Conductors, Semiconductors, Insulators, and Crystal-Growth Technology

Conductors, Semiconductors, Insulators, and Crystal-Growth Technology

A Practical Guide

Zeev Burshtein
Nes Ziona, Israel

WILEY

Published by John Wiley & Sons, Inc., Hoboken, New Jersey.
Published simultaneously in Canada.

Library of Congress Cataloging-in-Publication Data applied for:

Hardback ISBN: 9781394339853

Cover Design: Wiley

Set in 9.5/12.5pt STIXTwoText by Straive, Chennai, India

Contents

Introduction

For dozens of years, semiconductors and their applications are regularly studied in university undergraduate and graduate courses in departments of physics, electrical, and materials engineering. The specific subjects, their teaching order, and the depth of their analytical presentation, are naturally different in different departments, and also depend on the individual lecturer and his research experience. The subject of crystal growth technology is studied mostly in material engineering departments. These two subjects are linked strongly, because manufacturing of semiconductor devices strongly involve the growth of single crystals, and the crystal growing must be tailored to specific properties of the aimed-at applications. The present book is based on my own, originally Hebrew-written lecture notes of the said two subjects, which I used to teach separately for over 35 years in the Department of Materials Engineering at the Ben-Gurion University of the Negev, Israel. The analytical presentation adheres to formats used in teaching of experimental physics. I believe that many parts of the book are further suitable to provide a proper background to interested researchers. Parts dealing with conductors, semiconductors, and insulators provide a detailed background on electrical properties of such materials. Useful specific devices based on the different material properties are described. Subjects related to microelectronic devices and electronic integrated circuits, however, are not addressed; they rightfully deserve a separate attention! Parts related to the crystal growth technology provide a basic theoretical background of a solid formation effects. Detailed description of crystal growth means, especially furnaces and their specific designs per different candidate materials, is provided. Special specific examples of crystal growth are described in detail, including end-product mechanical manufacturing and quality control.

Quite simple numerical calculation exercises are included at the end of most chapters; they are aimed at assisting the reader in clarifying the meaning of the different described terms, their inter-relations, and the orders of magnitude of the various described physical effects. The given data are often presented in the mixed units that are common in the scientific literature. It is however strongly recommended that on solving the exercises numerically, the readers will convert all units into the SI (rationalized-MKS) system, and avoid using neither CGS nor any mixed units system for that purpose.

No references related to the different addressed subjects are provided throughout the text. A reader who wishes to delve, deepen, and expand on any subject can use the list of references for further reading, provided in Appendix **C**, or refer to scientific journals dealing with the subject. Referring to adequate commercial catalogues may also be found quite useful. A subject index and a table of physical constants are appended.

I am indebted to quite a number of colleagues with whom I shared many efforts over a full decade of research and development in the field of crystal growth, especially to Dr. Uri laor, Dr. Yehoshua Shimony, and Dr. Moshe Azulay.

I dedicate this book to the memory of my PhD instructor, the late Professor Abraham Many of the Hebrew University of Jerusalem, who was an example and a model in all his ways to me and to all his other students and colleagues. He introduced me to quite a number of subjects presented in this book, and I adopted his teaching spirit and methods in the writing of this present book.

Zeev Burshtein
Nes-Ziona, Israel

1 Solid State Structure

1.1 Crystalline Lattice

Solids are most often constructed of well-defined crystalline structures, whose essence is a periodic system of atoms or molecules spanned across a considerable range. The periodic structure comprises an imaginary crystalline lattice attached with an atomic basis. The lattice is a set of points in space from which the crystal looks the same. A crystalline lattice may be defined by a set of three **fundamental vectors**, $\underline{a}$, $\underline{b}$, and $\underline{c}$, such that the crystal appears identical when viewed from points $\underline{r}$ and $\underline{r}'$ satisfying

$$\underline{r}' = \underline{r} + n_1\underline{a} + n_2\underline{b} + n_3\underline{c}, \tag{1.1}$$

where n_1, n_2, and n_3 are arbitrary integers. The $\underline{L} \equiv n_1\underline{a} + n_2\underline{b} + n_3\underline{c}$ collection of points actually defines the crystalline lattice. The $\underline{L}$ vector is called **lattice vector** or **lattice translation vector**. The three-dimensional parallelepiped based on three fundamental vectors is called **unit cell**. A lattice and its constructing vectors are granted the adjective **primitive** if *all* lattice points are defined by Eq. (1.1). Otherwise, the unit cell is called **centered**. The aforesaid underlying ideas are displayed in Figure 1.1, describing rather a two-dimensional lattice. As the figure demonstrates, it is possible to choose different combinations of fundamental vectors to form different unit cells for the same lattice; in fact, an infinite number of such different unit cells can be chosen. A unit cell exhibits the property that it is possible to fill the entire space by its translations using all lattice vectors.

Notably, thus formed unit cells have identical volumes (identical areas in the case of the two-dimensional lattice of the figure). It is also possible to construct unit cells exhibiting the same property that are not necessarily a spatial parallelepiped of the fundamental lattice vectors. One important example is the **Wigner–Seitz unit cell** formed by the following procedure: initially, one of the lattice points is selected as an origin. Then, all out-pointing lattice vectors are sliced in two by adding perpendicular, infinitely large planes. The smallest enclosed polyhedron establishes the desired unit cell. It is quite obvious that its volume would be identical to that of the parallelepiped unit cell, because it spans the entire lattice volume with no voids, where each such cell is assigned to a different lattice point. Furthermore, by its definition, the Wigner–Seitz unit cell is independent of the selection of fundamental lattice vectors!

Conductors, Semiconductors, Insulators, and Crystal-Growth Technology: A Practical Guide,
First Edition. Zeev Burshtein.

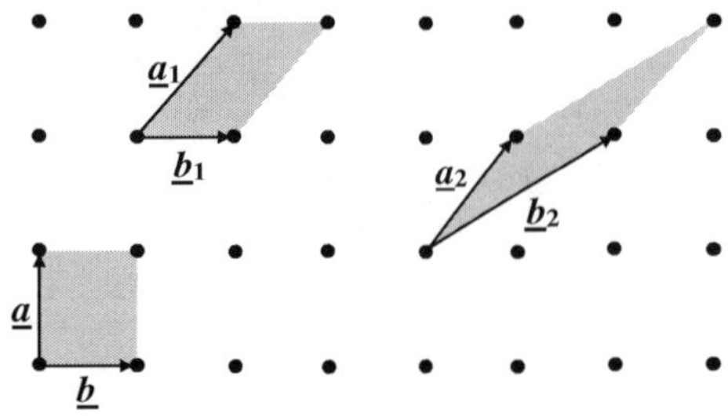

Figure 1.1 Schematic illustration of a two-dimensional lattice with three different selections of fundamental lattice vectors and the unit cells they generate (the shaded grey areas). The black small bullets indicate the lattice points, and the different choices are given different subscript numbering.

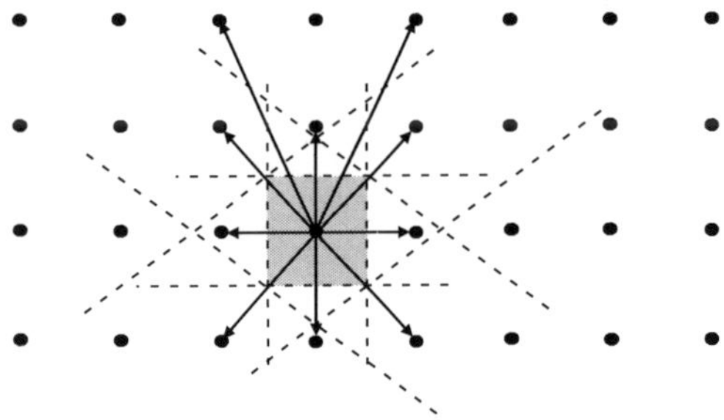

Figure 1.2 Illustration of the construction method of the Wigner–Seitz unit cell. The dashed lines mark the projections on the page plane of the perpendicular slicing planes of lattice vectors which outpoint from the lattice point selected as the origin. The gray-shaded area marks the Wigner–Seitz cell.

The formation of a two-dimensional Wigner–Seitz unit cell is illustrated in Figure 1.2.

The $\boldsymbol{V_C}$ unit cell volume is given by the absolute value of the determinant of the Cartesian coordinates of the $\underline{a}$, $\underline{b}$, and $\underline{c}$ lattice fundamental vectors:

$$\boldsymbol{V_C} = |(\underline{a} \times \underline{b} \cdot \underline{c})| = \textbf{abs} \begin{vmatrix} a_x & a_y & a_z \\ b_x & b_y & b_z \\ c_x & c_y & c_z \end{vmatrix} \tag{1.2}$$

As previously noted, the lattice is an abstract mathematical entity. The crystal itself is formed when an identical atom or an identical stiff collection of atoms (say a molecule) is attached to each lattice point. The attached entity is called **atomic basis**.

An attached atomic basis may sometimes reduce the lattice symmetry. For demonstration, we consider the two-dimensional square lattice of Figure 1.3, where each lattice point is attached with a couple of identical atoms such that the lattice point is exactly at the mid distance between the atoms. It is easily seen that while each lattice point exhibits a rotational symmetry of 90° about an axis crossing it perpendicularly to the page plane (a fourfold rotational symmetry denoted C_4), the rotational symmetry order of the crystal itself is only twofold (namely only a 180° rotation about each said axis yields the same crystal construction; it is denoted as C_2). Thus, different crystals may exhibit different symmetry orders even if they belong to the same crystalline lattice.

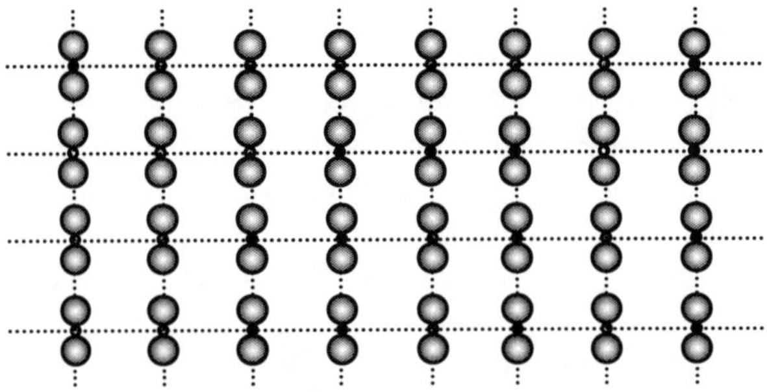

Figure 1.3 Schematic illustration of rotational symmetry reduction of a crystal lattice caused by addition of an atomic basis. The two-dimensional lattice of the small bullets has a fourfold rotational symmetry about axes perpendicular to the page plane crossing the lattice points. The symmetry is reduced to only a twofold rotational symmetry after addition of an atomic basis comprising two identical atoms (represented by balls) to each lattice point.

The various possible lattice shapes have been classified by Bravais into 14 different **families**, and subclassified as 7 crystalline **systems**. The crystalline systems are defined by the relationships among the lengths of the fundamental $\underline{a}$, $\underline{b}$, and $\underline{c}$ linearly independent (**not necessarily primitive**) vectors and their corresponding α, β, and γ inter angles. If a unit cell generated by these vectors is not primitive, it is then **centered**, comprising lattice points not coinciding with the cell corners. Bravais systems and the different families belonging to each system are listed and illustrated below.

The mathematical instrument used to present and analyze crystal symmetry is Group Theory. The symmetry of the different lattice structures is described by 32 groups of point-symmetry operations. The latter include inversion, mirror reflections, rotations (so-called "proper" rotations), and rotations followed by mirror reflections (so-called "improper" rotations), *that leave at least a single point in space unaltered*, and make the operated-upon object seem identical with the original structure. When other special symmetry operations are added – lattice translations and mirror reflections, each followed by a translation that differs from any lattice translation (called **glide-mirror**), and rotations, each followed by a translation that differs from any lattice translation (called **screw axis**) – 230 groups are obtained, called **space groups**. The space group properties cast conditions on the possible structure of atomic bases that could be attached to lattice points belonging to a certain space group.

(i) Cubic system $a = b = c$; $\alpha = \beta = \gamma = 90°$.

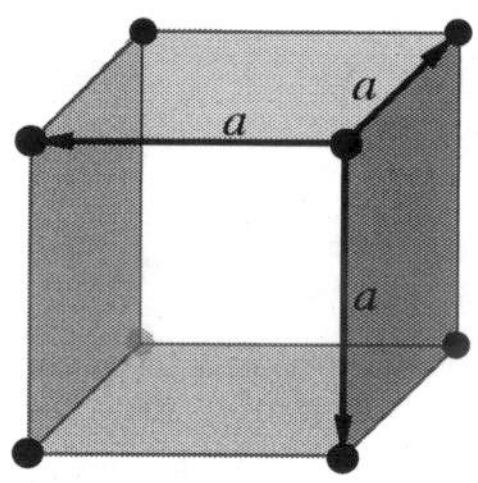

P

Primitive cubic

Lattice points at only cell vertices

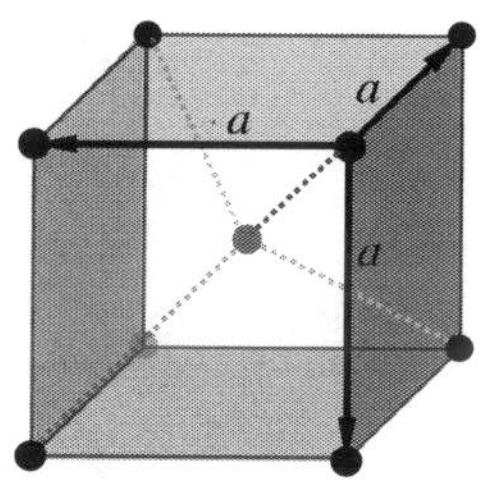

I

Body-centered cubic

Additional lattice point at body center

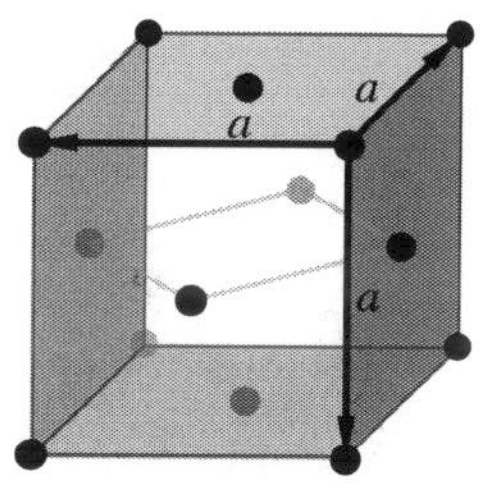

F

Face-centered cubic

Additional lattice point at each face center

(ii) Hexagonal system $a = b \neq c$; $\alpha = \beta = 90°$; $\gamma = 60°$.

P

Primitive hexagonal

Lattice points at only cell vertices

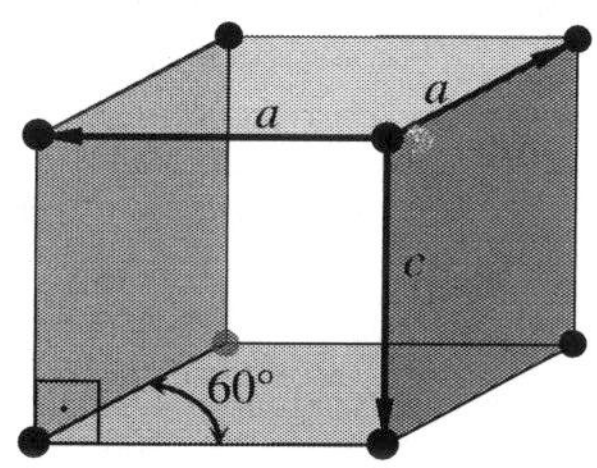

(iii) Rhombohedral (trigonal) system $a = b = c$; $\alpha = \beta = \gamma < 120°$ and $\neq 120°$ or $60°$.

P or R in different cases

Primitive Rhombohedral (trigonal)

Lattice points at only cell vertices

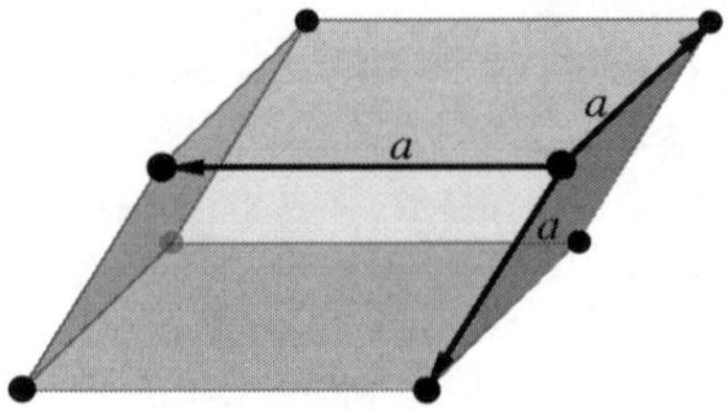

(iv) Tetragonal system $a = b \neq c$; $\alpha = \beta = \gamma = 90°$.

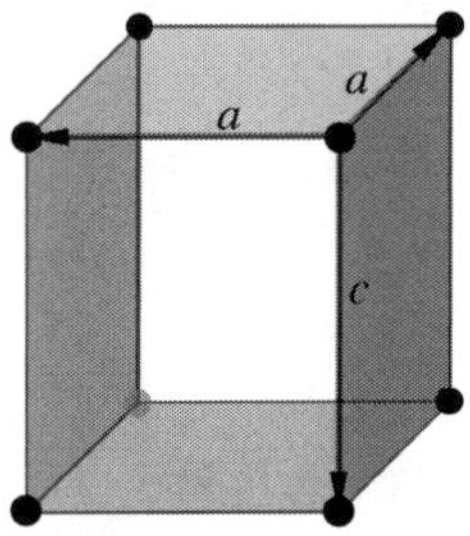

P

Primitive tetragonal

Lattice points at only cell vertices

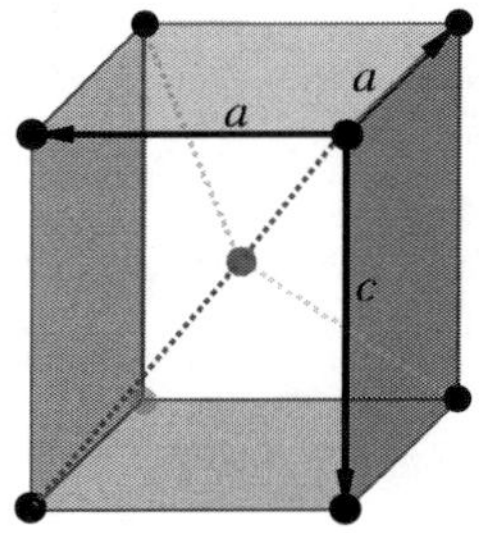

I

Body-centered tetragonal

Additional lattice point at body center

(v) Orthorhombic system $a \neq b \neq c$; $\alpha = \beta = \gamma = 90°$.

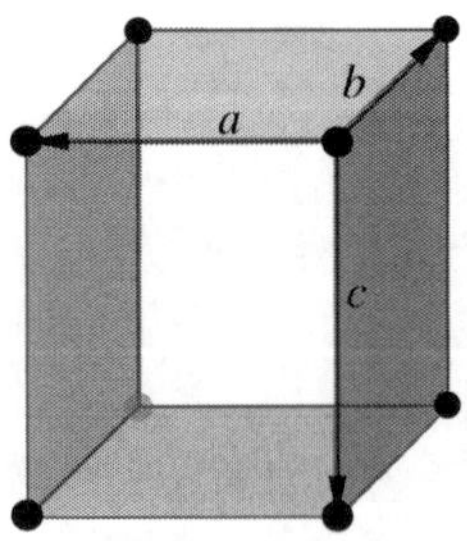

P

Primitive orthorhombic

Lattice points at only cell vertices

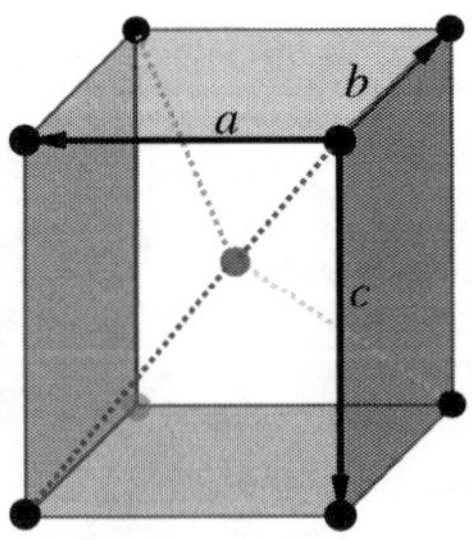

I

Body-centered orthorhombic

Additional lattice point at body center

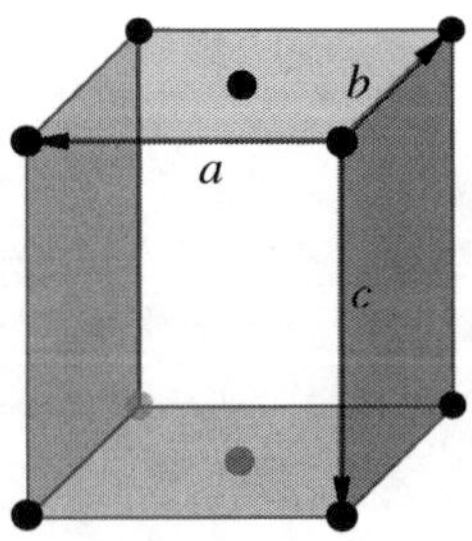

A or C in different cases

Base-centered orthorhombic

Additional lattice points at centers of one pair of opposite faces

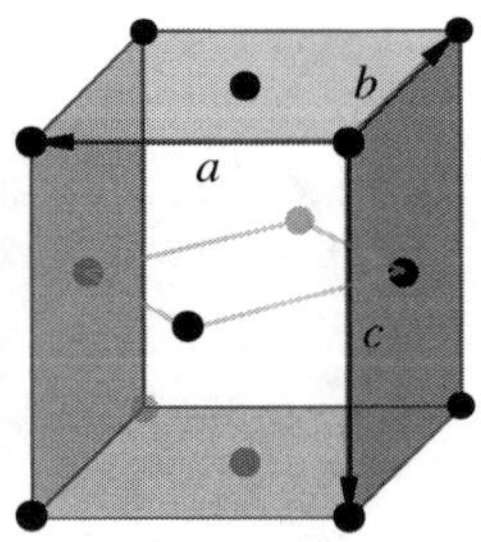

F

Face-centered orthorhombic

Additional lattice point at each face center

(vi) Monoclinic system $a \neq b \neq c$; $\alpha = \gamma = 90° \neq \beta$.

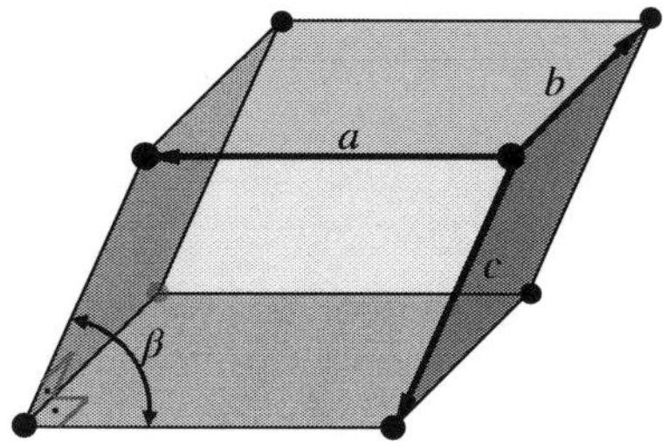

P
Primitive monoclinic
Lattice points at only cell vertices

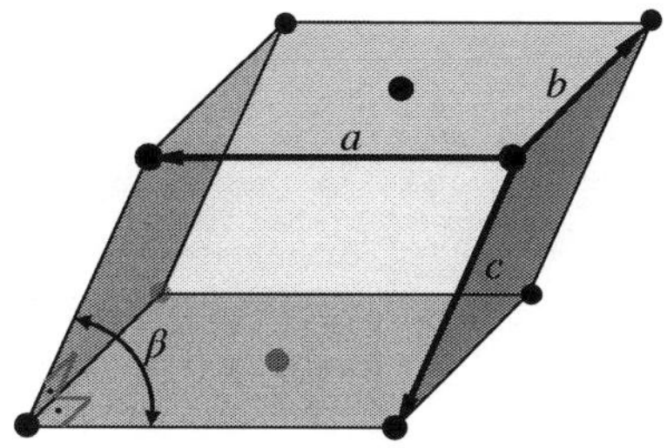

B or C in different cases
Base-centered tetragonal
Additional lattice points at centers of one pair of opposite rectangular faces

(vii) Triclinic system $a \neq b \neq c \neq a$; $\alpha \neq \beta \neq \gamma \neq \alpha$.

P
Primitive triclinic
Lattice points at only cell vertices

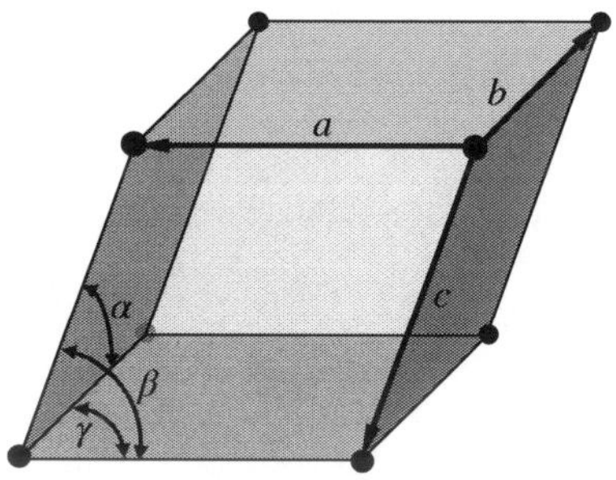

Solids may also appear as crystals exhibiting disorder at occasional sites or even in many sites; also as poly-crystalline forms, where a crystalline order is maintained at only tiny fractions of a millimeter; in glassy forms, where a crystalline order is only slightly larger than the size of a unit cell; and in an amorphous structure, where the order never exceeds the size of a single comprising molecule. In 1984, a special ordered structure of solids exhibiting a fivefold (C_5) rotational order was discovered, a symmetry that is inconsistent with a requirement of a perfect crystal. That discovery yielded an extended definition of the crystalline concept, and the person who made this discovery (D. Shechtman) was granted a Nobel Prize in chemistry in 2011. In our following discussions, however, we limit ourselves to consider only regularly ordered crystals, where needed.

1.2 Indication of Crystal Planes and Orientations

A **crystalline orientation** is an orientation of a lattice vector of the form $\underline{L} \equiv h\underline{a} + k\underline{b} + \ell\underline{c}$, where h, k, ℓ are integers that do not have a common factor. For Brevity, the $[h, k, \ell]$ symbol is used as the orientation index. (If initially the said three integers do have common factors, they are divided by the largest one).

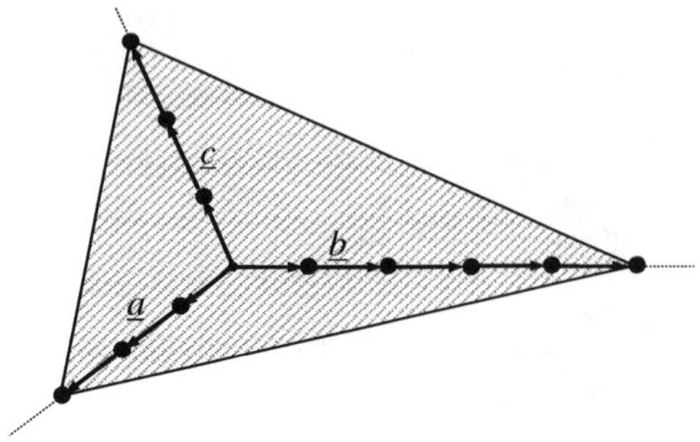

Figure 1.4 Demonstration of the definition method of a fundamental family of crystal plane. For more details, see the text.

A **crystalline plane** is a plane that contains lattice points. A **fundamental plane** is a plane defined by three lattice points placed along three fundamental vectors, namely one at $p_1\underline{a}$, another at $p_2\underline{b}$, and a third one at $p_3\underline{c}$ (p_1, p_2, and p_3 being integers), as described in Figure 1.4. A family of planes is the collection of all planes that are parallel to the fundamental plane. Each family of planes consumes the entire lattice points. In principle, the three p_1, p_2, and p_3 integers set may serve to define the planes family. Following Miller, however, the used method is as follows: Let M be the smallest common product among p_1, p_2, and p_3. Miller indices of the corresponding planes family are defined as

$$(h, k, \ell) \equiv \left(\frac{M}{p_1}, \frac{M}{p_2}, \frac{M}{p_3}\right). \tag{1.3}$$

For example, in the structure presented in Figure 1.4, $p_1 = 3$, $p_2 = 5$, and $p_3 = 3$. Therefore, $M = 15$. Miller indices of the planes family of Figure 1.4 is $(h, k, \ell) \equiv \left(\frac{15}{3}, \frac{15}{5}, \frac{15}{3}\right) = (5, 3, 5)$. In the special case of cubic unit cells, $[h, k, \ell]$ orientations are always perpendicular to the (h, k, ℓ) planes.

1.3 The Reciprocal Lattice

Vectors $\underline{A}$, $\underline{B}$, and $\underline{C}$ that form a lattice reciprocal to a one defined by $\underline{a}$, $\underline{b}$, and $\underline{c}$ vectors are defined as follows:

$$\underline{A} \equiv 2\pi\frac{\underline{b}\times\underline{c}}{\underline{a}\times\underline{b}\cdot\underline{c}}; \quad \underline{B} \equiv 2\pi\frac{\underline{c}\times\underline{a}}{\underline{b}\times\underline{c}\cdot\underline{a}}; \quad \underline{C} \equiv 2\pi\frac{\underline{a}\times\underline{b}}{\underline{c}\times\underline{a}\cdot\underline{b}}. \tag{1.4}$$

When simultaneously dealing with both crystal lattice and its reciprocal one, the former is often addressed as the **direct lattice**. Notably, the denominators in all above expressions are identical, and their size (their absolute value) is in fact the volume $\boldsymbol{V_C}$ of the direct lattice unit cell (Eq. [1.2]).

In the following, we summarize several basic properties of the reciprocal lattice:

(1) *Direct and crossed scalar products:*

$$\underline{a}\cdot\underline{A} = \underline{b}\cdot\underline{B} = \underline{c}\cdot\underline{C} = 2\pi;$$

$$\underline{a}\cdot\underline{B} = \underline{a}\cdot\underline{C} = \underline{b}\cdot\underline{A} = \underline{b}\cdot\underline{C} = \underline{c}\cdot\underline{A} = \underline{c}\cdot\underline{B} = 0.$$

In other words, the unit cell vectors of the reciprocal lattice are perpendicular to corresponding planes in the direct lattice: $\underline{A}$ is perpendicular to the $(\underline{b}, \underline{c})$ plane, $\underline{B}$ is perpendicular to the $(\underline{c}, \underline{a})$ plane, and $\underline{C}$ is perpendicular to the $(\underline{a}, \underline{b})$ plane.

(2) *The unit cell volume of the reciprocal lattice*:

$$\mathbf{V}_{rcp} = |(\underline{A} \times \underline{B} \cdot \underline{C})| = \frac{(2\pi)^3}{\mathbf{V}_C}.$$

(3) *Scalar product between direct and reciprocal lattice vectors*:

A reciprocal lattice vector: $\underline{G}(hk\ell) = h\underline{A} + k\underline{B} + \ell\underline{C}$;

A direct lattice vector: $\underline{L} = n_1\underline{a} + n_2\underline{b} + n_3\underline{c}$.

Their scalar product:
$$\left.\begin{aligned} \underline{G} \cdot \underline{L} &= (h\underline{A} + k\underline{B} + \ell\underline{C}) \cdot (n_1\underline{a} + n_2\underline{b} + n_3\underline{c}) \\ &= 2\pi(hn_1 + kn_2 + \ell n_3) = 2\pi n \end{aligned}\right\},$$

where $n \equiv (hn_1 + kn_2 + \ell n_3)$ is an integer.

(4) *Distance between adjacent crystalline planes*:
We state without a formal proof that the $d_{hk\ell}$ distance between adjacent (h, k, ℓ) crystalline planes is given by

$$d_{hk\ell} = \frac{2\pi}{|\underline{G}(hk\ell)|}. \tag{1.5}$$

(5) *The reciprocal lattice of the reciprocal lattice is the original direct lattice.*

1.4 X-ray Diffraction by Crystals

We consider now a parallel beam of X-rays striking a single crystal. The radiation $\underline{k}$ wave-vector size is defined as $|\underline{k}| = 2\pi/\lambda$, where λ is the radiation wavelength, and its orientation coincides with the beam propagation orientation. For simplicity and illustration, and without limiting the generality of the discussion, we assume that each lattice point is occupied by a single atom. Each such atom established a scattering center for the impinging radiation. The scattered radiation will exhibit a considerable intensity in orientations forming a constructive interference among the waves scattered off all crystal points. The combined effect of scattering and the following interference is termed **diffraction**. We proceed in our analysis by considering the optical waves path differences between two elastically scattered beams (namely with no wavelength changes) off two lattice points, A and B, as illustrated in Figure 1.5. Figure 1.6 provides a wave-vectors scheme related to the said process. The path difference δ_{BA} between beams scattered at the B point and the one scattered at the A point,

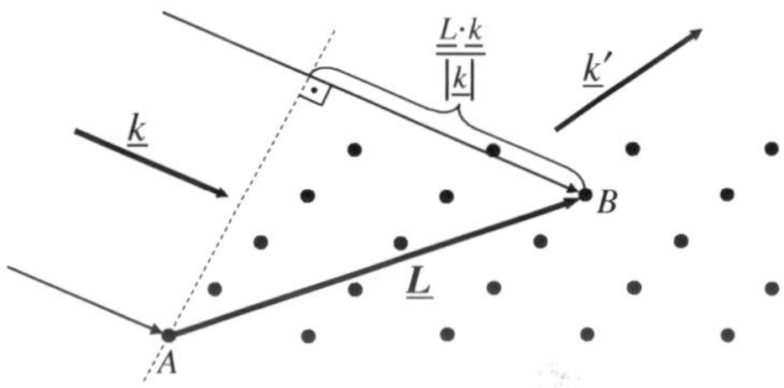

Figure 1.5 Auxiliary drawing for calculating the optical path difference between X-rays scattered off different A and B points of the crystal lattice. The beam path for reaching point B is longer relative to its path reaching point A. It is marked by a one-sided curled bracket.

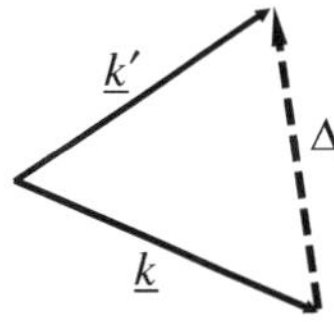

Figure 1.6 Wave-vectors scheme of the impinging X-ray beam $\underline{k}$, the scattered beam $\underline{k}'$, and the difference vector $\Delta\underline{k}$. In elastic scattering, the wavelengths of the impinging and scattered beams are identical, namely $|\underline{k}| = |\underline{k}'|$.

both in a $\underline{k}'$ orientation, is given by

$$\delta_{BA} = \frac{\underline{L} \cdot \underline{k}'}{|\underline{k}'|} - \frac{\underline{L} \cdot \underline{k}}{|\underline{k}|} = \frac{\underline{L} \cdot (\underline{k}' - \underline{k})}{k} = \frac{\underline{L} \cdot \Delta \underline{k}}{k}.$$

Constructive interference conditions are established when the optical paths differences we have calculated is an integral number of the radiation wavelength, namely when $\delta_{BA} = s\lambda$, where s is an integer. Then we get that

$$\underline{L} \cdot \Delta \underline{k} = \delta_{BA} \cdot k = s\lambda \cdot \frac{2\pi}{\lambda} = 2\pi s. \tag{1.6}$$

This condition should be valid for all lattice $\underline{L}$ vectors; therefore, $\Delta \underline{k}$ (see Fig. 1.6) must be a reciprocal lattice vector (Section 1.3): namely, a constructive interference occurs only in such specific cases where $\Delta \underline{k} = \underline{G}$. Experimental study of constructive orientations of X-rays scattering off a crystal provides a mapping of the reciprocal lattice vectors, hence of the direct lattice vectors as well. Intensity relations among beams diffracted in the different orientations provide information on the structure of the atomic basis attached to each lattice point. That issue, however, is beyond the scope of the present book. We just add that diffraction scattering patterns similar to those of X-rays are obtained for material particles scattering, like for electron and neutron beams. They are widely used in studies of solid state structure.

There exists an alternative, quite simplified way to address the X-ray diffraction effect. The crystal is described as a set of parallel planes containing lattice points. We have already indicated in Section 1.2 that such family of planes consumes all lattice points. The said planes are described as partially reflecting mirrors, as in Figure 1.7. The optical paths difference between the beam path marked II in Figure 1.7 and the one marked I there is given by

$$\delta_{II,I} = \delta_1 - \delta_2 = \frac{d}{\mathbf{sin}\,\theta} - \frac{d}{\mathbf{sin}\,\theta}\mathbf{cos}\,2\theta = 2d\,\mathbf{sin}\,\theta,$$

where θ is the grazing angle of the beam impingement relative to the crystal plane. Under constructive interference, this difference should be equal to a multiplication of the wavelength by an integer. Namely,

$$2d\,\mathbf{sin}\,\theta = n\lambda, \tag{1.7}$$

where n is an integer. This condition is known as **Bragg interference condition**.

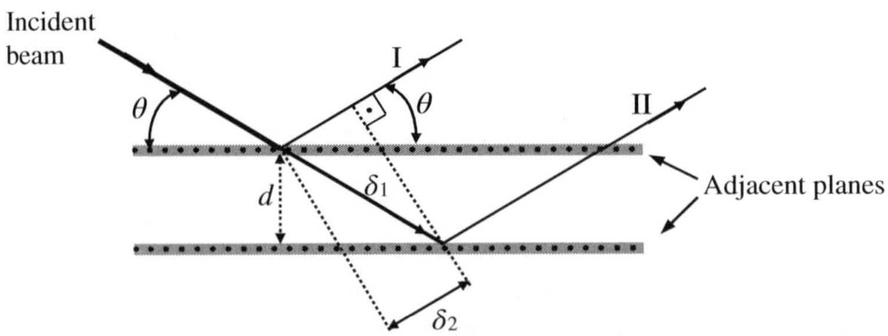

Figure 1.7 Auxiliary drawing for calculating the optical path difference between two X-ray beams partially reflected from two adjacent parallel planes of inter-distance *d*. The impinging beam is drawn using a thick line; the transmitted and reflected beams are drawn using thinner lines.

There is obviously a linkage between Bragg's condition and the previous result. We re-plot Figure 1.6 while adding the reflecting plane. It is readily seen that the θ angle of Figure 1.7 is half the angle between $\underline{k}'$ and $\underline{k}$ of Figure 1.8. Thus, we have that $|\Delta\underline{k}| = 2k\,\boldsymbol{sin}\,\theta$. As we showed earlier, under constructive interference $\Delta\underline{k}$ is necessarily some $\underline{G}$ vector of the reciprocal lattice. Namely,

$$|\underline{G}| = 2\frac{2\pi}{\lambda}\cdot\frac{n\lambda}{2d} \Rightarrow d = n\frac{2\pi}{|\underline{G}|}. \tag{1.8}$$

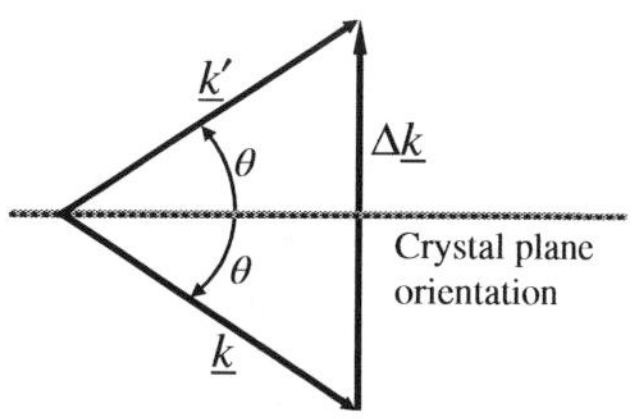

Figure 1.8 Auxiliary drawing of the impinging and scattered beam wave-number vectors, and their difference vector under constructive interference, to be used for calculating the size of the difference vector.

The n quantity is called **reflection order**. The $n = 1$ case is called **first-order reflection**, the $n = 2$ case is called **second-order reflection**, and so on. A high reflection order in fact expresses a situation where the constructive interference involves many planes, some imaginary, whose inter-distances are a division by n of the real $d_{hk\ell}$ planes inter-distance: $d_{hk\ell} = 2\pi/|G(hk\ell)|$ (per Eq. [1.5]). Further recall that the $\underline{G}$ reciprocal lattice vector is perpendicular to the Bragg plane (Fig. 1.8).

The Wigner–Seitz unit cell of the reciprocal lattice is called **first Brillouin zone**. To demonstrate the importance of using this concept, we consider the equation of its boundary surface. Let $\underline{G}$ be a vector whose perpendicular slicing plane establishes a surface of the Wigner–Seitz cell. Then, let $\underline{k}$ be a vector starting at the origin and ending at some point on the cell surface belonging to the said slicing plane, as drawn in Figure 1.9. It thus satisfies the equation

$$\frac{\underline{k}\cdot\underline{G}}{|\underline{G}|} = \frac{1}{2}|\underline{G}|.$$

Then

$$2k|\underline{G}|\,\boldsymbol{sin}\,\theta = |\underline{G}|^2 \Rightarrow 2k\,\boldsymbol{sin}\,\theta = \left(\frac{2\pi}{d}\right)n.$$

We set the relation $k = \dfrac{2\pi}{\lambda}$ and get $2d\,\boldsymbol{sin}\,\theta = n\lambda$, which again is Bragg's condition. In other words, **the boundary surface of the first Brillouin zone is the geometric position of all $\underline{k}$ vectors that satisfy Bragg's condition for the crystal family of planes which are parallel to that boundary surface**.

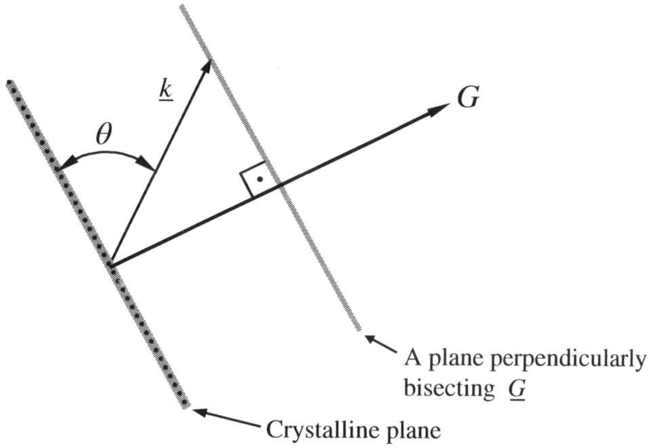

Figure 1.9 Auxiliary drawing used for finding the equation of the first Brillouin zone boundary surfaces. A boundary surface is defined by a plane bisecting perpendicularly the reciprocal lattice $\underline{G}$ vector.

Exercises for Chapter 1

1.1 The Cartesian coordinates of the fundamental vectors of the primitive unit cell of a given crystal in angstrom (Å) units are:

$$a(2.34,\ -10.14,\ 0);\quad b\left(0,\ 0.6,\ 6.72\right);\quad c\left(11.32,\ 5.28,\ 0\right).$$

(a) Plot the unit cell structure, and identify the crystalline system it belongs to.
(b) Calculate the volume of the unit cell, and the volume of its first Brillouin zone.
(c) Calculate the fundamental vectors of the reciprocal lattice, plot the unit cell structure, and determine the crystalline system it belongs to.
(d) Calculate the $d_{7,7,12}$ distance between the adjacent $(7,7,12)$ crystalline planes.

2 Crystal Lattice Vibrations

2.1 Dispersion Curves of Lattice Vibrations

In the previous chapter, the crystals were described as systems where the atoms or molecules belonging to each lattice point always remain steady. In reality, the said lattice points describe their equilibrium positions, while the atoms/molecules oscillate about these positions, thus distorting the crystal symmetry. (Such motions occur even at the absolute zero temperature – a quantum effect known as "zero-point fluctuations"). The oscillation amplitudes increase with temperature. When such distortion occurs locally, it spreads throughout the crystal due to coupling between the atoms. The oscillation propagation modes are classified as **longitudinal** and **transverse**. In a longitudinal oscillation, the orientation of the distortion wave propagation is *parallel* to the distortion orientation. In a transverse oscillation, the orientation of the distortion wave propagation is *perpendicular* to the distortion orientation. Transverse oscillation waves are also called **shear waves**.

To clarify these concepts, a temporary state of the crystalline whole plane position deviations is plotted schematically in Figure 2.1. The lattice points, that also mark the equilibrium positions of the atomic bases, are indicated by small bullets, while their belonging atomic bases are indicated by larger balls. A crystalline plane position deviation is marked u_s, where s is the plane ordinal number. The deviation sizes in the figure are grossly exaggerated; in reality, the deviations are extremely small compared to the interatomic distances.

In forming a wave equation analysis, we assume that the return force acting on each deviating atomic basis of mass M depends on its plane deviation from the equilibrium position relative to the deviation of the subsequent plane, namely

$$F_s = \sum_p C_p(u_{s+p} - u_s),$$

where F_s is the return force acting on an atomic basis of s ordinal number, C_p is the effective "spring constant" relevant to the deviation of an atomic basis in a plane whose ordinal number is $s + p$ relative to an atomic basis in a plane of s ordinal number. The equation of motion is

$$M\frac{d^2u_s}{dt^2} = \sum_p C_p(u_{s+p} - u_s). \tag{2.1}$$

We attempt a solution of the form $u_{s+p} = u_0 e^{i(s+p)ka} \cdot e^{-i\omega t}$, where a is the size of the lattice period (lattice constant) in the direction of the beam propagation, ω is the angular frequency, k is

Conductors, Semiconductors, Insulators, and Crystal-Growth Technology: A Practical Guide,
First Edition. Zeev Burshtein.

the size of the propagating wave vector, and u_0 is the oscillation amplitude. We then get

$$-\omega^2 M u_0 e^{iska} \cdot e^{-i\omega t} = \sum_p C_p \left(e^{i(s+p)ka} - e^{iska}\right) u_0 e^{-i\omega t};$$

$$\omega^2 M = -\sum_p C_p \left(e^{ipka} - 1\right).$$

Because $C_{-p} = C_p$, the latter equation assumes the form $\omega^2 M = -\sum_{p>0} C_p(e^{ipka} + e^{-ipka} - 2)$.

Then

$$\omega^2 = \frac{2}{M} \sum_{p>0} C_p(1 - \cos pka). \tag{2.2}$$

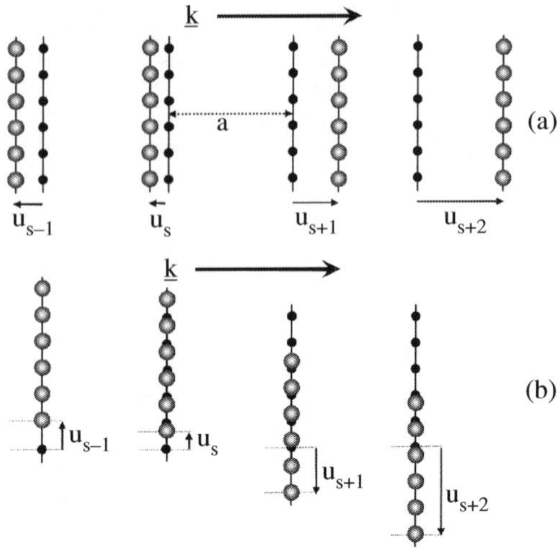

Figure 2.1 Schematic illustration of the propagation of longitudinal (a) and transverse (b) waves in a crystal. Small bullets indicate the equilibrium positions of the lattice points, and the larger balls indicate their attached atomic bases. $\underline{k}$-marked vectors indicate the propagation orientations of the oscillation waves. In order to emphasize the orientation of the positions deviations, the sizes of the molecule position deviations are grossly exaggerated, especially for the described transverse wave.

A $\omega(\underline{k})$ function is addressed as a **dispersion function** or as a **dispersion curve**.

We consider now a special simple case, where plane coupling is limited to only two adjacent planes, namely $C_p = 0$ for all $p > 1$ cases. The dispersion function then assumes the following form:

$$\omega = 2\sqrt{\frac{C_1}{M}} \cdot \left|\sin\frac{1}{2}ka\right|. \tag{2.3}$$

The last result is described graphically in Figure 2.2. As obvious by the function form and by the figure, the angular frequency ω is periodic as a function of k, the period being $2\pi/a$. However, the only region of k values that exhibits a physical meaning ranges between $-\pi/a$ and $+\pi/a$. Any oscillation of an actual atomic basis can be described by a k value of the said range as well as by a $k' = k + n(2\pi/a)$, where n is an integer. For proof, we consider the position-dependent part of the equation of motion:

$$u_0 e^{i(s+p)k'a} = u_0 e^{i(s+p)ka} e^{in(s+p)(2\pi/a)a} = u_0 e^{i(s+p)ka}; \quad \textbf{Q.E.D.}$$

As may be readily verified (Section 1.4), the range of k values between $-\pi/a$ and $+\pi/a$ is actually the first Brillouin zone. The physical meaning of the possibility to ignore k values outside that range is quite clear, because for a $k > |\pi/a|$ case, we would get that $\lambda < 2a$, namely an oscillation of an empty space between adjacent crystal planes; obviously physically meaningless.

To delve further into the physical meaning of the dispersion curve, we consider its shape in its first Brillouin zone limits. In the limit of very long waves, namely when $k \ll \pi/a$, we get that $\omega \cong \sqrt{\frac{C_1}{M}} \cdot ka$. The wave phase velocity v_φ, which actually is the speed of sound v inside the crystal, is given by

$$v = v_\varphi \equiv \frac{\omega}{k} = \sqrt{\frac{C_1}{M}} \cdot a, \tag{2.4}$$

which is a constant, independent of frequency close to that limit.

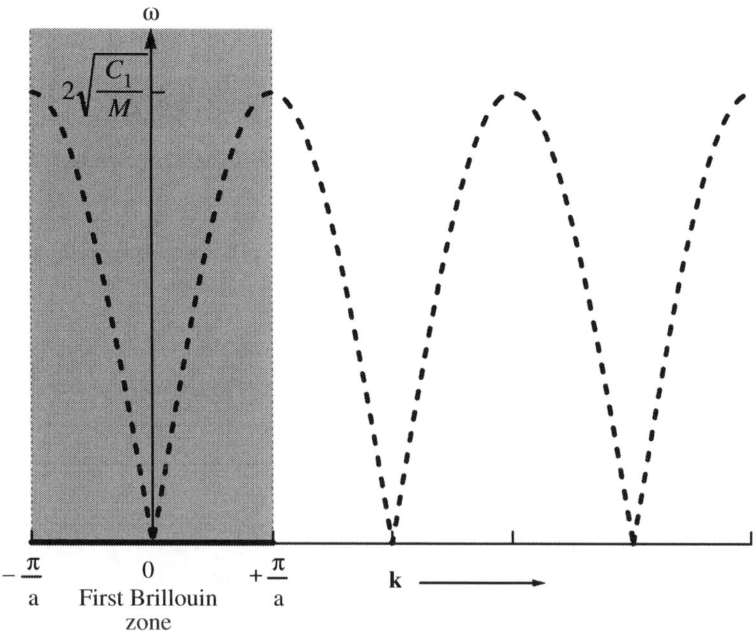

Figure 2.2 Shape of lattice vibrations dispersion curve in the special case where crystal planes coupling is limited to only between two adjacent planes (Eq. [2.2]).

Next we consider an edge of the Brillouin zone. In that limit, the wavelength is $\lambda = 2a$. Here, we rather seek to find the wave group velocity v_g. For such waves

$$v_g \equiv \frac{d\omega}{dk} = a\sqrt{\frac{C_1}{M}} \cdot \left| \cos\frac{1}{2}ka \right| \xrightarrow[k \to \pm\frac{\pi}{a}]{} 0.$$

The obtained vanishing of the group velocity means that the edge of the Brillouin zone describes a standing wave. This is a general result, not at all unique to the special case we addressed of assuming that the interaction is limited to only adjacent crystal planes. The reason is that the Brillouin zone edge is the spatial position of waves satisfying Bragg's interference conditions (Section 1.4). In other words, each lattice oscillation that propagates in the $\underline{k}$ direction is accompanied by an oscillation of the same wavelength that propagates in the $-\underline{k}$ direction. Interference between these two waves establishes a standing wave. The same consideration is approximately valid for waves belonging to regions close to the Brillouin zone edges; namely, their group velocity is very close to being zero. In Section 2.3, we provide a reminder about the **phase velocity** and **group velocity** terms, and to their physical meaning.

2.2 Solids Specific Heat

2.2.1 Energy Quantization of a Harmonic Oscillator

To demonstrate the effect of energy quantization, we consider a harmonic oscillator formed by a mass m hanging by a spring of a k constant (the latter property should not be confused as a wave-number, it is an entirely different physical quantity). The angular frequency of the mass oscillation is given by $\omega = \sqrt{k/m}$. Classically, the energy E stored by the mass oscillations is given by $E = \frac{1}{2}m\omega^2A^2$, where A is the oscillation amplitude. By quantum theory, however, when the system oscillates at its "self" (eigen) frequency, its energy may assume only discrete values, given by $E_n = \left(n + \frac{1}{2}\right)\hbar\omega$ for $n = 0, 1, 2\ldots\infty$, where $\hbar \equiv 1.054 \times 10^{-34}$ J · s is the reduced **Planck's constant**. Correspondingly, the classical amplitude may assume only a discrete set of values. That effect is presented schematically in Figure 2.3.

In cases of material oscillations, the basic $\hbar\omega$ quantum energy is called **phonon**; in view of the simple proportionality relation between the phonon energy and the angular frequency, they are often used as alternative symbols. The number of phonons describing the oscillator energy is called **occupation number**.

A harmonic oscillator under thermal equilibrium with a heat reservoir may oscillate at different amplitudes belonging to different occupation numbers. According to Boltzmann, the probability P_n for occupation of n phonons satisfies

$$P_n \propto e^{-n\hbar\omega/k_BT}, \tag{2.5}$$

where $k_B \equiv 1.38 \times 10^{-23}$ J/**deg** is the Boltzmann constant, and T is the absolute temperature. (Recall that the gas constant R relates to the Boltzmann constant according to $R = k_BN_A$, where $N_A = 6.022 \times 10^{23}$ mole^{-1} is Avogadro's number.) Using the Boltzmann relation, we get

$$P_n = \frac{e^{-n\hbar\omega/k_BT}}{\sum_{s=0}^{\infty} e^{-s\hbar\omega/k_BT}}. \tag{2.6}$$

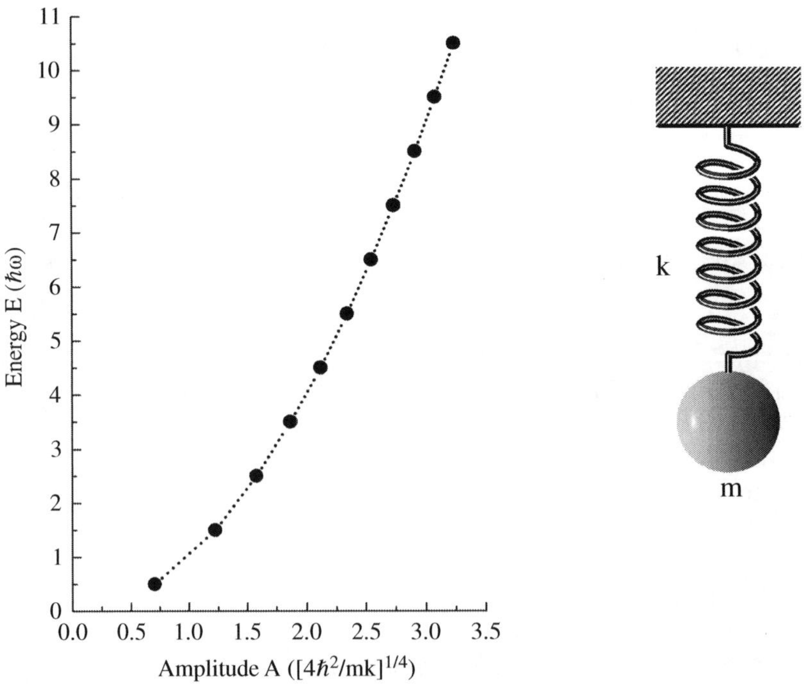

Figure 2.3 Energy of a harmonic oscillator (a mass m hanging by a spring of k constant) oscillating at its self ω angular frequency as a function of the oscillation amplitude. The quantization constraint, emphasized by the small full circles along the curve, dictates the existence of only discrete values of both energy and amplitude.

The average number $\langle n \rangle$ of phonons occupying the oscillator (also called **occupation number** or **Boltzmann distribution function**) would be given by

$$\langle n \rangle = \sum_{n=0}^{\infty} nP_n = \frac{ne^{-n\hbar\omega/k_BT}}{\sum_{s=0}^{\infty} e^{-s\hbar\omega/k_BT}} = \frac{1}{e^{\hbar\omega/k_BT} - 1}. \tag{2.7}$$

For the last mathematical transition, we used the following auxiliary calculations:

(i) For $x < 1$, we may write $\sum_s sx^s = x\frac{d}{dx}\sum_s x^s = x\frac{d}{dx}\left(\frac{1}{1-x}\right) = x\frac{1}{(1-x)^2}$;

(ii) $\sum_s e^{-s\hbar\omega/k_BT} = \frac{1}{1 - e^{-\hbar\omega/k_BT}}$.

We turn now to analyze the behavior of the occupation number in the limits of very low and very high temperatures:

$$\langle n \rangle \approx \begin{cases} e^{-\hbar\omega/k_BT} & \textit{for} \quad k_BT \ll \hbar\omega \quad (\textit{low temperature}) \\ \frac{k_BT}{\hbar\omega} & \textit{for} \quad k_BT \gg \hbar\omega \quad (\textit{high temperature}) \end{cases}. \tag{2.8}$$

In the low-temperature limit, the occupation number is very small compared to a unity ($\langle n \rangle \ll 1$), but it grows exponentially with the temperature. In the high-temperature range, the occupation number is very high compared to a unity ($\langle n \rangle \gg 1$), and it grows linearly with the temperature.

The average oscillator energy is given by

$$\langle E \rangle = \hbar\omega\langle n \rangle = \frac{\hbar\omega}{e^{\hbar\omega/k_BT} - 1}. \tag{2.9}$$

2.2.2 Einstein's Model of Solids Specific Heat

Einstein suggested to consider each atomic basis (molecule) in the solid as a single rigid oscillator exhibiting the same characteristic angular frequency ω for each degree of freedom of its motion. The molar thermal energy would then be given by

$$E = \frac{3N_A\hbar\omega}{e^{\hbar\omega/k_BT} - 1}, \tag{2.10}$$

where N_A is Avogadro's number. The molar specific heat at a constant volume C_V is obtained by the derivative of E per Eq. (2.10) according to the temperature T

$$C_V \equiv \left(\frac{\partial E}{\partial T}\right)_V = 3N_Ak_B\left(\frac{\hbar\omega}{k_BT}\right)^2 \frac{e^{\hbar\omega/k_BT}}{\left(e^{\hbar\omega/k_BT} - 1\right)^2}. \tag{2.11}$$

Accordingly, we turn now to analyze the behavior of the molar specific heat in the limits of very low or very high temperatures, as performed in Eq. (2.8):

$$C_V \approx \begin{cases} 3N_Ak_B\left(\frac{\hbar\omega}{k_BT}\right)^2 e^{-\hbar\omega/k_BT} & k_BT \ll \hbar\omega \quad (\textit{low temperature}) \\ 3N_Ak_B = 3R & k_BT \gg \hbar\omega \quad (\textit{high temperature}) \end{cases}. \tag{2.12}$$

It is readily seen that C_V saturates at a constant value in the high-temperature limit. On the other hand, at low temperatures, it drops down exponentially as the temperature drops and vanishes at the absolute zero. Experiments indeed show such behavior at both limits. However, especially the *shape* of the specific heat drop at low temperatures is not exponential, but rather proportional to T^3, namely as the third power of the absolute temperature. The Debye model of the specific heat to be hereby presented provides a result that assumes a better fit with experiment.

2.2.3 Debye Model of Solids Specific Heat

Debye model of solids specific heat is based on the existence of vibrational dispersion functions such as presented by Eq. (2.2). The first step is quantization of the said vibrations in a way analogous to the one performed for the harmonic oscillator. In addition, not all wave-vectors are allowed! Real solids have finite dimensions, hence boundary conditions that allow only discrete values for the vibrational modes. As an example, we consider a string of L length stretched between two ends, as demonstrated in Figure 2.4. The possible vibrational modes satisfy $\lambda = 2L \cdot \frac{1}{n}$, where λ is the wavelength and n is a positive integer called **vibrational order**. The belonging wave-numbers satisfy $|k| = \frac{2\pi}{\lambda} = n\frac{\pi}{L}$. Other series of allowed k values are obtained when one or two of the string ends are permitted to move freely in the vertical orientation. It is, however, assumed that due to the macroscopic dimensions of usual solids, details of their boundary conditions, and the corresponding wave-numbers of the allowed oscillations, bear no influence on conclusions related to their macroscopic properties.

The boundary conditions expressed in Figure 2.4 impose a situation of standing waves; however, as we shall see in Section 2.3, a standing wave is a combination of two same-wavelength waves propagating in opposite directions. For mathematical convenience, it is common to choose boundary conditions whose solutions provide directly the option of waves propagating in opposite directions. In the special case we consider, we choose **periodic boundary conditions** that satisfy $u_{s+\tilde{N}} = u_s$ for all s, where $L = \tilde{N}a$ is the crystal length, and a is the lattice period. Thus, we get

$$u_{s+\tilde{N}} = u_0 e^{-i\omega_k t} \cdot e^{i(s+\tilde{N})ka} = u_s = u_0 e^{-i\omega_k t} \cdot e^{iska}.$$

The latter is valid provided that

$$\tilde{N}ka = \pm r \cdot 2\pi$$

$$k = \pm r\frac{2\pi}{\tilde{N}a} = \pm r\frac{2\pi}{L},$$

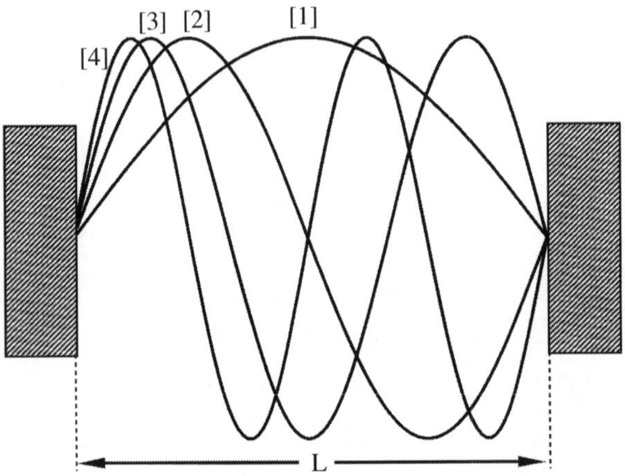

Figure 2.4 Schematic illustration of the outlines of the four lowest vibrational modes permitted for a stretched string of length L tied at both ends. The numbers next to each outline indicate the oscillation order.

where $r = 0, 1.2. \cdots \widetilde{N}(= L/a)$. There is no point in considering k values outside the first Brillouin zone (Fig. 2.2) because they describe an oscillation, which is identical to another k from within that zone. Thus, r is confined to $r = 0, 1.2. \cdots \frac{1}{2}\widetilde{N}$ values, namely a total of $\widetilde{N}$ different values of k. Recall that edge values of k in the first Brillouin zone in fact describe the same oscillation state, because their difference is a reciprocal lattice vector; for simplicity, we assumed that $\widetilde{N}$ is even. In a macroscopic crystal, $\widetilde{N}$ is a huge number, of the order of $10^7 - 10^8$ for a crystal being ~1 cm long. It is thus natural to consider the *density* of k values. The number of states per a unit wave-number in one dimension of k would be $\widetilde{D}_{1D} = 1/(2\pi/L) = (L/2\pi)$. In generalizing this result to a three-dimensional crystal, yet under a simplified case of a primitive cubic crystal whose lattice constant is a and is shaped as a rectangular box whose edges are L_1, L_2, and L_3, the obtained density of states would be $\widetilde{D}_{3D} = \left(\frac{L_1}{2\pi}\right)\left(\frac{L_2}{2\pi}\right)\left(\frac{L_3}{2\pi}\right) = \frac{\boldsymbol{V}}{(2\pi)^3}$, where $\boldsymbol{V} = L_1L_2L_3$ is the crystal volume.

Hereafter we will deal with the spatial density of unit cells $\widetilde{N} = 1/a^3$ (the number of primitive unit cells per unit crystal volume) and the density of wave-number states (the number of wave-number states per unit crystal volume), for which we get

$$\widetilde{D}(\underline{k}) = \frac{1}{(2\pi)^3}. \tag{2.13}$$

That density of states is constant, independent of $\underline{k}$. To simplify the equations, it is further assumed to be independent of details of a macroscopic crystal shape.

We next calculate the $\underline{k}$ states density along the *energy axis* of phonons (or alternatively the angular frequency ω axis), per a crystal unit volume. For that purpose, in Debye approximation, one assumes that the sound wave dispersion curves are the same in all crystal orientations (see Eq. [2.1]); in other words, that the angular frequency depends on the *size* of the $\underline{k}$ vector but not on its orientation. This assumption is indeed a gross approximation, because a distance between adjacent crystalline planes, hence their binding force, depends on the orientation. Still, this assumption provides a useful simplification of its dependent mathematical expressions. Under the said assumption, the energy surfaces (equi-energy surfaces) are spherical. A necessary correction, however, is to redefine the range of wave-vectors: instead of limiting them to inside the first Brillouin zone, we limit them to inside a sphere, such that their total number per unit crystal volume remains $\widetilde{N} = 1/a^3$. Namely

$$\frac{1}{(2\pi)^3} \cdot \frac{4}{3}\pi k_D{}^3 = \widetilde{N},$$

where the sphere radius k_D is the upper limit value for allowed $\underline{k}$ vectors. Then we get

$$k_D = (6\pi^2\widetilde{N})^{1/3} = \left(\frac{6}{\pi}\right)^{1/3} \cdot \left(\frac{\pi}{a}\right) = 1.24\left(\frac{\pi}{a}\right), \tag{2.14}$$

where the Debye sphere radius k_D is longer than the upper limit value of $\underline{k}$ at the edge of the first Brillouin zone along the $\underline{a}$ direction (Fig. 2.1). We further assume that the oscillation frequency is identical for both longitudinal and transverse modes belonging to the same $\underline{k}$ wave-vector (this assumption is an approximation as well; in fact, in most cases, the frequencies of longitudinal waves are considerably larger than of transverse waves). The density of ω states per a unit crystal volume satisfies

$$D(\omega)d\omega = 3\widetilde{D}(\underline{k})d^3\underline{k}.$$

The multiplication by 3 at the right hand side of the equation is because there exist three oscillation modes of the crystal per each value of k : one longitudinal and two transverse, which, according to our assumption, are of an identical frequency. Therefore

$$D(\omega)d\omega = 3\widetilde{D}(\underline{k})4\pi k^2\frac{dk}{d\omega} = \frac{3}{(2\pi)^3}4\pi k^2\frac{1}{v_g(\omega)},$$

where v_g is the sound wave group velocity at ω angular frequency. In addition to the previous approximations, in the Debye approximation, the real dispersion curve, such as calculated in Eq. (2.2), is replaced by an approximated one that exhibits no change of slope up to the maximal allowed value of k. In other words, there is no curving like the one causing the group velocity to vanish at the Brillouin zone edge (Fig. 2.2). The dispersion curve keeps the same slope as the real one occurring at the near-zero values of k (namely $v_g(k) = v_g(0) = v$). The dispersion curve assumes a simple form: $\omega = v \cdot k$ for both transverse and longitudinal waves. Accordingly, the density of angular frequency states is given by

$$D(\omega) = \frac{3}{(2\pi)^3}\cdot\frac{\omega^2}{v^3}. \tag{2.15}$$

A schematic description of the shape of a real dispersion curve and the one obtained by the Debye approximation is provided in Figure 2.5.

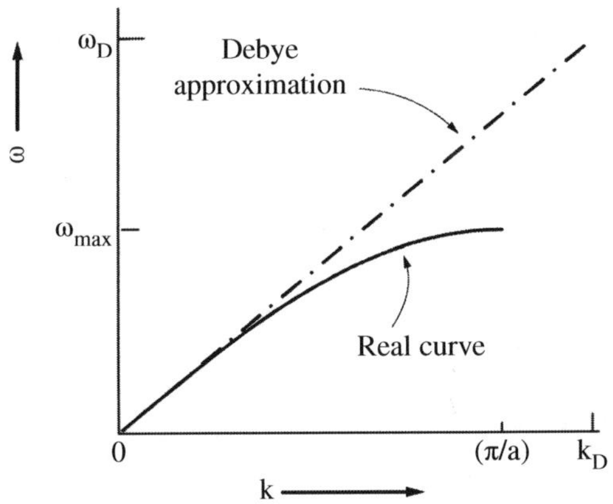

Figure 2.5 Debye approximation for a dispersion curve. The solid continuous line describes a real dispersion curve in the crystal orientation of the fundamental $\underline{a}$ vector of a cubic unit cell. The curve ends at the π/a limit of the first Brillouin zone, with a maximal ω_{max} frequency. The dash-dotted line is the Debye approximation.

The maximal value of the angular momentum given by $\omega_D = v \cdot k_D$ is called **Debye frequency**. The justification for adopting that result is the fact that at low temperatures, practically only low-frequency vibrations are excited, for which the real and Debye curves coincide.

For the specific heat calculation, we start by calculating the thermal energy stored in a crystal *unit volume*. We denote that energy as $\overline{E}$ to be distinguished from the molar thermal energy E as we previously denoted (see Eq. [2.10]).

$$\overline{E} = \int_0^{\omega_D} d\omega \cdot D(\omega)\cdot\langle n(\omega)\rangle\cdot\hbar\omega$$

$$= \int_0^{\omega_D} d\omega\frac{3}{2\pi^2}\cdot\frac{\omega^2}{v^3}\cdot\frac{1}{e^{\hbar\omega/k_BT}-1}\cdot\hbar\omega;$$

$$\overline{E} = \frac{3\hbar}{2\pi^2 v^3}\int_0^{\omega_D}\frac{\omega^3}{e^{\hbar\omega/k_BT}-1}d\omega.$$

As a change of variable, we denote $x = \hbar\omega/k_BT$, then $\omega = \frac{k_BT}{\hbar}x$, $d\omega = \frac{k_BT}{\hbar}dx$, and $x_D = \hbar\omega_D/k_BT$. We get the expression

$$\overline{E} = \frac{3\hbar}{2\pi^2 v^3}\cdot\left(\frac{k_BT}{\hbar}\right)^4\cdot\int_0^{x_D}\frac{x^3}{e^x-1}dx.$$

We are interested in the case of low temperatures, where $k_BT \ll \hbar\omega_D$. In that limit, $x_D \gg 1$. We may then practically take the upper limit of the above integral as ∞, then the integral assumes the numerical $\pi^4/15$ value. Thus, we get

$$\overline{E}\xrightarrow[T\ll\hbar\omega_D/k_B]{}\frac{\hbar\pi^2}{10\text{v}^3}\left(\frac{k_BT}{\hbar}\right)^4. \tag{2.16}$$

We further denote $\theta_D \equiv \hbar\omega_D/k_B$. That quantity is called **Debye temperature**.

$$\theta_D = \frac{\hbar\omega_D}{k_B} = \frac{\hbar\text{v}k_D}{k_B} = \frac{\hbar\text{v}}{k_B}(6\pi^2\widetilde{N})^{\frac{1}{3}}. \tag{2.17}$$

From this result, we extract the speed of sound v, set in Eq. (2.16), and receive

$$\overline{E} = \frac{8\pi^4}{5}\theta_D(k_B\widetilde{N})\left(\frac{T}{\theta_D}\right)^4. \tag{2.18}$$

For obtaining the molar thermal energy, we have to divide the latter expression by $\widetilde{N}$ and multiply by Avogadro's number N_A. We get

$$E = \frac{8\pi^4}{5}\theta_D(k_BN_A)\left(\frac{T}{\theta_D}\right)^4. \tag{2.19}$$

The molar specific heat in the limit of small temperatures would thus be given by

$$C_V = \left.\frac{\partial E}{\partial T}\right]_V = \frac{12\pi^4}{5}(k_BN_A)\left(\frac{T}{\theta_D}\right)^3 \quad \text{for } T \ll \theta_D. \tag{2.20}$$

Indeed, the Debye model for the low temperatures specific heat is consistent with the general experimental results of it being proportional to the third power of the absolute temperature. Carrying out of actual measurements allows obtaining the actual values of the characteristic Debye temperature of each material. Such results exhibit very good agreement with calculated values of that parameter using Eq. (2.17) based on the known speed of sound and the unit cells density. That fit justifies the different approximations that establish the Debye model of specific heat.

Experimental Debye temperatures of different solids are of the order of 100–400 K. Table 2.1 provides a list of Debye temperatures measured for different solid chemical elements.

Einstein's model is still partially suitable for solids whose building blocks are molecules rather than simple atoms. Such molecules vibrate and rotate about their equilibrium posture, whose frequencies exhibit only a small change as a function of the wave-vector size. Each such oscillation contributes to the thermal energy of the crystal, hence to the specific heat, according to the basic principles of Einstein's theory. Those contributions are extremely small in the low-temperature limit, where each molecule could be justifiably approximated as a rigid body.

2.3 A Reminder: Standing and Propagating Waves; Group and Phase Velocity

A wave propagating in an x direction is a function of the form $\psi(\omega t - kx)$. The $\varphi = \omega t - kx$ quantity is the wave phase. We seek the properties of points in space where the phase remains the same, namely satisfying $\frac{d\varphi}{dt} = 0$:

$$\frac{d\varphi}{dt} = \omega - k\dot{x} = 0 \quad \Rightarrow \quad \text{v}_\varphi \equiv \dot{x} = \frac{\omega}{k}.$$

The v_φ quantity is called **phase velocity**.

Table 2.1 Debye temperatures of different elements in their solid phase.

Element name (*Z* number)	Symbol	Debye temperature (*K*)	Element name (*Z* number)	Symbol	Debye temperature (*K*)
Beryllium (4)	Be	1440	Rhenium (45)	Re	300
Carbon[a)] (6)	C	2230	Palladium (46)	Pd	275
Sodium (11)	Na	150	Silver (47)	Ag	215
Magnesium (12)	Mg	400	Cadmium (48)	Cd	172
Aluminum (13)	Al	390	Indium (49)	In	100
Silicon (14)	Si	645	Tin (50)	Sn	260
Potassium (19)	K	100	Antimony (51)	Sb	140
Calcium (20)	Ca	230	Barium (56)	Ba	110
Titanium (22)	Ti	350	Lanthanum (57)	La	150
Chromium (24)	Cr	485	Hafnium (72)	Hf	213
Manganese (25)	Mn	350	Tantalum (73)	Ta	245
Iron (26)	Fe	420	Tungsten (74)	W	310
Cobalt (27)	Co	385	Rhenium (75)	Re	300
Nickel (28)	Ni	375	Osmium (76)	Os	250
Copper (29)	Cu	315	Iridium (77)	Ir	285
Zinc (30)	Zn	250	Platinum (78)	Pt	225
Gallium (31)	Ga	125	Gold (79)	Au	170
Germanium (32)	Ge	374	Mercury (80)	Hg	96
Strontium (38)	Sr	147	Thallium (81)	Tl	100
Zirconium (40)	Zr	280	Lead (82)	Pb	105
Molybdenum (42)	Mo	380	Bismuth (83)	Bi	100
Ruthenium (44)	Ru	400			

a) In the diamond phase.

A group velocity is an effect resulting from interference among different waves of similar frequency and wavelength. For demonstration, we consider the interference result between two such sinusoidal waves of same amplitude

$$\begin{aligned}\psi &= \psi_1 + \psi_2 = A\,\boldsymbol{sin}(\omega_1 t - k_1 x) + A\,\boldsymbol{sin}(\omega_2 t - k_2 x) = \\ &= 2A\,\boldsymbol{cos}\left(\frac{\omega_1 - \omega_2}{2}t - \frac{k_1 - k_2}{2}x\right)\cdot\boldsymbol{sin}\left(\frac{\omega_1 + \omega_2}{2}t - \frac{k_1 + k_2}{2}x\right) \\ &= \left[2A\,\boldsymbol{cos}\left(\frac{\Delta\omega}{2}t - \frac{\Delta k}{2}x\right)\right]\cdot\boldsymbol{sin}(\omega t - kx),\end{aligned}$$

where $\omega \equiv \dfrac{\omega_1 + \omega_2}{2}$, $\Delta\omega \equiv \omega_1 - \omega_2$, $k \equiv \dfrac{k_1 + k_2}{2}$, and $\Delta k \equiv k_1 - k_2$.

The resulted function may be described as a *nearly sinusoidal* wave, whose angular frequency is an average between the frequencies of its two components, its wave number is an average between the wave-numbers of its two components, and its amplitude varies relatively slowly as a function of time and space. The latter said amplitude modulation establishes a propagating wave. The phase velocity of the *amplitude modulation* is the **group velocity** of the composed wave; it is given by

$$\mathrm{v_g} = \frac{\Delta\omega/2}{\Delta k/2} = \frac{\Delta\omega}{\Delta k} \xrightarrow[\substack{continuously\ composed\\ group}]{} \frac{d\omega}{dk}.$$

A **standing wave** is a group of waves that exhibits points in space where the amplitude is steadily zero. For demonstration, we consider the interference result between two sinusoidal waves of equal amplitude, frequency, and wave-vector, yet propagating in *opposite* directions:

$$\begin{aligned}\psi &= \psi_1 + \psi_2 = A\,\mathbf{sin}(\omega t - kx) + A\,\mathbf{sin}(\omega t + kx)\\ &= [2A\,\mathbf{cos}(kx)]\mathbf{sin}(\omega t).\end{aligned}$$

It is readily seen that the group amplitude in that particular case is steadily zero at points satisfying $\cos(kx) = 0$, namely where

$$x_n = \frac{1}{2}\left(\frac{\pi}{k}\right) + n\left(\frac{\pi}{k}\right),$$

for an integer n.

Exercises for Chapter 2

2.1 Assume that crystalline aluminum *Al* is primitive cubic with a lattice constant $a = 2.55$ Å. Its atomic mass is $A = 27$. It is given that the return force constant between adjacent planes is 11.5 N/m; calculate numerically and plot a graph of the dispersion curve of longitudinal sound waves of aluminum, calculate the sound velocity in this material and the energy of an excited phonon belonging to the Brillouin zone surface. Compare the obtained value with the characteristic $k_B T$ thermal energy of a degree of freedom at room temperature (300 K).

2.2 Calculate the occupation number of phonons having $\omega = 10^{13}$ rad/s angular frequency at room temperature (300 K).

2.3 Calculate the contribution of these phonons to the molar specific heat at room temperature (300 K) and at 30 K.

2.4 Calculate the Debye temperature of silicon *Si* based of the following given data: Atomic weight $A = 28.1$; material density $\rho = 2.33\ \mathrm{g/cm^3}$; sound velocity $\mathrm{v} = 5.75 \times 10^5$ cm/s.

2.5 Calculate the temperature dependence of the molar specific heat of silicon *Si* at the low-temperature limit. Base your calculations on data provided in the previous (2.4) exercise.

3 Free Electrons in Metals

3.1 Free Electron Density of States

Much like phonons, electrons also exhibit dispersion curves that relate their kinetic energy to their wave vector. The said dispersion curves derive from de Broglie's relation

$$p = \frac{h}{\lambda}, \text{ or alternatively } \underline{p} = \hbar \underline{k}, \tag{3.1}$$

where $\underline{p} = m\underline{v}$ is the electron linear momentum, $h = 6.626 \times 10^{-34}$ J · s is the **Planck constant**, $\hbar \equiv h/2\pi$ is the **reduced Planck constant** (yet also often addressed as just "Planck constant"), λ is the central wavelength of the wave group that determines the electron motion, and m is the electronic mass. In view of the simple proportionality relation between the electronic linear momentum and the $\underline{k}$ wave vector, the latter is often addressed as a "linear momentum."

The electron kinetic energy E is given by

$$E = \frac{1}{2}mv^2 = \frac{p^2}{2m} = \frac{\hbar^2 k^2}{2m}. \tag{3.2}$$

The last expression is addressed as the dispersion curve of free electrons. The $E(k)$ relation given in Eq. (3.2) is continuous, because the energy may assume any value. However, when electrons are contained inside a solid metal, some of their properties are similar to those of other waves contained inside a closed box. We have already seen in relation to crystal lattice oscillations, that boundary conditions required by the finite nature of the crystal dimensions allow the occurrence of only discrete values for the oscillations wave vectors. Such conclusion is valid for any wave effect. Here again, it is assumed that in macroscopic crystals, the details of the boundary conditions and the different conclusions concerning their belonging allowed k states have no impact on the crystal macroscopic properties. Similarly to the case of crystal inner vibrations described in Section 2.2.3, and for the sake of mathematical convenience, we use the periodic boundary conditions. A further issue causing a difference between free electrons and electrons contained inside a metal is that the latter are exposed to the atomic potential, which exhibits the lattice symmetry. According to **Bloch theorem**, eigenfunctions of the electronic states under such conditions (combination of a crystalline potential symmetry with periodic boundary conditions) exhibit the form

$$\psi_{\underline{k}}(\underline{x}, t) = u_{\underline{k}}(\underline{x}) e^{-i(\omega t - \underline{k} \cdot \underline{x})}, \tag{3.3}$$

where the $u_{\underline{k}}(\underline{x})$ function satisfies the periodic lattice translational symmetry.

Conductors, Semiconductors, Insulators, and Crystal-Growth Technology: A Practical Guide,
First Edition. Zeev Burshtein.

The density of k-space states in one dimension under periodic lattice translations is $(L/2\pi)$, where L is a one-dimensional size of the crystal. In three dimensions, it is $\boldsymbol{V}/(2\pi)^3$, where $\boldsymbol{V}$ is the crystal volume. Because each "allowed" $\underline{k}$ state could occupy two electrons of opposite spins, the density of electron states in k-space per unit crystal volume would be

$$\widetilde{D}(k) = \frac{2}{(2\pi)^3}. \tag{3.4}$$

We turn now to calculate the density of electron states $D(E)$ along the electron energy E axis. Assuming that energy surfaces (equi-energy surfaces) in the k-space are spherical as in Eq. (3.2), and that the energy is independent of the spin state, we get

$$D(E)dE = \widetilde{D}(k)d^3\underline{k} = \frac{2}{(2\pi)^3}4\pi k^2 dk.$$

Using Eq. (3.2), we get $\frac{dE}{dk} = \frac{\hbar^2 k}{m}$ and $k = \frac{\sqrt{2mE}}{\hbar}$. We set it in the above relation and get the following expression for the density of electron states per unit energy per unit crystal volume:

$$D(E) = \frac{\sqrt{2m^3}}{\pi^2\hbar^3}E^{1/2}. \tag{3.5}$$

Despite the fact that we are dealing with electrons confined inside a crystal, it is customary to use the **free electrons** term when the electrons contribute to electrical conductivity (see Section 3.6). Electrons that don't contribute to the electrical conductivity are called **bound electrons**.

3.2 Fermi–Dirac Distribution

In this section, we calculate the probability of an allowed energy level E to occupy an electron. For that purpose, we use the Boltzmann generalized relation, stating that the probability $P_n(E)$ that a thermodynamic system would contain n particles having a total energy E satisfies the relation

$$P_n(E) = Ce^{-\frac{E-\mu n}{k_B T}}, \tag{3.6}$$

where μ is the chemical potential, and the C constant is a normalization factor adjusted to satisfy $\sum_n P_n(E) = 1$. It is readily obvious that Eq. (2.5), which we used for calculating the phonon distribution function, is a special case of Eq. (3.6) for a system whose chemical potential is zero. When applied to electrons, the μ quantity is called **electrochemical potential**.

In the special case of phonons, there was no limit to the number of $\hbar\omega$ phonons that a harmonic oscillator may occupy. Particles of such nature are called **bosons**. In terms of the mass bonded to a spring of Figure (2.3), it means that there is no upper limit to the size of the oscillation amplitude. However, in relation to electrons, the situation is fundamentally different. According to **Pauli exclusion principle**, only a single electron may occupy an allowed quantum state (the electron spin is included as one of the state-characterizing quantum numbers). Particles having such nature are called **Fermions**. It means that n in Eq. (3.6) may assume only one of two values: either "0" or "1". Obviously, the energy and electrochemical potential of an empty state are both zero. Therefore

$$P_1(E) = \frac{Ce^{-\frac{E-\mu n}{k_B T}}}{\sum\limits_{all\ states} P_n(E)} = \frac{e^{-\frac{E-\mu}{k_B T}}}{e^0 + e^{-\frac{E-\mu}{k_B T}}} = \frac{1}{1 + e^{-\frac{E-\mu}{k_B T}}}. \tag{3.7}$$

The occupation probability function given by Eq. (3.7) applies to all free electrons in the metallic solid, because under thermal equilibrium they all belong to the same electrochemical potential μ. In both metals and semiconductors, the electrochemical potential is called **Fermi energy** (or quite commonly **Fermi level**, although it is not necessarily an allowed energy level). It is usually denoted as E_F. Additionally, $P_1(E)$ is rather written as f(E). Rewriting Eq. (3.7), we get

$$\mathrm{f}(E) = \frac{1}{1 + e^{-\frac{E-E_F}{k_B T}}}. \tag{3.8}$$

In Eq. (3.9), we provide an analysis of the behavior of the electron occupation probability f(E) in extreme cases of very low and of very high energies, and at the special $E = E_F$ point.

$$\mathrm{f}(E) = \begin{cases} 1 - e^{-\frac{E_F-E}{k_B T}} & \mathit{for} \quad E < E_F \ \mathit{and} \ E_F - E \gg k_B T \\ \frac{1}{2} & \mathit{for} \quad E = E_F \\ e^{-\frac{E-E_F}{k_B T}} & \mathit{for} \quad E > E_F \ \mathit{and} \ E - E_F \gg k_B T \end{cases}. \tag{3.9}$$

In the limit of low energies (top line in Eq. [3.9]), the probability is very close to 1. It means that practically all allowed levels up to almost E_F will be occupied. In the limit of high energies (bottom line in Eq. [3.9]), the probability turns close to zero. It means that practically only a tiny part of the allowed states in that limit will be occupied. Large change in the occupation probability exists over an energy range that is approximately $3k_BT$ around E_F. The probability there drops from a value close to a unity, down to a value close to being zero.

At room temperature, the $k_BT \cong 0.026$ eV thermal energy is very small compared to typical Fermi energy values in metals (to be specifically calculated later). In the low-temperature limit, the said range of steep probability drops is obviously very narrow. Thus, near the absolute zero temperature, all allowed states will be occupies up to the Fermi energy, each with a single electron, and all allowed states above Fermi energy will be unoccupied. For $E = E_F$, the probability is 1/2, namely the probability for a state to be either occupied or unoccupied is the same. All above-indicated features are expressed in Figure 3.1.

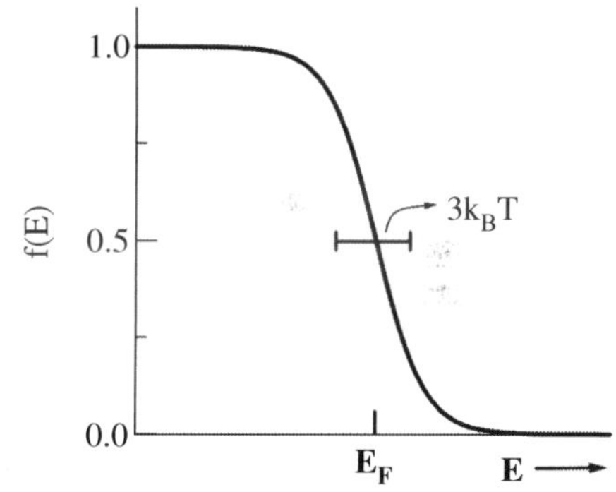

Figure 3.1 Fermi–Dirac distribution function.

3.3 Fermi Energy in Metals

We start by considering the electronic states occupation in the limit of very low temperatures that are close to the absolute zero $T = 0$. The electrons occupy the lowest possible energy states, where each state is occupied by two electrons of opposite spins (see Section 3.1). At the absolute zero temperature, the highest occupied state energy coincides with the **Fermi energy** E_F according to its thermodynamic definition (Section 3.2), and it is correspondingly also denoted as E_F. The linear momentum (wave-number vector) of an electron at that state (Fermi momentum) is denoted as k_F, and its velocity (Fermi velocity) is denoted as v_F. These features are demonstrated schematically in Figure 3.2

An electronic linear momentum satisfies the requirement that a sphere of k_F radius in the linear momentums space includes all N electrons contained in a unit crystal volume. Namely

$$\frac{4}{3}\pi k_F^3 \cdot \frac{2}{(2\pi)^2} = N.$$

Thus, we get:

$$\left.\begin{aligned} k_F &= (3\pi^2 N)^{1/3} \\ E_F &= \frac{\hbar^2}{2m}(3\pi^2 N)^{2/3} \\ v_F &= \frac{\hbar\, k_F}{m} = \frac{\hbar}{m}(3\pi^2 N)^{1/3} \end{aligned}\right\}. \tag{3.10}$$

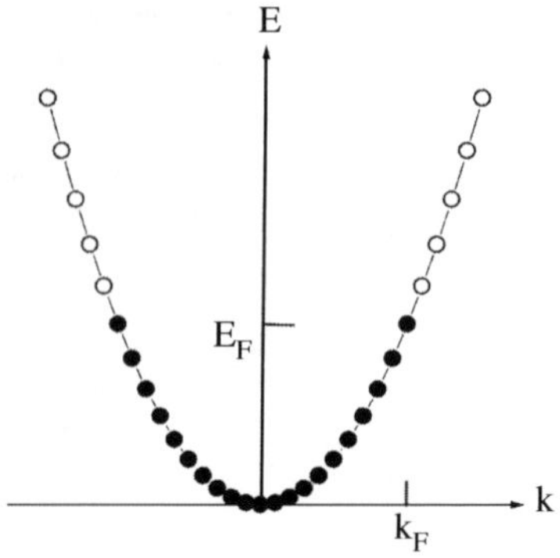

Figure 3.2 One-dimensional dispersion (energy versus wave number) curve for electrons in a metal. The hollow circles describe allowed yet unoccupied energy states; the full circles describe allowed energy states occupied by electrons at the absolute zero temperature.

It is worthwhile knowing the orders of magnitude of the Fermi energy and Fermi velocity in metals. A basic characteristic of metals is that the number of electrons contributing to the electrical conductivity is of the same order of magnitude as the number of the unit cells (the number of atoms in the simple cases). Representative given numerical data would be

$$m = 9.1 \times 10^{-31}\ \text{kg};$$
$$N = 3 \times 10^{22}\ \text{cm}^{-3} = 3 \times 10^{28}\ \text{m}^{-3};$$
$$\hbar = 1.054 \times 10^{-34}\ \text{J} \cdot \text{s}.$$

First basic equation:

$$E_F = \frac{\hbar^2}{2m}(3\pi^2 N)^{2/3}.$$

Setting:

$$E_F = \frac{1.054^2 \times 10^{-68} \times (3 \times 3.14^2 \times 3 \times 10^{28})^{2/3}}{2 \times 9.1 \times 10^{-31}} = 5.75 \times 10^{-19}\ \text{J}$$
$$= 5.75 \times 10^{-19} \frac{\text{J}}{1.6 \times 10^{-19} \frac{\text{J}}{\text{eV}}} = 3.59\ \text{eV}.$$

Second basic equation:

$$v_F = \frac{\hbar}{m}(3\pi^2 N)^{1/3}.$$

Setting:

$$v_F = \frac{1.054 \times (3 \times 3.14^2 \times 3 \times 10^{22})^{1/3}}{9.1 \times 10^{-31}} = 1.1 \times 10^6\ \frac{\text{m}}{\text{s}}.$$

As seen, the Fermi energy size is of the order of several electron-volts, and electrons of such energy reach a velocity of the order of a third of a percent of the speed of light!

Fundamentally, the size of the Fermi level may change as a function of temperature. Qualitatively, its size reduces upon increase of temperature, because the probability for electron occupation at higher energies increases, where the density of states is also larger. Correspondingly, the probability for electron occupation at lower energies decreases, where the density of states is also smaller. The total number of electrons is unchanged; therefore, the Fermi energy will become smaller than given by Eq. (3.10). An explicit calculation of the change ΔE_F (not detailed here) yields the expression $\Delta E_F \cong -0.8k_BT(k_BT/E_F)$. Because $k_BT \ll E_F$, we get that $|\Delta E_F| \ll k_BT$. It is thus a tiny change, and usually ignored. For all practical purposes, Fermi energies in metals are calculated using Eq. (3.10). In semiconductors, on the other hand, the changes are large, exhibiting gross impacts on their electrical conduction properties. Such issues will be dealt with in detail in Section 5.4.

3.4 Thermionic Emission

The Fermi–Dirac distribution shows that at a finite temperature, there is a probability for a small part of the electrons to occupy energy states that are considerably higher than the Fermi energy. Some of these electrons could overcome their binding to the metal and become free in the empty space. Such emission of electrons by a metal is called **thermionic emission**. The energy difference Φ between the lowest energy of a free electron in the vacuum space, and the Fermi energy, is called **work function** of the metal. It is a quantity characteristic of each metal, and is different in different metals. An energy scheme related to the thermionic emission effect is provided in Figure 3.3.

Consider now an electron of energy $E = \frac{\hbar^2}{2m}\left(k_x^2 + k_y^2 + k_z^2\right)$ that overcomes the potential barrier and crosses the metal surface. The k_y and k_z values remain unchanged on crossing the barrier. Therefore, an electron energy crossing the barrier must satisfy the inequality

$$E \geq E_F + \Phi + \frac{\hbar^2}{2m}\left(k_y^2 + k_z^2\right);$$

Or alternatively,

$$E_x \equiv \frac{\hbar^2 k_x^2}{2m} \geq E_F + \Phi.$$

We turn now to calculate $n(k_x)$, defined as the density of electrons per unit wave vector per unit crystal volume, which have sufficient energy to cross the crystal surface into vacuum.

$$n(k_x)dk_x = \frac{2}{(2\pi)^3}dk_x \iint dk_y dk_z.$$

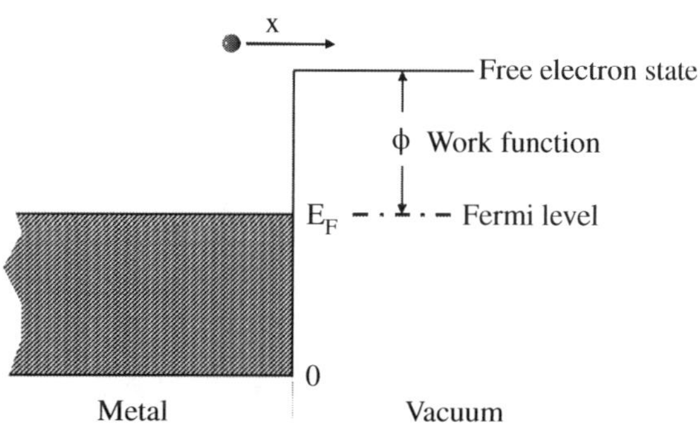

Figure 3.3 Energy scheme in the vicinity of a metal surface. An electron having sufficient energy can overcome the ϕ surface potential barrier (work function) and become free. X marks a Cartesian coordinate perpendicular to the surface.

We denote by $dk_r^2 = k_y^2 + k_z^2$ and form the integration over a circular surface in the (k_y, k_z) plane: we replace the $dk_y dk_z$ infinitesimal segments by $2\pi k_r dk_r$ (see Fig. 3.4). Then

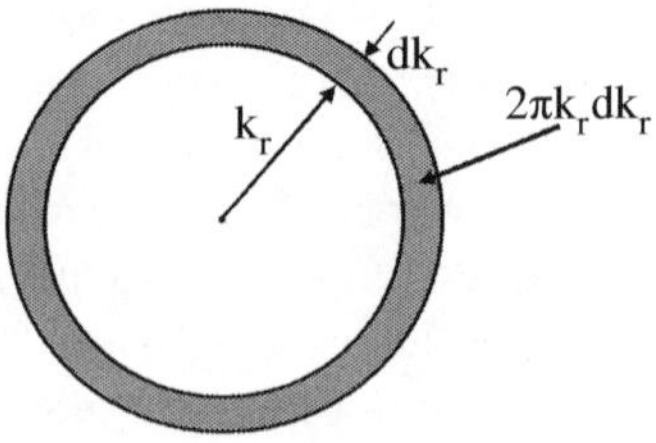

Figure 3.4 Auxiliary drawing to the conversion of the infinitesimal segments in the above integrals.

$$n(k_x)dk_x = \frac{2}{(2\pi)^3} dk_x \int_0^\infty k_r dk_r \mathrm{f}(E).$$

The density of the thermionic emission current is

$$dJ = qv_x n(k_x)dk_x,$$

where q is the electronic elementary charge. Thus

$$dJ = q\frac{\hbar k_x}{m} \cdot \frac{2}{(2\pi)^2} dk_x \int_0^\infty k_r dk_r \ \mathrm{f}(E).$$

$$J = \frac{2q\hbar}{(2\pi)^2 m} \int k_x dk_x \int_0^\infty k_r dk_r \ \mathrm{f}(E).$$

We now set the specific form of Fermi–Dirac distribution function in the $E - E_F \gg k_B T$ limit (Eq. [3.9]). Further recalling that $E = \frac{\hbar^2}{2m}\left(k_x^2 + k_y^2 + k_z^2\right) = \frac{\hbar^2}{2m}\left(k_x^2 + k_r^2\right)$, we get:

$$J = \frac{2q\hbar}{(2\pi)^2 m} \int k_x dk_x \int k_r dk_r e^{-\frac{E-E_F}{k_B T}}$$

$$= \frac{2q\hbar}{(2\pi)^2 m} e^{\frac{E_F}{k_B T}} \int k_x e^{-\frac{\hbar^2 k_x^2}{2mk_B T}} dk_x \int_0^\infty k_r e^{-\frac{\hbar^2 k_r^2}{2mk_B T}} dk_r.$$

We perform a variable change: $\xi \equiv k_r^2$, hence $d\xi = 2k_r dk_r$. We set it above, integrate, and get

$$J = \frac{2q\hbar}{(2\pi)^2 m} \cdot \frac{2mk_B T}{\hbar^2} e^{\frac{E_F}{k_B T}} \int dk_x k_x e^{-\frac{\hbar^2 k_x^2}{2mk_B T}}.$$

We perform an additional variable change: $E_x \equiv \frac{\hbar^2 k_x^2}{2m}$, hence $dE_x \equiv \frac{\hbar^2}{m} k_x dk_x$. We set it in the latter integral, specify the integration limits, and integrate as below:

$$\frac{m}{\hbar^2} \int_{E_F+\Phi}^\infty dE_x e^{-\frac{E_x}{k_B T}} = \frac{mk_B T}{\hbar^2} e^{-\frac{E_F+\Phi}{k_B T}}.$$

Combining all above calculations, we get

$$J = \left[\frac{2qmk_B^2}{(2\pi)^3 \hbar^2}\right] T^2 e^{-\frac{\Phi}{k_B T}} = \boldsymbol{A} T^2 e^{-\frac{\Phi}{k_B T}}. \tag{3.11}$$

The above $\boldsymbol{A} = \left[\frac{2qmk_B^2}{(2\pi)^3 \hbar^2}\right]$ constant is known as **Richardson–Dushman constant**. Its numerical value is 1.2×10^6 A/m^2**deg**^{2} = 120 A/cm^2**deg**^{2}.

According to the last result, the thermionic emission is exponentially larger in metals of smaller work functions, and it increases as a function of temperature.

Experimental measurement of thermionic emission is usually made inside vacuum chambers containing two electrodes connected to a variable high-voltage source. The cathode is made of the studied metal, whose temperature may be varied using a proper heating element. The voltage is applied between the electrodes, and the generated current is measured by a serially connected ammeter, as described schematically in Figure 3.5.

As to the current versus voltage shape, notably, some current through the electrical circuit still exists under a zero external voltage. The reason is that the thermionic emission of electrons turns the cell into an electromotive source. The current initially grows under low external voltages. The reason is that initially emitted electrons accumulate near the cathode, thus forming a negative space charge that prevents the emission of additional electrons. Thus, the zero-voltage current is much smaller than the thermionic emission one. Application of the external voltage reduces the said space charge while increasing the current reaching the anode. On further increase of voltage, the current increase moderates, and eventually the current value saturates. The saturated value is identified as the cathode thermionic emission current. In studying of metal properties, the saturated thermionic emission is measured as a function of temperature. In the results analysis, the ratio J/T^2 is plotted versus $1/k_BT$ on a semi-logarithmic scale. The obtained curve is linear; its slope yields the value of the metal work function Φ, and its ordinate crossing value then yields the Richardson–Dushman $\boldsymbol{A}$ constant. For alkali metals such as cesium, rubidium, and potassium, work functions are in the 2.1–2.3 eV range; work functions of noble metals such as gold and platinum are in the 5.1–5.6 eV range. Table 3.1 provides a list of experimental work functions of many metals. Values of the Richardson–Dushman constant depend on the metal surface cleanliness; however, even for metals that underwent careful surface cleaning procedures, the experimental value may be up to 5 times smaller than the theoretical one expressed by Eq. (3.11). It is usually attributed to intrinsic processes of reflection and scattering of electrons at the materials surfaces.

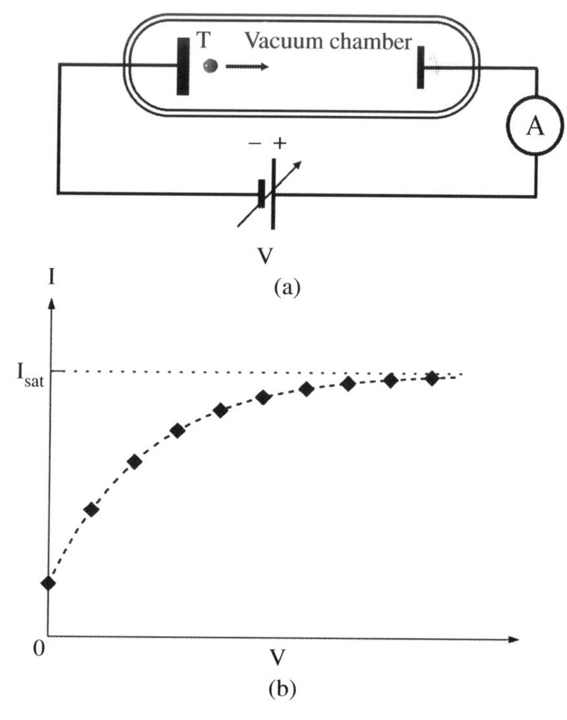

Figure 3.5 Schematic illustration of an experimental setup used in measuring of thermionic electron emission. (a) A cathode heated to a T temperature inside a vacuum chamber emits electrons collected by the anode under an external V voltage. An ammeter connected in series measures the collected current. The dependence of the I current on the collecting V voltage exhibits a characteristic form described in (b). The full rhombus-shaped symbols indicate experimentally obtained results.

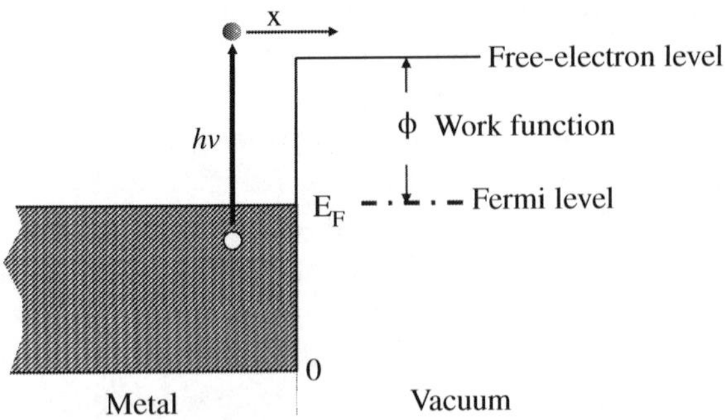

Figure 3.6 Energy scheme in the vicinity of a metal surface. A photon of $h\nu$ energy that is greater than the metal ϕ work function is absorbed inside the metal, thus granting an electron sufficient energy to cross the work function surface barrier and become free outside.

3.5 The Photoelectric Effect

The photoelectric effect is the emission of electrons out of a metal surface resulting by a partial light absorption by the metal. Such emission is enabled only when an absorbed photon energy $h\nu$, where h is the Planck constant and ν is the light frequency, is greater than the metal work function Φ. It should be emphasized that if the photons' energy is smaller than the work function, no emission occurs even if the incident light intensity is very high. The process energy scheme is presented in Figure 3.6.

The maximal kinetic energy that an emitted electron could have as a result of illumination by monochromatic light is given by

$$E_{kin} = \hbar\nu - \Phi. \tag{3.12}$$

It is possible to stop emitted electrons from being collected by a counter electrode by applying a stopping voltage instead of a collecting voltage, namely reversing the voltage polarity compared to that described in Figure 3.5(a) (see Fig. 3.7). Complete stopping of all emitted electrons obtains when the maximal kinetic energy equals qV_S, where q is the electron elementary charge and V_S is the applied stopping voltage measured when the current through the ammeter becomes zero. Namely, $qV_S = \hbar\nu - \Phi$, or

$$V_S = \frac{\hbar}{q}\nu - \frac{1}{q}\Phi. \tag{3.13}$$

It is possible to carry out an experiment of measuring the stopping voltage of photoelectrically emitted electrons under

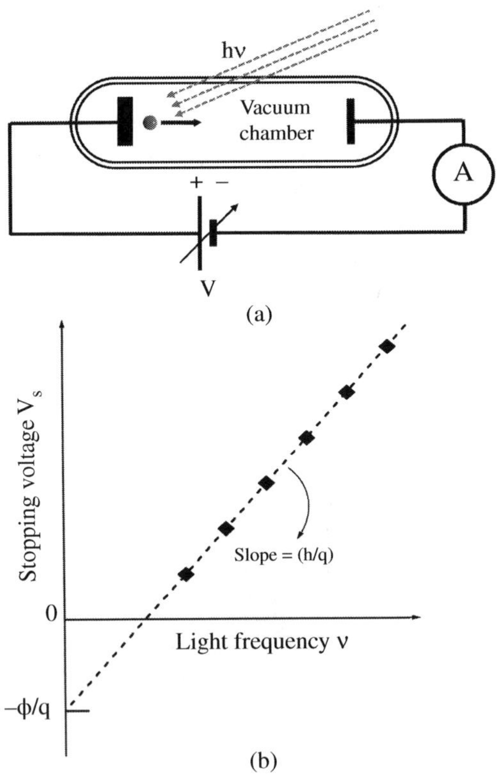

Figure 3.7 (a) Schematic illustration of experimental setup used in measuring the maximal kinetic energy of photoelectrically emitted electrons from a metal surface by determining the stopping voltage. (b) Method of results analysis by plotting the stopping voltage as a function of the incident monochromatic light frequency, allowing determination of both Planck constant and the metal work function. Full rhombus-shaped symbols indicate the experimental results; the dashed line provides their linear fit.

impinging monochromatic light beams of different frequencies. Plotting a graph of the stopping voltage as a function of the light frequency yields a straight line of h/q slope and $-\Phi/q$ intercept on the ordinate axis. The electron elementary charge q is a known quantity, thus such an analysis allows determining both Planck constant h and the metal work function Φ. In all made experiments, the obtained Planck constant is identical, coinciding with the value provided by the historic Planck model of the black-body radiation. Work functions of chemically different metals are different, being a characteristic property of the metal. They turn out to be identical with values obtained by measuring of thermionic emission of the same metals (Section 3.4). The above interpretation of the photoelectric effect was suggested by Einstein in 1921, for which he was awarded the 1928 Nobel prize of physics, because by its essence it provided a direct proof for the validity of quantization of electromagnetic radiation.

We turn now to carry out a detailed calculation of the saturated photoemission. We consider an electron whose energy is given by

$$E = \frac{\hbar^2}{2m}\left(k_x^2 + k_y^2 + k_z^2\right).$$

The k_y and k_z values remain unchanged on crossing the surface potential barrier. Thus, an electron crossing the barrier needs to have an energy that is greater than the barrier height $E_F + \Phi$ plus the part that remains unchanged upon that crossing: $E_r = \frac{\hbar^2}{2m}\left(k_y^2 + k_z^2\right) = \frac{\hbar^2}{2m}k_r^2$. We denote $E_x = \frac{\hbar^2}{2m}k_x^2$. Thus, the condition for being able to cross the barrier is $E_x \geq E_F + \Phi$.

To actually calculate the emission current, we need to express the energy distribution of electrons excited by the incident light absorption near the surface. The distribution is obviously different from that of a thermodynamic equilibrium, and the Fermi–Dirac distribution is no more valid. We then replace the Fermi distribution by an assumed new one, we denote by $f^*(E)$:

$$f^*(E) = \begin{cases} 1 & \text{for} \quad 0 < E < E_F - h\nu \\ 1-\theta & \text{for} \quad E_F - h\nu < E < E_F \\ \theta & \text{for} \quad E_F < E < E_F + h\nu, \end{cases} \tag{3.14}$$

where $\theta \ll 1$ is a quantity proportional to the light intensity, whose essence will be discussed later in detail. The $f^*(E)$ functional shape is based on the limit of Fermi–Dirac distribution at very low temperatures. It is also assumed that the probability of exciting a low-energy electron is independent of its original energy. Thus, under steady light illumination, an electron occupation probability below the Fermi energy reduces by a constant factor θ between E_F and $E_F - h\nu$, and increases by the same factor between E_F and $E_F + h\nu$. The overlooking of temperature effects on the occupation probability is justified as long as $(h\nu - \Phi) \gg k_B T$. Figure 3.8 displays the form of the newly assumed distribution probability function.

We continue by calculating $n(k_x)$, the concentration of electrons satisfying the conditions that enable their crossing the potential barrier

$$n(k_x)dk_x = \frac{2}{(2\pi)^3}dk_x \int dk_y dk_z f^*(E).$$

Their contribution to the current density J would be given by

$$dJ = qn(k_x)dk_x\frac{\hbar k_x}{m} = q\frac{2}{(2\pi)^3}dk_x\frac{\hbar k_x}{m}\int dk_y dk_z f^*(E),$$

then

$$J = q\frac{2}{(2\pi)^3}\frac{\hbar}{m}\int k_x dk_x \int dk_y dk_z f^*(E)$$

Intermediate calculations

(i) $E_x = \dfrac{\hbar^2 k_x^2}{2m} \Rightarrow k_x dk_x = \dfrac{m}{\hbar^2} dE_x;$

We integrate now over a circular surface as per Figure 3.4:

(ii) $\begin{cases} \text{we change } dk_y dk_z \dfrac{\hbar^2 k_x^2}{2m} \text{ into } 2\pi k_r dk_r \\ E_r = \dfrac{\hbar^2 k_x^2}{2m} \Rightarrow 2\pi k_r dk_r = \dfrac{2\pi m}{\hbar^2} \end{cases}$

$$J = q\frac{2}{(2\pi)^3} \cdot \frac{\hbar}{m} \cdot \frac{m}{\hbar^2} \int_{E_F+\phi}^{\infty} dE_x \frac{2\pi m}{\hbar^2} \int_0^{\infty} dE_r \cdot f^*(E).$$

We carry out a change of variable in the latter integral: E instead of E_r. Because $E_r = E - E_x$, then $dE_r = dE$, yet the integration limits should also be correspondingly changed, to receive

$$J = q\frac{2}{(2\pi)^3} \cdot \frac{2\pi m}{\hbar^3} \int_{E_F+\Phi}^{\infty} dE_x \int_{E_x}^{\infty} dE \cdot f^*(E).$$

Intermediate calculation of the latter integral

$$\int_{E_x}^{\infty} f^*(E) dE = \begin{cases} \theta(E_F + h\nu - E_x) & \text{for } E_F < E_x < E_F + h\nu \\ 0 & \text{for } E_F + h\nu < E_x \end{cases}$$

$$J = q\frac{2}{(2\pi)^3} \cdot \frac{2\pi m}{\hbar^3} \cdot \theta \int_{E_F+\Phi}^{E_F+h\nu} (E_F + h\nu - E_x) dE_x = -q\frac{1}{(2\pi)^3} \cdot \frac{2\pi m}{\hbar^3} \cdot \theta (E_F + h\nu - E_x)^2 \Big|_{E_F+\Phi}^{E_F+h\nu}$$

$$J = \theta \frac{2\pi q m}{h^3} (h\nu - \Phi)^2 \quad \text{for} \quad h\nu > \Phi. \tag{3.15}$$

This last result is called **Fowler equation**.

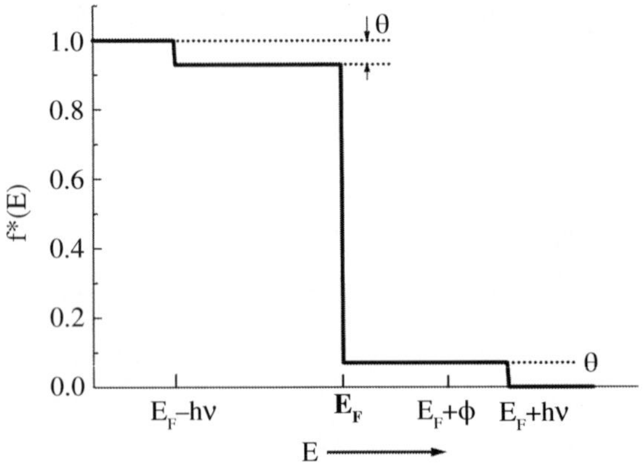

Figure 3.8 Electron occupation probability distribution as a function of energy close to the metal surface at a very low temperature under illumination with a monochromatic light of $h\nu$ photons energy per Eq. (3.14).

We turn to perform a quantitative calculation of the θ parameter. Let F be the photons flux impinging the metal surface. The units of F are photons per unit area per unit time. Then:

$F(1-r)$ is the number of photons absorbed near the metal surface per unit area per unit time, where r is the optical reflectance of the metal surface.

$F(1-r)A$ is the total rate of photons absorbed below the surface of an illuminated A area.

The effective volume where the said photons were absorbed is $A\alpha^{-1}$, where α is the light absorption coefficient. The rate of electron excitation per unit volume would be $\frac{F(1-r)A}{A\alpha^{-1}} = F(1-r)\alpha$. Then $F(1-r)\alpha \cdot \tau$ would be the average concentration of excited electrons near the metal surface up to an α^{-1} depth below the surface, where τ is the effective lifetime of an excited electron.

We further assume that near the metal surface, the dominant said lifetime-determining factor is scattering at the surface. Such assumption is not always valid, but such a mechanism is always an important component in determining the excited electron lifetime. In cases where this approximation is valid, we may write $\tau \approx \frac{\alpha^{-1}}{\langle |v_x| \rangle}$, where $\langle |v_x| \rangle$ is a representative average of an excited electron velocity oriented toward the surface. The density of excited electrons would be $F(1-r)/\langle |v_x| \rangle$. (It is interesting to notice that under dominance of surface scattering assumption, the density of excited electrons is independent of the light absorption depth.) To assess the electron velocity of the excited electrons, we choose $E_F + \Phi$ as a representative energy. Then

$$E_F + \Phi = \frac{\hbar^2}{2m}\left(k_x^2 + k_y^2 + k_z^2\right) = \frac{3\hbar^2}{2m}\left\langle k_x^2 \right\rangle \cong \frac{3\hbar^2}{2m}\langle |k_x| \rangle^2.$$

We then extract $\langle |k_x| \rangle$ from the two sides of the equation, use the relation $\langle |v_x| \rangle = \hbar \langle |k_x| \rangle$, and get

$$\langle |v_x| \rangle \cong \left[\frac{2(E_F + \Phi)}{3m}\right]^{1/2}.$$

The concentration of excited electrons would then be given by

$$N_{exc} \cong F\frac{(1-r)\sqrt{3m/2}}{(E_F + \Phi)^{1/2}}. \qquad \text{(i)}$$

Independently calculated (see Eqs. [3.5] and [3.14]), the excited electron concentration is given by

$$N_{exc} = \int_{E_F}^{E_F+h\nu} f^*(E)D(E)dE = \theta\frac{\sqrt{2m^2}}{\pi^2\hbar^3}\int_{E_F}^{E_F+h\nu} E^{1/2}dE.$$

We approximate the latter integral by assuming that $(h\nu - \Phi) \lesssim 0.2\,\Phi \lesssim 0.1E_F$; the above integral upper limit may be replaced by $E_F + \Phi$; then, the integral is approximated as $\Phi(E_F + \Phi)^{1/2}$. Under such approximation

$$N_{exc} \approx \theta\frac{\sqrt{2m^2}}{\pi^2\hbar^3}\Phi(E_F + \Phi)^{1/2} \qquad \text{(ii)}$$

Comparing Eqs. (i) and (ii), we get $\theta \approx F \cdot \frac{\sqrt{3}\pi^2\hbar^3(1-r)}{2m\Phi(E_F + \Phi)}$. We set this result in the Fowler equation (see Eq. [3.15]) and get

$$\frac{J}{qF} \approx \frac{\sqrt{3}(1-r)}{8\Phi(E_F + \Phi)}(h\nu - \Phi)^2 \equiv C^{(F)}(h\nu - \Phi)^2. \qquad \text{(3.15-a)}$$

The $C^{(F)} \equiv \sqrt{3}(1-r)/8\Phi(E_F+\Phi)$ quantity appearing in Eq. (3.15-a) is called **Fowler coefficient**. Notably, the $\eta \equiv J/qF$ on the left-hand side is actually the quantum efficiency of the photoelectric emission. In experiments, the saturated photoemission current is measured as a function of the incident light frequency. It is analyzed by plotting the square root of the quantum efficiency $(J/qF)^{1/2}$ as a function of $h\nu$. The slope of the obtained linear fit yields the value of $[C^{(F)}]^{1/2}$, and its intersection on the abscissa yields the work function Φ (Fig. 3.9). An excellent agreement exists between work function values determined that way (photoelectric effect) and work functions of same metals obtained via analysis of the thermionic emission effect (Section 3.4). A reasonable agreement is also obtained with the theoretical Fowler coefficient $C^{(F)}$ (Eq. [3.15-a]), although the experimental values are usually smaller than the theoretical ones. The differences are attributed to the existence of light absorption processes that don't involve electron excitations, but mostly to the fact that a considerable fraction of excited electrons are either back-reflected at the surface, or lose their energy by scattering at surface imperfections. The deviation from the theoretical prediction is often used to assess the latter effects.

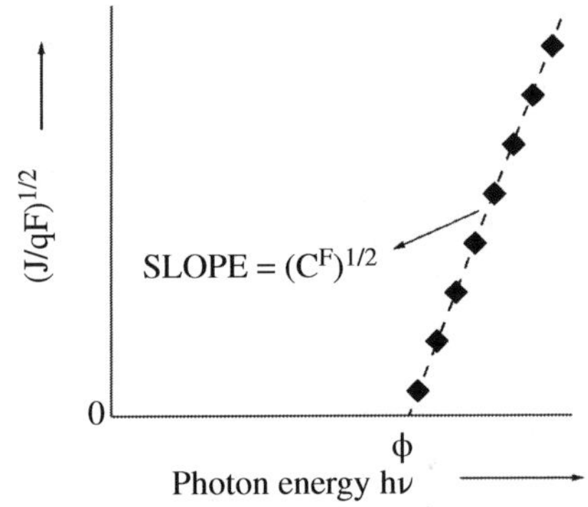

Figure 3.9 Schematic illustration of an analysis method of experimental electron photoemission current results from a metal into vacuum to obtain the metal work function and assess the electron surface scattering. Full rhombus-shaped symbols indicate the experimental results; the dashed line indicates their linear fit.

An energy scheme similar to the one presented in Figure 3.6 also occurs in contacts between a metal and an insulator, or between a metal and a semiconductor, namely an energy barrier forms for transition of electrons from the metal into the other material. In such constructions, it is also possible to achieve photoelectric transmission of electrons from the metal into the other material. That process is known as **internal photoemission**. Equations developed for electron photoemission into vacuum may also be used to describe the internal photoemission, with only minor changes. Additional discussion related to contact effects between metals and semiconductors is provided in Section 8.1.2. Table 3.1 in Section 3.8 provides a work functions list of many metals measured by the different methods described in this section.

3.6 Electrical Conductivity and Mobility

When an external electric field $\underline{\mathcal{E}}$ is applied across a material containing free electrons, a force $-q\underline{\mathcal{E}}$ acts on each electron. The field application accelerates the electron velocity. According to Newton's second law, the acceleration $d\underline{\mathrm{v}}/dt$ satisfies

$$-q\underline{\mathcal{E}} = m\frac{d\underline{\mathrm{v}}}{dt} = m\frac{d}{dt}\left(\frac{\hbar \underline{k}}{m}\right) = \hbar\frac{d\underline{k}}{dt},$$

$$\frac{d\underline{k}}{dt} = -\frac{q}{\hbar}\underline{\mathcal{E}}.$$

The meaning of the last result is that under an external electric field, the entire sphere of electron-occupying states (called **Fermi sphere**) moves at a constant speed in the k space. At time t after the electric field onset, each and every electron experiences a coherent change of its $\underline{k}$

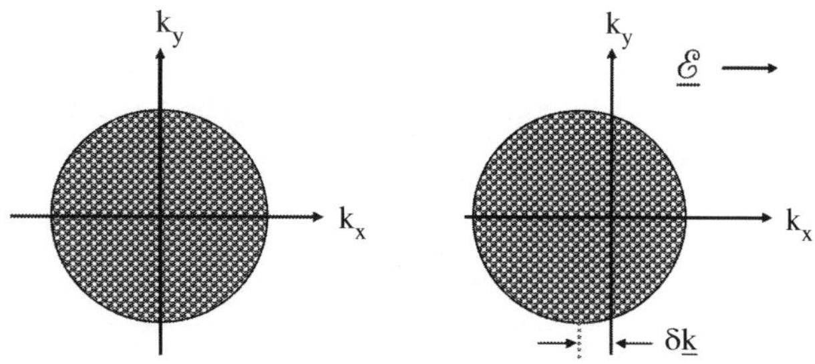

Figure 3.10 Illustration of the average shift of the Fermi sphere of electrons in a metal in the linear momentum space under application of an external electric field. For details, see the text.

state by the amount $\delta \underline{k} = \frac{-q\underline{\mathcal{E}}}{\hbar} t$ (see Fig. 3.10). Due to various scattering processes, such as by lattice vibrations, structural faults, and presence of foreign atoms, the described coherent process is interrupted after some time. We denote by τ the average duration until the said scattering occurs for all moving electrons. τ is usually called **scattering lifetime**. The average linear momentum (wave number) change acquired by each electron is given by $\langle \delta \underline{k} \rangle = \frac{-q\underline{\mathcal{E}}}{\hbar} \tau$. The velocity of each electron will change by $\delta \underline{\mathrm{v}}$, whose average value is given by

$$\langle \delta \underline{\mathrm{v}} \rangle = \frac{\hbar}{m} \langle \delta \underline{k} \rangle = -\left(\frac{q\tau}{m}\right) \underline{\mathcal{E}} \equiv -\mu \underline{\mathcal{E}}. \tag{3.16}$$

The average change in the free electron velocity $\langle \delta \underline{\mathrm{v}} \rangle$ is called **drift velocity**. The $\mu \equiv \frac{|\langle \delta \underline{\mathrm{v}} \rangle|}{\mathcal{E}}$ quantity is called **electron mobility**, and it expresses the value of the drift velocity per unit electric field. The mobility units are $[\mu] = \mathrm{m}^2/\mathrm{V} \cdot \mathrm{s}$. In the above-described case, namely in an isotropic metal, we also have the relation $\mu = \frac{q\tau}{m}$. The mobility size is a characteristic property of each material. The first reason is that the scattering lifetime is a characteristic property of the material. Further, in many materials, the conducting electrons behave as if they exhibit a mass that is different than the mass of a free electron; that **effective mass** is also a characteristic property of each material. The "effective mass" issue will be addressed at length in the next chapter. In all materials, it is possible to increase the mobility of electrons by improving the crystal structure perfection and the chemical purity; in many materials, lowering of temperature also increases the mobility due to lengthening of the scattering lifetime.

Characteristic values of free electron mobilities in different types of materials are listed below:

(i) In liquids: 10^{-5}–$10^{-1}\ \mathrm{cm}^2/\mathrm{V} \cdot \mathrm{s}$
(ii) In molecular solids: $10^{-3} - 10\ \mathrm{cm}^2/\mathrm{V} \cdot \mathrm{s}$
(iii) In metals: 1–$100\ \mathrm{cm}^2/\mathrm{V} \cdot \mathrm{s}$
(iv) In semiconductors: 10–$10^6\ \mathrm{cm}^2/\mathrm{V} \cdot \mathrm{s}$

The electrical current density is given by

$$\underline{J} = -qn\delta \underline{\mathrm{v}} = -q\mu n \underline{\mathcal{E}} \equiv -\sigma \underline{\mathcal{E}}, \tag{3.17}$$

where n is the density of electrons, and $\sigma \equiv q\mu n$ is the **conductivity**. The conductivity units are $[\sigma] = \Omega^{-1}\mathrm{m}^{-1}$. It should be noticed that the density of electrons denoted here by a lowercase n has been denoted by the uppercase N in previous sections; use of the lowercase denotation is the accepted practice in the subject we are dealing with now.

The issue of electron motion under the action of an external electric field may be approached differently. The effective results that were obtained for the distribution of all free electrons may be interpreted as a property of each individual electron. Such imaginary electron is addressed as an

effective electron. For example, in the absence of an external electric field, the mean velocity of all free electrons is zero. Thus, each electron is considered as if at rest in the absence of any electric field. Scattering is regarded as an **effective friction force**, being proportional to the velocity change $\delta\underline{v}$ by the $-\frac{m}{\tau}\delta\underline{v}$ relation. According to Newton's second law, the balance of forces acting on an individual effective electron thus becomes

$$m\frac{d}{dt}(\delta\underline{v}) = -q\underline{\mathcal{E}} - \frac{m}{\tau}\delta\underline{v}. \tag{3.18}$$

Under a steady state, $\delta\underline{v} = \text{const.}$, whose full time derivative vanishes. Then $-q\underline{\mathcal{E}} = \frac{m}{\tau}\delta\underline{v}$, or

$$\delta\underline{v} = -\left(\frac{q\tau}{m}\right)\underline{\mathcal{E}}, \tag{3.19}$$

a result identical to the previously obtained one (Eq. [3.16]), which justifies the said approach. In future discussions, we will treat free electrons in their role as electrical conducting charge carriers as the said effective electrons, unless specifically stated otherwise.

Quite importantly, a minus "−" sign given to the electron elementary charge is written in proper places of all latter expressions. This is because the electron charge is indeed negative and correspondingly is the orientation of its motion relative to the orientation of the applied external field. When a combination of both positive and negative charge carriers will be considered, negative and positive signs will be attached to the q-symbol according to the charge carrier nature. Positive and negative charge carriers coexist in electrolytic solutions, where positive ions play the role of positive charge carriers, and negative ions play the role of negative charge carriers. Similar cases also occurs in solids, as will be described and discussed in detail in different following sections.

3.7 Galvanomagnetic Effects: Cyclotron Motion and Hall Effect

Under a combined application of external electric field $\underline{\mathcal{E}}$ and a magnetic induction field $\underline{\mathcal{B}}$, an electron experiences two forces: $-q\underline{\mathcal{E}}$ resulting from the electric field and a Lorentz force $-q\delta\underline{v}\times\underline{\mathcal{B}}$ resulting from the magnetic field. The balance of forces acting on an electron yields

$$m\frac{d}{dt}(\delta\underline{v}) = -q(\underline{\mathcal{E}} + \delta\underline{v}\times\underline{\mathcal{B}}) - \frac{m}{\tau}\delta\underline{v}. \tag{3.20}$$

Like Eq. (3.18), this equation is expressed for an individual effective electron, and we have demonstrated its relevance to the electric field part. Obviously, the force acting on the real electrons under a magnetic field depends on their real velocity and not on the average velocity change $\delta\underline{v}$. A static magnetic field that acts on the Fermi sphere of electrons "resting" in the k-space distorts its spherical shape, yet by a first-order approximation, the average electron velocity remains zero! As we shall see later, changes in the average velocity are effected only if $\delta\underline{v} \neq 0$. That justifies the writing of the effective electron equation of motion in the form appearing in Eq. (3.20). That equation is written for the MKS units system (also known as SI units system). It may be transformed into the cgs units system by replacing $\mathcal{B}$ by $\frac{1}{c}\mathcal{B}$, where c is the speed-of-light constant, and expressing $\mathcal{B}$ with "gauss" rather than with "Tesla" units (recall that 1 Tesla ≡ 1 Weber/m^2 ≡ 10,000 gauss).

3.7.1 Cyclotron Motion

The cyclotron motion is a particular case obtained under the following conditions:

(i) $\underline{\mathcal{E}} = 0$
(ii) $\tau \to \infty$
(iii) $\underline{\mathcal{B}}$ is oriented in the z direction

The equation of motion thus reduces to the form

$$\left.\begin{aligned} m\frac{d}{dt}\delta \mathrm{v}_x &= -q\mathcal{B}\delta \mathrm{v}_y \\ m\frac{d}{dt}\delta \mathrm{v}_y &= +q\mathcal{B}\delta \mathrm{v} \end{aligned}\right\},$$

whose steady-state solution is

$$\left.\begin{aligned} \delta \mathrm{v}_x &= (\delta \mathrm{v})_0 \boldsymbol{cos}\omega_c t \\ \delta \mathrm{v}_y &= (\delta \mathrm{v})_0 \boldsymbol{sin}\omega_c t \\ \omega_c &= \frac{q\mathcal{B}}{m} \end{aligned}\right\}, \tag{3.21}$$

where $(\delta \mathrm{v})_0$ is some initial velocity granted to the electrons cloud of the Fermi sphere. Thus, the effective electrons exhibit a circular motion of ω_c angular frequency, called **cyclotron rotation**. As to the real electrons, it means the addition of a circular rotation component to their original individual fixed velocity. The (ii) condition means that $\tau \gg \omega_c^{-1}$.

Let us now consider a numerical example. The given data are:

$$\mathcal{B} = 1.0 \text{ Tesla}; \quad m = 9.1 \times 10^{-31} \text{ kg}; \quad q = 1.6 \times 10^{-19} \text{ Coulomb}; \quad \omega_c = ?.$$

Basic equation:

$$\omega_c = \frac{q\mathcal{B}}{m}.$$

Setting:

$$\omega_c = \frac{1.6 \times 10^{-19} \times 1}{9.1 \times 10^{-31}} \cong 1.8 \times 10^{11} \text{ rad/s}.$$

The obtained frequency is of a similar order of magnitude as lattice vibrations. As a result, a strong interaction exists between the electron circular motion components and the lattice vibrations, causing considerable scattering of the electrons. In order to realize the $\tau \gg \omega_c^{-1}$ condition required to render this effect noticeable, studied metals are cooled down to very low temperatures to reduce the phonon population, and simultaneously use very high magnetic fields to correspondingly increase the cyclotron frequency. The effect is measured by exposing the material to a high-frequency (r.f. ≡ radio-frequency) circularly polarized electromagnetic radiation. In resonance, when the r.f. frequency equals ω_c, enhanced absorption of the electromagnetic radiation takes place, whose occurrence indicates an increases in the effective electron rotation radius.

3.7.2 Hall Effect

The Hall effect is a particular case observed under the following conditions:

(i) $\underline{\mathcal{E}} \neq 0$
(ii) $\tau \ll \omega_c^{-1}$
(iii) $\underline{\mathcal{B}}$ is oriented in the z direction

The equations of motion then assume the form

$$\left.\begin{aligned} m\left(\frac{d}{dt}+\frac{1}{\tau}\right)\delta \mathrm{v}_x &= -q(\mathcal{E}_x+\mathcal{B}\delta \mathrm{v}_y) \\ m\left(\frac{d}{dt}+\frac{1}{\tau}\right)\delta \mathrm{v}_y &= -q(\mathcal{E}_y-\mathcal{B}\delta \mathrm{v}_x) \\ m\left(\frac{d}{dt}+\frac{1}{\tau}\right)\delta \mathrm{v}_z &= -q\mathcal{E}_z \end{aligned}\right\}.$$

Under steady-state conditions, the time derivatives are zero. Recalling the $\mu = q\tau/m$ relation, the equations reduce to the form

$$\left.\begin{aligned} \delta \mathrm{v}_x &= -\mu\mathcal{E}_x - \mu\mathcal{B}\delta \mathrm{v}_y \\ \delta \mathrm{v}_y &= -\mu\mathcal{E}_y - \mu\mathcal{B}\delta \mathrm{v}_x \\ \delta \mathrm{v}_z &= -\mu\mathcal{E}_z \end{aligned}\right\},$$

whose solutions for $\delta \mathrm{v}_x$ and $\delta \mathrm{v}_y$ is

$$\left.\begin{aligned} \delta \mathrm{v}_x &= \frac{-\mu}{1+(\mu\mathcal{B})^2}(\mathcal{E}_x - \mu\mathcal{B}\mathcal{E}_y) \\ \delta \mathrm{v}_y &= \frac{-\mu}{1+(\mu\mathcal{B})^2}(\mathcal{E}_x + \mu\mathcal{B}\mathcal{E}_x) \end{aligned}\right\}.$$

We include the relation $\underline{J} = -qn\delta\underline{\mathrm{v}}$ and get

$$\left.\begin{aligned} J_x &= \frac{qn\mu}{1+(\mu\mathcal{B})^2}(\mathcal{E}_x - \mu\mathcal{B}\mathcal{E}_y) \\ J_y &= \frac{-\mu}{1+(\mu\mathcal{B})^2}(\mathcal{E}_y + \mu\mathcal{B}\mathcal{E}_x) \end{aligned}\right\}. \tag{3.22}$$

We now look at a particular case of a conducting material designed in the form of a long and thin rectangular plane-parallel plate whose faces are oriented parallel to the X-, Y-, and Z-axes, as described in Figure 3.11. A current of J_x density is made to flow in the X-axis direction by added

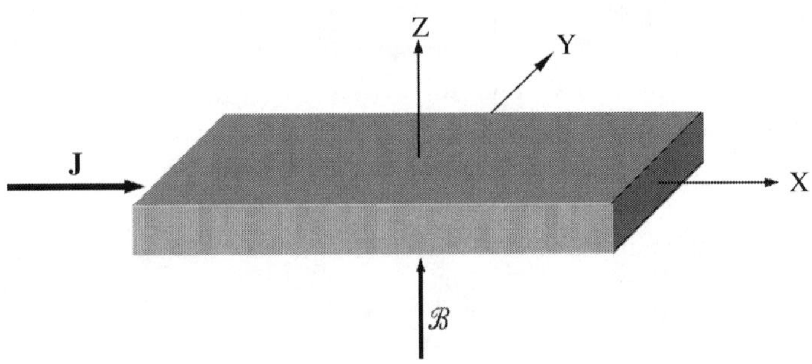

Figure 3.11 Auxiliary drawing demonstrating the orientations of the magnetic induction field $\mathcal{B}$, the electrical current density J, and the rectangular plane-parallel conducting material, in relation to the Cartesian axis system used in the discussion of the Hall effect.

electrical contacts and application of an external $\mathcal{E}_x$ electric field in that direction. The current in the Y direction is necessarily zero. Setting these conditions in Eq. (3.22), we get

$$\frac{\mathcal{E}_y}{\mathcal{E}_x} = -\mu\mathcal{B}; \quad J_x = q\mu n\mathcal{E}_x. \tag{3.23}$$

The appearance of an $\mathcal{E}_y$ electric field resulting by applications of the $\mathcal{B}$ magnetic induction field in the Z direction is called **Hall effect**. Recall condition (ii) in the beginning of this subsection; it readily implies that $\mu\mathcal{B} \ll 1$, then also $|\mathcal{E}_y| \ll |\mathcal{E}_x|$. To understand the physical mechanism responsible for the appearance of $\mathcal{E}_y$, we consider an electron moving in the plate whose coordinated motion with all other free electrons generates the X-oriented current. The electron, actually carrying a negative charge, moves in the opposite direction. On application of the magnetic field, a Lorentz force given by $\underline{F} = (-q)\delta\underline{v} \times \underline{\mathcal{B}}$ is generated, which acts on the electron. In the present configuration, it is oriented oppositely to the Y-axis orientation. Before a steady state forms, the moving electrons will deviate a bit from their original track toward −Y direction and accumulate on the corresponding plate side-wall. A small electric field then forms, compensating the Lorentz force caused by the magnetic field.

Measuring the Hall electric field $\mathcal{E}_y$ generated between the two plate side-walls allows calculating the mobility μ by using Eq. (3.23). A surprising result is that in many solids, the Hall voltage polarity is opposite to the expected polarity presented in our previous discussion. It indicates the presence of positive charge carries in those solids. (It should be noticed that the orientation of the Lorenz force is the same for positive or negative charge carriers that form a current in the same direction; thus, positive charges will accumulate at the same side-wall indicated for accumulation of negative charge carriers in our previous discussion; therefore, the polarity of the Hall effect reverses). The existence of positive charge carriers is a basic property of semiconductors and will be discussed extensively in our following chapters.

A combined measurement of a Hall effect $\mathcal{E}_y$ and the current density J_x, and knowing the size of the magnetic field $\mathcal{B}$ allows calculating the electron density n. For that purpose, a **Hall coefficient** R_H quantity is defined by

$$R_H \equiv \frac{\mathcal{E}_y}{J_x\mathcal{B}} \cong \frac{\mathcal{E}_y}{q\mu n\mathcal{E}_x \cdot \mathcal{B}} = -\frac{1}{qn}. \tag{3.24}$$

Measuring of the Hall coefficient allows a straightforward extraction of the electron density n.

3.8 Appendices on Metals' Physical Properties and Applications

3.8.1 Work Functions of Metals

Table 3.1 Work functions of different metals.

Name	Chemical symbol	Work function (eV)	Name	Chemical symbol	Work function (eV)
Cesium	Cs	2.10	Aluminum	Al	4.30
Rubidium	Rb	2.23	Silver	Ag	4.30
Potassium	K	2.32	Titanium	Ti	4.30
Barium	Ba	2.61	Tin	Sn	4.35
Strontium	Sr	2.62	Iron	Fe	4.46

(Continued)

Table 3.1 (Continued)

Name	Chemical symbol	Work function (eV)	Name	Chemical symbol	Work function (eV)
Calcium	Ca	2.91	Mercury	Hg	4.46
Lithium	Li	2.91	Tungsten	W	4.47
Lanthanum	La	3.20	Antimony	Sb	4.47
Thorium	Th	3.50	Molybdenum	Mo	4.67
Scandium	Sc	3.51	Ruthenium	Ru	4.68
Thallium	Tl	3.80	Copper	Cu	4.68
Magnesium	Mg	3.80	Rhenium	Re	4.95
Hafnium	Hf	3.90	Cobalt	Co	4.95
Zirconium	Zr	4.10	Beryllium	Be	4.95
Tantalum	Ta	4.17	Rhodium	Rh	4.95
Lead	Pb	4.17	Tellurium	Te	4.98
Bismuth	Bi	4.17	Nickel	Ni	5.05
Cadmium	Cd	4.17	Gold	Au	5.05
Gallium	Ga	4.26	Palladium	Pd	5.07
Zinc	Zn	4.28	Iridium	Ir	5.18
Niobium	Nb	4.30	Platinum	Pt	5.59
Vanadium	V	4.30			

3.8.2 Photomultipliers

A most important application of the photoelectric effect is the measuring of visible and near-visible electromagnetic radiation using a device called **photomultiplier**. The device is based on a metallic cathode that emits electrons upon incidence by light whose photons energy is larger than the metal work function. The photon wavelength whose photon energy is equal to the cathode work function is called **cutoff wavelength** λ_{co}. The initial emission quantum efficiency (namely the number of emitted electrons per number of absorbed photons) obviously depends on the photons' energy, and its maximal obtained value is about $\eta = 10\%$. The emitted electrons are accelerated while impacting through a series of about ten electrodes called **dynodes**. The voltage between adjacent dynodes is typically about 100 V. An accelerated electron of about 100 eV kinetic energy impinging a dynode causes a secondary emission of approximately 4 additional electrons, thus inducing an exponential increase of the electrical current reaching the anode. The said construction is hosted inside a high-vacuum tube. The entire device amplification, namely the number of electrons collected at the anode relative to the number of electrons emitted at the cathode, is about $g = 5^{10} \approx 10^7 \frac{collected\ electrons}{emitted\ electrons}$. A typical maximal amplification of a photomultiplier, namely the number of electrons collected at the anode per incident photon, is of the order of $\eta \cdot g \approx 0.1 \times 10^7 = 10^6 \frac{electrons}{incident\ photon}$.

Figure 3.12 presents the construction principle of a photomultiplier. The output resistor R is external to the circuit of the light detection. The C_p capacitor in most devices is the electrical capacitance between the anode and ground connection (thus granted the "parasitic" adjective).

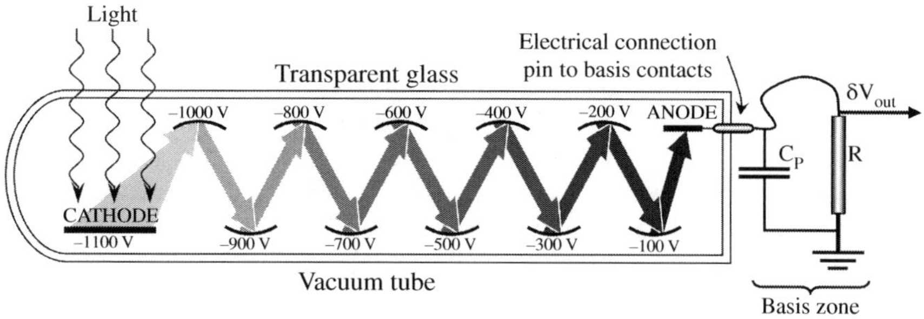

Figure 3.12 Schematic illustration of the construction and operation principle of a photomultiplier tube. The darkening of the arrows coming out of each dynode illustrates the increase in the intensity of the electron current as a result of secondary emissions of electrons caused by former accelerated electrons impinging the dynode.

In actual engineering, the dynodes are arranged spirally around the longitudinal axis of the photomultiplier, and not on two opposite sides as described in the diagram. This ensures that the direction of movement of the electrons is maintained until very close to the collecting dynode, thus optimizing the rate of electron impact on each dynode. The typical distance between dynodes is about 1 cm. The total transition time of the electron beam between the cathode and anode is about 20 ns, and it obviously depends on the distance between adjacent dynodes, their total number, and the inter-dynode voltage. The transition time that determines the voltage signal rise over the output resistance R (and its parasitic parallel capacitance C_p) is that between the last dynode and the anode, and is about 2 ns.

The voltages across the different dynodes are supplied by a voltage divider comprising serially connected resistors described in Figure 3.13. Its inner electrical connections are included in the tube basis, usually positioned close to the anode (right side in Fig. 3.12). Thin metallic wires correspondingly connect to each target dynode inside the tube (only the penetrating pin connected to the anode is depicted in Fig. 3.12). Sometimes parallel capacitors C are connected across the several last resistors to compensate against effective resistance reduction due to large parallel currents developing between adjacent dynodes inside the tube. Typical selected resistors resistances may vary between several dozen kΩ up to about a MΩ. It depends on the prospected light intensities, and the cooling means required to minimize heating up of the individual resistors entailing subsequent heat flow toward the dynodes.

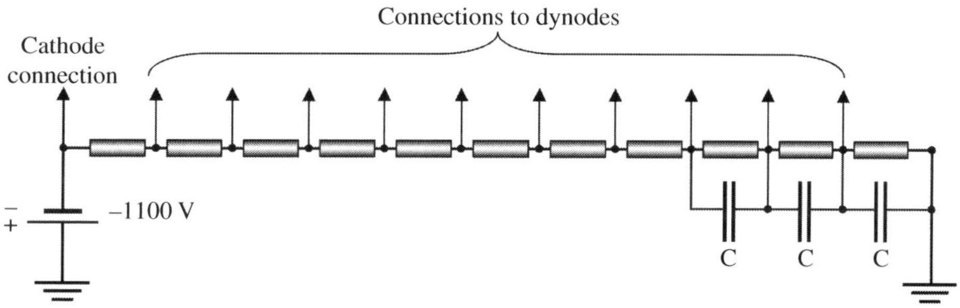

Figure 3.13 Scheme of an electrical circuit that distributes the voltage among the different photomultiplier dynodes.

One of the main reasons for noise in a photomultiplier electrical circuit is thermionic emission, especially at the cathode, but also at the different dynodes. It is possible to reduce such emissions by cooling the entire photomultiplier device, while simultaneously reducing the cathode and dynodes heating due to heat generated in the voltage divider resistors.

The spectral response range of a photomultiplier depends on two factors: one is the work function of the metal serving as cathode, which determines the long wavelength cutoff λ_{co}. The other is the optical transparency range of the entrance window (often just the vacuum tube glass material); it usually determines the short wavelength limit $\lambda_{g\ell}$. A most common cathode material is cesium antimonide Cs_3Sb, which is also used to coat the dynodes, and is wavelength sensitive up to slightly over 700 nm. Sensitivity of up to 1700 nm is achieved by using indium–gallium–arsenide $InGaAs(Cs)$-coated indium phosphide InP substrates. The coating material is a degenerate n-type semiconductor, and the photoemitted electrons initially occupy its conduction band. The long-wave cutoff photon energy reflects the chemical affinity of that semiconductor, being about 0.73 eV. Extended discussions on the matter of semiconductors are provided in the following chapters; as to the specific affinity issue, see Section 8.1.2. The practical response limit on the short wavelength (ultraviolet) side is about 170 nm, and as indicated, it is determined by the tube glass material.

Light intensity measurements are formed by either of two methods: one is measuring of separate single-photon absorption events, or of discrete, almost simultaneously absorbed cluster of photons; the other is a continuous measurement of the light intensity. The latter is used when durations between arrival of single photons, or of single clusters of photons, are too short to be distinguished between.

3.8.2.1 Single-Photon Counting

We calculate now the voltage signal developed over the output resistor R following emission of a single electron resulted by absorption of a single photon. As previously calculated, a sum of $N = 10^7$ electrons would reach the anode. The parasitic capacitance of the anode coupling to ground is $C_p \cong 10\,\text{pF} = 10^{-11}$ F. The signal height is given by

$$\delta V_{out} = \frac{qN}{C_p} = \frac{1.6 \times 10^{-19} \times 10^7}{10 \times 10^{-12}} = 0.16\text{ V} = 160\text{ mV},$$

where q is the electron elementary charge. Such size of signals allows registering and counting the appearance of single photons. In a photon-counting mode, it is necessary that the C_p parasitic capacitance won't discharge during the signal development duration ($t_t \cong 2$ ns), but on the other hand, be fully discharged between two successive signals. Namely, one needs that $RC_p \gg 2$ ns:

$$R \gg \frac{t_t}{C_p} = \frac{2 \times 10^{-9}}{10^{-11}} = 200\ \Omega.$$

As an example, a common choice is $R = 20 \times 200\,\Omega = 4\,\text{k}\Omega$. In such case, the counting rate f is limited to a case where a typical duration between successive pulses is at least 5 durations of the capacitor charge decay: $f \precsim 5 \times 10^6$ Hz. Figure 3.14 shows a voltage versus time trace developed across an R resistor resulted by a single electron photoemitted at the cathode, as displayed on an oscilloscope screen. The trace clearly shows the voltage rise across the capacitance during the flow of electrons from the last dynode toward the anode, and its following decay due to the capacitor C_p discharge through the R resistor.

A common commercial application of a photomultiplier rather relates to detection and spectroscopic analysis of X-ray radiation fields. A device called scintillation counter comprises a

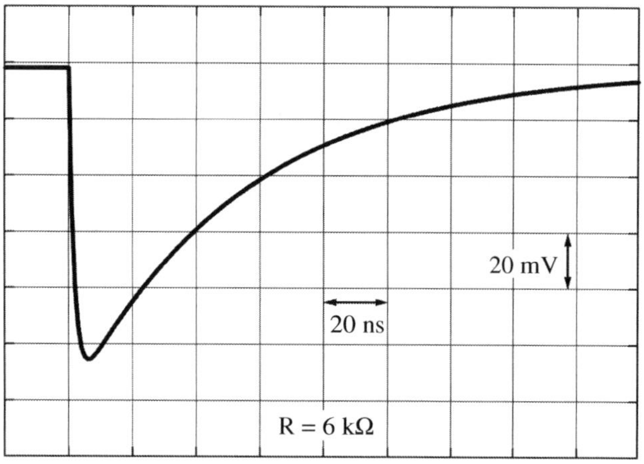

Figure 3.14 A typical voltage versus time trace resulting from emission of a single electron at the cathode as displayed on an oscilloscope screen developed over an $R = 6\ k\Omega$ resistor external to a photomultiplier tube.

photomultiplier where a special type crystal is placed close to the glass tube in front of the cathode. The said (scintillating) crystal emits a visible light pulse immediately after absorption of a gamma or an X-ray photon. Stopping and absorbing mechanisms of such high-energy photons are described in more detail in Section 9.1.2. Some of the different scintillating crystals used for that purpose are thallium-doped sodium iodide $NaI(Tl)$, bismuth germanate $Bi_4Ge_3O_{12}$ (abbreviated as BGO), and cadmium tungstate $CdWO_4$. The scintillation intensity (namely the number of emitted visible photons) is usually proportional to the energy of the absorbed gamma or X-ray photon. In most scintillating materials, the emitted light pulse duration (called "lifetime") is of the order of a tenth microsecond, and in other scintillators, it is of several dozen microseconds. In $NaI(Tl)$, the lifetime is about 0.23 μs; in $Bi_4Ge_3O_{12}$, it is about 0.3 μs; and in $CdWO_4$, it may range between 5 and 20 μs. A part of the emitted photons are oriented toward the cathode, eventually inducing the said voltage signal across the R resistor. The mere appearance of the signal indicates the occurrence of the gamma or X-ray photon absorption event, and its height is proportional to the photon energy. In that way, it is possible to obtain a spectroscopic mapping of the photon energies that existed in the radiation field. Aside, on using a sodium iodide scintillating crystal, care should be taken to prevent exposure to free air; being hygroscopic, it tends to absorb water vapor that then dissolves its surface, causing scattering of the light originally intended to pass through toward the photomultiplier cathode.

3.8.2.2 Photomultiplier Responsivity Under Continuous Light Intensity

Under light intensities that are considerably larger than addressed in the previous subsection, the light intensity may be simply assessed by measuring the voltage developed across the R resistor, namely measuring of $\delta V = \delta V_{out}$ (see Fig. 3.12).

The device **responsivity** is defined as the ratio between the voltage signal δV recorded at the device output and the power of light P incident on its sensitive surface. The formal definition is $\mathscr{R} \equiv \delta V/P$. For obtaining a general expression based on the photomultiplier properties, we denote by A the cathode surface area and by F the incident photons flux density (in units of photons per unit area per unit time). For simplicity, we assume that all photons are of same λ wavelength. The photomultiplier plays a role of a current source relative to the output resistor R. The electrical current would be

$$I = q \cdot \eta F \cdot A \cdot g. \tag{3.25}$$

The voltage between the resistor edges would be

$$\delta V = IR = q\eta FAgR. \tag{3.26}$$

The power of the incident light is given by $P = FA \cdot h\nu$; therefore, the responsivity would be

$$\mathscr{R} \equiv \frac{\delta V}{P} = \frac{q\eta FAgR}{FA \cdot h\nu} = \frac{q\eta gR}{h\nu} = qgR \cdot C^{(F)} \frac{(h\nu - \Phi)^2}{h\nu}, \tag{3.27}$$

where the cathode $C^{(F)}$ Fowler constant is given by (see the text that follows Eq. [3.15-a])

$$C^{(F)} = \sqrt{3}(1 - r)/8\Phi(E_F + \Phi), \tag{3.28}$$

where r is the reflection factor of the incident light, Φ is the cathode material work function, and E_F is its Fermi energy.

We denote by $\lambda_{co} = hc/\Phi$ (the cutoff wavelength), and in addition $h\nu = hc/\lambda$, and get

$$\mathscr{R}(\lambda) = qgRC^{(F)} \frac{hc}{\lambda} \left[1 - \frac{\lambda}{\lambda_{co}}\right]^2 \textit{for } \lambda_{g\ell} < \lambda < \lambda_{co}. \tag{3.29}$$

3.8.3 Electron Guns

One of the most common applications of thermionic emission is a device called **electron gun**. It comprises a vacuum tube (not necessarily a transparent glass) containing an electron emitting cathode heated to a high temperature. These electrons are accelerated by an electric field and, in many cases, shaped into a quasi-parallel beam by combinations of grids and electric and magnetic fields. The cathode material is almost always tungsten W by virtue of its endurance up to temperatures as high as 2400 °C. The wire forming the heating element is usually also tungsten made. A schematic illustration of the basic electron gun construction is presented in Figure 3.15

Until about 75 years ago, electron guns had a widespread use as tubes for amplification, processing, and shaping of electronic signals in radio and television transmitters and receivers. However, they have been almost entirely replaced by semiconductor devices.

Until about two decade ago, they were a source for receiving images on television and oscilloscopes screens, where the impact orientation of the beams producing those screen images were controlled by inner capacitor plates that applied electric fields perpendicularly to the beam orientation. They are still used for X-rays producing tubes, where the electrons are accelerated by voltages of tens of thousands up to millions volts. When the electrons impact a metal anode, the

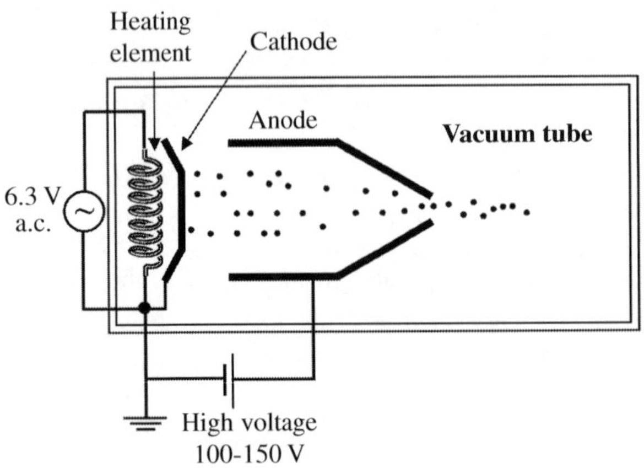

Figure 3.15 Schematic illustration of a basic electron gun construction. It displays a case where electrons emitted from a heated cathode are accelerated and shaped to form a narrow, quasi-parallel beam propagating in a high-vacuum container.

anode emits the X-rays. For acceleration voltages V_p ranging between tens of thousands and hundreds of thousands volts, the emitted radiation spectrum exhibits a smooth continuous spectrum up to a maximal photon energy of qV_p, where q is the electronic elementary charge. The physical mechanism responsible for creation of the said continuum is attributed to **deceleration** of the electron velocity while penetrating the anode material (the German **Bremsstrahlung** term, meaning "stopping radiation" is often used). On top of this continuum, narrow emission peaks are obtained, emanating from the inner target electron transitions while returning to fill emptied shells of the anode atoms, mainly the K atomic shells that were emptied by the impact of the high-energy electrons. For example, on impact of a tungsten W anode, characteristic photon energies are $K_{\alpha 1} = 59.319$ keV; $K_{\alpha 2} = 57.982$ keV; $K_{\beta 1} = 67.245$ keV; $K_{\beta 2} = 66.952$ keV. On impact of a copper Cu anode, the characteristic photon energies are $K_{\alpha 1} = 8.048$ keV; $K_{\alpha 2} = 8.028$ keV; $K_{\beta} = 8.905$ keV. These generated X-rays are widely used in various medical, industrial, and scientific research applications.

Exercises for Chapter 3

3.1 The following table shows the results of a set of measurements of thermionic emission from a certain metal as a function of temperature

Temperature (°C)	330	475	530	625
Current density (A/cm^2)	7×10^{-9}	10^{-5}	7×10^{-5}	1.4×10^{-3}

Calculate the metal work function in units of electron-volt (eV), and its Richardson–Dushman constant.

3.2 The following table summarizes experimental results of photoelectric electron emission current densities obtained under illumination by different light wavelengths

Wavelength (Å)	5390	4860	4070	3750
Light intensity (photons/cm$^2 \cdot$ s)	10^{16}	8×10^{14}	4.2×10^{15}	6×10^{15}
Emission current density (A/cm^2)	3.6×10^{-8}	9.5×10^{-9}	1.65×10^{-7}	4.05×10^{-7}

The density of electrons in the metal is 3.8×10^{22} cm^{-3}, and the light reflection coefficient at all wavelengths is 0.95.

(a) Calculate the metal work function and Fermi energy in units of electron-volt (eV).
(b) Calculate the experimental value of the Fowler coefficient and compare it with the theoretically expected value.
(c) What should be the metal temperature at which the thermionic emission would be the same as the specific photoelectric current measured at 5390 Å.

3.3 (a) Calculate the electron mobility of a metal whose resistivity is $1.5 \times 10^{-6}\ \Omega \cdot$ cm, and its electron density is 5×10^{22} cm^{-3}.
(b) Calculate the electrons' scattering lifetime in that metal.

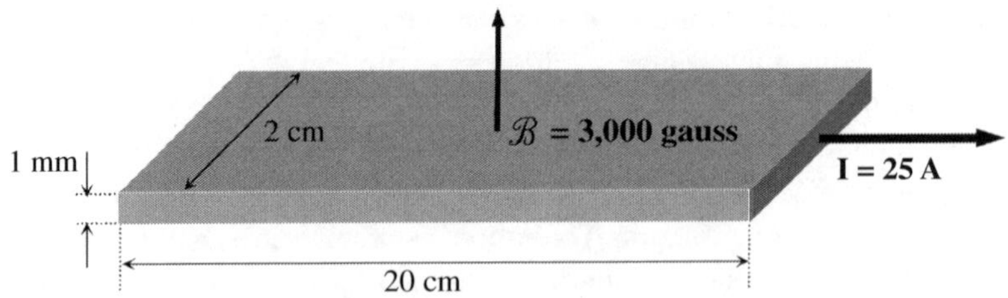

Figure 3.16 Auxiliary drawing demonstrating the orientation and strength of the magnetic induction field $\mathscr{B}$, the electrical current I, and the rectangular plane-parallel conducting material dimensions, in relation to the given data of Exercise 3.3.

(c) Figure 3.16 displays a rectangular plane-parallel plate of the above given metal while carrying a current under a magnetic field perpendicular to the large plate face. Calculate the developed Hall voltage, and indicate its orientation. (A reminder: 1 Tesla = 10,000 gauss).

3.4 Figure 3.17 presents four oscilloscope traces displaying the time dependence of the voltage developed over the R exit resistor of a photomultiplier for different values of its resistance, related to discrete voltages appearing as a result of single-photon absorption events at the cathode. Calculate the value of the parasitic electrical circuit capacitance C_p and the photomultiplier gain g. Given that the photomultiplier comprises 10 dynodes, find out the average number of electrons emitted by a single dynode following an impact by an accelerated electron.

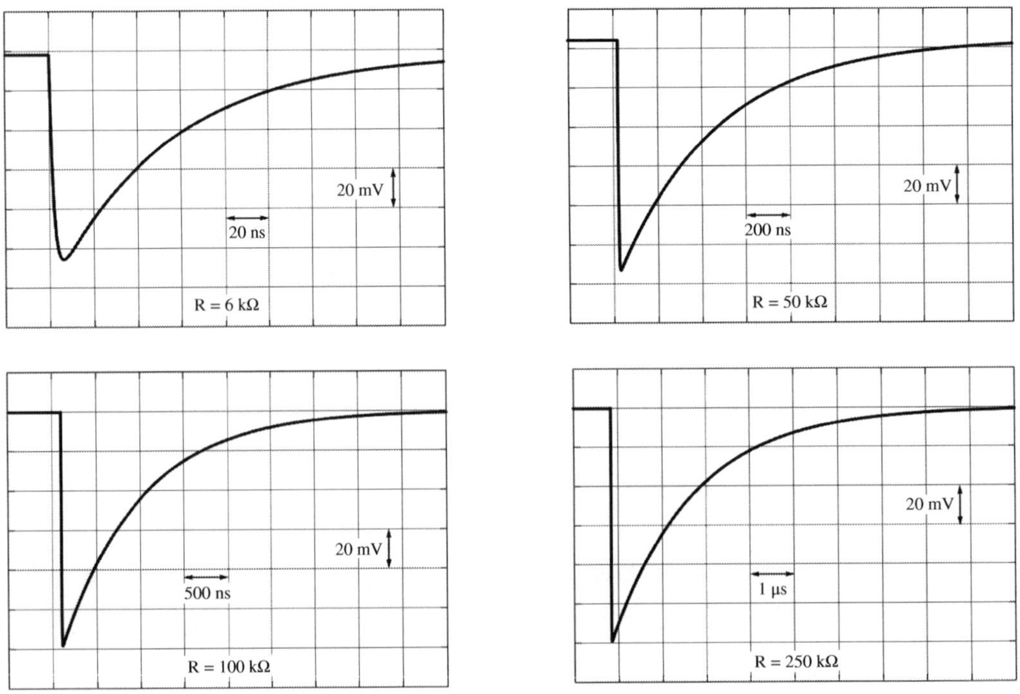

Figure 3.17 Voltage versus time traces displayed on an oscilloscope screen as developed over an exit resistor of a photomultiplier tube; detailed related questions are given in the text of Exercise 3.4.

3.5 Given that the voltage between the last dynode and the anode in the photomultiplier described in exercise 3.4 is 100 V, and the transit time between the two is 1.8 ns, calculate their distance.

3.6 It is given that the work function of the said photomultiplier cathode metal coating is 2.1 eV, the Fermi energy is 4.0 eV, and the reflectance of an incident light is 91.5%, calculate its cutoff response wavelength, the Fowler coefficient for photoelectric emission, and the quantum yield for electron emission by impact of 500 nm wavelength photons (green light).

3.7 It is given that the glass forming the said photomultiplier tube is opaque to wavelengths shorter than 200 nm. Calculate numerically and plot a graph of the responsivity curve $\mathscr{R}$ as a function of wavelength λ between 200 nm and the cutoff wavelength, for an output resistor of 20 kΩ.

4 Preliminary Concepts Regarding Semiconductors

4.1 Dispersion Curves of Electrons in Solids

We start by repeating and delving into the issue of dispersion curves of electrons in solids (Section 3.1). The crystal lattice is considered as a potential perturbation on the free electron state whose dispersion curve is given by (Eq. [3.2])

$$E = \frac{\hbar^2 k^2}{2m}. \tag{4.1}$$

An electronic wave under the influence of the crystal periodic potential undergoes diffraction processes similar to those discussed in relation to X-rays (Section 1.4). Particularly, an electron whose wave vector $\underline{k}$ is located at an edge of a Brillouin zone satisfies a Bragg reflection condition. Interference between the original wave and its reflected one creates a standing wave. It means that the group velocities at the Brillouin zone edge are zero (see also Section 2.1). It further means that on approaching the Brillouin zone edge, the dispersion curve of Eq. (4.1) bends toward becoming strictly flat at the zone edge itself, as in Figure 4.1. A resulting effect is the creation of discontinuities in the energy E dependence on k. Differently stated, separate allowed and forbidden energy zones are created. The electronic waves are no longer simple plane waves, where each belongs to a single wave vector, but rather wave groups comprising a distribution of $\underline{k}$ vectors belonging to higher-order Brillouin zones. The $\underline{k}$ vectors (presented one dimensionally in Fig. 4.1) are thus a representative of the sum of contributing wave vectors. For simplicity, the representative wave vector is chosen to belong to the first Brillouin zone. That form of representation is called **reduced representation**. In Figure 4.1, they are presented as dotted curves.

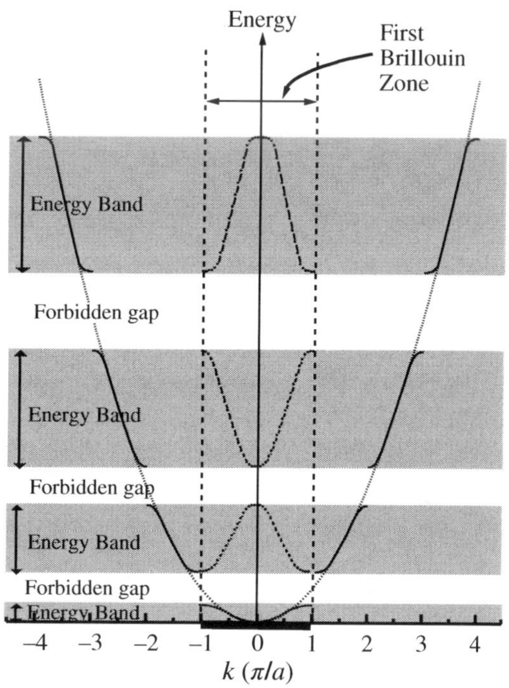

Figure 4.1 Electron energy band structure in a crystal, a result of the crystalline periodic potential. For details, see the text.

Conductors, Semiconductors, Insulators, and Crystal-Growth Technology: A Practical Guide,
First Edition. Zeev Burshtein.

The electronic wave functions are chosen to satisfy periodic boundary conditions (Sections 2.1, 2.2.3, 3.1, and Eq. [3.3]). Namely

$$\psi(\underline{x}+\underline{L}) = \psi(\underline{x}) \Rightarrow e^{-i(\omega t-\underline{k}\cdot\underline{x})} = e^{-i(\omega t-\underline{k}\cdot\underline{x})}e^{i\underline{k}\cdot\underline{L}},$$

where $\underline{L}$ is a crystal length dimension vector. When $\underline{k}$ and $\underline{L}$ are parallel, we get that

$$e^{ik\cdot L} = 1 \Rightarrow kL = n\cdot 2\pi$$

$$k = n\frac{2\pi}{L};\quad n = 0, \pm 1, \pm 2, \ldots \pm \frac{L}{2a},$$

where a is the lattice period. The density of wave vectors in the k-space is $\frac{L}{2\pi}$ in one dimension and $\frac{L^3}{(2\pi)^3}$ in three dimensions; considering the two allowed spin states per each allowed k state, the allowed density of electron states in three dimensions per unit crystal volume is

$$\widetilde{D}(\underline{k}) = \frac{2}{(2\pi)^3}. \tag{4.2}$$

The total density of states in an allowed energy band is $2N$, where N is the density of primitive unit cells in the crystal.

Consider some energy band around the $k = 0$ point. Usually, it exhibits either a maximum or a minimum there. Thus, in expanding the energy into a Taylor series in k, the linear addend is zero, and the first non-zero addend belongs to the squared k:

$$E(k) = \frac{1}{2}\frac{\partial^2 E}{\partial k^2}\bigg]_{k=0}\cdot k^2 + \cdots$$

For sufficiently small k values, higher-order addend contributions may be neglected. Then

$$E(k) = \frac{\hbar^2 k^2}{2\dfrac{\hbar^2}{\partial^2 E/\partial k^2]_{k=0}}} \equiv \frac{\hbar^2 k^2}{2m^*}, \tag{4.3}$$

where the $m^* \equiv \dfrac{\hbar^2}{\partial^2 E/\partial k^2]_{k=0}}$ quantity is called **effective mass**. Use of the effective mass concept allows writing the electron dispersion function in the low-energy range in a form similar to that of free electrons. Obviously, for energies belonging to higher k values, additional addends of the Taylor expansion need to be written. Yet the quantity $m^* \equiv \hbar^2/\frac{\partial^2 E}{\partial k^2}$ is always addressed as the "effective mass." An interesting property of such effective mass is that the sum of its inverse in one-dimensional wave number across a full energy band is zero

$$\sum \frac{1}{m^*} = \frac{1}{\hbar^2}\int_{-\pi/a}^{+\pi/a} \frac{\partial^2 E}{\partial k^2}\frac{2}{2\pi}dk = \frac{1}{\pi\hbar^2}\frac{\partial E}{\partial k}\bigg]_{-\pi/a}^{+\pi/a} = 0. \tag{4.4}$$

Figure 4.2 provides a plot of the effective mass dependence on k along an entire first Brillouin zone.

Due to the curved nature of the electron dispersion, it is not anymore possible to interpret an electron energy as due to a kinetic energy. The electron is considered in terms of a matter wave, whose velocity is described by the group velocity of its wave function $\omega(k) \equiv \hbar^{-1}E(k)$. Its velocity is then given by the group velocity $v_g = \partial\omega/\partial k = \hbar^{-1}\partial E/\partial k$. Figure 4.3 demonstrates the k dependence of the electron velocity $v \equiv v_g$ throughout the entire first Brillouin zone.

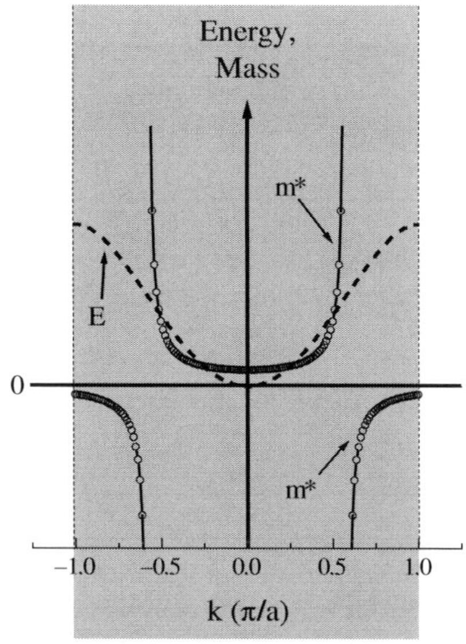

Figure 4.2 Wave-vector *k* dependence of the effective mass *m** and of an electron energy *E* relative to the band bottom, for an electron energy band of the first Brillouin zone (emphasized by a gray background).

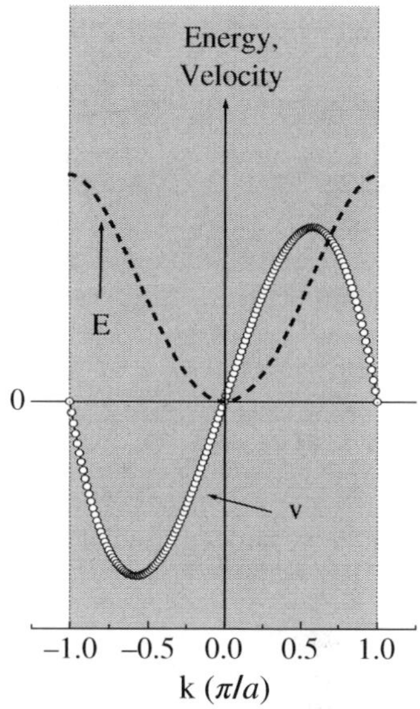

Figure 4.3 Wave-vector *k* dependence of the electron velocity *v* and its energy *E* relative to the band bottom at the first Brillouin zone (emphasized by a gray background).

4.2 Electron Motion Under External Electric Fields

An external electric field $\mathcal{E}$ exerts a force on the electron. The power of the work done by this force increases the electron energy according to

$$\frac{dE}{dt} = -q\mathcal{E}\mathrm{v}_g, \tag{4.5}$$

where v_g is the electron (group) velocity. On the other hand, we may write

$$\frac{dE}{dt} = \frac{dE}{dk} \cdot \frac{dk}{dt} = \hbar \mathrm{v}_g \frac{dk}{dt}.$$

Comparing the above two relations, we get

$$\frac{dk}{dt} = -\frac{q\mathcal{E}}{\hbar}. \tag{4.6}$$

This last result is identical with the one obtained at the beginning of Section 3.6. For simplicity, in following expressions we avoid adding the subscript "g" to the group velocity and just use "v".

The physical meaning of Eq. (4.6) is that the wave vector *k* representing the state of the electron changes at a constant rate under an external electric field. When the wave vector reaches the edge of the first Brillouin zone, the opposite zone edge becomes the representing wave vector *k*. The electron state continues to change at the same rate and same orientation. In that way, the electron *k* state changes periodically, thus forming a periodic motion. Figure 4.3 describes the dependence of the electron velocity on the wave vector *k*.

The theoretically expected electron periodical motion never occurs in reality, because it is interrupted by scattering, mostly by lattice vibrations, but also by crystalline imperfections. Yet it is interesting to assess the expected amplitude A had that motion not been interrupted. Twice the amplitude is the motion between two points where the velocity becomes zero: an edge of the Brillouin zone (one edge of the energy band) and the $k = 0$ state (the second edge of the energy band).

$$2A = \left|\int \mathrm{v}dt\right| = \left|\int \mathrm{v}\frac{dt}{dk}dk\right| = \left|\int \frac{1}{\hbar}\frac{dE}{dk}\frac{-\hbar}{q\mathcal{E}}dk\right| = \frac{1}{q\mathcal{E}}\int_{E_0}^{E_{\mathbf{max}}} dE = \frac{1}{q\mathcal{E}}(E_{\mathbf{max}} - E_0).$$

$$A = \frac{1}{2q\mathcal{E}}\Delta E, \tag{4.7}$$

where $\Delta E \equiv (E_{max} - E_0)$ is the energy band width. Notably, the expected amplitude is rather inversely proportional to the applied electric field. The oscillation time period T is the time required to scan through the entire Brillouin zone between its two edges:

$$T = \left(\frac{2\pi}{a}\right) \Big/ \left(\frac{q\mathcal{E}}{\hbar}\right) = \frac{2\pi\hbar}{q\mathcal{E}a}. \tag{4.8}$$

Based on the above results, we provide a numerical assessment of the amplitude and time period under a typically used external electric field in a typical semiconductor.

Given:

$$\Delta E = 3\ \text{eV} = 3 \times 1.6 \times 10^{-19} = 4.8 \times 10^{-19}\ \text{J};\quad a \cong 1.5 \times 10^{-10}\ \text{m};$$

$$\mathcal{E} \approx 100\ \frac{\text{V}}{\text{cm}} = 10^4\ \frac{\text{V}}{\text{m}};\quad \hbar = 1.06 \times 10^{-34}\ \text{J} \cdot \text{s}.$$

First setting (Eq. [4.7]):

$$A = \frac{4.8 \times 10^{-19}}{2 \times 1.6 \times 10^{-19} \times 10^4} = 1.5 \times 10^{-4}\ \text{m} = 1.5 \times 10^{-2}\ \text{cm}.$$

Second setting (Eq. [4.8]):

$$T = \frac{6.28 \times 1.06 \times 10^{-34}}{1.6 \times 10^{-19} \times 10^4 \times 1.5 \times 10^{-10}} = 2.8 \times 10^{-9}\ \text{s}.$$

In reality, the mean free time till scattering of an electron is of the order of 10^{-13}–10^{-11} s. Therefore, electrons could never perform a full periodic motion, not even a considerable part of it. Only in extreme cases of extremely pure materials having a high electron mobility, and under very high external electric fields, can a dependence of the mobility on the magnitude of the electric field be observed, which then indicates that the electrons were able to perform a significant part of the periodic oscillation amplitude.

4.3 Electrical Conductivity and Charge Carrier Mobility

We first show that an energy band, whose allowed states are fully occupied, can never contribute an electrical current. The reason is that even under an external electric field, the distribution of electrons among the states cannot be different than the distribution under thermodynamic equilibrium conditions. Under either an external electric field or its absence, the electrical current density generated by the electrons will always be given by

$$J_x = -q\int \mathrm{v}(\underline{k})dn(\underline{k}) = -q\left[\int_{-\pi/a}^{0} \mathrm{v}(\underline{k})n(\underline{k})dk_x + \int_{0}^{+\pi/a} \mathrm{v}(\underline{k})n(\underline{k})dk_x\right]. \tag{4.9}$$

In the above expression, we chose x to be oriented along one of the fundamental lattice orientations. We change the variable in the second integral to read $\underline{\kappa} \equiv -\underline{k}$. By symmetry also $\mathrm{v}(\underline{k}) = -\mathrm{v}(-\underline{k})$. In a symmetric electron distribution relative to the $k = 0$ point, we also have $n(\underline{k}) = n(-\underline{k})$. Then

$$J_x = -q\left[\int_{-\pi/a}^{0} \mathrm{v}(\underline{k})n(\underline{k})dk_x + \int_{0}^{-\pi/a} \mathrm{v}(-\underline{\kappa})n(-\underline{\kappa})(-d\kappa_x)\right].$$

In the second integral, we change the variable name $\underline{\kappa}$ into $\underline{k}$ and reverse its limits. Then

$$J_x = -q\left[\int_{-\pi/a}^{0} \mathrm{v}(\underline{k})n(\underline{k})dk_x - \int_{-\pi/a}^{0} \mathrm{v}(-\underline{\kappa})n(-\underline{\kappa})(-d\kappa_x)\right] = 0. \tag{4.10}$$

Identical results are valid for any symmetric occupation of electrons in the band, a situation that obviously exists in a filled band.

When an external electric field is applied over a material whose band of electron states is not fully occupied, the occupation symmetry is distorted. Each electron changes its k state compared to its state under thermodynamic equilibrium by an average value given by

$$\Delta k_x = -\frac{q\mathcal{E}_x}{\hbar}\tau, \tag{4.11}$$

where τ is the average scattering time, that for simplicity we assume to be independent of k. The average added velocity experienced by an electron is given by

$$\Delta \mathrm{v}_x = \frac{d\mathrm{v}_x}{dt}\tau = \tau\frac{d}{dt}\left(\frac{-1}{q\mathcal{E}_x}\cdot\frac{dE}{dt}\right) = \frac{-\tau}{q\mathcal{E}_x}\cdot\frac{d}{dt}\left(\frac{dE}{dk}\cdot\frac{dk}{dt}\right) = \frac{-\tau}{q\mathcal{E}_x}\cdot\frac{-q\mathcal{E}_x}{\hbar}\cdot\frac{dk}{dt}\cdot\frac{d}{dk}\left(\frac{dE}{dk}\right)$$
$$= \frac{-\tau}{q\mathcal{E}_x}\cdot\frac{-q\mathcal{E}_x}{\hbar}\cdot\frac{-q\mathcal{E}_x}{\hbar}\cdot\frac{d^2E}{dk^2} = -\frac{q\tau}{m^*(k)}\mathcal{E}_x.$$

In the course of the above development, we used Eqs. (4.5) and (4.6), and the definition of the effective mass $m^*(k)$, which follows Eq. (4.3). In the last right transition at the top line, we switched the derivation order between the time t and the wave vector k. The electrical current density is thus given by

$$J_x = -q\int_{-\pi/a}^{+\pi/a} \Delta\mathrm{v}_x(\underline{k})dn(\underline{k}) = q^2\tau\mathcal{E}_x\int_{-\pi/a}^{+\pi/a}\frac{1}{m^*(k)}dn(\underline{k}). \tag{4.12}$$

We now consider two special cases; one case is of a nearly empty band, whose very small number of electrons accumulates at its bottom. In that case, we may assume the effective mass m^* to be constant, take it out of the integration symbol, and denote it rather as m_n. We get

$$J_x = q\left(\frac{q\tau}{m_n}\right)\mathcal{E}_x \cdot n_{tot} = q\mu_n n_n^{tot}\mathcal{E}_x, \tag{4.13}$$

where $\mu_n \equiv \frac{q\tau}{m_n}$ is called **electron mobility** and $n_n^{tot} \equiv n_{tot}$ is the total concentration of electrons per unit volume occupying the band. In later expressions, we deleted the upper "tot" superscript.

The second special case is of a nearly full band. The belonging electrons obviously occupy all available lower-energy states, while the highest-energy states remain unoccupied. Recall that the sum through the entire band of all inverse effective masses is zero (see Eq. [4.4]); namely

$$\int_{\substack{\text{all}\\\text{states}}} \frac{1}{m^*(k)} dn(\underline{k}) = 0.$$

Therefore

$$\int_{\substack{\text{filled}\\\text{states}}} \frac{1}{m^*(k)} dn(\underline{k}) = - \int_{\substack{\text{empty}\\\text{states}}} \frac{1}{m^*(k)} dn(\underline{k}).$$

We define $m_p^*(k) \equiv -m^*(k)$ for every unoccupied (empty) state. The m_p^* quantity is called **hole effective mass**. Because the effective mass belonging to the top energy band is negative (see Fig. 4.2), the hole effective mass is positive. The current density would then be given by (Eq. [4.12]),

$$J_x = -q^2\tau\mathcal{E}_x \int_{\substack{\text{empty}\\\text{states}}} \frac{1}{m^*} dn(\underline{k}) = q^2\tau\mathcal{E}_x \int_{\substack{\text{empty}\\\text{states}}} \frac{1}{m_p^*} dn(\underline{k}).$$

Because all empty states belong to a very narrow energy range near the top of the energy band, it is possible to assume that the effective hole mass is constant. Then it may be taken out of the integration symbol and placed before it. We denote it now as just m_p and get

$$J_x = q\left(\frac{q\tau}{m_p}\right)\mathcal{E}_x n_p^{tot} = q\mu_p n_p^{tot}\mathcal{E}_x, \tag{4.14}$$

where n_p^{tot} is the density of the total number of empty states, to be called **hole density**, and $\mu_p \equiv \dfrac{q\tau}{m_p}$ is called **hole mobility**. In later expressions, we discard the "tot" superscript related to the hole density. The accepted symbols used to indicate the different above-discussed quantities are μ_n and μ_p to the electron and hole mobilities, respectively; n and p (or alternatively n_n and n_p) to their densities; and m_n and m_p to their effective masses, respectively.

We have already established that in the above-presented extreme limits, the hole contribution to an electrical current has the same direction as electrons under the same electric field orientation. Because each hole describes a quantum state that misses an electron, they are considered as electrons carrying a positive charge, yet a different effective mass. That interpretation is supported by Hall effect measurements, as mentioned in Section 3.7. Semiconductors may simultaneously occupy holes and electrons at the same place in the crystal, yet only while belonging to different bands. Thus, an electrical current in a semiconductor is a sum of separate contributions of these two-type charge carriers, which are denoted by J_n and J_p for electrons and holes, respectively. These subjects are addressed extensively, at large detail, in the following chapters.

Exercises for Chapter 4

4.1 The dispersion curve in the first Brillouin zone of allowed electron states in a primitive cubic crystal whose lattice constant is $a = 5.7$ Å is given by

$$E = Ak^2 - Bk^4,$$

where E is the energy, k is the wave vector, and the constants are $A = 2.7 \times 10^{-38}\ \mathrm{J \cdot m^2}$ and $B = 4.45 \times 10^{-58}\ \mathrm{J \cdot m^4}$. Plot a graph of the dispersion curve, and calculate the energy band width in electron-volt (eV) units, the effective electron mass had the band been occupied by very few electrons, and the effective hole mass had the band been slightly less than fully occupied by electrons (namely occupied by very few holes). Compare the obtained effective masses with the size of the free electron rest mass.

4.2 Calculate the amplitude and time period motion of a single electron occupying the energy band described in Exercise 4.1 under an external electric field of 100 V/cm in the absence of any scattering. Compare the obtained amplitude with the size of the lattice constant.

4.3 The actual charge carrier scattering time of few electrons occupying the energy band described in Exercise 4.1 at room temperature is 3×10^{-13} s and that of few occupying holes is 2×10^{-12} s. Calculate the characteristic mobility of few occupying electrons and of few occupying holes, and express them in units of $\mathrm{cm^2/V \cdot s}$.

4.4 Calculate the electrical current density that develops in the same crystal, when the energy band given in the above exercises occupies electrons at a density of $4 \times 10^{13}\ \mathrm{cm^{-3}}$, and otherwise when it occupies holes at the same density, under an external electric field of 100 V/cm.

5 Charge Carriers Under Thermal Equilibrium

5.1 Fundamentals of Energy Band Structure

As seen in the previous chapter, electron energy states are clustered in bands of dense energy states with energy separations between. At the absolute zero temperature, the electrons occupy the lowest possible energy states. In a special case, where the highest occupied band is fully occupied, the material is called **semiconductor**. That full band is then called **valence band**. The next higher-energy band, which at the absolute zero temperature is entirely unoccupied, is called **conduction band**. In a simplistic model, any material exhibiting two valence electrons would be a semiconductor, because the density of allowed states in the valence band is $2N$, where N is the molecules density (the factor of 2 emanates from the two allowed spin states per every valence electron). The reality is often more complicated, for example because of overlaps between energy bands of electrons belonging to different electronic configurations. The energy separation between the valence band top (denoted by E_V) and the conduction band bottom (denoted by E_C) is called **energy gap**, and is denoted as E_g. The energy gap is almost entirely independent of temperature, and may be regarded as constant for most practical needs. The said energy scheme is presented in Figure 5.1.

In contrast to Figure 4.1, where the horizontal axis described the electronic wave vector, the horizontal axis in the present figure relates to a geometric position in the crystal; this is despite the fact that energy states addressed in Chapter 4 described an electron as being simultaneously present throughout the entire crystal volume. As to be shown later in detail, the crystal properties may vary as a function of the geometric position in the crystal. Thus, any energy attributed to an electron at a certain geometric place expresses the energy of a group of waves comprising a large number of wave vectors. Typical energy gaps in semiconductors range between 0.2 eV and 2 eV, while typical widths of energy bands is several electron-volts. Notably, the energy bands model shown in Figure 5.1 is very simplistic; for example, there are crystalline materials exhibiting partial overlap between the valence and conduction bands. Such materials are called **semimetals**. We will address that issue briefly in Section 9.2.2, but won't elaborate on that issue further.

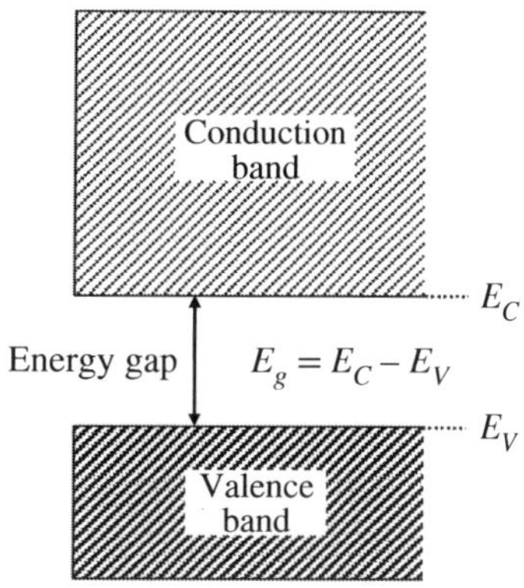

Figure 5.1 Schematic illustration of the fundamental energy band structure and naming in a semiconductor.

Conductors, Semiconductors, Insulators, and Crystal-Growth Technology: A Practical Guide, First Edition. Zeev Burshtein.

Temperature increase above the absolute zero induces excitation of electrons from the valence band top to the conduction band bottom. The electrons thereby residing in the conduction band behave like free electrons exhibiting an effective mass m_n, which differs from the truly free electron rest mass m. They are thus sometimes addressed as **quasi-free** electrons. The unoccupied (empty) states at the valence band top (holes) correspondingly behave like quasi-free **positive electrons**, exhibiting an effective mass m_p. In low-symmetry crystals, the m_n and m_p quantities are tensors. For simplicity, however, they are often regarded as simple scalar quantities.

In crystals that are pure and free of structural defects, all electrons and holes emanate from thermal excitation between the bands. Such crystals are called **intrinsic**, and the electron and hole densities are equal: $n_0 = p_0$, where n_0 and p_0 are the electron and hole densities under thermal equilibrium, respectively (the "0" subscript label relates the quantities to the thermal equilibrium; in the absence of thermal equilibrium, that subscript is often omitted). The situation is different in crystals called **extrinsic**, where the different charge carrier densities are influenced by the presence of impurities. In many such cases, $n_0 \neq p_0$ even under thermal equilibrium.

The electron and hole effective masses are usually smaller than the free electron rest mass. A typical value of an effective mass is $m_{n,\,p} \cong 0.2m$, where m is the free electron rest mass. Usually, the effective electron mass is smaller than the free hole mass, a relation related to the fact that conduction bands are broader than valence bands. At a finite T temperature, the charge carriers exhibit thermal equilibrium with lattice vibrations, thereby carrying a kinetic energy. When imagined to behave like an ideal gas, each degree of freedom carries an average of $\frac{1}{2}k_BT$ energy, and a total kinetic energy of $\frac{1}{2}m^* \left\langle \mathrm{v}_{th}{}^2 \right\rangle = \frac{3}{2}k_BT$, where m^* is the charge carrier effective mass and v_{th} is its average thermal velocity. At room temperature (300 K), a thermal velocity order of magnitude is $\mathrm{v}_{th} \approx 10^7$ cm/s. For comparison, recall that in metals, electron velocity may reach velocities as high as 10^8 cm/s (see Eq. [3.10] and its following text).

To conclude this section, we provide an energy gaps table of several known semiconductors (Table 5.1).

Table 5.1 Partial compilation of room temperature energy gaps of several semiconductor, and their gap type: **d** – direct; **i** – indirect. **Direct gap** and **indirect gap** concepts are discussed in Section 6.2.

Material	Chemical formula	Energy gap (eV)	Gap type
Silicon	Si	1.14	i
Germanium	Ge	0.67	i
Indium antimonide	InSb	0.18	d
Indium arsenide	InAs	0.35	d
Gallium arsenide	GaAs	1.43	d
Indium phosphide	InP	1.35	d
Cadmium telluride	CdTe	1.45	d
Zinc oxide	ZnO	3.2	d

5.2 Electrical Conduction and Hall Effect

Semiconductors differ from metals in the sense that they exhibit two types of charge carriers; each provides its different contribution to the electrical conductivity. It is so because the different charge carriers occupy *different energy bands*. The different current densities J_p and J_n of holes and electrons, respectively, are given by

$$\left.\begin{aligned} J_p &= q\mu_p p_0 \mathcal{E} \\ J_n &= q\mu_n n_0 \mathcal{E} \end{aligned}\right\}, \tag{5.1}$$

where μ_p and μ_n are the hole and electron mobilities, respectively. The total current density is the sum of the two separate contributions

$$J = J_p + J_n = q(\mu_p p_0 + \mu_n n_0)\mathcal{E}. \tag{5.2}$$

The $\sigma \equiv q(\mu_p p_0 + \mu_n n_0)$ quantity is the **specific conductivity** ($\sigma = 1/\rho$, where ρ is the specific resistivity) whose units are $\Omega^{-1}\,\mathrm{m}^{-1}$. The relation $J = \sigma\mathcal{E}$ is called **Ohm's law**.

The Hall effect caused by application of an external magnetic field on a current carrying material (Section 3.7.2) is also a sum of separate contributions by holes and electrons. When velocity changes occurring to each type carrier under the external magnetic field are calculated separately, and their separate contributions are summed, we get (see also Eq. [3.22])

$$J_x = J_{px} + J_{nx} = \frac{qp_0\mu_p}{1+(\mu_p\mathcal{B})^2}(\mathcal{E}_x + \mu_p\mathcal{B}\mathcal{E}_y) + \frac{qn_0\mu_n}{1+(\mu_n\mathcal{B})^2}(\mathcal{E}_x - \mu_n\mathcal{B}\mathcal{E}_y)$$

$$J_y = J_{py} + J_{ny} = \frac{qp_0\mu_p}{1+(\mu_p\mathcal{B})^2}(\mathcal{E}_y - \mu_p\mathcal{B}\mathcal{E}_x) + \frac{qn_0\mu_n}{1+(\mu_n\mathcal{B})^2}(\mathcal{E}_y + \mu_n\mathcal{B}\mathcal{E}_x).$$

Setting the condition $J_y = 0$, and also the $(\mu_p\mathcal{B})^2 \ll 1$ and $(\mu_n\mathcal{B})^2 \ll 1$ conditions, we get

$$\frac{\mathcal{E}_y}{\mathcal{E}_x} = \frac{p_0{\mu_p}^2 - n_0{\mu_n}^2}{\mu_p p_0 + \mu_n n_0}\mathcal{B}; \quad R_H \equiv \frac{\mathcal{E}_y}{J_x\mathcal{B}} = \frac{p_0{\mu_p}^2 - n_0{\mu_n}^2}{q(\mu_p p_0 + \mu_n n_0)^2}. \tag{5.3}$$

Using Eq. (5.3), the statement made in Section 3.7.2, that the Hall effect polarity is opposite in crystals of opposite-type charge carriers, may be readily verified.

5.3 Impurities and Crystalline Defects

Introduced impurities or other crystal defects involve the appearance of additional allowed states in the forbidden energy gap. Advanced materials purification procedures allow obtaining materials with impurity densities that are as low as 10^9–$10^{10}\,\mathrm{cm}^{-3}$ (unit cell densities in crystals are of the order of $10^{22}\,\mathrm{cm}^{-3}$). The said materials purity is thus impressively high.

Impurities and crystalline defects exhibit a considerable influence on the material electrical properties. First, they act as scattering centers, causing shortening of the scattering lifetime τ appearing in charge carrier mobility equations (Eqs. [4.13] and [4.14]). The mobility obtained in a perfectly pure material is determined solely by the lattice vibrations, and is often addressed as **lattice mobility**. Still, the major effect of impurities presence in semiconductors is their contribution to the charge carrier densities.

Consider for example a germanium *Ge* crystal where an arsenic *As* atom replaces a "legitimate" germanium atom that was properly placed in the lattice. The crystalline bonding of germanium is based on covalent bonds at a tetrahedral symmetry, where each germanium atom contributes four electrons. The valence of an arsenic atom is 5, and therefore, it has an excess bond over the required four covalent bonds. The fifth electron involved in the bonding is in fact very weakly bonded, and a tiny added energy would excite it into the conduction band of the germanium crystal. Each replacing arsenic atom in fact generates a discrete energy level E_D below the conduction band. Typical depths of levels generated by arsenic *As* and by other similar impurities are of the order of 0.1 – 0.01 eV (Fig. 5.2).

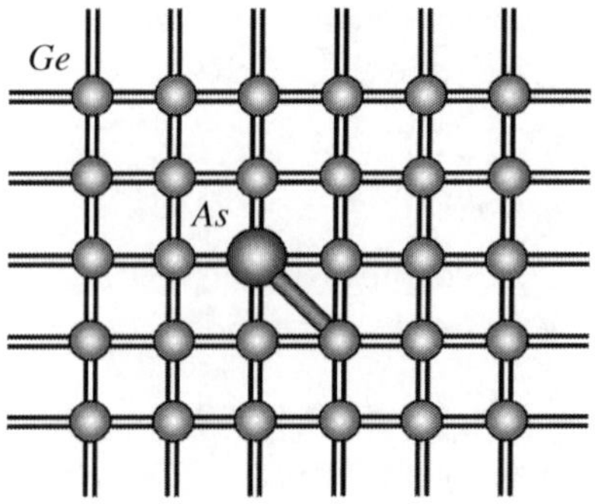

Figure 5.2 Schematic illustration of the generation of a conducting electron-donating level by replacing a lattice germanium atom with an arsenic *As* atom of valency 5. A covalent bond is indicated by a straight line. The excess bond provided by the arsenic is indicated by a broadened line.

It is possible to obtain a qualitative assessment of the reason why such generated states are very shallow (namely so close below the conduction band). Consider a valence electron of the arsenic separating from it, and then orbiting it in a mode similar to an electron orbiting a hydrogen nucleus. If the orbit radius is sufficiently large, it is possible to adopt a macroscopic model, whereby the germanium host plays the role of a medium between the positive arsenic ion and the orbiting electron; the attraction between the two is ε_r times smaller than between an electron and a proton of the hydrogen atom, where ε_r is the medium relative dielectric constant. All orbits are then ε_r times longer, and the binding energies are ${\varepsilon_r}^2$ times smaller than in the hydrogen atom. The hydrogen ionization energy is $E_H^i = 13.6$ eV; For example, in germanium, $\varepsilon_r \cong 15$; therefore

$$\Delta E_D \equiv E_C - E_D \approx \frac{E_H^i}{{\varepsilon_r}^2} = \frac{13.6\text{ eV}}{15^2} = 0.06\text{ eV}.$$

Fivefold valence impurity (dopant) atoms contributing a quasi-free electron in germanium are called **donors**. Threefold valence impurity (dopant) atoms may contribute a hole by capturing an electron that belonged to a neighbor germanium atom; they are called **acceptors**. The generated energy level is denoted by E_A. An example of such case is the indium *In*. Figure 5.3 demonstrates the method by which an energy levels scheme of a semiconductor is presented when it contains impurity levels located in the forbidden gap. For simplicity, the full widths of the valence and conduction band are not presented, and only the valence band top E_V and the conduction band bottom E_C are shown. Table 5.2 provides a list of different dopant atoms in germanium and the depth of the energy levels they produce in the germanium forbidden energy gap. Controlled addition of dopants to semiconductor materials allows control of the electrical conduction properties of these materials.

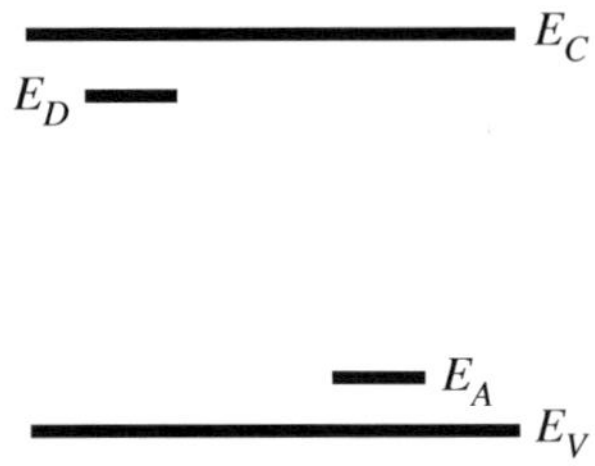

Figure 5.3 Energy scheme of a semiconductor containing donor and acceptor dopant states in different crystal areas.

Table 5.2 Energy depth of different impurities (dopants) in germanium crystals.

Acceptor impurities			Donor impurities		
Element	**Chemical symbol**	**$E_A - E_V$ (eV)**	**Element**	**Chemical symbol**	**$E_C - E_D$ (eV)**
Magnesium	Mg	0.092	Phosphorus	P	0.012
Copper	Cu	0.042	Arsenic	As	0.0127
Gold	Au	0.5	Antimony	Sb	0.0096

5.4 Fermi–Dirac Distribution and Charge Carrier Concentrations

In Chapter 3, we presented the Fermi–Dirac distribution function, which describes the occupation probability of an electron at an allowed state of E energy.

$$\mathrm{f}_n(E) = \frac{1}{1 + e^{\frac{E-E_F}{k_BT}}}, \tag{5.4}$$

where E_F is the Fermi energy. The value $\mathrm{f}_p \equiv 1 - \mathrm{f}_n$ describes the probability for not occupying the same state, thus describes a **hole** occupation probability. In a direct calculation, one obtains

$$\mathrm{f}_p(E) = \frac{1}{1 + e^{\frac{E_F-E}{k_BT}}}. \tag{5.5}$$

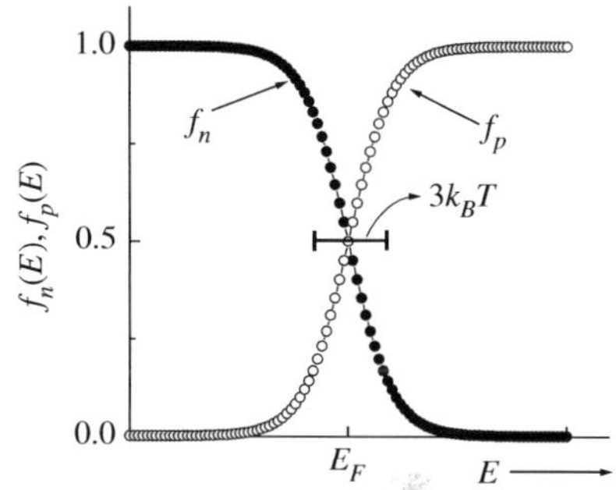

Figure 5.4 Fermi–Dirac distribution (occupation distribution as a function of energy) for electrons (f_n) and holes (f_p).

Figure 5.4 describes the shape of the two functions, f_n and f_p.

We refer to a semiconductor crystal as **non-degenerate** if all conduction and valence states satisfy the condition $|E - E_F| \gg k_BT$, because then is becomes possible to use the approximations presented in Eq. (3.9). Otherwise, the semiconductor is referred to as a **degenerate semiconductor**. A degenerate semiconductor contains a relatively high concentration of charge carriers of one type, and its properties are quite similar to those of a metal. In our future discussions, we deal mostly with non-degenerate semiconductors.

The electron concentration in the conduction band per unit crystal volume is given by

$$n_0 = \int_{E_C}^{\infty} n(E)dE = \int_{E_C}^{\infty} D(E)\mathrm{f}_n(E)dE, \tag{5.6}$$

where $n(E)$ is the occupation density of electrons per unit energy, and $D(E)$ is the density of allowed states for electrons per unit energy in the conduction band. For expressing the density of allowed electron states, we adopt the expression provided in Eq. (3.5) for free electrons in a metal, and for the Fermi–Dirac distribution, we adopt the approximation for a non-degenerate crystal (see Eq. [3.9]). Namely

$$D(E) = \frac{\sqrt{2m_n{}^3}}{\pi^2\hbar^3}(E - E_C)^{1/2}; \quad \mathrm{f}_n(E) \cong \boldsymbol{exp}\left(-\frac{E - E_F}{k_BT}\right). \tag{5.6-a}$$

The occupation density of electrons per unit energy as a function of the energy distance from the conduction band bottom is presented in Figure 5.5. The density is zero at the band bottom, yet it grows as a function of energy due to the increase in the allowed states density. Then, a peak is obtained followed by a decrease, which is a result of an exponential reduction in the occupation probability (Eq. [5.6-a]). We set Eq. (5.6-a) in Eq. (5.6) and get

$$n_0 = \frac{\sqrt{2m_n{}^3}}{\pi^2\hbar^3}\int_{E_C}^{\infty}(E - E_C)^{1/2}e^{-\frac{E-E_F}{k_BT}}dE$$

$$= \frac{4\pi(2m_n)^{3/2}}{h^3}e^{-\frac{E_C-E_F}{k_BT}}\int_{0}^{\infty}(E - E_C)^{1/2}e^{-\frac{E-E_C}{k_BT}}d(E - E_C).$$

We perform a variable change: we denote by $\frac{1}{2}\xi^2 \equiv (E - E_C)/k_B T$ and get

$$n_0 = \frac{4\pi(2m_n)^{3/2}}{h^3} e^{-\frac{E_C - E_F}{k_B T}} \times \sqrt{\pi}\left[\frac{1}{\sqrt{2\pi}}\int_0^\infty \xi^2 e^{-\xi^2} d\xi\right].$$

The value in the square brackets is 0.5 (half the variance in the standard Gaussian bell shape). Thus, we get

$$n_0 = 2\left(\frac{2\pi m_n k_B T}{h^2}\right)^{3/2} e^{-\frac{E_C - E_F}{k_B T}} \equiv N_C e^{-\frac{E_C - E_F}{k_B T}}.$$

An analogous expression obtains for holes (also under non-degeneracy conditions). In summary

$$\left.\begin{aligned} n_0 &= N_C e^{-\frac{E_C - E_F}{k_B T}}; N_C = 2\left(\frac{2\pi m_n k_B T}{h^2}\right)^{3/2} \\ p_0 &= N_V e^{-\frac{E_F - E_V}{k_B T}}; N_V = 2\left(\frac{2\pi m_p k_B T}{h^2}\right)^{3/2} \end{aligned}\right\}. \tag{5.7}$$

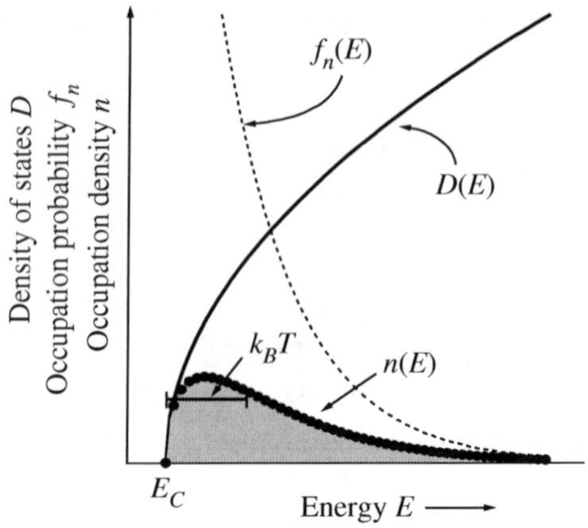

Figure 5.5 Energy dependence of the density of allowed electron states $D(E)$, the occupation probability $f_n(E)$, and the occupation probability $n(E)$ in the vicinity of the conduction band bottom. Notably, the occupation probability $n(E)$ exhibits the shape of a single peak, with an effective width of approximately $k_B T$. The gray area below the $n(E)$ curve is its integral, which provides the density of the total number of occupying electrons.

The N_C and N_V quantities are called **effective concentration of electron states** and **effective concentration of hole states**, respectively. They are written in a form that provides an impression that the conduction band is a discrete state of density N_C, separated by $E_C - E_F$ from the Fermi state, and that the valence band is a discrete state of density N_V, separated by $E_F - E_V$ from the Fermi state. That approach is quite justified, because the temperature dependence of the effective N_C and N_V band densities are very small compared to the exponential factor in the theoretical expressions for the charge carrier concentrations in Eq. (5.7). In correspondence to that approach, in plotting of energy states diagrams, the conduction and valence bands of Figure 5.1 are replaced by discrete E_C and E_V states, as done for example in plotting of Figure 5.3.

An interesting result emanates from the expressions in Eq. (5.7): consider the product between the two charge carrier concentrations. We get

$$n_0 p_0 = N_C N_V e^{-\frac{E_g}{k_B T}}. \tag{5.8}$$

The expression on the right-hand side depends solely on the absolute temperature T, and on neither n_0 nor p_0. That result is called **law of mass action**, in analogy to a similar relation between chemically reacting materials. That result is further interpreted as a statement on dynamic equilibrium relation between free electrons and holes. The $n_0 p_0$ product on the left-hand side is proportional to electron and hole recombination rate; under thermodynamic equilibrium, that rate equals their generation rate. Correspondingly, the $N_C N_V \exp(-E_g/k_B T)$ product on the right-hand side – which depends only on the temperature – is proportional to the thermal generation rate of electrons and holes. That particular issue will be discussed in detail in

Section 6.2. It is further worth noting that while the effective band concentrations N_C and N_V are (moderately) temperature dependent, the energy gap E_g exhibits only negligible temperature dependence. For most practical purposes, it may be considered a constant, characteristic of the specific semiconductor at all temperatures.

We consider now the product definition $n_0p_0 \equiv n_i{}^2$, where n_i denotes the intrinsic density of charge carriers, namely the density obtained under thermal equilibrium in an absolutely pure crystal, where each hole is a result of a thermally excited electron out of the valence band into the conduction band; namely, the number of free holes is exactly the same as the number of free electrons. In an intrinsic crystal, the intrinsic conductivity σ_i is given by (see Eq. [5.2] and the following text)

$$\sigma_i = q\mu_p n_i + q\mu_n n_i = q(\mu_p + \mu_n)(N_C N_V)^{1/2} e^{-\frac{E_g}{2k_BT}}.$$

Quite often in intrinsic crystals, mobilities of both electrons and holes are proportional to $T^{-3/2}$ over a considerable range of temperatures. It is then possible to write $\mu_p = \mu_p^{(0)} T^{-3/2}$ and $\mu_n = \mu_n^{(0)} T^{-3/2}$, where $\mu_p^{(0)}$ and $\mu_n^{(0)}$ are constants. Setting it in the above expression for the intrinsic conductivity we get

$$\sigma_i = q\left(\mu_p^0 + \mu_n^0\right) \cdot 2(m_p m_n)^{3/4} \left(\frac{2\pi k_B}{h^2}\right)^{3/2} e^{-\frac{E_g}{2k_BT}}. \tag{5.9}$$

Thus, the intrinsic conductivity σ_i varies exponentially with temperature. In experimental studies, the specific electrical conductivity of an intrinsic semiconductor is measured as a function of the absolute temperature. The obtained specific conductivity is plotted as a function of $1/k_BT$ on a semi-logarithmic scale. The slope of the obtained linear fit yields the half-energy-gap $E_g/2$, and its intercept on the ordinate (Y-axis) yields the product $q\left(\mu_p^0 + \mu_n^0\right) \cdot 2(m_p m_n)^{3/4} \left(\frac{2\pi k_B}{h^2}\right)^{3/2}$. The above-described procedure is one of the experimental methods used for finding out the size of energy gaps in semiconductors.

An extrinsic semiconductor crystal containing impurities usually exhibits different concentrations of holes and electrons. Still, under thermal equilibrium conditions, the product between their concentrations remains the same, as given by Eq. (5.8). We thus further consider the connection between the position of the Fermi level and the presence of charge carrier donating impurities.

We start by considering rather an intrinsic crystal, for which we have $n_0 = p_0$:

$$\frac{p_0}{n_0} = \frac{N_V}{N_C} \exp\left(\frac{E_V + E_C - 2E_F}{2k_BT}\right) = 1,$$

whence

$$E_F = \frac{1}{2}(E_C + E_V) + \frac{1}{2}k_BT\boldsymbol{ln}\left(\frac{N_V}{N_C}\right) = \frac{1}{2}(E_C + E_V) + \frac{3}{4}k_BT\boldsymbol{ln}\left(\frac{m_p}{m_n}\right). \tag{5.10}$$

Apparently, in an intrinsic crystal, the Fermi level position is located at approximately the middle between the valence and conduction band. Its deviation from the mid-gap is very small, of an order not greater than k_BT; at room temperature, $k_BT \cong 0.026$ eV. That deviation obviously depends on the ratio between the effective hole and electron masses. An extreme case of that ratio occurs in the indium antimonide *InSb* semiconductor, where $m_p/m_n \cong 20$. The deviation from mid-gap center per Eq. (5.10) would then be

$$\frac{3}{4}k_BT\boldsymbol{ln}(m_p/m_n) \underset{300\,\mathrm{K}}{=} 0.75 \times 0.026 \times \boldsymbol{ln}20 \cong 0.06\ \mathrm{eV}.$$

That deviation is still smaller than the band gap of indium antimonide, in itself quite small (about 0.18 eV). The deviation from mid-gap in other known semiconductors is considerably smaller.

We turn to consider the expected change of Fermi energy position as a function of added impurities. Suppose they are electron donors of N_D density and E_D energy, satisfying

$$\text{(i) } n_D = N_D e^{-\frac{E_D - E_F}{k_B T}}; \quad \text{(ii) } n_0 = N_V e^{-\frac{E_C - E_F}{k_B T}}, \tag{5.11}$$

where n_D is the density of donors still occupying electrons. Dividing equation (i) by (ii):

$$\frac{n_D}{N_D} = \frac{n_0}{N_C} e^{\frac{\Delta E_D}{k_B T}}, \tag{5.12}$$

where $\Delta E_D \equiv E_C - E_D$ is the energy depth of the donor impurities below the conduction band. Room temperature values of N_C are of the order of $N_C \approx 10^{19}\ \text{cm}^{-3}$. Limiting ourselves to cases where $n_0 \ll N_C$, say $n_0 < 10^{17}\ \text{cm}^{-3}$, and to shallow donor impurities, say $\Delta E_D \leq k_B T$, we always get that $n_D/N_D \ll 1$. That means that practically all donor impurities are ionized. If the donor densities are considerably greater than the intrinsic electron density at the same temperature, then $n_0 \cong N_D$, which is valid only if $n_i \ll N_D \ll N_C$. Under such conditions, we may write

$$n_0 = N_C e^{\frac{E_C - E_F}{k_B T}} \cong N_D,$$

yielding

$$E_F \cong E_C - k_B T \boldsymbol{ln}(N_C/N_D). \tag{5.13}$$

We thus see that the Fermi energy lowers upon increase of temperature. That tendency continues even when the underlying assumptions for obtaining Eq. (5.13) are no more valid, namely when the density of charge carriers resulted by thermal excitation from the valence band exceeds those contributed by the donor impurities. In other words, the charge carriers approach their (higher temperature) intrinsic values. The Fermi energy then settles near the mid position between the valence and conduction bands. On the other hand, if the temperature is rather lowered toward a range where $k_B T \ll \Delta E_D$, most donor impurities become occupied. Thermal excitation from the valence band is also negligible. The Fermi energy will then be located between E_C and E_D. We then replace Eq. (5.11) by

$$\begin{aligned} &\text{(i) } N_D - n_D = N_D e^{-\frac{E_F - E_D}{k_B T}}; \\ &\text{(ii) } n_0 = N_C e^{-\frac{E_C - E_F}{k_B T}}, \end{aligned} \tag{5.14}$$

Instead of the previous approximation, we write $n_0 = N_D - n_D$ and get

$$\begin{aligned} E_F = &\frac{1}{2}(E_C + E_D) \\ &+ \frac{1}{2} k_B T \boldsymbol{ln}(N_D/N_C). \end{aligned} \tag{5.15}$$

This expression is very similar to the one obtained for the intrinsic crystal (Eq. [5.10]), where the donor impurities "band" in the present case replaces the valence band role of the intrinsic case.

A schematic description of the change of the Fermi energy position as a function of the temperature in the case of a crystal containing donor impurities is provided in Figure 5.6. There are two regions displaying a steep change in the Fermi energy position upon increase of temperature. One is in the low-temperature range, where $k_B T$ is close in value to the donors impurity state depth below the conduction band ΔE_D; the second is in the high-temperature range, where n_i assumes a value that is similar to the donors volume density N_D.

In the case of acceptor-type impurities (*hole-donating* ones), their energy is denoted by E_A, and their density by N_A. When they are the only ones present in the crystal, conclusions symmetric to those concerning electron and donor impurities would be drawn. However, if acceptors would be added to an extrinsic *n*-type crystal, the Fermi energy would lower. When their density becomes equal to the donors density, the crystal assumes a condition similar to that of an intrinsic crystal, and it is then addressed as a **compensated intrinsic crystal**. The Fermi energy position is then very close to the mid band gap. By further addition of acceptor impurities, the Fermi energy further lowers toward the valence band. The crystal becomes a *p*-type semiconductor. Thus, indicating that the Fermi energy closeness to a certain band is equal to stating its electrical conductivity type: in an *n*-type crystal, the Fermi energy is close to the conduction band; in a *p*-type crystal, the Fermi energy is close to the valence band! Figure 5.7 provides energy scheme drawings of *n*-type and *p*-type crystals that include the position of the Fermi energy.

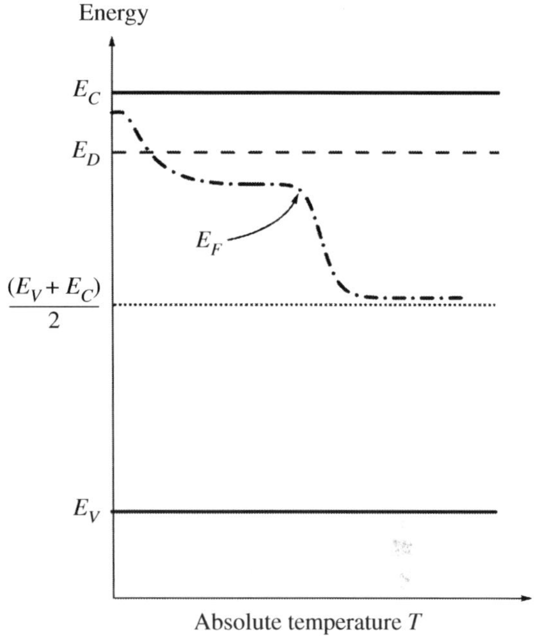

Figure 5.6 Schematic illustration of a Fermi energy position variation in an *n*-type crystal as a function of temperature.

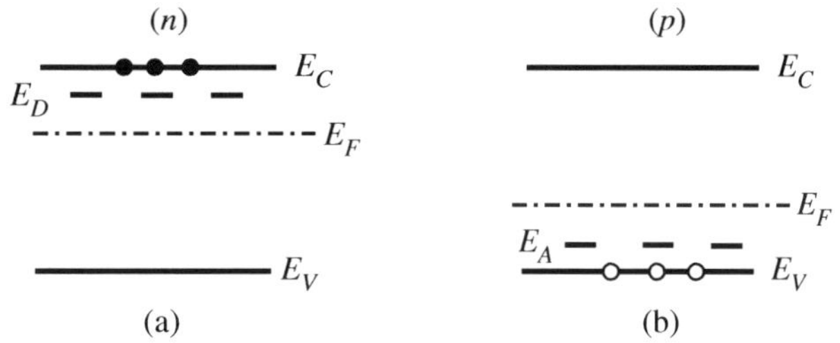

Figure 5.7 Illustration of typical energy schemes in (a) an *n*-type crystal and in (b) a *p*-type crystal. The essence of the majority type charge carriers is accentuated by drawing full circles for electrons, and open circles for holes, in the appropriate bands.

Exercises for Chapter 5

5.1 The room temperature (300 K) electron mobility in a certain semiconducting crystal is $3900\,\text{cm}^2/\text{V}\cdot\text{s}$, and it is proportional to $T^{-3/2}$, where T is the absolute temperature. Formulate a technical expression that describes the electron mobility as a function of the absolute temperature, given that the effective mass of free electrons in the conduction band is 0.15 of a free electron rest mass. Calculate the conducting electron mean scattering time at room temperature and at −200 °C.

5.2 The room temperature free hole density in the semiconductor of Exercise 5.1 is 1.7×10^{14} cm^{-3}, while the free electron density is 3.4×10^{12} cm^{-3}. The hole mobility is 1.9×10^{3} cm^2/V · s. What is the crystal conductivity type? Calculate the crystal electrical conductivity in Ω^{-1} cm^{-1} units, and the current density in A/cm^2 units when under an external electric field of 15 V/cm. Calculate the electron and hole densities that would exist if there were no impurities in the crystal. What would then be the conductivity and the current density?

5.3 The hole effective mass in the semiconductor described in Exercise 5.2 is 0.6 of a free electron rest mass m. Calculate the effective states density of the valence and conduction bands at room temperature and at 80 K. Also calculate the energy gap in electron-volt units (eV). In your calculations, use data and results belonging to the previous exercises.

5.4 Calculate the separation of the Fermi energy from the valence band under conditions given in Exercise 5.2. What would be the separation at a temperature of 80 K, assuming that all acceptors remain as well ionized at that temperature? Plot the crystal energy schemes for both temperatures.

5.5 The following table summarizes results of conductivity measurements of an intrinsic semiconductor at different temperatures:

Temperature (°C)	15	30	80	110	125	160	275	195
Resistivity (Ω · cm)	7180	3170	345	120	75	28.5	20	12.5

(a) Calculate the semiconductor energy gap.
(b) It is given that the effective electron mass is 0.18 of a free electron rest mass, and the effective hole mass is 0.32 of the free electron rest mass. Further, the room temperature hole mobility is 1100 cm^2/V · s. Calculate the room temperature electron mobility, and formulate technical equations for the temperature dependence of hole and electron mobilities.
(c) Calculate the effective room temperature states densities of the valence and conduction bands.

6 Charge Carrier Dynamics

6.1 Charge Carrier Lifetime

As seen in the previous chapter, free electron and hole concentrations under thermal equilibrium satisfy the **law of mass action**:

$$p_0 n_0 = e^{-\frac{E_g}{k_B T}}. \tag{6.1}$$

The "0" subscript attached to both type carriers' concentrations indicate their relevance to the thermal equilibrium condition. There are different means to induce changes in the carriers' concentrations. For example, it is possible to excite electrons from the valence bands into the conduction band by illuminating the crystal with light whose photon energies $h\nu$ are larger than the energy gap E_g, namely $h\nu > E_g$. The valence band is then enriched with holes, and the conduction band is enriched with electrons. It is also possible to expose the crystal to nuclear α, β, or γ radiation, or inject excess charge carriers as in a p–n junction; the latter will be thoroughly discussed in Section 7.1.

For demonstration, assume that electron transition from the valence band into the conduction band is induced by a very short light pulse. Each absorbed photon generates a pair of charge carriers: a hole and an electron. Therefore, at the $t = 0$ moment, $\delta p = \delta n$, where $\delta p = p - p_0$ and $\delta n = n - n_0$ are the concentrations of excess holes and electrons, respectively. As off this moment, enhanced recombination between electrons and holes set in, until their concentrations return to their thermal equilibrium values. The disappearance rates of the excess charge carriers are expressed as

$$U_p \equiv \frac{d(\delta p)}{dt}; \quad U_n \equiv \frac{d(\delta n)}{dt}. \tag{6.2}$$

The recombination processes could be either direct or indirect. Indirect processes involve lattice imperfections or impurities, which generate allowed energy states inside the forbidden gap, called **recombination centers**. The process involves an initial trapping of one-type carrier, then its recombination with the opposite-type carrier. If only direct recombination occurs, the situation is often addressed as a **trapping-absence condition**, satisfying $\delta p(t) = \delta n(t)$ at any t moment. The disappearance rates of the excess charge carriers are then also identical, namely $U_p = U_n$.

Recombination lifetimes of electrons and holes are defined as

$$\tau_p(\delta p) \equiv \frac{\delta p}{U_p}; \quad \tau_n(\delta n) \equiv \frac{\delta n}{U_n}. \tag{6.3}$$

Obviously, in the absence of trapping, $\tau_p(\delta p) = \tau_n(\delta n) \equiv \tau(\delta p = \delta n)$.

Conductors, Semiconductors, Insulators, and Crystal-Growth Technology: A Practical Guide,
First Edition. Zeev Burshtein.

Excess carrier lifetimes obtained in different crystals, under various conditions, span a very broad range of orders of magnitude, between 10^{-2} s and 10^{-9} s.

When the concentration of the excess charge carriers are relatively small, usually such satisfying δp, $\delta n \ll n_0 + p_0$, their lifetimes are constants, which are independent of neither δp nor δn. For example, the rate equation describing the excess hole concentration after the end of an exciting light pulse is given by

$$\frac{d(\delta p)}{dt} = -\frac{\delta p}{\tau_p},$$

A differential equation whose solution is

$$\delta p(t) = \delta p(0) e^{-t/\tau_p}. \tag{6.4}$$

If the crystal is exposed to a continuous generation of holes at a rate G per unit volume (say by a continuous illumination), the rate equation determining their excess concentration is given by

$$\frac{d(\delta p)}{dt} = G - \frac{\delta p}{\tau_p}.$$

If illumination starts at the $t = 0$ moment, and $G = G_0$ is steady for every $t \geq 0$, the solution is

$$\delta p(t) = G_0 \tau_p \left(1 - e^{-t/\tau_p}\right). \tag{6.5}$$

Equation (6.5) portrays a linear increase of the excess hole concentration as a function of time for $t \ll \tau_p$, then a gradually moderating increase rate, asymptotically reaching a saturated $\delta p(\infty) = G_0 \tau_p$ value. For practical purposes, it is possible to assume the achievement of a saturated value for a time t satisfying $t \geq 3\tau_p$. If the illumination stops abruptly later on, an exponential reduction of the excess concentration occurs according to Eq. (6.4). The form of the concentration of excess holes as a function of time under constant illumination for a limited time duration is presented in Figure 6.1.

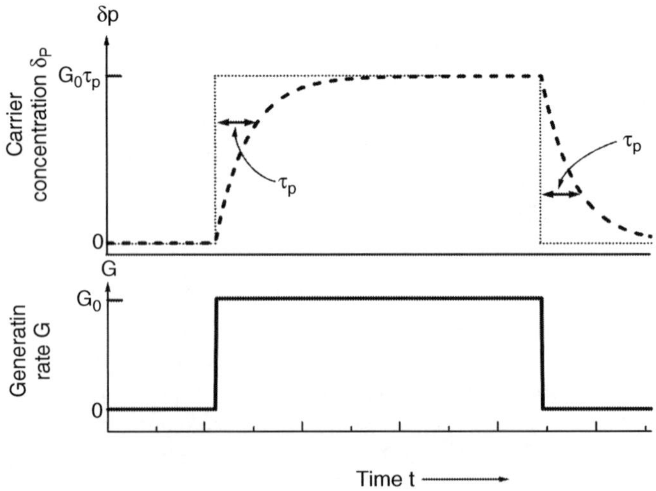

Figure 6.1 Illustration of the change in the hole excess concentration as a function of time under a constant illumination for a limited duration. The hole generation rate caused by the illumination is displayed by a thick line in the lower graph; the hole excess concentration is displayed by a dashed thick line in the upper graph.

Changes in charge carriers' concentrations express themselves in changes of the material electrical conductivity. Under thermal equilibrium, the electrical conductivity is given by

$$\sigma_0 = q(\mu_p p_0 + \mu_n n_0);$$

but under illumination:-

$$\sigma = q(\mu_p p + \mu_n n).$$

Thus

$$\delta\sigma = q(\mu_p \delta p + \mu_n \delta n). \tag{6.6}$$

In the absence of trapping, namely when $\delta p = \delta n$, we get $\delta\sigma = q(\mu_p + \mu_n)\delta p$.

6.2 Trapping and Recombination Cross Sections

The physical **interaction cross-section** S_{AB} between two A and B particles is formally defined as

$$S_{AB} \equiv \frac{P_{AB}}{\mathrm{v}_{AB} N_B}, \tag{6.7}$$

where P_{AB} is the probability per unit time for the interaction to occur for a single A particle, v_{AB} is the average relative velocity between the A particle and all other B particles, and N_B is the concentration (density) of the B particles. The "cross-section" concept has a simple geometric image. We describe a single A particle crossing perpendicularly the basis of a cylinder of S base area, whose length is $\mathrm{v}_{AB}t$ (see Figure 6.2). The cylinder contains B particles at an N_B concentration, all at rest. The B particles are assumed to behave like stiff spheres, where each sphere projection area is S_{AB}. The A particle is assumed to behave like an arrow, whose cross section in its velocity orientation is zero. The probability that the A particle will hit a B particle (namely realize the interaction) before crossing the entire cylinder length is given by the ratio between the sum of all B particles cross sections and the cylinder base area, namely $S_{AB} \cdot N_B \cdot (S \cdot V_{AB} \cdot t)/S$. Dividing by the time t yields the interaction probability per unit time P_{AB}. Setting in Eq. (6.7) yields the required cross-section definition per Eq. (6.7). It may be readily verified that $S_{AB} = S_{BA}$ (Figure 6.2).

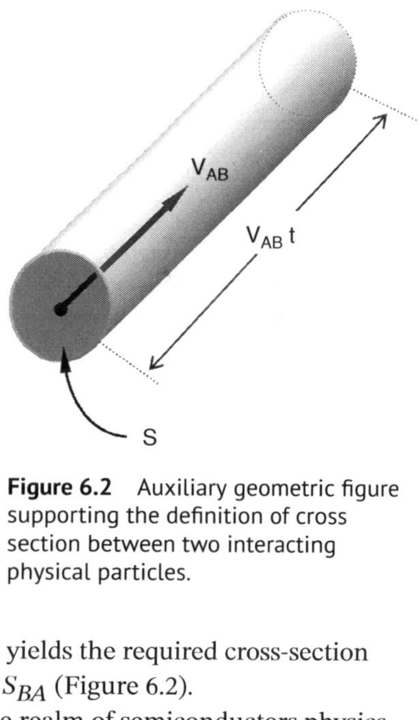

Figure 6.2 Auxiliary geometric figure supporting the definition of cross section between two interacting physical particles.

To clarify the application of the cross-section concept in the realm of semiconductors physics, we consider a case where n_0 electrons in the conduction band are in thermal equilibrium with an electron trapping level E_t residing below the conduction band having a density N_t. We denote by n_t the density of trapped electrons. At a certain moment, the conduction band level was added with excess δn electrons excited out of the trapping level. The thermal equilibrium was violated, and a trapping process starts. The trapping rate is proportional to the temporal summed electron density n and to the density of the vacant trapping levels $N_t - n_t$. The trapping rate is then given by

$$\frac{d(\delta n)}{dt} = -K_n \cdot n(N_t - n_t),$$

or

$$\frac{1}{n} \cdot \frac{d(\delta n)}{dt} = -K_n(N_t - n_t), \tag{6.8}$$

where K_n is a proportionality coefficient.

The quantity on the left-hand side of Eq. (6.8) expresses the trapping probability of a single excess electron per unit time. Turning to the cross-section concept defined by Eq. (6.7), we choose the thermal velocity of the free electrons v_{th} to represent the average relative velocity between a free electron and the trapping levels. We then get the following relation between the trapping cross section S_n and the proportionality coefficient K_n

$$K_n = v_{th} \cdot S_n. \tag{6.9}$$

In the above discussion, we used the "trap" and "trapping" terms in order to delve into familiarity with the effect. Still, the obtained conclusions are quite general. Particularly, instead of referring to the unoccupied traps density $(N_t - n_t)$, we may equally refer to the hole density p, as follows.

We discuss now the relation between the interaction cross-section and lifetime concepts. We assume the absence of trapping conditions, denote by G_{th} the rate of thermal generation of electron–hole pairs, and denote by R the rate of their mutual recombination rate. The time-dependent rate of charge carrier density reduction would then be $U = R - G_{th}$. The thermal generation process may be assumed to be independent of the carriers' densities, especially in non-degenerate crystals. The reason is that the effective densities of the valence and conduction bands are of the order of $10^{19} - 10^{20}$ cm^{-3}, while in non-degenerate semiconductors, typical carriers' occupation of the bands rarely exceeds an order of 10^{17} cm^{-3}. We may thus write $U = R - G_0$, where G_0 is the thermal generation rate of electron–hole pairs under thermal equilibrium. However, under thermal equilibrium, U is zero, therefore $G_0 = R_0 = K_{np}n_0p_0$. On the other hand, in the absence of thermal equilibrium, we get

$$U = K_{np}np - K_{np}n_0p_0 = K_{np}(np - n_0p_0) = v_{th}S_{np}(n_0 + p_0 + \delta p)\delta p.$$

Therefore

$$\tau \equiv \frac{\delta p}{U} = \frac{1}{v_{th}S_{np}(n_0 + p_0 + \delta p)}. \tag{6.10}$$

The last result shows that the lifetime of excited excess carriers depends on the concentration (density) of the excess carriers. In a limit of very small perturbations satisfying $\delta p \ll n_0 + p_0$, the lifetime is approximately constant, given by

$$\tau \cong \frac{1}{v_{th}S_{np}(n_0 + p_0)} \quad \text{for} \quad \delta p = \delta n \ll n_0 + p_0. \tag{6.11}$$

However, under large perturbations, and as long as $\delta p \gg n_0 + p_0$, the rate equation governing the decay of the excess charge carriers' concentration would be

$$\frac{d(\delta p)}{dt} = -v_{th}S_{np}(\delta p)^2,$$

whose solution is

$$\delta p(t) = \frac{\delta p(0)}{1 + v_{th}S_{np}\delta p(0) \cdot t} \quad \text{for} \quad \delta p(t) \gg n_0 + p_0. \tag{6.12}$$

For very large perturbations, we thus expect a very large decay rate of the excess carriers' concentration according to Eq. (6.12). Notably, the $t_{1/2}$ time duration where the perturbation δp

decays to half of its initial $\delta p(0)$ value is given by $t_{1/2} = 1/v_{th}S_{np}\delta p(0)$. In other words, the said time duration becomes shorter as a function of the initial perturbation size. However later on, after the excess charge carriers' density reduces to below $n_0 + p_0$, the decay continues exponentially according to Eq. (6.4). Figure 6.3 portrays, on a semi-logarithmic scale, the temporal evolution of the excess charge carriers' concentration for a case where $\delta p(0)/(n_0 + p_0) = 100$.

Quite often, trapping and recombination cross sections of charge carriers measured in semiconductors exhibit magnitudes of an order of atomic or molecular geometric cross sections, namely about 10^{-15} cm^2. However, considerably other orders of magnitude also occur, ranging between 10^{-12} cm^2 and 10^{-35} cm^2. The decisive factor regarding the cross-section values is the mechanism by which the charge carriers dissipate their excess energy. There are three known major mechanisms:

(1) **Radiative transitions**, namely recombination by emission of a photon.
(2) **Vibrational transition**, namely recombination by excitation of lattice vibrations (phonons).
(3) **Auger process**, namely recombination by giving up the excess energy to a close free electron.

6.2.1 Radiative Transitions

Consider a semiconductor under thermal equilibrium, where electron–hole recombination is dominantly radiative. The law of mass action given by Eq. (6.1) is then relevant. The left-hand side of the equation is the n_0p_0 product. When multiplied by $v_{th}S_{np}$, it provides the electron–hole recombination rate. Under thermal equilibrium, thermal recombination and generation rates are equal. Therefore, the thermal generation rate of electron–hole pairs is given by

$$G_{th} = v_{th}S_{np}N_CN_Ve^{-E_g/k_BT}. \tag{6.13}$$

Notably, the generation rate in a given semiconductor depends only on the material temperature. Its units in the MKS (SI) units system is m^{-3} s^{-1}.

Photons generated sufficiently close to the crystal surface by the radiative recombination mechanism will be emitted out of the crystal. We are then able to estimate the above-discussed

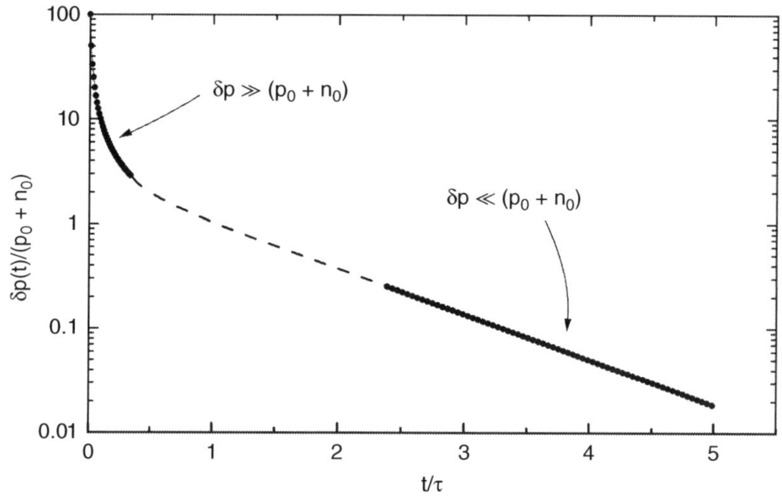

Figure 6.3 Semi-logarithmic plot of excess charge carrier concentration in a semiconductor after an initial excitation as a function of time, in the absence of trapping.

generation rate by assessing the radiation emission rate. We assume that the emitted photon spectral distribution within their specific range satisfy Planck's equation of the theoretical black body

$$\boldsymbol{\Phi}_{BB}(h\nu) = \frac{2\pi}{h^3c^2} \cdot \frac{(h\nu)^2}{\boldsymbol{exp}\left(\frac{h\nu}{k_BT}\right) - 1}, \tag{6.14}$$

where $\boldsymbol{\Phi}$ expresses the rate of photon emission per unit area per unit time (proof of Planck's equation is beyond the scope of the present book). The semiconductor crystal differs from a theoretical black body by the fact that its emitted photon energy is limited to an effective k_BT width about the band gap E_g (see Fig. 5.5). The emitted photons emanate from a near-surface layer whose depth is $\alpha^{-1}(h\nu)$, where α is the $h\nu$ photon absorption coefficient. The total emission rate per unit volume of the effective emitting region is obtained by multiplying the expression in Eq. (6.14) by k_BT, setting $h\nu \cong E_g$, and by dividing the effective thickness of the said layer. Thus, we get

$$G_{th} \cong \frac{2\pi \cdot k_BT \cdot {E_g}^2}{h^3c^2} \cdot \frac{\alpha(E_g)}{\boldsymbol{exp}\left(\frac{E_g}{k_BT}\right) - 1} \underset{E_g \gg k_BT}{\cong} \frac{2\pi \cdot k_BT \cdot {E_g}^2\alpha(E_g)}{h^3c^2}\boldsymbol{exp}\left(-\frac{E_g}{k_BT}\right).$$

We compare the last result with Eq. (6.13): we set there $v_{th} = (3k_BT/m)^{1/2}$, where m is the free electron rest mass, and also $N_C = 2(2\pi m_n k_BT/h^2)^{3/2}$ and $N_V = 2(2\pi m_p k_BT/h^2)^{3/2}$ (see Eq. [5.7]), and solve for S_{np}

$$S_{np} = \frac{1}{4(2\pi)^2} \cdot \frac{h^3{E_g}^2}{c^2m^{5/2}{k_B}^{5/2}}\left(\frac{1}{T}\right)^{5/2}\left(\frac{m^2}{m_nm_p}\right)^{3/2}\alpha(E_g). \tag{6.15}$$

As an example, we calculate the room temperature (T = 300 K) electron–hole recombination cross section of an imaginary semiconductor whose energy gap is 1.0 eV, its electron and hole effective masses are both equal to the free electron rest mass m, and the light absorption coefficient at the energy gap is 1.0 cm^{-1}.

Given numerical data are:

$h = 6.626 \times 10^{-34}\ \text{J}\cdot\text{s}$	$m_p = m_n = m$
$c = 3 \times 10^8\ \text{m/s}$	$E_g = 1.0\ \text{eV} = 1.6 \times 10^{-19}\ \text{J}$
$m = 9.11 \times 10^{-31}\ \text{kg}$	$\alpha(E_g) = 1.0\ \text{cm}^{-1} = 100\ \text{m}^{-1}$
$k_B = 1.38 \times 10^{-23}\ \text{J/K}$	

Setting the data in Eq. (6.15), we get

$$S_{np} = \frac{1}{4 \times (6.28)^2} \cdot \frac{6.626^3 \times 10^{-102} \times 1.6^2 \times 10^{-38}}{3^2 \times 10^{16} \times (9.11 \times 1.38)^{5/2} \times 10^{-54\times 2.5}}\left(\frac{1}{300}\right)^{5/2} 100;$$

$$S_{np} = 6 \times 10^{-29}\ \text{m}^2 = 6 \times 10^{-25}\ \text{cm}^2.$$

The last result allows expressing a technical equation for the radiative recombination cross section, which is adequate for a general semiconductor:

$$S_{np}[\text{cm}^2] = 6 \times 10^{-25} E_g^2\ [\text{eV}]\left(\frac{300}{T[\mathbf{deg}\ \text{K}]}\right)^{5/2}\left(\frac{m^2}{m_pm_n}\right)^{3/2} \cdot \alpha[\text{cm}^{-1}]. \tag{6.16}$$

More examples of technical equations to be used in other physical problems are presented in Section 9.2.

Semiconductors whose valence band top and conduction band bottom occur at different wave vectors are said to have an **indirect band gap** (see Table 5.1). A radiative recombination between an electron and a hole then must involve a phonon absorption or emission to match the wave vectors of both carriers. For that reason, the transition is weak, and both absorption coefficient and radiative recombination cross section are small. On the other hand, in crystals having a **direct band gap**, the radiative transition is strong, and both absorption coefficient and radiative recombination cross section are large. For example, germanium *Ge* has an indirect band gap. Its optical absorption coefficient at a wavelength corresponding to its energy gap is about 100 cm^{-1}. The energy gap itself is $E_g = 0.67$ eV, and the hole and electron effective masses are $m_p = 0.3\,m$ and $m_n = 0.082\,m$, respectively. Setting these values in Eq. (6.16), we get $S_{np}(Ge) = 7 \times 10^{-21}$ cm^2. As already seen (Eq. [6.11]), the recombination lifetime is given by

$$\tau = \frac{1}{v_{th} S_{np}(p_0 + n_0)}.$$

Setting $v_{th} \cong 10^5$ m/s, $S_{np}(Ge) = 7 \times 10^{-25}$ m^2, and $(p_0 + n_0) = 2n_i = 4.8 \times 10^{-19}$ m^{-3} (as for an intrinsic germanium at room temperature), we get $\tau \cong 0.3$ s. Actual lifetimes measured in germanium are never longer than 100 μs; this difference results from the involvement of non-radiative recombination processes competing with the radiative one, correspondingly dominating the lifetime. The gallium arsenide *GaAs* semiconductor, on the other hand, exhibits a direct band gap. The radiative lifetime of excess charge carriers in that material is of the order of 1 ns, thus dominating the recombination process. This material is used for manufacturing of diode lasers that are based on exploiting the emitted radiation, as will be described in detail in Section 7.4.

6.2.2 Vibrational Transitions

A vibrational transition is an electron–hole recombination event enabled by dissipating their entire energy to lattice vibrations (phonons). The maximal phonon energy in a material is given by $E^{max}_{phonon} = \hbar\omega_D \cong k_B T_D$, where ω_D is the material characteristic Debye angular frequency, and T_D is its Debye temperature (see Section 2.2). For example, in germanium *Ge*, the Debye temperature is 374 K (see Table 2.1). Thus, $E^{max}_{phonon} = 0.032$ eV. As indicated above, the germanium energy gap is $E_g = 0.67$ eV. The recombination thus requires a simultaneous emission of at least 21 phonons, and the probability of its occurrence is therefore very small. A vibrational electron–hole recombination in most semiconductors is thus quite negligible.

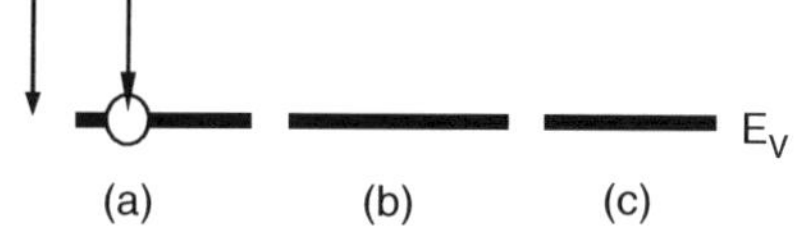

Figure 6.4 Auxiliary drawing demonstrating the Auger process in an *n*-type semiconductor: (a) electron–hole recombination is followed by excitation of a nearby electron into a high kinetic energy state. (b) the high-energy electron decays non-radiatively down to the band bottom (the energy gradually given to lattice vibrations). (c) The excited Auger electron remains at the conduction band bottom.

6.2.3 Auger Process

Auger process relates to an interaction ending by a simultaneous excitation of a mobile (free) electron, granting it a kinetic energy. In our case, it addresses an electron–hole recombination that excites a neighbor charge carrier, which thereby acquires kinetic energy. Valence and conduction bands in semiconductors are usually much broader than the energy gap, and in that aspect, the Auger process is possible. The cross section for that process is obviously proportional to the carriers' density in the band, and it is thus quite important, and indeed dominant in degenerate semiconductors. Due to the high states density in valence and conduction bands, a carrier acquiring the kinetic energy decays very quickly to the lowest available unoccupied state by exciting lattice vibrations (the above "high-state density" description means that the energy separation between allowed states within a band is tiny compared to $k_B T_D$). A typical time duration for a carrier to lose its kinetic energy is of the order of 10^{-12} s. In Figure 6.4, we present a drawing describing the process. A theoretical analysis of the Auger process is beyond the scope of this book and won't be presented here. In Section 9.2, we present a specific case where the Auger mechanism is the one dominating the charge-carriers recombination process.

6.3 Charge-Carriers Drift and Diffusion

The process called **diffusion** is a result of a random motion of physical objects that are not bound to fixed positions. A trivial statistical consideration suggests that the objects diffusion expresses itself by a direct streaming upon occurrence of a gradient in their concentrations. The reason is that statistically, the number of objects moving randomly from the dense region to the sparse one is larger than those moving randomly from the sparse region to the dense one. A concentration gradient of charge carriers may be a result of an internal structure, such as non-uniformity in the concentration of the carrier donor impurities, or a result of an external reason such as non-uniform heating or illumination. In a first-order approximation, the diffusion current density is assumed to be proportional to the carriers' concentration gradient, an assumption verified experimentally. Any movement of a charge carrier entails an electrical current. There are thus two factors causing electrical current – an external electric field, and diffusion. Expressed formally, the hole and electron currents are given by

$$\left.\begin{aligned} \underline{J}_p &= q\mu_p p\underline{\mathcal{E}} - qD_p\,\mathbf{grad}\,p \\ \underline{J}_n &= q\mu_n n\underline{\mathcal{E}} + qD_n\,\mathbf{grad}\,n \\ \underline{J} &= \underline{J}_p + \underline{J}_n \end{aligned}\right\}, \tag{6.17}$$

where $\underline{J}_p$ and $\underline{J}_n$ are the current densities carried by the holes and electrons, respectively; D_p and D_n are positive constants called **diffusion coefficients** of the holes and electrons, respectively; $\underline{\mathcal{E}}$ is the macroscopic local electric field; and $\underline{J}$ is the total electrical current density. The first addend in each of the two top equations is the electrical current density contributed by the carrier drift, and the second one is the electrical current density contributed by the carriers' diffusion. It is important to notice the plus and minus signs preceding the different contributions: the particle's streaming orientation is always opposite to the sign of their concentration gradient; the holes are electrically positive particles; therefore, their diffusion contribution to the electrical current is given the "–" minus sign. This sign reverses when electron diffusion is considered. The units of the diffusion coefficients are m^2/s in the MKS units system and cm^2/s in the CGS units system.

The electrical current is zero in a crystal under thermal equilibrium. According to the principle of detailed equilibrium, both hole and electron electrical currents are then separately zero, namely $\underline{J}_p = 0$ and $\underline{J}_n = 0$. Consider for example the conclusion drawn by the $\underline{J}_n = 0$ condition, more specifically regarding its x-component.

$$\underline{J}_{nx} = q\mu_n n_0 \mathcal{E}_x + qD_n \frac{dn_0}{dx} = 0.$$

Therefore

$$\frac{dn_0}{n_0} = -\frac{\mu_n}{D_n}\mathcal{E}_x dx = \frac{\mu_n}{D_n} dV,$$

where V is the local macroscopic potential at the mobile (free) electrons in the crystal. Integration of the two equation sides yields

$$n_0 = Const \cdot \boldsymbol{exp}\left(\frac{\mu_n}{D_n} V\right). \tag{6.18}$$

The *Const* quantity value depends on the selection of a reference electrical potential. For that purpose, we turn now to compare the latter result with another expression for n_0 given by Eq. (5.7), which is $n_0 = N_C \cdot \boldsymbol{exp}\left(-\frac{E_C - E_F}{K_B T}\right)$. We denote by $E_C - E_F = -qV$, whose meaning is that we choose as a reference potential, the potential of an electron at the Fermi level. In other words, the electrical potential energy of an electron residing at the conduction band bottom is defined by its energy separation from the Fermi level. Then, we get $n_0 = N_C \cdot \boldsymbol{exp}\left(\frac{qV}{K_B T}\right)$. We compare this expression with the one of Eq. (6.18) and get the identity $q/k_B T = \mu_n/D_n$. A similar expression is obtained for holes, and with a trivial change we get

$$\left.\begin{aligned} D_n &= \frac{k_B T}{q}\mu_n \\ D_p &= \frac{k_B T}{q}\mu_p \end{aligned}\right\}. \tag{6.19}$$

These relations between diffusion coefficients and carrier mobilities are called **Einstein relations**.

In our above discussion, we chose the Fermi energy to play the role of reference energy for the electron potential. It is worthwhile delving into the meaning and implications of that choice. Recall that choosing a reference potential is quite arbitrary, and for convenience we may choose such in a different way (the Einstein relations are independent of the particular potential energy reference). The former choice may also be represented in the form

$$E_C = E_F - qV.$$

The Fermi energy is a common quantity of the entire crystal volume. On the other hand, the electron potential energy usually varies as a function of position. Therefore, the changing form of the conduction band bottom E_C follows that of the declared V potential. The E_g gap between the conduction and valence bands is constant, independent of the macroscopic V potential. The reason is that E_g is determined by inter atomic potential gradients, whose size is of the order of 10^8 V/cm (a potential change of 1 V across atomic dimensions of ~1 Å). On the other hand, the electric fields generated by the macroscopic potential changes never exceed the order of 10^5 V/cm. The latter fields generation relates to the electrical diffusion currents: if for example the crystal contains a large concentration of donor impurities at a certain (limited) region, the concentrations gradient would cause the mobile electrons at the band bottom to shift from the high concentration to the lower concentration, thus generating local space charges: a positive charge where the donor impurities remain unoccupied and a negative charge where the mobile electrons reach. These space charges are responsible for the generation of an internal electric field. In Figure 6.5 we present a possible structure of valence and conduction bands energy scheme in a crystal under a position-dependent macroscopic potential.

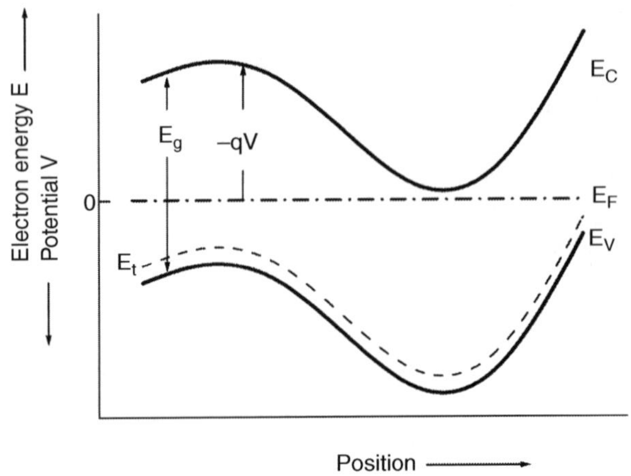

Figure 6.5 Demonstration of a possible energy bands scheme in an extrinsic crystal of a non-uniform distribution of donor and/or acceptor impurities, thus exhibiting changes of position. The conduction and valence bands, as well as carriers trapping bands, vary without change of their energy separations. The Fermi energy is identical throughout the entire crystal volume. *V* is the electrical potential of a mobile (free) electron residing at the bottom of the conduction band relative to the Fermi energy. Importantly, the orientation of the electron increase of energy is opposite to the orientation of electrical potential increase.

6.4 Quasi-Fermi Energy

In the previous section, we emphasized that a reference electrical potential for the electron potential energy may be selected quite arbitrarily "for convenience". Our selection then was the crystal Fermi energy (level) E_F, whence the electron potential energy has been attributed to a mobile (free) electron residing at the conduction band bottom (namely $-qV = E_C - E_F$). In analyses to be performed in the present section, we attribute the electronic potential energy to a function that relates to the two bands system, defined as $E_i \equiv \frac{1}{2}[E_C + E_V + k_BT\mathbf{ln}(N_V/N_C)]$. Then, the electronic potential V is defined to satisfy the relation $-qV = E_C - E_i$. In an intrinsic crystal, such defined E_i function coincides with the Fermi energy throughout its entire volume (see Eq. [5.10]). In an extrinsic crystal, on the other hand, where the concentration of impurities is not necessarily uniform, that energy differs from the Fermi energy, and it curves rigidly with the curvature of the energy bands, as demonstrated in Figure 6.6.

For a crystal under thermal equilibrium, we may now write

$$n_0 = N_C e^{-\frac{E_C - E_F}{k_BT}} = N_C e^{-\frac{(E_C - E_i) + (E_i - E_F)}{k_BT}} = N_C e^{-\frac{(E_C - E_i)}{k_BT}} \cdot e^{-\frac{(E_i - E_F)}{k_BT}};$$

$$n_0 = n_i e^{\frac{qV + E_F}{k_BT}}.$$

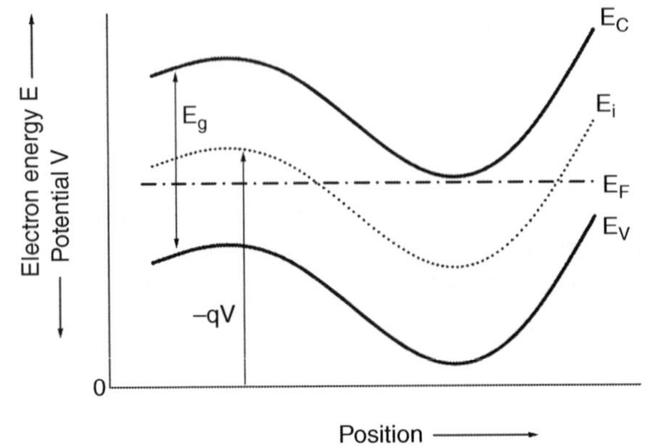

Figure 6.6 Curving of energy bands in an extrinsic crystal where the donor and/or acceptor density of impurities is not uniform. Also shown is the *Ei* energy defined in the text. Worth noticing is that the orientation of increase in the electron energy is opposite to the orientation of increase in the electrical potential.

An analogous calculation for holes yields a similar expression; in summary, we have

$$\left.\begin{aligned} p_0 &= n_i e^{-\frac{qV(\underline{x})+E_F}{k_B T}} \\ n_0 &= n_i e^{+\frac{qV(\underline{x})+E_F}{k_B T}} \end{aligned}\right\},$$

where $\underline{x}$ is the position vector in the crystal. The above result is valid for a non-degenerate semiconductor under thermal equilibrium, and is not different from the relations given in Eq. (5.7). If the crystal is illuminated, or exposed to the action of an external electric field, the Fermi energy concept is no more valid. Therefore, for mathematical convenience, new functions are defined to enable expressing the charge carrier densities when thermal equilibrium conditions are violated. They are called **quasi-Fermi state for holes** and **quasi-Fermi state for electrons,** denoted by $F_p(\underline{x})$ and $F_n(\underline{x})$ for holes and electrons, respectively. They are defined to satisfy the relations

$$\left.\begin{aligned} p(\underline{x}) &= n_i e^{-\frac{qV(\underline{x})+F_p(\underline{x})}{k_B T}} = N_V e^{-\frac{F_p(\underline{x})-E_V(\underline{x})}{k_B T}} \\ n(\underline{x}) &= n_i e^{\frac{qV(\underline{x})+F_n(\underline{x})}{k_B T}} = N_C e^{-\frac{E_C(\underline{x})-F_n(\underline{x})}{k_B T}} \end{aligned}\right\}. \tag{6.20}$$

A direct equivalent formal definition of $F_p(\underline{x})$ and $F_n(\underline{x})$ would be

$$\left.\begin{aligned} F_p &\equiv E_V + k_B T\,\mathbf{ln}\left(\tfrac{N_V}{p}\right) \\ F_n &\equiv E_C - k_B T\,\mathbf{ln}\left(\tfrac{N_C}{n}\right) \end{aligned}\right\}. \tag{6.21}$$

For brevity in writing Eq. (6.21), we refrained from including the $\underline{x}$ dependencies of the E_V, E_C, p, n, F_p, and F_n quantities.

Under thermal equilibrium, both quasi-Fermi levels coincide with the Fermi level. In the absence of thermal equilibrium, they are two different quantities, and they vary as functions of position in the crystal. There are two factors that determine their position dependences: one is variations in the charge carrier concentrations, namely $p = p(\underline{x})$ and $n = n(\underline{x})$. These variations are not straight forwardly connected, and they depend on the mechanisms responsible for the generation of excess charge carriers. The other is position changes of the valence and conduction bands, namely $E_V = E_V(\underline{x})$ and $E_C = E_C(\underline{x})$, among other things caused by application of an external electric field. The latter are always interconnected, because at each geometric $\underline{x}$ position, their energy separation is equal to the fundamental material energy gap: $E_C(\underline{x}) - E_V(\underline{x}) = E_g$.

To demonstrate the usefulness of the "quasi-Fermi states" quantities, we convert the form of the current density Eq. (6.17) as follows

$$\begin{aligned} \underline{J}_p &= q\mu_p p \underline{\mathcal{E}} - qD_p\,\mathbf{grad}\,p \\ &= -q\mu_p p\,\mathbf{grad}\,V - q\frac{k_B T}{q}\mu_p\left(p\frac{-q\,\mathbf{grad}\,V - \mathbf{grad}\,F_p}{k_B T}\right) = \mu_p p\,\mathbf{grad}\,F_p. \end{aligned}$$

On performing a similar procedure concerning the electron current density, we get

$$\left.\begin{aligned} \underline{J}_p &= \mu_p p\,\mathbf{grad}\,F_p \\ \underline{J}_n &= \mu_n n\,\mathbf{grad}\,F_n \end{aligned}\right\}. \tag{6.22}$$

An advantage of using the latter expressions is that they are much simpler than those of Eq. (6.17).

We turn now to consider the shape of quasi-Fermi levels under different conditions, which actually cause violation of thermal equilibrium.

(a) The semiconductor is under external excitation of charge carriers: Under such conditions, the hole thermal equilibrium concentration changes from p_0 into $p_0 + \delta p$ and the electron concentration changes from n_0 into $n_0 + \delta n$. The hole quasi-Fermi level F_p, which coincided with the Fermi level E_F before excitation, shifts downward, becoming closer to the valence band top (Eq. [6.21]). The electron quasi-Fermi level F_n, which coincided with the Fermi level E_F before excitation, shifts upward, becoming closer to the conduction band bottom (Eq. [6.21]). The said changes appear as a kind of "splitting" of the Fermi level into two separate ones, F_p and F_n (Figure 6.7).
(b) The semiconductor is under the influence of an external electric field without external excitation of charge carriers: Due to absence of external charge carrier excitation, we may conclude that $F_p(\underline{x}) = F_n(\underline{x})$, and that their separations from the carrier energy bands is the same as the Fermi level E_F had before application of the external electric field. Thus, their slope is the same as that of the carriers' energy bands. We express it also by a direct calculation, for example via the hole current density: $J_p = q\mu_p p_0 \underline{\mathcal{E}} = \mu_p p_0 \boldsymbol{grad}(-qV)$, but on the other hand, according to Eq. (6.22), we have $J_p = \mu_p p_0 \, \boldsymbol{grad} \, F_p$. Therefore, $F_p = F_n = -qV$. As expected, under no external carriers excitation, both $F_p(\underline{x})$ and $F_n(\underline{x})$ behave like the macroscopic potential energy caused by the external electric field (Figure 6.8).

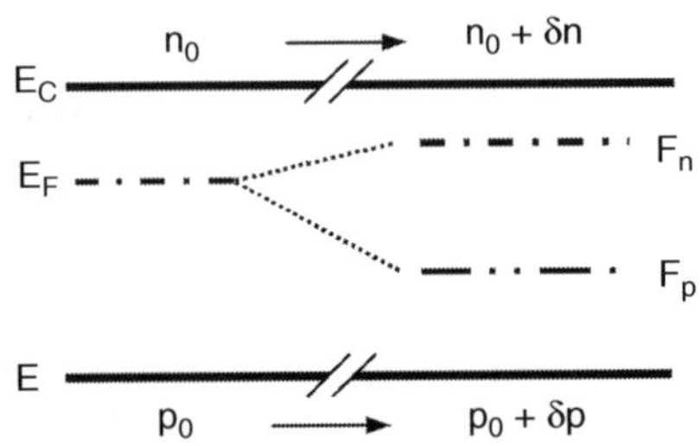

Figure 6.7 "Splitting" of quasi-Fermi levels due to excitation of excess holes and electrons (right side).

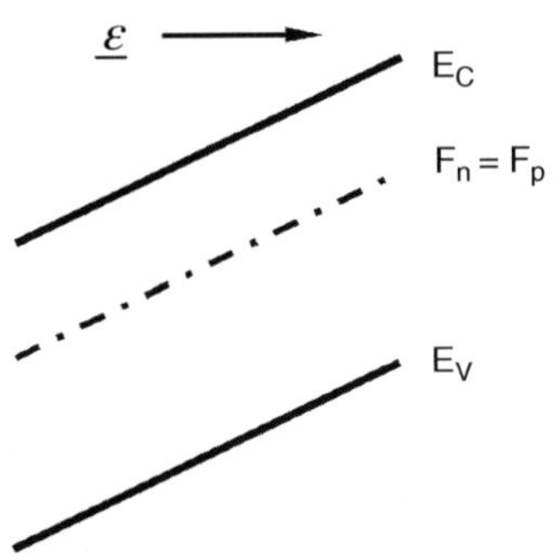

Figure 6.8 "Bending" of quasi-Fermi levels due to application of an external electric field.

6.5 Transient Currents

6.5.1 The Continuity Equations

Recall the electrical charge continuity equation in free space given by

$$\frac{\partial \rho}{\partial t} = -\boldsymbol{div} \, \underline{J}, \tag{6.23}$$

where ρ here denotes the electrical charge density. The physical meaning of Eq. (6.23) is that the electrical charge is fundamentally conserved, meaning that any change in the electrical charge density within a certain space, expressed at the left side of the equation, is caused by an electrical current expressed at the right side of the equation. In shaping of a continuity equation for charge carriers in a semiconductor, additional effects must be taken into account, meaning disappearance of charge carriers by trapping and recombination, as well as by their generation via a variety of processes, some of which have already been described. Continuity equations for charge carrier concentrations accounting for such processes would be

$$\left.\begin{aligned}\frac{\partial p}{\partial t} &= G_p - U_p - \frac{1}{q}\mathbf{div}\,\underline{J}_p \\ \frac{\partial n}{\partial t} &= G_n - U_n + \frac{1}{q}\mathbf{div}\,\underline{J}_n\end{aligned}\right\}, \tag{6.24}$$

where G and U are the generation and recombination rates of the charge carriers, respectively. We subtract the second equation from the first, recall that p_0 and n_0 are time-independent, and get

$$\frac{\partial}{\partial t}(p-n) = (G_p - G_n) - (U_p - U_n) - \frac{1}{q}\mathbf{div}(\underline{J}_p + \underline{J}_n);$$

$$\frac{\partial}{\partial t}(\delta p - \delta n) = (G_p - G_n) - (U_p - U_n) - \frac{1}{q}\mathbf{div}\,\underline{J}. \tag{6.24-a}$$

Poisson equation also participates in the present issue; in MKS (SI) units system, it is written as

$$\mathbf{div}\,\underline{\mathcal{E}} = \frac{1}{\varepsilon_r\varepsilon_0}\rho, \tag{6.25}$$

where ε_0 is the permittivity of free space, and ε_r is the material relative dielectric constant. In a uniform semiconductor and the absence of trapped charge, we may write the space-charge density in the form $\rho = q(\delta p - \delta n)$. Poisson equation may be then written as

$$\mathbf{div}\,\underline{\mathcal{E}} = \frac{q}{\varepsilon_r\varepsilon_0}(\delta p - \delta n). \tag{6.26}$$

6.5.2 Dielectric Relaxation

We consider now a uniform semiconductor, where at a certain moment $t = 0$, an excess density of charge carriers is introduced, which forms an electrical space charge. As a result of mutual repulsion, the excess charge will gradually migrate toward the crystal surface. It disappears entirely, if the crystal has electrical contacts. That described process is called **dielectric relaxation**. For a qualitative analysis, we put forward several assumptions that emphasize the effect. We assume the absence of any trapped (immobile) charge, and the absence of any generation or recombination of charge carriers (namely $U_p = U_n = 0$ and $G_p = G_n = 0$). We further assume that the sample is under a sufficiently high external electric field such that all diffusion currents may be neglected compared to drift currents (see Eq. [6.17] and text). For further simplification, we assume only a very small addition of charge carriers, satisfying $\delta p + \delta n \ll p_0 + n_0$ allowing a $\underline{J} = \sigma\underline{\mathcal{E}} \cong \sigma_0\underline{\mathcal{E}}$ approximation. Then

$$-\frac{1}{q}\,\mathbf{div}\,\underline{J} = -\frac{1}{q}\sigma_0\,\mathbf{div}\,\underline{\mathcal{E}} = -\frac{1}{q}\sigma_0\frac{q}{\varepsilon_r\varepsilon_0}(\delta p - \delta n) = -\frac{\sigma_0}{\varepsilon_r\varepsilon_0}(\delta p - \delta n).$$

On setting the expression for $-\frac{1}{q}\mathbf{div}\,\underline{J}$ appearing in the continuity equation (6.24-a), we get

$$\frac{\partial(\delta p - \delta n)}{\partial t} = -\frac{\sigma_0}{\varepsilon_r\varepsilon_0}(\delta p - \delta n).$$

The latter is a differential equation for $(\delta p - \delta n)$, whose solution under the above initial condition is

$$(\delta p - \delta n) = (\delta p - \delta n)_0 e^{-t/\tau_{DR}} \quad \text{for} \quad t > 0, \quad \text{where} \quad \tau_{DR} = \frac{\varepsilon_r\varepsilon_0}{\sigma_0}. \tag{6.27}$$

The last result may also be expressed in the form

$$\rho = \rho_0 e^{-t/\tau_{DR}},$$

where ρ is the electrical space-charge density.

We are thus considering a case of a space charge that decays exponentially over time. Its characteristic decay time τ_{DR} is called **dielectric relaxation time**, and it depends on the material properties. In insulators, that time is very long, because the specific conductivity σ_0 is extremely small. To demonstrate the effect order of magnitude, we calculate its size in a typical semiconductor like germanium. Given data would be $\sigma_0 = 0.1\,\Omega^{-1}\text{cm}^{-1} = 10\ \Omega^{-1}\text{m}^{-1}$, $\varepsilon_r = 15$, and $\varepsilon_0 = 8.85 \times 10^{-12}\frac{\text{Coulomb}}{\text{V·m}}$. Setting in Eq. (6.27), we get

$$\tau_{DR} = \frac{\varepsilon_r \varepsilon_0}{\sigma_0} = \frac{15 \times 8.85 \times 10^{-12}}{10} = 1.32 \times 10^{-11}\text{s}.$$

In metals, where the specific conductivity is greater by several orders of magnitude, the dielectric relaxation time may be as short as 10^{-16} s.

We turn to analyze the meaning of the assumption that diffusion currents are negligible compared to drift currents. For example, consider the hole current component in an x orientation. Following Eq. (6.17), we get the condition

$$D_p \frac{\partial(\delta p)}{\partial x} \ll \mu_p p \mathcal{E}_x.$$

Setting Einstein's relations (Eq. [6.19]) and solving for E_x, we get

$$\mathcal{E}_x \gg \frac{k_B T}{q} \cdot \frac{1}{p} \cdot \frac{\partial(\delta p)}{\partial x}.$$

We assume that the small added concentration of charge carriers satisfies $\delta p < 10^{-2} p$ over a range not smaller than 10^{-2} cm. At room temperature, $k_B T/q = 0.026$ V. Setting the values in the above equation, we get the requirement that $\mathcal{E}_x \gg 0.026$ V/cm. For added electrons, we get a similar result. We still have to remember that the requirement concerning the minimal size of the external electric field depends on the disturbance size δp and on the range δx over which it occurs. As much as the disturbance size grows, and its range narrows, the stricter would be the condition regarding the size of the electric field for which it would be justified to ignore diffusion currents compared to drift currents.

6.5.3 Charge Carrier Transit

Let us consider a uniform semiconductor shaped in the form of a long rectangular prism of length L under an electric field $\mathcal{E}_x$, where x is co-oriented with the prism length. A disturbance of thermal equilibrium is created in a form of added holes at one edge of the prism, near the positive electrical connection, as shown in Figure 6.9. Under proper conditions to be specified below, the disturbance will gradually drift toward the opposite prism end, and disappear at the counter electrical contact. Such an effect is called **charge carrier transit**. To simplify the analysis, and for the purpose of illustration, we assume the crystal to be p-type where $n_0 = 0$, and further neglect all recombination effects (namely $\tau = \infty$), and that the width of the region where the initial disturbance occurred is small compared to the prism length L. We make an important change of denotations: we denote by j_c the electrical current density that emanates from a real motion of charge-carrying particles. Thus far in our discussions, the possibility that an electrical current may arise by a mechanism different than just a motion of charge-carrying particles has not been mentioned. The latter mechanism will be discussed later. The dynamic equations describing the obtained electrical current for the said one-dimensional model along the x-axis would be

$$\left.\begin{aligned}&\text{(i) holes current density: } j_c = q\mu_p p(x.t)\mathcal{E}(x.t);\\&\text{(ii) continuity equation: } \frac{\partial j_c}{\partial x} = -q\frac{\partial (p-p_0)}{\partial t};\\&\text{(iii) Poisson equation: } \frac{\partial \mathcal{E}}{\partial x} = q\frac{(p-p_0)}{\varepsilon_r\varepsilon_0}.\end{aligned}\right\}. \tag{6.28}$$

We take a time t derivative of both sides of Poisson's Eq. (6.28)-(iii), reverse the order of $\mathcal{E}$ derivative taken between x and t, and get

$$\frac{\partial p}{\partial t} = \frac{\varepsilon_r\varepsilon_0}{q}\frac{\partial}{\partial x}\left(\frac{\partial \mathcal{E}}{\partial t}\right).$$

We then rewrite equation (ii) in the form

$$\frac{\partial p}{\partial t} = -\frac{1}{q}\frac{\partial j_c}{\partial x}.$$

Comparing the right-hand sides of the above two equations, we get

$$\frac{\partial}{\partial x}\left(j_c + \varepsilon_r\varepsilon_0\frac{\partial \mathcal{E}}{\partial t}\right) = 0.$$

We must then conclude that the quantity $J(t) \equiv j_c + \varepsilon_r\varepsilon_0\frac{\partial \mathcal{E}}{\partial t}$ ***is a constant independent of*** x.

The obtained result is general, because it refrained from neglecting the diffusion currents (that we may rather do in various following cases). Maxwell interpreted the $J(t)$ constant of the left-hand side as an actual total electrical current in any electrical circuit, which is the same at each point along an electrical circuit. The first addend on the right-hand side is a current contributed by the actual motion of electrically charged particles; the second addend he named as a **"displacement current"**. A displacement current may even provide the major contribution to the total current. For example, the electrical current in a medium between a capacitor plates is entirely a displacement current under charging or discharging. Summarizing, the electrical current $J(t)$ would always be given by

$$J(t) \equiv j_c(x,t) + \varepsilon_r\varepsilon_0\frac{\partial \mathcal{E}(x,t)}{\partial t}. \tag{6.29}$$

Recalling that the right-hand side is independent of x, we calculate an average of Eq. (6.29) by integrating along the semiconductor length, and dividing by that length L:

$$\begin{aligned}J(t) &= \frac{1}{L}\int_0^L J(t)dx = \frac{1}{L}\int_0^L q\mu_p p(x,t)\mathcal{E}(x,t)dx + \frac{\varepsilon_r\varepsilon_0}{L}\cdot\frac{\partial}{\partial t}\int_0^L \mathcal{E}(x,t)dx;\\J(t) &= \frac{q\mu_p}{L}\int_0^L p(x,t)\mathcal{E}(x,t)dx - \frac{\varepsilon_r\varepsilon_0}{L}\cdot\frac{\partial V}{\partial t},\end{aligned} \tag{6.30}$$

where V is the external voltage supplied by the electrical voltage supply (see Fig. 6.9). If that voltage is constant, its partial time derivative is zero, and we may write

$$J(t) = q\mu_p\frac{1}{L}\int_0^L p(x,t)\mathcal{E}(x,t)dx = q\mu_p\overline{p(x,t)\mathcal{E}(x,t)}, \tag{6.31}$$

where $\overline{p(x,t)\mathcal{E}(x,t)}$ is the average of the product $p(x,t)\mathcal{E}(x,t)$ along the sample. That average is independent of x, and it obviously depends only on time. In a uniform semiconductor, where the position dependence of the electrical field is small (an approximately constant field), the said average may be approximated as a product of the two separate averages

$$J(t) = q\mu_p \overline{p}(t)\overline{\mathcal{E}}(t), \tag{6.32}$$

where $\overline{p}(t) \equiv (1/L)\int_0^L p(x,t)dx$ is the average hole density in the crystal, and $\overline{\mathcal{E}}(t) = V/L$ is the average electrical field. The latter is relevant as long as the external voltage variations over time are sufficiently slow to justify neglecting of displacement currents. Under the said assumptions, the time variation of the electrical currents is practically determined by the time dependence of the average concentration of the charge carriers.

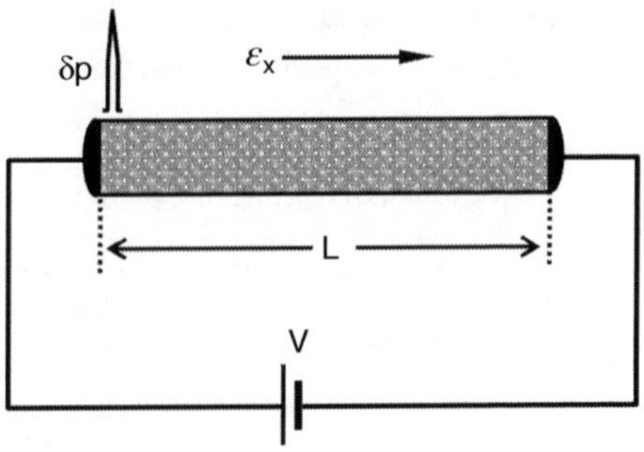

Figure 6.9 Schematic illustration of a uniform p-type semiconductor under an electric field when a sudden short-time narrow region near the positive electrode is added with excess holes.

We now delve further into the issue of neglecting the displacement currents, and see that it is quite justified over a broad range of practical conditions even if the voltage source exhibits a considerable alternating component. Eq. (6.29) suggests that displacement currents may be neglected when

$$\frac{\partial V}{\partial t} \ll \frac{q\mu_p L}{\varepsilon_r \varepsilon_0}\overline{p(x,t)\mathcal{E}(x,t)} \approx \frac{q\mu_p \overline{p}(t)V}{\varepsilon_r \varepsilon_0},$$

or alternatively presented via defining a characteristic frequency f of the voltage variations

$$\mathrm{f} \equiv \frac{1}{V}\frac{\partial V}{\partial t} \ll \mathrm{f}_{\boldsymbol{max}} \equiv \frac{q\mu_p \overline{p}(t)}{\varepsilon_r \varepsilon_0}. \tag{6.33}$$

The above-defined f_{max} provides an upper limit estimate of external voltage variations' frequency, for which it would be justified to ignore the contribution of displacement currents. As a numerical example, we choose specific properties of a typical p-type semiconductor in a range where f_{max} would assume a relatively low value:

Given

$$\mu_p = 1000\ \mathrm{cm^2/V\cdot s} = 0.1\ \mathrm{m^2/V\cdot s};\quad \overline{p} = 10^{13}\ \mathrm{cm^{-3}} = 10^{19}\ \mathrm{m^{-3}};$$

$$\varepsilon_r = 15;\quad \varepsilon_0 = 8.85\times 10^{-12}\ \mathrm{Coulomb/V\cdot m}$$

Setting in Eq. (6.33)

$$\mathrm{f} \equiv \frac{1}{V}\frac{\partial V}{\partial t} \ll \mathrm{f}_{\boldsymbol{max}} \equiv \frac{1.6\times 10^{-19}\times 0.1\times 10^{-19}}{15\times 8.85\times 10^{-12}} = 1.2\times 10^{9}\ \mathrm{Hz}.$$

A conclusion drawn on the basis of the last result is that in semiconductors, it is usually safe to ignore the contribution of displacement currents even if an external voltage supply exhibits some noise, as long as the noise frequency never exceeds a GHz range; say if it is only several dozen MHz. The problem arises strongly when dealing with insulating materials, where the density of mobile (free) charge carriers is close to zero.

We return to the issue presented at the beginning of the present subsection (Figure 6.9). A very short light pulse (or some other means) has generated an addition of holes whose electrical charge is not compensated by neither free nor trapped electrons at a region whose width is $\delta x(0)$

and is located at the very left end of the crystal (connected to the positive electrical contact). We mathematically define an additional axis system that moves from left to right at a velocity $\frac{dx}{dt} = \mu_p \mathcal{E}(x, t)$. We correspondingly analyze the full time differential of the hole density at the moving axis system:

$$\frac{dp(x(t), t)}{dt} = \frac{\partial p}{\partial x} \cdot \frac{dx}{dt} + \frac{\partial p}{\partial t} = \frac{\partial p}{\partial x} \cdot \mu_p \mathcal{E} - \frac{1}{q}\frac{\partial j_c}{\partial x} = \frac{\partial p}{\partial x}\mu_p \mathcal{E} - \mu_p \left[\frac{\partial p}{\partial x}\mathcal{E} + p\frac{\partial \mathcal{E}}{\partial x}\right].$$

Opening the square brackets, and setting Poisson's equation (Eq. [6.28]-(iii)), we get

$$\frac{d(\delta p)}{dt} = -\frac{q\mu_p}{\varepsilon_r \varepsilon_0} p \cdot \delta p. \tag{6.34}$$

This differential equation describes the time variation of added hole density in an axis system moving at a $\mu_p \mathcal{E}(x, t)$ velocity in the X-axis direction. In the following sub-sections, we consider its solution under various boundary conditions.

6.5.3.1 Excess Charge Carrier Drift in an Insulating Semiconductor

An insulator obviously satisfies the condition $p_0 = 0$. Then $p = \delta p$ and Eq. (6.34) becomes

$$\frac{d(\delta p)}{dt} = -\frac{q\mu_p}{\varepsilon_r \varepsilon_0} (\delta p)^2,$$

Whose solution is (see also Eq. [6.12])

$$\delta p(t) = \frac{\delta p(0)}{1 + \frac{q\mu_p}{\varepsilon_r \varepsilon_0} \delta p(0) \cdot t}. \tag{6.35}$$

The solution in fact expresses a dielectric relaxation, whose effective constant τ_{DR} depends on the gradually changing charge carrier concentration. This equation is valid as long as the origin of the moving axis is still contained inside the crystal. In a case of a uniform crystal and a constant voltage source, it means for $0 < t < t_t \equiv L/\mu_p \mathcal{E} = L^2/\mu_p V$. The t_t quantity is called **transit time**, and it describes a frontal drift of the excess charge carriers generated at one edge of the crystal till reaching the counter edge. If $\delta p(0)$ is sufficiently small, such that $\frac{q\mu_p \delta p(0) \cdot t_t}{\varepsilon_r \varepsilon_0} \ll 1$, then $\delta p(t) \cong \delta p(0)$ during the entire time duration where the equation is valid. It means that the front width δx of the excess charge carriers remains the same throughout the entire drift duration, until exiting the crystal. If that condition is not satisfied, it means that $\delta p(0)$ reduces over time. Because charge carriers don't disappear, it means that the carriers front δx broadens. The shape of the obtained current trace in the two cases, assuming a uniform electrical field across the crystal volume, is described in Figure 6.10. The current exhibits a constant value $J(0) = q\mu_p \delta p(0)\delta x(0)V/L^2$ as long as the entire front remains inside the crystal. When it reached the other edge, past the transit time, the current reduces to zero. The shape of the current reduction reflects the shape of the front width upon reaching the other end: if the front has broadened, the current reduction to zero lengthens.

Uniformity of the external electric field is a valid assumption up to a limited size of the excess electrical charge added into the crystal. With larger perturbations, the electrical field inside the crystal deforms by values that exhibit the same order as the originally applied field; it then strongly affects the current time dependence. It is called **space-charge limited current**. Addressing that issue is quite complex, requiring acquaintance with effects and concepts to be presented in Chapter 8. Section 6.6 provides a detailed discussion of space-charge limited currents with proper cross-referencing to Chapter 8.

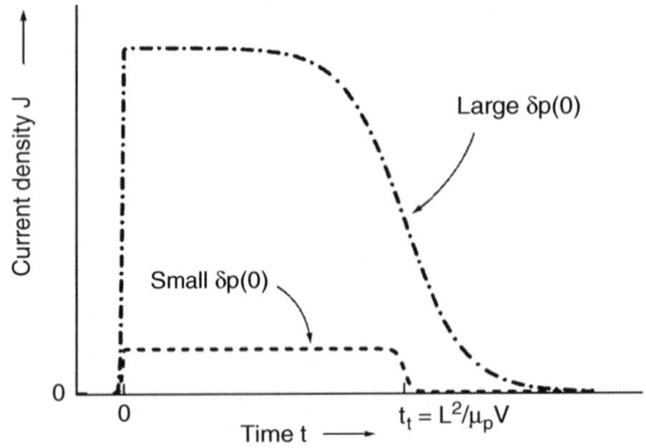

Figure 6.10 Typical current density versus time dependence following a short pulse of excess hole generation at the positive edge of an insulating semiconductor under an external electric field. Two cases are described: a small front that remains narrow, and a large front that broadens during its drifting toward the opposite edge.

6.5.3.2 Small Excess Charge Carrier Drift in a Conducting Material

For analysis of a conducting material case, we assume that $p_0 > 0$ while $\delta p(0) \ll p_0$. Eq. (6.34) then reduces to the form

$$\frac{d(\delta p)}{dt} = -\frac{q\mu_p p_0}{\varepsilon_r \varepsilon_0} \delta p,$$

whose solution is

$$\delta p(t) = \delta p(0) e^{-t/\tau_{DR}},$$

where $\tau_{DR} = \varepsilon_r \varepsilon_0 / q\mu_p p_0 = \varepsilon_r \varepsilon_0 / \sigma_0$ is the dielectric relaxation time (Eq. [6.27]). The shape of the initially generated excess carriers front remains unchanged, but its size, and obviously the total number of excess carriers, decays exponentially. Excess charge carriers exit both sides of the crystal. The time dependence of the obtained current is presented in Figure 6.11. The increased current density at $t = 0$, given by $\delta J(0) = q\mu_p \delta p(0) \delta x(0) V/L^2$, decays exponentially immediately after its start, in the form $\delta J(t) = \delta J(0) e^{-t/\tau_{DR}}$ until the reduced-in-size excess charge carrier front reaches the edge side of the crystal, getting out of it, and disappear into the electrical contact.

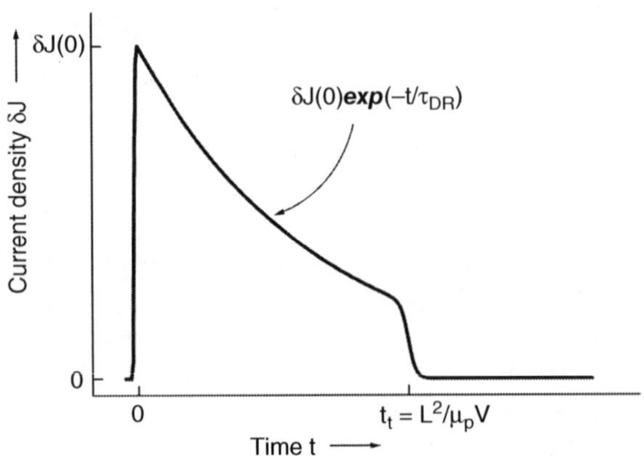

Figure 6.11 Typical increased current density versus time dependence following a short, small pulse of excess hole generation at the positive edge of a conducting material under an external electric field. A uniform dielectric size reduction of the excess carrier front occurs during its drift toward the opposite electrode, until exiting the crystal at the negative electrode.

6.5.3.3 Small Excess Charge Carrier Drift in a Semiconductor

In the case of a semiconductor, the dynamics of the two types of charge carriers needs to be accounted for. The equations of motions would be

$$\left.\begin{aligned} &(i)\ \underline{J}_p = q\mu_p p\underline{\mathcal{E}} - qD_p\,\textbf{grad}\,(\delta p)\\ &(ii)\ \underline{J}_n = q\mu_n n\underline{\mathcal{E}} + qD_n\,\textbf{grad}\,(\delta n)\\ &(iii)\ \frac{\partial p}{\partial t} = G - \frac{\delta p}{\tau_p} - \frac{1}{q}\,\textbf{div}\,\underline{J}_p\\ &(iv)\ \frac{\partial n}{\partial t} = G - \frac{\delta n}{\tau_n} - \frac{1}{q}\,\textbf{div}\,\underline{J}_n \end{aligned}\right\}. \qquad (6.36)$$

In addition, we also use Einstein's relations of Eq. (6.19). Notably, we returned to use the capital "*J*" letter for the position-dependent electrical current emanating from the motion of electrically charged particles. It means that we ignore the displacement currents (see Eq. [6.29]). That would be justified if the external electric field remains unchanged over time. It, however, doesn't imply that it is the same as a function of position; in principle, it may even be grossly different in different places.

The dielectric relaxation time in a semiconductor is usually very short compared to typical transition times. Therefore, if at a certain moment an excess density of charge carriers are generated, say δp, it will be balanced by opposite-type charge carriers δn within a time duration that is of the order of the dielectric relaxation time. It thus implies that $\delta p \cong \delta n$. Any deviation from electrical balance (the difference between δp and δn) must be extremely small, because otherwise it would generate a very large internal electric field. To estimate an allowed upper limit for deviation from electrical charge balance, let us assume a deviation that distorts a typical external electrical field of, say 10 V/cm, applied over a sample of 1 mm length, meaning a distortion of $d\mathcal{E}_x/dx = 10^6$ V/mm^2. According to Poisson's equation (Eq. [6.26]), we would have

$$\frac{d\mathcal{E}}{dx} = \frac{q(\delta p - \delta n)}{\varepsilon_r\varepsilon_0} \Rightarrow (\delta p - \delta n) = \frac{\varepsilon_r\varepsilon_0}{q}\frac{d\mathcal{E}}{dx}.$$

Setting the assumed values, we get

$$(\delta p - \delta n) = \frac{15 \times 8.85 \times 10^{-12} \times 10^6}{1.6 \times 10^{-19}} = 8.3 \times 10^{14}\ \text{m}^{-3} = 8.3 \times 10^8\ \text{cm}^{-3}.$$

The obtained concentration is in fact tiny compared to actual charge carrier concentrations in semiconductors. For example, the room temperature intrinsic concentrations of holes and electrons in germanium *Ge* is 2.4×10^{13} cm^{-3}. Thus for germanium and other similar semiconductors, it is quite justified to assume that electrical charge balance is maintained as long as the external electrical fields are not extremely high. Namely, it is assumed that quite accurately, $\delta p = \delta n$.

We thus turn to solving the dynamic equations of Eq. (6.36) under the following assumptions:

(1) The crystal is a uniform semiconductor.
(2) There exists a non-uniform δp perturbation to the charge carrier concentrations.
(3) The perturbation is small.
(4) The crystal is under the action of a small external electric field.
(5) There is no trapping, namely always $\delta p = \delta n$, therefore also $\tau_p = \tau_n \equiv \tau$.

For solving Eq. (6.36), we set Eq. (6.36)-(i) in Eq. (6.36)-(iii) and get

$$\frac{\partial p}{\partial t} = G - \frac{\delta p}{\tau} - \mu_p p\,\textbf{div}\,\underline{\mathcal{E}} - \mu_p(\underline{\mathcal{E}} \cdot \textbf{grad}(\delta p)) + D_p\,\textbf{div}\,\textbf{grad}\,(\delta p);$$

Similarly, we set Eq. (6.36)-(ii) in Eq. (6.36)-(iv) and get

$$\frac{\partial n}{\partial t} = G - \frac{\delta n}{\tau} + \mu_n n\, \mathbf{div}\, \underline{\mathcal{E}} + \mu_p(\underline{\mathcal{E}} \cdot \mathbf{grad}\,(\delta n)) + D_p \mathbf{div}\, \mathbf{grad}(\delta n);$$

We multiply the first equation by $\mu_n n$, the second by $\mu_p p$, sum the two results, and divide the sum by $\mu_n n + \mu_p p$. We get

$$\frac{\partial(\delta p)}{\partial t} = G - \frac{\delta p}{\tau} - \mu(\underline{\mathcal{E}} \cdot \mathbf{grad}\,(\delta p)) + D\, \mathbf{div}\, \mathbf{grad}(\delta p), \tag{6.37}$$

where

$$\mu \equiv \frac{n-p}{\frac{p}{\mu_n} + \frac{n}{\mu_p}}; \quad D \equiv \frac{n+p}{\frac{p}{D_n} + \frac{n}{D_p}}.$$

The above μ quantity is called **ambipolar mobility**, and the D quantity is called **ambipolar diffusivity**. Under sufficiently small perturbations such that $\delta p \ll n_0 + p_0$, we get

$$\mu \equiv \frac{n_0 - p_0}{\frac{p_0}{\mu_n} + \frac{n_0}{\mu_p}}; \quad D \equiv \frac{n_0 + p_0}{\frac{p_0}{D_n} + \frac{n_0}{D_p}}. \tag{6.38}$$

In that case, the ambipolar mobility and ambipolar diffusivity are characteristic constants of the conditions-abiding crystal. An important effect is the limit values of the ambipolar mobility and diffusivity in semiconductors of a distinct conductivity type: if $n_0 \gg p_0$, we get $D \rightarrow D_p$ and $\mu \rightarrow \mu_p$. On the other hand, if $p_0 \gg n_0$, we get $D \rightarrow D_n$ and $\mu \rightarrow (-\mu_n)$. Thus, under those distinct cases, the dominating ambipolar diffusivity and mobility are those of the minority carriers! It should be clarified that while the excess minority carriers indeed drift and diffuse in the form described according to Eq. (6.38) – to be later analyzed in considerable detail, the majority carriers behave differently – they accumulate at the region where the excess minority carriers drifted to, thus effectively balancing the space charge that would have been otherwise generated. Considering the meaning of these dynamic equations from the "point of view" of the majority carriers, they describe the drift of the position where they accumulate to screen the said developing space charge.

Equation (6.37) in a one-dimensional case has the form

$$\frac{\partial(\delta p)}{\partial t} = G - \frac{\delta p}{\tau} - \mu\mathcal{E}\frac{\partial(\delta p)}{\partial x} + D\frac{\partial^2(\delta p)}{\partial x^2}. \tag{6.39}$$

We turn now to discuss the solution of Eq. (6.39) for several different special cases.

6.5.3.4 Steady Excitation of a Narrow Region with no External Electric Field

As above stated, we set $\mathcal{E} = 0$ in Eq. (6.39), and $G = 0$ except for a very narrow region at the crystal center $x = 0$ position, as portrayed in Figure 6.12 below. Under steady excitation, $\frac{\partial(\delta p)}{\partial t} = 0$, and Eq. (6.39) reduces to the form

$$\frac{d^2(\delta p)}{dx^2} = \frac{\delta p}{D\tau} \quad \text{for } |x| > 0,$$

whose solution under the said boundary conditions is

$$\delta p(x) = \delta p(0) e^{-|x|/\sqrt{D\tau}}. \tag{6.40}$$

The $L_{Df} \equiv \sqrt{D\tau}$ quantity is called **diffusion range**. The spatial distribution of the excess, oppositely charged carrier pairs is displayed in Figure 6.12.

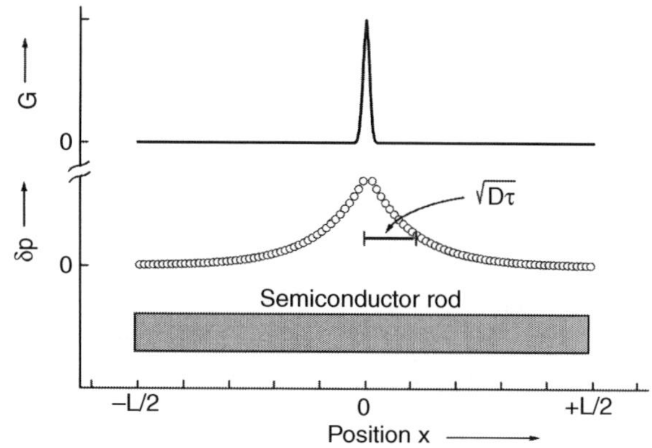

Figure 6.12 Schematic illustration of the distribution of excess opposite-type charge carrier pairs (bottom) under a steady excitation at a very narrow region (top) in the absence of any external electrical field.

6.5.3.5 Steady Excitation of a Narrow Region Under a High External Electric Field

It is thus basically assumed that $G = 0$ over the entire crystal volume, except for a very narrow region at the crystal center $x = 0$ position as described in Figure 6.13. In contrast to the previous case, it is assumed that the external electric field is sufficiently high to allow neglecting of the contribution of diffusion currents compared to drift currents. Under steady excitation, $\frac{\partial(\delta p)}{\partial t} = 0$, and Eq. (6.39) reduces to the form

$$\frac{d(\delta p)}{dx} = -\frac{\delta p}{\mu \mathcal{E} \tau},$$

whose solution is

$$\delta p(x) = \delta p(0)e^{-x/\mu\mathcal{E}\tau} \quad \text{for} \quad x/\mu > 0. \tag{6.41}$$

The $L_{\mathcal{E}} \equiv \mu\mathcal{E}\tau$ quantity is called **drifting range** (in the professional literature, the originally used German **Schub-weg** term is quite often used). The spatial distribution of the excess oppositely charged carrier pairs is displayed in Figure 6.13.

It should be emphasized again that ignoring of the diffusion effect upon which the present development was based, is justified as long as the $\mu\mathcal{E}\tau \gg \sqrt{D\tau}$ condition is satisfied. In a case of an intrinsic semiconductor, where $n_0 = p_0$, we always have $\mu = 0$. Then, the opposite of the above condition prevails irrespective of the external electric field, namely always $\mu\mathcal{E}\tau \ll \sqrt{D\tau}$.

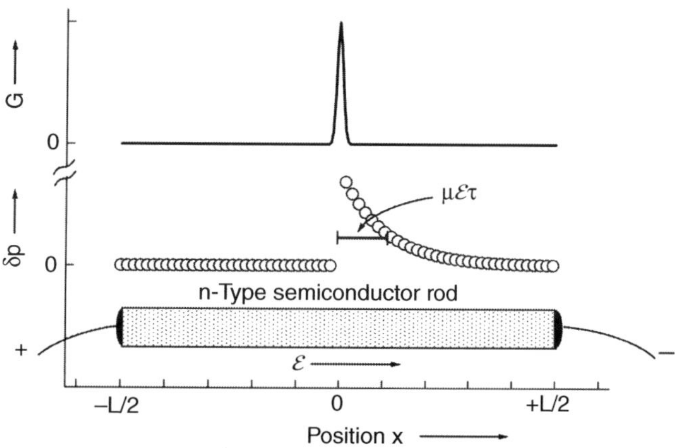

Figure 6.13 Schematic illustration of the distribution of excess opposite-type charge carrier couples in an *n*-type semiconductor (bottom) under a steady excitation at a very narrow region (top) under a very high external electrical field. In a *p*-type crystal, the charge carriers would drift in the opposite direction.

6.5.3.6 Transient Excitation in a Narrow Region Under a High External Electric Field

We assume the external electric field to be sufficiently high such that $\mu\mathcal{E}\tau \gg \sqrt{D\tau}$. We may then ignore the last addend in Eq. (6.39). After the end of excitation, we have $G = 0$, and Eq. (6.39) assumes the form

$$\frac{\partial(\delta p)}{\partial t} + \mu\mathcal{E}\frac{\partial(\delta p)}{\partial x} = -\frac{\delta p}{\tau}.$$

We perform a variable change as follows: we mathematically define an additional axis system moving at a constant velocity $dx/dt = \mu\mathcal{E}$, in which $\delta p = \delta p(x(t), t)$, where $x(t) = \mu\mathcal{E}t$. The left-hand side of the above equation is the full differential of δp according to t. Then

$$\frac{d(\delta p)}{dt} = -\frac{\delta p}{\tau},$$

whose solution is

$$\delta p(t) = \delta p(0)e^{-t/\tau} \text{ for } x = \mu\mathcal{E}t, \text{ and } \delta p = 0 \text{ for any } x \neq \mu\mathcal{E}t. \tag{6.42}$$

If we had included the diffusion effect in our equation, we would have obtained a broadening of the moving charge-carriers front. The physical image is thus of a narrow front of charge carriers drifting in the electrical field direction (or in the opposite direction for the other type of minority carriers), gradually broadening with time.

Figure 6.14 relates to an experiment that demonstrates the effect, named after **Shockley and Haynes**. The subject of the experiment is an n-type semiconductor shaped as a long rectangular

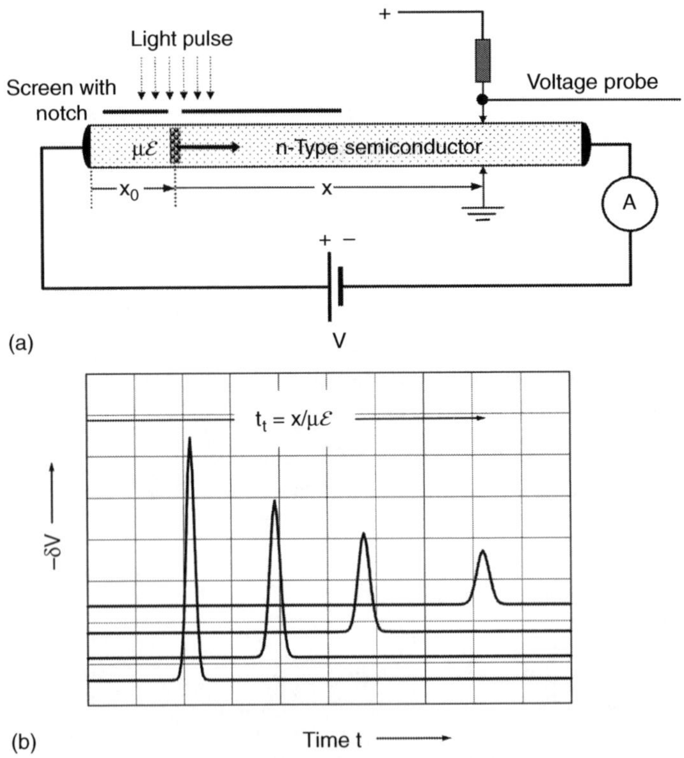

Figure 6.14 Description of an experiment and its results aimed at measuring the excess carrier lifetime and ambipolar diffusivity in an n-type semiconductor. In the experiments, the position of the exciting light pulse is changeable in order to vary its distance x from the electrical probe. (a) The array of electrical connections. (b) Shape of voltage traces measured at the probe for different x distances, as displayed on an oscilloscope screen.

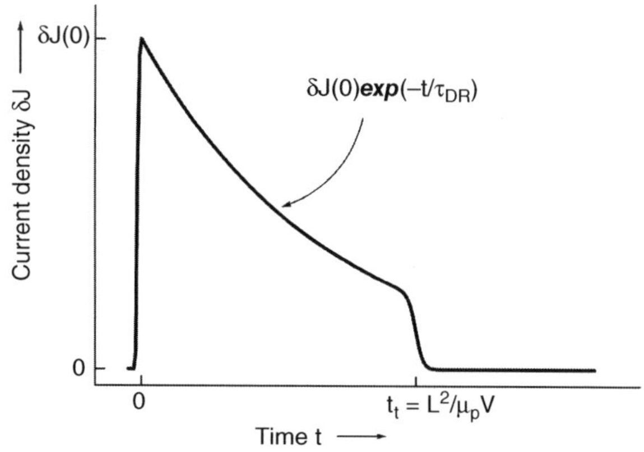

Figure 6.15 Description of the added current density as a function of time after a short-time excitation of excess electron–hole pairs at a narrow region in an *n*-type semiconductor, as described in Figure 6.14a. Electron–hole pairs' recombination occurs while the excess carrier front drifts, becoming lower, till reaching and exiting the crystal end connected to the negative electrical contact of the voltage supply.

prism under a high longitudinal electric field. A short light pulse having a very small geometric cross section is targeted to hit the surface, to form a very narrow front of excess hole and electron pairs in the crystal cross section close to the positive contact. A probe that measures changes in the prism transverse conductivity is placed close to the negative contact, and uses a relatively small cross electric field. The transverse conductivity increases when the excess carriers' front traverses the probe position. It is thus possible to record the front height, width, and arrival time on an oscilloscope screen, thereby calculate the ambipolar mobility μ, diffusivity D, and lifetime τ. In the original experiment by Shockley and Haynes, the excess minority carriers were obtained by contact injection, an effect to be described in detail in Chapter 7 in relation to a p–n junction.

The time dependence of the total added current that follows a light pulse, as measured by an ammeter **A** connected in series to the semiconductor (Fig. 6.14a), is presented schematically in Figure 6.15. The added current density at the light pulse absorption moment is given by $\delta J(0) = q(\mu_p + \mu n)\delta p(0)\delta x(0)V/L^2$, where $\delta x(0)$ is the generated front-width at $t = 0$ moment, V is the voltage between the contacts, and L is the crystal total length. Later on, the added current density reduces with time according to $\delta J(t) = \delta J(0)e^{-t/\tau}$, until the reduced-in-size excess carriers' front reaches the crystal edge connected to the negative electrode, where they exit the crystal. On exiting the crystal, the added current density drops to zero. The time duration between the front generation and its exiting the crystal (called **transit time**) depends on its generation position. In the case shown in Figure 6.14a, it is given by $t_t = (L - x_0)L/\mu V$, where μ is the ambipolar mobility. Because we considered an n-type crystal, the ambipolar diffusivity satisfies $\mu \cong \mu_p$.

Notably, the obtained current density versus time trace is quite similar to the one described in Figure 6.11. However, the current decay in that case emanated from a dielectric relaxation, while in the present considered case, it emanates from recombination of the excess hole–electron pairs.

6.6 Space-Charge Limited Currents

The system we now consider is an insulating semiconductor crystal (namely one of a high-energy gap) shaped in the form of a thin slab of thickness L, with electrical contacts on opposite faces, and a voltage V applied between the contacts. The system is quite similar to a two-plates capacitor whose electrical capacitance density (electrical capacitance per unit area) is $\boldsymbol{C} = \varepsilon_r\varepsilon_0/L$, and the electrical charge density on each plate is $\boldsymbol{Q} = \varepsilon_r\varepsilon_0 V/L$. Initially, the crystal itself remains free of charge carriers (zero electrical current). However, under certain conditions, some electrical charge still penetrates the crystal, generating a charge distribution in the space between the electrodes. When there are no carrier traps in the semiconductor, the penetrated space charge assumes the form of free charge carriers, expressing itself by an electrical current through the crystal.

An electrical contact that allows penetration of space charge into an insulating material is called "injecting contact". In the context of our discussed subject, an injecting contact is characterized by the fact that it induces a very large reservoir of charge carriers of a single specific type very close to the contact inside the crystal (called "accumulation layer") and persists in maintaining that reservoir for a large range of external voltages. Therefore, there is a point in the crystal near the injecting contact, where the electrical field assumes a zero value for all used external voltages. Chapter 8 provides a broad and in-depth discussion on the influence of electrical contacts on the bands structure of semiconductors. **Prior reading of particularly Section 8.1 would be quite useful for readers of the following discussion!** A metallic contact may function as a hole-injecting contact if it is of a sufficiently large work function, or as an electron-injecting contact if it is of a sufficiently small work function. Figure 6.16 displays an energy states' scheme obtained close to a hole-injecting contact under a high external electric field. The thickness of the hole accumulation layer (Debye range L_D) is very small compared to the crystal thickness L. The hole current in the region where the electric field is close to zero, are dominated by diffusion, while deeper into the crystal volume, where the electric field is high, the hole current are dominated by the drift.

An alternative means of forming an injecting electrical contact is to illuminate the near-contact material surface with a strongly absorbed light. The practical way is to illuminate the crystal through the electrical contact by making the contact partially transparent. Use of strongly absorbed light guaranties that the absorption is limited to a very narrow region adjacent to the contact, whose thickness is small compared to the entire crystal length. The absorbed photons

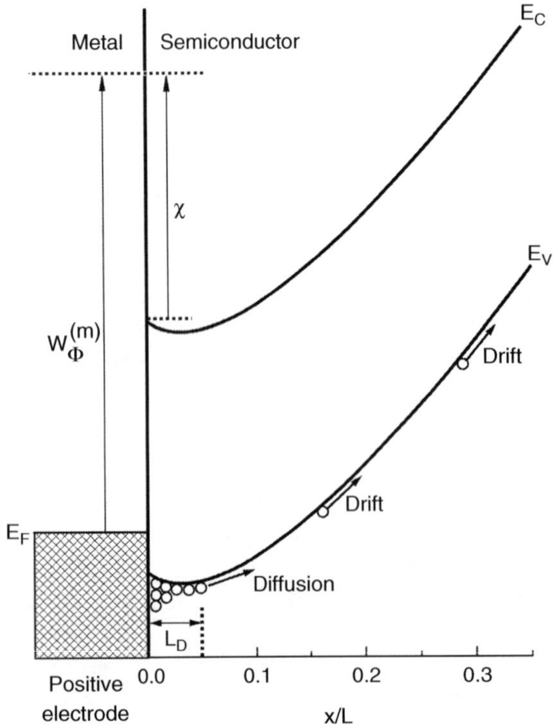

Figure 6.16 Energy scheme of an insulating semiconductor under the action of a high electric field near a hole-injecting electrode at a positive polarity. $w_{\Phi}^{(m)}$ is the metallic electrode work function, and χ is the semiconductor affinity. For additional details and discussion, see the text.

generate many electron–hole pairs, and the applied electrical field separates between the two-type carriers. If the illuminated contact is positive, the electrons will move a very short distance and be "absorbed" by the electrode, while the holes will be "injected" into the crystal. If the contact is negative, the holes will be the ones absorbed by the contact, and the electrons will be the ones injected into the crystal.

In our following discussion, we assume that the positive electrode (anode) is hole injecting, while the negative electrode (cathode) is not. We analyze two cases: one where the applied voltage is constant, and the currents reached a steady state, the other is transient, occurring immediately after the voltage onset (or immediately after the positive contact becomes "injecting"), before reaching the steady state. To simplify the analysis, we assume an absence of carriers trapping.

6.6.1 Constant Space-Charge Limited Current

When a constant current flows through the crystal, the electric field is also kept time-independent throughout the entire crystal volume, and displacement currents are zero. Thus, the equations to be solved are (see Eqs. [6.28])-(i) and [6.28])-(iii))

$$\left.\begin{array}{ll}\text{(i) Holes current density:} & J = q\mu_p p(x)\mathcal{E}(x);\\ \text{(ii) Poisson equation:} & \dfrac{\partial \mathcal{E}}{\partial x} = \dfrac{q}{\varepsilon_r\varepsilon_0}p(x).\end{array}\right\}, \tag{6.43}$$

and the addition of the boundary conditions $\mathcal{E}(0) = 0$ and $\int_0^L \mathcal{E}(x)dx = V$, where x is the distance from the anode (see Fig. 6.16). For punctuality, it should be pointed out that in writing the current density Eq. (6.43)-(i), we ignored all diffusion currents (Section 6.3), which is justified where the electrical voltage is very high. Rather adjacent to the injecting electrode, as displayed in Figure 6.16, the electrical field is zero, and the currents generating the carriers' injection are dominated by diffusion. It is, however, a very narrow region compared to the entire crystal length, and we ignore it for simplicity. It may be further concluded that because the current J is constant, the free hole density becomes smaller as the distance from the anode increases and the electric field grows. To continue the process of solving the equations (6.43), we extract the hole density $p(x)$ using Poisson's equation, and set it in the current equation to get $J = \varepsilon_r\varepsilon_0\mu_p\mathcal{E}(x)\frac{\partial \mathcal{E}}{\partial x}$. Then

$$J = \frac{1}{L}\int_0^L Jdx = \frac{\varepsilon_r\varepsilon_0\mu_p}{L}\int_0^L \mathcal{E}(x)\frac{d\mathcal{E}(x)}{dx}dx = \frac{\varepsilon_r\varepsilon_0\mu_p}{L}\int_{\mathcal{E}(0)}^{\mathcal{E}(L)} \mathcal{E}(x)d\mathcal{E}(x) = \frac{\varepsilon_r\varepsilon_0\mu_p}{2L}\mathcal{E}^2(L). \tag{6.44}$$

For obtaining a specific expression for the current density, we adopt **Child hypothesis** stating that the position x dependence of the electric field follows the relation $\mathcal{E}(x) = L^{-1/2}\mathcal{E}(L)x^{1/2}$. That hypothesis is consistent with the boundary condition that the electric field is zero adjacently to the injecting anode at $x = 0$, while assuming the proper value at $x = L$. Setting it at the second boundary condition that the voltage between the electrodes is V, we get $\mathcal{E}(L) = \frac{3}{2}(V/L)$. That size of the electric field is 50% larger than the V/L electric field that would have prevailed without the carriers' injection at the anode. We set the obtained value in Eq. (6.44) and get

$$J = \frac{9}{4}\frac{\varepsilon_r\varepsilon_0\mu_p}{2L^3}V^2. \tag{6.45}$$

That result is known as **Child equation**; the current density is proportional to the squared voltage between the electrodes and is inversely proportional to the third power of the crystal thickness.

We turn to calculate the total space charge $\boldsymbol{Q}_{SC}$ contained inside the crystal. We compare two expressions used for the current, one given by Eq. (6.44) and the other is $J = q\mu_p p(L)\mathcal{E}(L)$ obtained by using Eq. (6.43)-(i). Solving for $p(L)$, we get $p(L) = \frac{\varepsilon_r \varepsilon_0}{2qL}\mathcal{E}(L)$. We know that the product between the hole density and electric field are the same for any x. By using Child's hypothesis concerning the electric field dependence on x, we get $p(x) = \frac{\varepsilon_r \varepsilon_0 \mathcal{E}(L)}{2qL^{1/2}}x^{-1/2}$. Then

$$\boldsymbol{Q}_{SC} = q\int_0^L p(x)dx = \varepsilon_r\varepsilon_0\mathcal{E}(L) = \frac{3}{2}\varepsilon_r\varepsilon_0\frac{V}{L}. \tag{6.46}$$

The space charge contained in the crystal is thus 1.5 times greater than the charge that would have been contained on each electrode (per unit area) without charge injection into the insulating crystal.

The above discussion and equation developments were limited to no-trapping conditions. If carrier trapping is significant, the material can still carry electrical currents, where a part of the injected space charge is trapped, and the other part is mobile. Assuming that the fraction θ of mobile charge is independent of the position x and the voltage V, Child's equation assumes a simple form

$$J = \frac{9}{4}\theta\frac{\varepsilon_r\varepsilon_0\mu_p}{2L^3}V^2. \tag{6.47}$$

Experiments performed using various insulating semiconductors show that values of θ may be as small as $10^{-2} - 10^{-9}$, and in the same material, being different between cases of hole or electron injection. Under very high voltages, a situation develops that carrier traps become close to be fully filled. In that limit $\theta \to 1$, and a further increase of voltage also increases the free charge carriers concentration.

Regarding the essence of Child's hypothesis, it is worthwhile noting that a different form of the electrical field dependence on position could be hypothesized, say $\mathcal{E}(x) = L^{-\alpha}\mathcal{E}(L)x^{\alpha}$, where $0 < \alpha < 1$. Choosing $\alpha \geq 1$ is impossible because it implies an infinitely large space charge inside the crystal. Choosing an $\alpha \neq \frac{1}{2}$ will only change the numerical factor appearing in Child's equation. Experimental results usually support Child's choice of $\alpha = \frac{1}{2}$.

6.6.2 Transient Space-Charge Limited Current

Suppose that at a certain $t = 0$ moment, the anode contact with the insulating semiconductor suddenly becomes hole injecting. Such situation can be realized by a strongly absorbed light pulse, whose rise-time is very short, or alternatively by a sudden application of a voltage between the electrodes, provided of course that the anode is fundamentally hole injecting into the said material. As a result, at the $t = 0$ moment, holes will start drifting toward the negative electrode (cathode). The electrical current at each point x and time $t > 0$ is given by the sum of conduction and displacement currents (see also Eq. [6.29])

$$J(t) = q\mu_p p(x,t)\mathcal{E}(x,t) + \varepsilon_r\varepsilon_0\frac{\partial\mathcal{E}(x,t)}{\partial t}, \tag{6.48}$$

where in expressing the conduction current, we ignored a possible contribution of diffusion currents. In addition, Poisson equation should also be included

$$\frac{\partial\mathcal{E}(x,t)}{\partial x} = \frac{q}{\varepsilon_r\varepsilon_0}p(x,t). \tag{6.49}$$

Maintaining continuous injection requires having $\mathcal{E}(0, t) = 0$ for $t \geq 0$ as a boundary condition. Solving for $p(x, t)$ in Eq. (6.49) and setting it in Eq. (6.48) yields

$$J(t) = \varepsilon_r \varepsilon_0 \left[\frac{1}{2} \mu_p \frac{\partial \mathcal{E}^2(x,t)}{\partial x} + \frac{\partial \mathcal{E}(x,t)}{\partial t} \right]. \tag{6.50}$$

Averaging Eq. (6.50) along the crystal, we get

$$J(t) = \frac{1}{L} \int_0^L J(t)dx = \frac{\varepsilon_r \varepsilon_0 \mu_p}{2L} \mathcal{E}^2(L,t) + \varepsilon_r \varepsilon_0 \frac{\partial}{\partial t} \int_0^L \mathcal{E}(x,t)dx, \tag{6.51}$$

where $\mathcal{E}(L, t)$ is the electric field at the cathode. Recall that $\int_0^L \mathcal{E}(x,t)dx = V = constant$. Therefore, its partial derivative according to time is zero, and so is the second addend in Eq. (6.51). Thus

$$J(t) = \frac{\varepsilon_r \varepsilon_0 \mu_p}{2L} \mathcal{E}^2(L,t). \tag{6.52}$$

Because the electrical current is independent of x, we may set in Eq. (6.50) any desired value of x, say $x = L$. Doing that, and comparing with Eq. (6.52), we get

$$\frac{1}{2} \mu_p \frac{\partial \mathcal{E}^2(L,t)}{\partial x} + \frac{\partial \mathcal{E}(L,t)}{\partial t} = \frac{\mu_p}{2L} \mathcal{E}^2(L,t). \tag{6.53}$$

We denote by t_{sc} the time at which the first front of injected space charge reaches the cathode at $x = L$. Inspecting Eq. (6.49), we trivially conclude that <u>before</u> t_{sc}, when $p(x, t) = 0$, also $\partial\mathcal{E}(x, t)/\partial x]_{x=L} = 0$. Thus, for $t < t_{sc}$, the first addend on the left-hand side of Eq. (6.53) is zero! Then this equation reduces to

$$\frac{\partial \mathcal{E}(L,t)}{\partial t} = \frac{\mu_p}{2L} \mathcal{E}^2(L,t). \tag{6.54}$$

We obtained a regular differential equation for $\mathcal{E}(L, t)$; our boundary condition is $\mathcal{E}(L, 0) = V/L$ meaning that before any holes injection into the crystal has started, the electrical field across the crystal was constant, having the V/L value throughout the entire crystal volume, including the specific $x = L$ point. The obtained solution for $\mathcal{E}(L, t)$ is (see also Eqs. [6.12] and [6.35])

$$\mathcal{E}(L,t) = \frac{V}{L} \cdot \frac{1}{1 - t/2t_t} \quad \text{for} \ \ t < t_{sc}, \tag{6.55}$$

where $t_t = L^2/\mu_p V$ is the transition time of a narrow small front of holes between the two crystal sides (Fig. 6.10 and related text). Setting it in Eq. (6.52), we get

$$J(t) = \frac{\varepsilon_r \varepsilon_0 \mu_p V^2}{2L^3} \frac{1}{(1 - t/2t_t)^2}, \quad \text{for} \ \ t \leq t_{sc}. \tag{6.56}$$

According to Eq. (6.56), at the $t = 0$ moment, the current is $4/9 \cong 0.45$ of the steady space-charge-limited current given by Eq. (6.45). After that moment, the current rises until the t_{sc} moment when the injected holes front reaches the cathode. Recall that at any given moment shorter that t_{sc}, the space between the leading front and the cathode is entirely free of charge carriers; therefore, the electrical field there doesn't change as a function of x, maintaining the initial $\mathcal{E}(L, t)$ value. Thus, the leading hole front drifted at a velocity given by $\mu_p\mathcal{E}(L, t)$, and its arrival time at the cathode t_{sc} is given by solving the equation

$$L = \int_0^{t_{sc}} \mu_p \mathcal{E}(L,t)dt. \tag{6.57}$$

Setting the specific expression for $\mathcal{E}(L, t)$ from Eq. (6.55) and solving for t_{sc}, we get

$$t_{sc} = 2(1 - e^{-1/2})\, t_t \cong 0.78\, t_t. \tag{6.58}$$

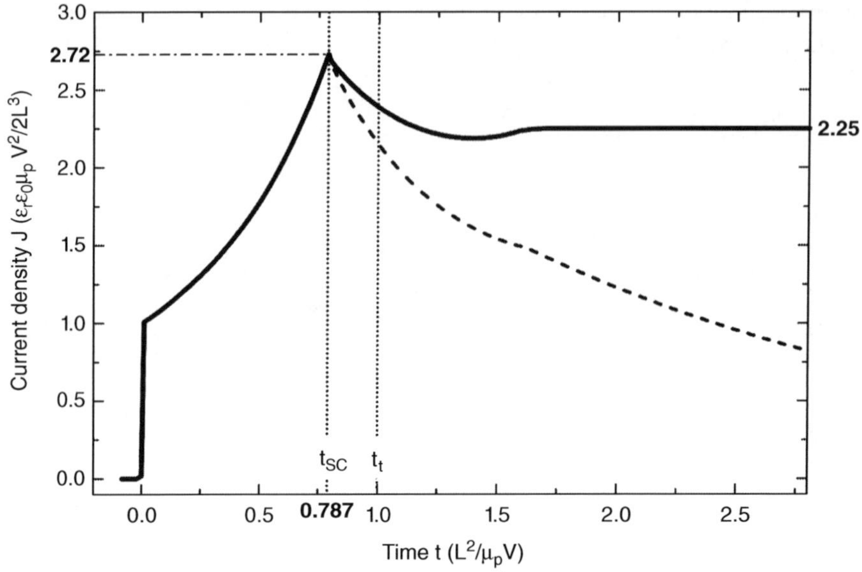

Figure 6.17 Time dependence of the current density in an insulating semiconductor after a sudden transformation of the positive electrode into an unlimited hole injector starting at the t = 0 moment. A solid line shows a case where the hole injection is continual. A dashed line shows a special case, where the hole injection lasts for only a short time of the order of t_{SC}.

The current density equation exhibits a peak value at the t_{SC} moment, which is about 20% higher than the current calculated for the steady current case (Child Eq. [6.45]). Later on, the current reduces gradually, with a general tendency toward the steady value. The process is quite complicated, involving the exiting of some space charge out of the crystal and its redistribution inside the crystal, as in the steady current state. The time required for reaching the steady current state is of the order of t_{SC}. A specific calculation involves some numerical computation and won't be presented here. Figure 6.17 provides a drawing of the discussed current versus time trace. It also shows an additional trace of a case where the current reduces gradually toward zero. Such cases are typical when the electrical contact transforms into a carrier-injecting one by a short light pulse, and it loses its injecting capability within a duration of the order of t_{SC}. The current size reduction reflects the emptying process of both the injecting region and the entire crystal volume from the space charge. A common physical process involved in a current reduction is a gradual trapping of the injected charge carriers, a common case that has not been addressed by our present analysis.

Exercises for Chapter 6

6.1 A semiconductor is shaped as a thin plate of thickness 0.3 mm, width 1.5 mm, and length 20 mm. Electrical contacts are soldered at the far two ends of the plate, and an external voltage is applied. The (dark) room-temperature free hole density is 1.7×10^{14} cm^{-3}, and the free electron density is 3.4×10^{12} cm^{-3}. The hole mobility is 1.9×10^3 cm^2/V · s, and the electron mobility is 3.9×10^3 cm^2/V · s. The sample is exposed to a constant illumination, causing excitation of holes and electrons at a 2×10^{16} cm^{-3} s^{-1} rate throughout its whole volume. The excess carriers lifetime is 22 μs. The electrical field is 45 V/cm. Calculate the change in the specific conductivity and in the current density under the constant illumination compared to their dark condition values. The effective hole mass is 0.6, and of the electron mass is 0.15 compared to the free electron rest mass. Calculate the energies of the hole and electron quasi-Fermi states, and their splitting under the illumination.

6.2 (a) The semiconductor described in Exercise 6.1 is exposed to an extremely short light pulse generating an excess of 2×10^{16} cm^{-3} holes and electrons throughout its entire volume. Calculate the change in the current density at immediately after the light pulse ends and the time until it reduces to half that value.

(b) The same light pulse is collimated to a circle of 1.2 mm diameter on the sample face center, generating a 2×10^{16} cm^{-3} excess of holes and electrons in the volume under the said collimator. Calculate the change in the current density immediately after the end of the light pulse.

6.3 Temperature dependencies of the carriers mobility of the semiconductor of Exercise 6.1 are

$$\mu_p(T)[\mathrm{cm}^2/\mathrm{V}\cdot\mathrm{s}] = 1.9 \times 10^3 \left(\frac{300}{T[°\mathrm{K}]}\right)^{3/2};$$

$$\mu_n(T)[\mathrm{cm}^2/\mathrm{V}\cdot\mathrm{s}] = 3.9 \times 10^3 \left(\frac{300}{T[°\mathrm{K}]}\right)^{3/2}.$$

Calculate the carriers diffusion constants. Then, assume that the hole concentrations remain the same between room temperature and 80° K, and calculate the ambipolar diffusivities and mobilities at room temperature and at 80° K. Calculate the diffusivity and the drift ranges under an external electric field of 45 V/cm.

6.4 As complementary to Figure 6.16, draw an energy level scheme of an insulating semiconductor under the action of a high voltage close to the negative, electron-injecting contact.

6.5 The electron mobility in a certain insulating semiconductor is 1000 cm^2/V · s, and the hole mobility is 90 cm^2/V · s; the relative dielectric constant is $\varepsilon_r = 7.7$. The material is shaped in a form of a thin plate, 0.8 mm thick and 2.2 cm^2 face area. Electrical contacts are placed on the two opposite large faces and an external voltage is applied.

(a) Calculate the electrical current obtained under a voltage of 100 V if the positive electrode is hole injecting, or alternatively if the negative electrode is electron injecting.

(b) Given that the material contains deep hole-traps of $N_t = 5 \times 10^{10}$ cm^{-3} density, calculate the voltage under which all the traps are filled when the positive electrode is hole injecting.

7 p–n Junction-Based Devices

7.1 The p–n Junction

A p–n junction of a semiconductor is a region where the conductivity type changes from being *p*-type to *n*-type, while the other material properties are identical. Forming of such junction involves a controlled addition of impurities to the different regions. A common technical procedure is growing the crystal as entirely of one specific conductivity type, and then depositing on its surface a layer of impurities that potentially act as opposite-type charge carriers (donors or acceptors). Eventually the crystal is heated, thus inducing diffusion of the deposited impurities into the crystal. Initially, the added impurities compensate the initial conductivity type, and then they reverse the type at a limited region near the surface. For simplicity, the situation is always considered as if the junction is formed by a tight contact between two separate crystals, one of type p and the other of type n.

A p–n junction in fact exhibits three regions: a *p*-type region, a transition region, and an *n*-type region. Obviously, it is the transition region that is addressed as a **junction**. The junction "border" at the *p*-type region side is denoted by X_p, and at the *n*-type region side, the "border" is denoted by X_n. Because by its definition the junction region contains charge carriers of varying concentrations, the charge carriers diffuse from their high concentration range to their low concentration range. Holes of the p-range diffuse toward the n-range, and electrons from the n-range diffuse toward the p-range. As a result, the n-range becomes positively charged, and the p-range becomes negatively charged. The thus created electrical field causes the charge carriers to drift in the opposite direction until these two opposite currents balance (see also Section 6.3). A schematic illustration of the space-charge distribution about a typical p–n junction is presented in Figure 7.1.

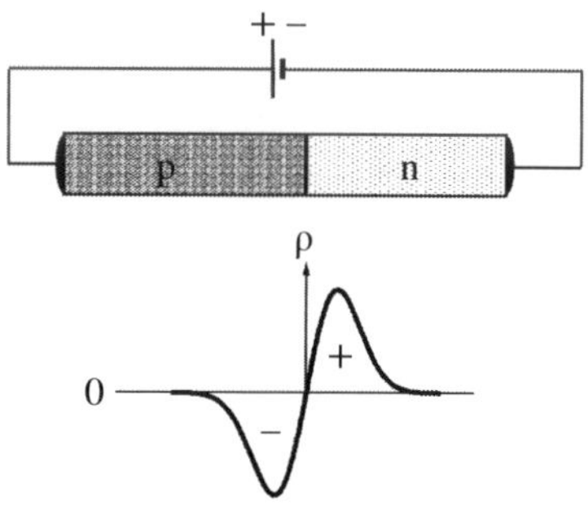

Figure 7.1 Schematic illustration of a p–n junction, and the space-charge distribution around the junction area. Also shown are electrical contacts to an external DC voltage source at a **forward** polarity.

The generation of the internal electrical field also involves bending of the energy bands, as described schematically in Figure 7.2. The potential difference between the p-and n-regions is often addressed as a **potential barrier**. The reason for that naming is that it is the cause for balancing the diffusion. As seen in the figure, the potential barrier is quite close in size to the energy gap, yet always smaller.

Conductors, Semiconductors, Insulators, and Crystal-Growth Technology: A Practical Guide,
First Edition. Zeev Burshtein.

As will be discussed later in detail, a p–n junction exhibits a rectification property. Namely, in a certain polarity of the external voltage, it allows obtaining a high current, while in the opposite polarity, the allowed current is tiny. To simplify the theoretical analysis, we put forward several simplifying assumptions:

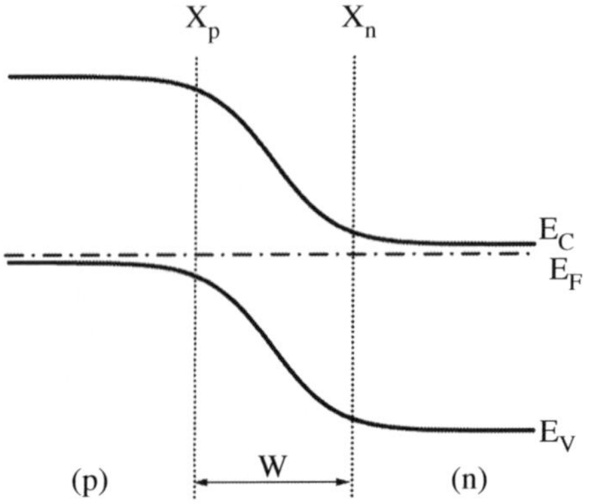

Figure 7.2 Demonstration of energy band bending generated in the transition region between p and n type conduction in a semiconductor (p–n junction).

(1) The entire applied external voltage falls across the junction.
(2) The applied voltage V_a is smaller than the potential barrier.
(3) The electrical metallic contacts are "ohmic" (meaning that they don't cause any bands bending), and their distance from the junction is large compared to the $\sqrt{D_p \tau_p}$ and $\sqrt{D_n \tau_n}$ diffusion ranges.
(4) The effective junction width $W \equiv |X_n - X_p|$, namely the width of the region where the potential barrier occurs, is rather narrow compared to the $\sqrt{D_p \tau_p}$ and $\sqrt{D_n \tau_n}$ diffusion ranges.

The above first assumption (1) is based on the fact that the density of charge carriers in the junction itself is extremely small. Therefore, its electrical resistance is very large compared to those of the p- and n-regions connected in series. Thus, an externally applied voltage will fall mostly across the junction region.

The bending of the energy bands actually changes under an external electric field. For example, under application of a **forward voltage** (when the contact to the p-region is positive and the contact to the n-region is negative), the potential barrier becomes smaller. The concept of a Fermi level is no more valid; it has to be replaced by two quasi-Fermi levels of holes and electrons, F_p and F_n, respectively. Figure 7.3 displays the variation of the quasi-Fermi levels about the junction region. In the figure, they are plotted as if they don't change until crossing through the junction; specifically, the hole quasi-Fermi level in the junction is the same as in the p-region, and it changes steeply only in the n-region adjacent to the junction. Similarly, the electron quasi-Fermi level in the junction is as in the n-region, and it changes steeply only in the p-region adjacent to the junction. We will prove that behavior to be a good approximation of the actual case.

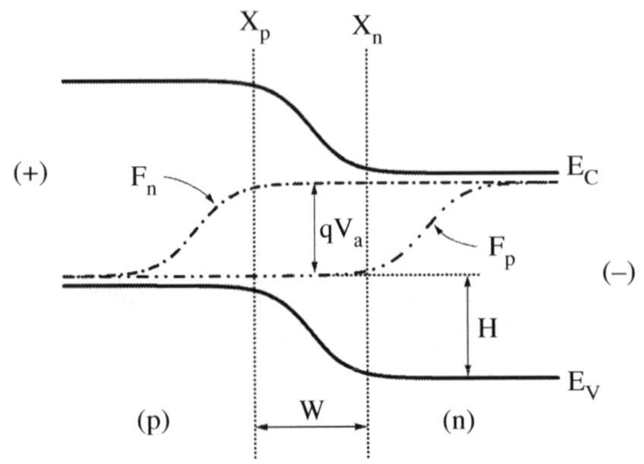

Figure 7.3 Energy states scheme in the vicinity of a p–n junction after application of an external *Va* voltage in a forward polarity.

We denote:

1. n_p – Electron density in the p-region
2. n_n – Electron density in the n-region
3. p_p – Hole density in the p-region
4. p_n – Hole density in the n-region

The hole current density is given by $J_p = p\mu_p \frac{dF_p}{dx}$ (Eq. [6.22]). Assuming that the junction is very narrow (assumption **(4)**), hole recombination in the junction itself is negligible. Thus, hole current calculated at X_p is practically the same as its calculated value at X_n. Namely

$$J_p(X_p) = p_p\,\mu_p \left.\frac{dF_p}{dx}\right|_{X_p} \cong p_n\,\mu_p \left.\frac{dF_p}{dx}\right|_{X_n} = J_p(X_n).$$

Therefore,

$$\frac{\left.\frac{dF_p}{dx}\right|_{X_p}}{\left.\frac{dF_p}{dx}\right|_{X_n}} \cong \frac{p_n}{p_p} \ll 1.$$

The $p_n \ll p_p$ condition is valid as long as the external voltage applied over the junction is not too large (assumption **(2)**). The conclusion drawn by the last result is that the slope of the hole quasi-Fermi state on the p-side of the junction crossing is very small compared to its slope on the n-side of the junction crossing. For simplicity, it is assumed that the slope of F_p is completely zero throughout the entire junction width. Such conclusion is consistent with the previous one, and this is how it is drawn in Figure 7.3. A symmetric conclusion would be drawn regarding F_n.

The total current is given by

$$J = J_p(X_p) + J_n(X_p) \cong J_p(X_n) + J_n(X_p). \tag{7.1}$$

To obtain an expression for the total current, we calculate the separate contributions of the holes and electrons at the indicated positions. To obtain the hole current, we start by calculating $p_n(X_n)$:

$$p_n(X_n) = N_V e^{-\frac{H}{k_BT}} = N_V e^{-\frac{H-qV_a+qV_a}{k_BT}} = N_V e^{-\frac{H-qV_a}{k_BT}} \cdot e^{\frac{qV_a}{k_BT}} = p_{n0} e^{\frac{qV_a}{k_BT}},$$

where p_{n0} is the hole concentration (density) at the n-region, far from the junction. As to the meaning of the H quantity and other quantities appearing in the expressions, see the energy scheme of Figure 7.3. The hole concentration (density) at the junction n-side thus increases after the onset of the external voltage. The added concentration is given by

$$\delta p(X_n) \cong p_n(X_n) - p_{n0}(X_n) = p_{n0}(X_n)[e^{\frac{qV_a}{k_BT}} - 1]. \tag{7.2}$$

This effect of growth in minority carriers' concentration due to application of an external voltage is called **charge carrier injection**. That added concentration diffuses into the n-region further from the junction limit, and under a steady state will be given by (see Eq. [6.40])

$$\delta p(x - X_n) = \delta p(X_n) \cdot e^{-\frac{x-X_n}{\sqrt{D_p\tau_p}}}. \tag{7.3}$$

Using Eq. (7.3), we may derive the size of the hole current

$$J_p(X_n) = -qD_p \left.\frac{d(\delta p_n)}{dx}\right|_{X_n} = qD_p \frac{\delta p(X_n)}{\sqrt{D_p \tau_p}} = \frac{qD_p}{\sqrt{D_p \tau_p}} p_{n0}[e^{\frac{qV_a}{k_B T}} - 1].$$

And symmetrically for the electron current

$$J_n(X_p) = \frac{qD_n}{\sqrt{D_n \tau_n}} n_{p0}[e^{\frac{qV_a}{k_B T}} - 1].$$

The total current would thus be

$$J = \left(\frac{qD_p}{\sqrt{D_p \tau_p}} p_{n0} + \frac{qD_n}{\sqrt{D_n \tau_n}} n_{p0} \right) [e^{\frac{qV_a}{k_B T}} - 1] = (J_{ps} + J_{ns})[e^{\frac{qV_a}{k_B T}} - 1],$$

or alternatively expressed as

$$J = J_s[e^{\frac{qV_a}{k_B T}} - 1] \quad \text{where } J_s = \frac{qD_p}{\sqrt{D_p \tau_p}} p_{n0} + \frac{qD_n}{\sqrt{D_n \tau_n}} n_{p0}. \tag{7.4}$$

Analysis of Eq. (7.4) shows that there is a very small range of voltages $|V_a| \ll k_B T/q$, which exhibits a linear dependence of the electrical current as a function of voltage. For negative voltages, namely voltages of a **reverse polarity** satisfying $|V_a| \gg k_B T/q$, the reverse current saturates at the $J = -J_s$ value. For positive voltages, namely of a **forward polarity**, the electrical current density increases exponentially. Figure 7.4 provides a schematic plot of the current density dependence on the voltage (called "current/voltage characteristic") in a p–n semiconductor device at room temperature (where $k_B T/q = 0.025$ eV).

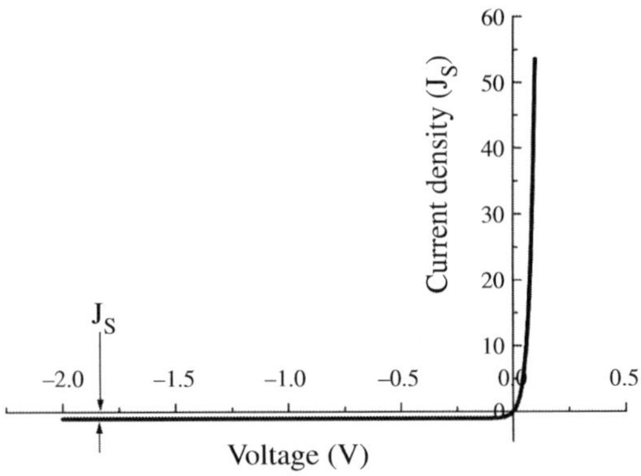

Figure 7.4 Room temperature current density as a function of voltage in a p–n junction. Notably, at already a voltage Va = 0.1 V, the forward voltage current density is 50 times larger than the saturated reverse-voltage current density.

It is possible to understand the saturation effect at the reverse voltage by the fact that under the reverse polarity, both p- and n-regions adjacent to the junction are depleted from their minority carriers, instead of being enriched by injection under a forward voltage. While the concentration of injected minority carriers can grow many times larger than without the voltage, their concentration may reduce by the limited amount of depletion to zero. In other words, the maximal obtained depletions obtained under a reverse voltage polarity is n_{p0} and p_{n0}. In other words, $\delta n_p(X_p) = -n_{p0}$ and $\delta p_n(X_n) = -p_{n0}$. Correspondingly, the saturated currents then result from diffusion of the minority carriers from the depths of the p- and n-regions toward the junction.

7.2 A Rectifying Diode

A diode is an electronic device containing two external electrical contacts. A p–n junction having two electrical contacts, one to the p-side and the other to the n-side, as in Figure 7.1, has rectification capabilities. It means that the current under a reverse polarity is almost zero compared to the current under a forward polarity (see Fig. 7.4). For simplicity, it is assumed that the rectifying diode in its forward polarity exhibits a zero electrical resistance, while in the reverse polarity, the electrical resistance is infinite. A device based on such property has quite a number of applications. The standard symbol of a rectifying diode in an electrical circuit is shown in Figure 7.5, where the orientation of the triangle apex indicates the high conductivity orientation (namely the high conductivity orientation in the figure is upward). A most common use of rectifying diodes is in electrical circuits transforming an alternating voltage (AC) into a direct voltage (DC). Figure 7.6 presents two types of rectifying electrical circuits: one is called **half-wave rectifying**; the other is called **full-wave rectifying**. The half-wave rectifying circuit uses a single diode connected in series to both the alternating voltage source and the load resistor R_L. During the half wave where the alternating voltage polarity coincides with the direction of the diode electrical conduction, the diode resistance is negligibly small, and almost the entire voltage develops across the load resistor. In the opposite polarity, the diode resistance is extremely large compared to that of the load resistance, and the voltage across the load resistance is practically zero. Thus, the voltage across the load resistance lasts only during half of the alternating voltage wave duration; that is the source of the adjective granted to that circuit arrangement. The full-wave rectifying circuit uses four diodes as displayed in Figure 7.6, such that the load resistor R_L is always connected in series to two of the diodes at their conducting polarity.

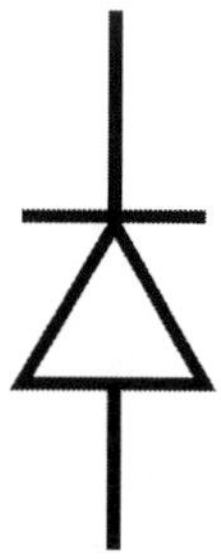

Figure 7.5 Standard symbol shape of a rectifying diode in plotting of an electrical circuit.

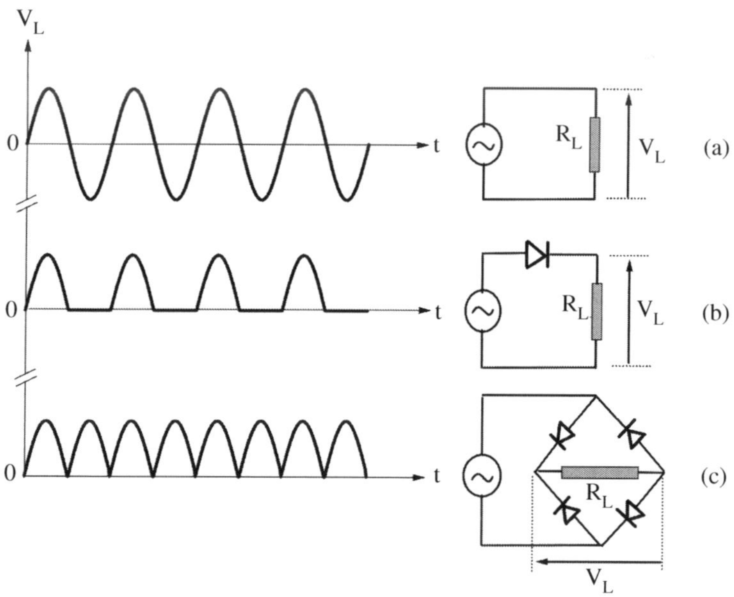

Figure 7.6 The V_L voltage obtained over the R_L load resistor in different electrical circuits, two of which include rectifying diodes. In the top circuit (a), the load resistor is connected directly to the alternating (AC) voltage source, and no rectification occurs. In the mid circuit (b), a single rectifying diode is connected in series to the load resistor, yielding a half-wave rectification. In the bottom circuit (c), the load resistor is connected to a 4-diodes "bridge", yielding a full-wave rectification.

7.3 Current Breakdown Under Reverse Voltage

7.3.1 Introduction

According to our previous analysis of current/voltage characteristics in a p–n junction, the reverse current is very small and voltage-size independent. That property is maintained up to a certain reverse voltage V_{br}, which depends on particular details of the junction construction. At that voltage, a process called **current breakdown** occurs. The current thus grows to very high levels, at a rate that grossly exceeds the growth rate of the forward current. A schematic illustration of a current versus voltage characteristic that includes the breakdown effect is displayed in Figure 7.7. It must be emphasized that the essence of the breakdown current in itself doesn't have anything destructive to the diode. An only factor that might cause destruction is excess overheating, because the large reverse current also involves a relatively high voltage, hence a high heating power compared to a forward current obtained at a low voltage. Such overheating might induce diffusion of the donor and acceptor impurities that form the p–n junction, thus impairing the junction properties. In extreme cases, it might also cause a junction melting. In electrical circuits, such cases may be readily prevented, for example by limiting the current by adding a proper series resistor, adding some active cooling of the diode, or by optimizing the heat removal from the junction area.

There are two known physical mechanisms causing a reverse voltage current breakdown in p–n junctions: a **Zener breakdown**, and an **avalanche breakdown**, to be hereby described.

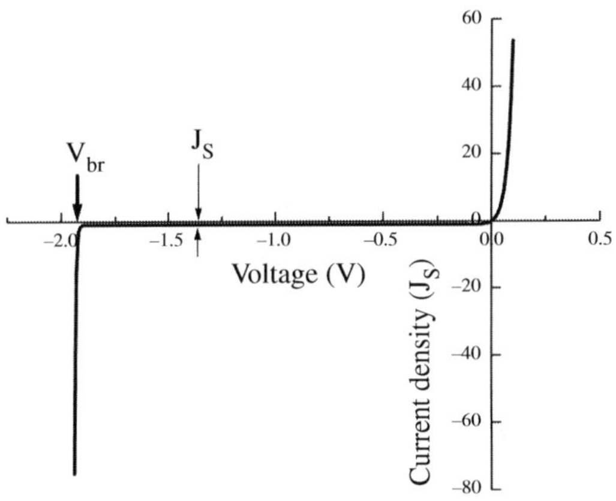

Figure 7.7 Current versus voltage characteristic in a p–n junction where the reverse voltage breakdown occurs at a reverse voltage of 1.9 V.

7.3.2 Zener Breakdown

Figure 7.8 provides a schematic structure of the energy bands in a reverse-biased junction. Energetic overlap between the full valence band on the p-side and the empty conduction band on the n-side is obtained at already quite low voltages. Thus, electrons start to cross from the p-side to the n-side by the **tunnelling** quantum mechanical mechanism. Such process is negligible is cases of a broad junction, yet becomes very large in a narrow junction, occurring when the crystals forming the two junction sides contain very large dopant concentrations. A breakdown voltage incurs when the tunnelling current turns greater than the saturated reverse current. The large steepness of the current versus voltage rise stems from the steep rise of the number of electrons having the same energy as empty states, and simultaneously the distance between them and the empty states becomes shorter. The above-described process is called **Zener effect**. If the dopant concentrations are quite large, the Zener current breakdown may be achieved at voltages of only several volts. If, however, the junction is broad, the Zener current breakdown will be preceded by an avalanche mechanism, to be discussed below. Quantitative analysis of the Zener effect is beyond the scope of the present book and won't be presented here.

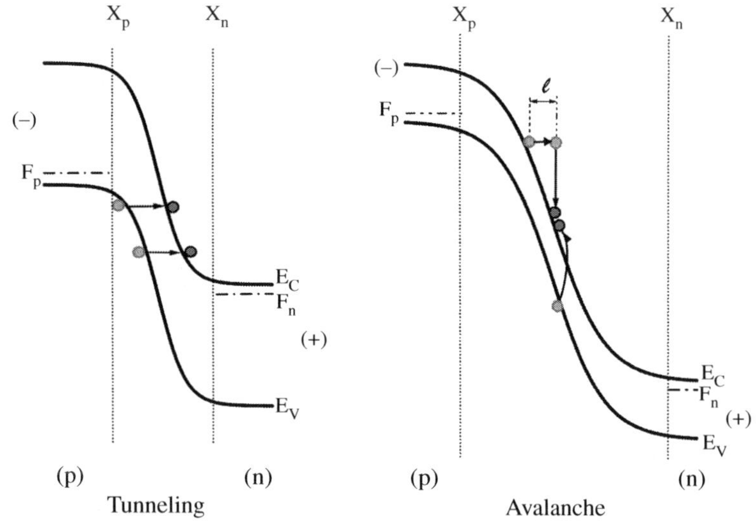

Figure 7.8 Energy state structure in a p−n junction under a reverse voltage (negative polarity to p-region and positive to the n-region). The left part describes electrons tunnelling from filled valence states on the p-side of the junction into empty states in the junction conduction band, which are distant but unoccupied. The right side describes an electron in the junction conduction band acquiring kinetic energy by the high electrical field, then dissipating it by exciting an additional electron from the junction valence band into the junction conduction band.

7.3.3 Avalanche Breakdown

In cases where dopant concentrations that form the p- and n-regions of a p–n junction are relatively small, the junctions' widths are quite large. A Zener current breakdown is then negligible. Instead, an avalanche breakdown occurs at considerably higher voltages. The process involves impact ionizations of the semiconductor atoms by energetic charge carriers. Namely, a charge carrier is accelerated by the electric field, acquires a kinetic energy, eventually losing it by exciting a valence-band electron into the conduction band. The ionization generates an electron–hole couple that may even participate in additional impacts. The current thus grows very steeply as a function of voltage.

We turn to express the breakdown voltage V_{br} as a function of the crystal and junction characteristic properties. To achieve an impact ionization, the macroscopic internal electrical field must be sufficiently large to grant the accelerated electrons a kinetic energy that is larger than the energy gap E_g within a single mean free path ℓ. In other words, the condition $q\mathcal{E}_{br}\ell \geq E_g$ is necessary to initiate the impact ionization; or

$$\mathcal{E}_{br} \geq \frac{E_g}{q\ell}. \tag{7.5}$$

It is important to emphasize that the charge carriers we now consider are the true particles and not the effective charge carriers as presented and discussed in Sections 3.6. and 4.1.

The electrical field inside the junction is not uniform. Therefore, the condition for the avalanche breakdown is determined by the region where the internal electric field assumes its maximal value. To assess the size of the maximal field, we consider a simplistic model of the space-charge distribution described in Figure 7.9. The junction is described as a region entirely depleted of all mobile (free) charge carriers, such that its space charge originates from the ionized

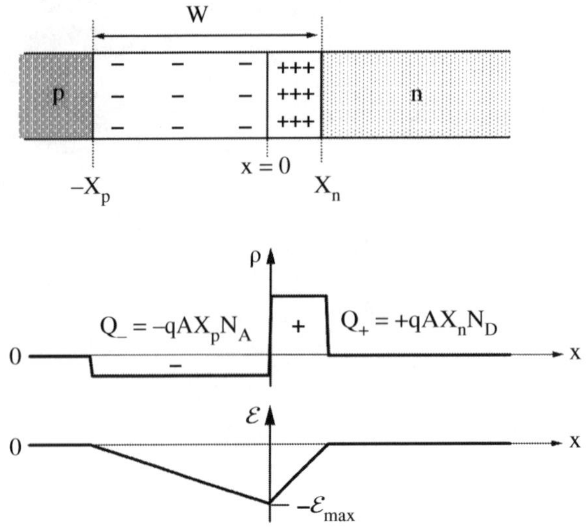

Figure 7.9 A model of the space-charge distribution in a p–n junction, and the resulting space charge and electric field oriented perpendicularly to the junction cross section.

hole and electron contributing dopants (acceptors and donors, respectively). We further assume that the dopant concentrations, N_A in the p-region and N_D in the n-region, are uniform up to assumed X_p and X_n "border" planes of the p and n junction sides, respectively. The connecting plane area is designated as A. Uniformity of the donor and acceptor impurity concentrations is assumed separately for the space-charge in each of the junction sides. The junction actually exhibits a distribution of electrical dipoles such that the positive space-charge Q_+ in the n-side of the junction is the same as the negative space-charge Q_- in the p-side of the junction.

Following Poisson Eq. (6.25), we write

$$\left.\begin{aligned}\frac{d\mathcal{E}}{dx} &= \frac{q}{\varepsilon_r\varepsilon_0}N_D \qquad \text{for} \quad < x < X_n;\\ \frac{d\mathcal{E}}{dx} &= -\frac{q}{\varepsilon_r\varepsilon_0}N_A \qquad \text{for} \quad -X_p < x < 0.\end{aligned}\right\}. \tag{7.6}$$

The variation of the electric field as a function of position perpendicularly to the junction cross section (x-orientation) is displayed in Figure 7.9. Notably, outside the junction itself (namely in the distinct p- and n-regions) the fields are tiny, practically zero due to the high electrical conductivities in those regions. The maximal electric field $\mathcal{E}_{max}$ may be obtained by integration in Eq. (7.6)

$$-\mathcal{E}_{max} = \int_{-X_p}^{0} \frac{d\mathcal{E}}{dx}dx = \frac{-q}{\varepsilon_r\varepsilon_0}N_AX_p = \frac{-q}{\varepsilon_r\varepsilon_0}N_DX_n. \tag{7.7}$$

An additional integration provides the voltage applied across the junction:

$$V = \int_{-X_p}^{X_n} -\mathcal{E}(x)dx = \frac{1}{2}W\mathcal{E}_{max} = \frac{1}{2}\frac{q}{\varepsilon_r\varepsilon_0}N_DW,$$

where $W \equiv X_p + X_n$ is the junction width. Using the relation $X_pN_A = X_nN_D$, we get $X_n = \frac{WN_A}{N_A + N_D}$. We set it in the above expression for V and get

$$V = \frac{1}{2}\frac{q}{\varepsilon_r\varepsilon_0}\left(\frac{N_AN_D}{N_A + N_D}\right)W^2.$$

Extracting the junction width W, we get

$$W = \left[\frac{2\varepsilon_r\varepsilon_0 V}{q}\left(\frac{1}{N_A} + \frac{1}{N_D}\right)\right]^{1/2}. \tag{7.8}$$

According to Eq. (7.8), the junction width increases by application of an external voltage ($V = V_a + E_g/q$). Also seen is that the junction width decreases by increasing the dopant concentrations (the donors N_D and acceptors N_A).

We can focus our attention now on the avalanche breakdown issue. We assume that avalanche triggering occurs at the very place where the electrical field attains its maximal value. The required voltage is given by $V_{br} = \frac{1}{2}W\mathcal{E}_{br} = \frac{1}{2}WE_g/q\ell$. We set the junction width according to Eq. (7.8) and get

$$V_{br} = \frac{1}{2}\frac{E_g}{q\ell}\left[\frac{2\varepsilon_r\varepsilon_0 V_{br}}{q}\left(\frac{1}{N_A} + \frac{1}{N_D}\right)\right]^{1/2}.$$

Extracting V_{br} that appears on both sides of the equation, we get

$$V_{br} = \frac{\varepsilon_r\varepsilon_0 E_g^2}{2q^3\ell^2}\left(\frac{1}{N_A} + \frac{1}{N_D}\right). \tag{7.9}$$

Already at the present stage of analysis, one may infer some important qualitative conclusions. For example, the avalanche breakdown voltage increases as a function of the semiconductor energy gap. Correspondingly, breakdown voltages in silicon (Si) diodes would be larger than in germanium (Ge) diodes (see Table 5.1). Further, the breakdown voltage is dominated by the part of the junction whose dopant density is lowest! In other words, it is sufficient that only one of either N_A or N_D is small to turn the avalanche breakdown voltage very high.

The electron carrier mean free path till scattering ℓ is usually longer than the holes' one. That quantity is connected to the mean scattering time τ appearing in the mobility expressions (see Eqs. [3.16], [4.13], and [4.14]), and it may be approximately expressed as $\ell \approx \mathrm{v}_{th}\tau_n = \tau_n\sqrt{3k_BT/m_n}$ (see Section 5.1). A representative expression for τ_n would be $\tau_n \cong m_n\mu_n/q$ (see Eqs. [3.16] and [4.13]). Setting in the previous expression, we get $\ell^2 \approx \frac{3m_nk_BT}{q^2}\mu_n^2$. Setting in Eq. (7.9), we get

$$V_{br} = \frac{\varepsilon_r\varepsilon_0 E_g^2}{6qm_nk_BT\mu_n^2}\left(\frac{1}{N_A} + \frac{1}{N_D}\right).$$

This result indicates that the avalanche breakdown voltage is inversely proportional to the squared mobility. In most semiconductors, the mobility reduces upon increase of temperature. When the dominant scattering mechanism limiting the carrier mobility is dominated by the interaction with lattice oscillations, we get that the mobility behaves as $\mu_n = \mu_n^{(0)}T^{-3/2}$ (see the text in Section 5.4, which precedes Eq. [5.9], Exercise 5.1, and Exercise 6.3). Setting in the above equation, we get

$$V_{br} = \frac{\varepsilon_r\varepsilon_0 E_g^2 T^2}{6qm_nk_Bm_n\left(\mu_n^{(0)}\right)^2}\left(\frac{1}{N_A} + \frac{1}{N_D}\right). \tag{7.10}$$

We thus obtained that the avalanche breakdown voltage is proportional to the squared absolute temperature. In conclusion, however, it should be noted that the actual external voltage that triggers the avalanche is smaller than the value expressed by Eq. (7.10) by approximately E_g/q, because the p–n junction already exhibits such an internal voltage at thermal equilibrium.

7.3.4 Zener Diode

A **Zener diode** is a p–n junction diode where the reverse voltage current breakdown is used in an electrical circuit. The same name is used even if the current breakdown mechanism is the electron avalanche (Section 7.3.3). The feature utilized is the fact that even tiny voltage changes around V_{br} cause very large changes in the current. In that way, it is possible to stabilize a voltage across a resistor in an electrical circuit. Figure 7.10 displays an electrical circuit where a Zener diode is used to stabilize the voltage across a load resistor R_L. The voltage supplier suspected as exhibiting voltage variation across R_L is simplistically described as a series connection of a stable voltage battery and an alternating voltage. In the absence of the Zener diode, the voltage across R_L is determined by the ratio between it and an additional series resistor R_i. Therefore, the voltage across R_L will then include an alternating component. On addition of the Zener diode parallel to R_L, each time the voltage across it could differ from V_{br}, the effective resistance of the Zener diode would accordingly change (decrease or increase), and the voltage between their common ends would settle as V_{br}. In designing of such electrical circuits, care is taken to assure that the minimal supplied voltage will never be smaller than V_{br}. That way the voltage across R_L will always be V_{br}, and the remainder of the source voltage supply will appear fully across R_i.

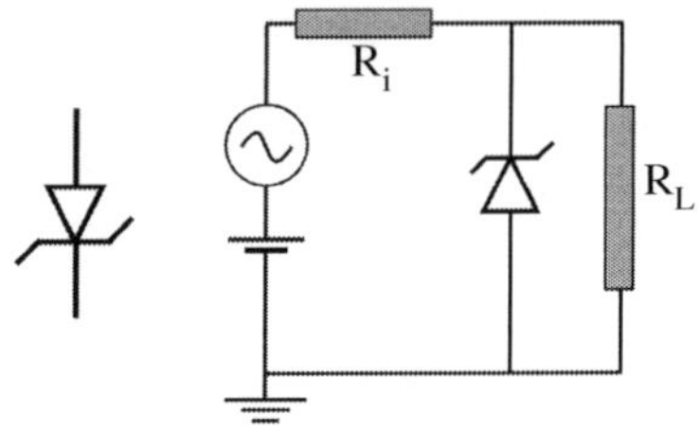

Figure 7.10 Left side shows the standard symbol shape of a Zener diode used in plotting of an electrical circuit. The right side displays a basic electrical circuit that includes a Zener diode stabilizing the voltage across a load resistor *RL*. For explanation, see the text.

7.4 Diode Lasers

7.4.1 Introduction

Minority charge carriers injection occurring in a p–n junction under a forward voltage (see Eqs. [7.2] and [7.3]) could involve a radiative recombination of the excess injected minority carriers with the majority charge carriers present in that region. Namely, electrons injected into the p-region will recombine with the local holes, and holes injected into the n-region will recombine with the local electrons. Light emission from the junction region will appear especially in semiconductors of a direct band gap, such as gallium arsenide (*GaAs*), where radiative recombination is dominant (see Section 6.2.1). To increase the recombination rate, the p- and n-regions are designed to have large concentrations of carriers-contributing impurities, which brings each region to have degeneracy conditions. In other words, the quasi-Fermi energies be placed inside the corresponding bands. For demonstration, in Figure 7.11, we describe the band structure in the momentum space (the *k*-space), with an indication of a possible radiative transition in the p-region into which a large concentration of electrons were injected. As illustrated, the radiative recombination is direct in the crystalline *k*-space, and the photon energy is slightly larger than the band gap. The light emission spectrum extends from E_g as the lower energy limit, till $(F_n - F_p)$ as the upper energy limit. Symmetric conditions obviously exist in the n-side of the junction under injection of holes.

The geometric width of the region where the radiative recombination occurs is quite similar to the diffusion range ℓ_D of the injected charge carriers. We thus perform a numerical assessment of the size of that range for electrons injected into the p-region in a gallium arsenide (*GaAs*) crystal in the vicinity of room temperature.

Given data are:

$$k_BT/q = 0.025\,\text{V};$$

$$\mu_n \cong 10,000\,\frac{\text{cm}^2}{\text{V}\cdot\text{s}} = 1.0\,\frac{\text{m}^2}{\text{V}\cdot\text{s}};$$

$$\tau_n \cong 10^{-9}\,\text{s}.$$

Basic equation:

$$\ell_D = \sqrt{D_n\tau_n} = \sqrt{(k_BT/q)\mu_n\tau_n}.$$

Setting:

$$\ell_D = (0.025 \times 1 \times 10^{-9})^{1/2} = 5 \times 10^{-6}\,\text{m}.$$

Thus: $\ell_D = 5\,\mu\text{m}$.

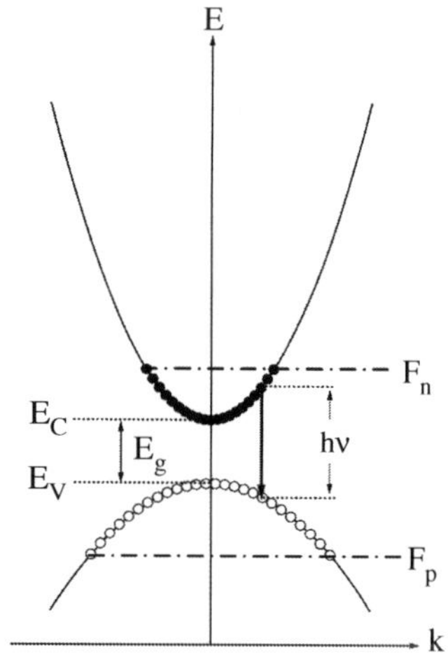

Figure 7.11 Charge carrier distribution plotted on dispersion curves of the valence and conduction bands. Especially indicated is a direct radiative transition in the *k*-space of a *p*-type region into which high concentration of electrons have been injected.

There are two kinds of devices based on the described effect: a **light-emitting diode** (LED) and a **diode laser**. The diode laser is in fact an extreme case of the LED, where stimulated emission of light (see Section 7.4.2) takes control of the electron–hole recombination process. In the following section we explain the physical principles underlying a laser operation, and delve on the example of a gallium arsenide (*GaAs*) diode laser.

7.4.2 Physical Principles of Laser Operation

Light crossing through matter might cause electron transitions from occupied states into unoccupied ones, when the photon energy equals the difference between the state energies. If the energy of the occupied state is lower than the unoccupied state energy, the process is called **absorption**. The reason for that naming is that the energy subtracted from the radiation field was added to the matter. If the energy of the occupied state is higher than the unoccupied state energy, the process is called **stimulated emission**. The "stimulated" adjective stems from the fact that while such process occurs even in the dark, the radiation field presence increases the probability of its occurrence.

Notably, absorption and stimulated emission are in fact entirely symmetric, because both occur only in the presence of the radiation field; therefore, both are entitled to the "stimulated" adjective. Still that adjective is attached only to the emission process, in order to discriminate it from the rather common process called **spontaneous emission**. That emission from a high-energy level to a lower one occurs independently of the presence of any radiation field (in quantum mechanics, the spontaneous emission is also described as stimulated by the vacuum electromagnetic radiation, often addressed as "zero-point fluctuations"). The spontaneous emission is so common that people tend to forget that is was very difficult to be theoretically explained. The spontaneous emission is consistent with the second law of thermodynamics. Recall that the law states that thermodynamic equilibrium between equi-energy systems prefers the one of higher entropy. An atom at a low state plus a photon at a random propagation orientation in space, is less ordered – then of higher entropy than the system of an atom at the higher state, while both systems obviously contain the same energy.

A schematic illustration of the three radiative mechanisms is presented in Figure 7.12.

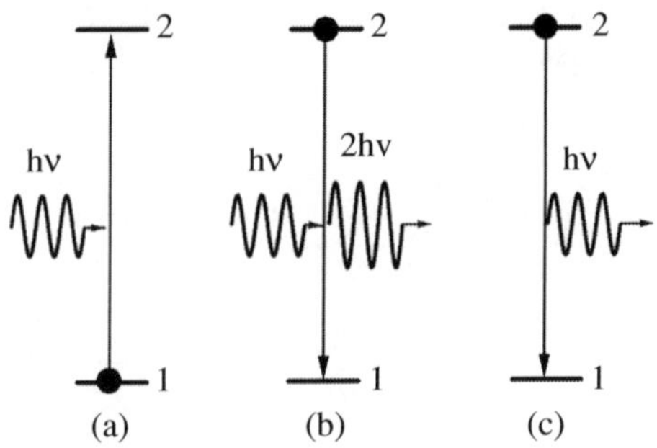

Figure 7.12 Schematic illustration of radiative transitions between energy states in matter. (a) Absorption: a photon of the radiation field is absorbed by an atom at a low-energy state (1), and the radiation intensity reduces. (b) Stimulated emission: The radiation field stimulates an atom at a high-energy state (2) to transform to a lower-energy state (1), and the radiation field is enriched with a single photon *whose* phase coincides with that of the stimulating field. It is important to notice that the photon addition is expressed by increase of the field amplitude and not by a frequency change. (**c**) Spontaneous emission: an atom at a high-energy state (2) performs a sudden radiative transition to the lower-energy state (1) and enriches the radiation field with a single photon.

A monochromatic light beam of wavelength λ and intensity F crossing a uniform matter of thickness L, where only the (a) and (b) mechanisms are involved, changes according to the **Beer–Lambert** law

$$F(L) = F(0)\boldsymbol{exp}(\gamma L),$$

where $F(0)$ is the light intensity on entering the matter. The amplification factor γ is given by

$$\gamma(\lambda) = g(\lambda)\frac{\lambda^4 \Delta N}{8\pi n^2 c\tau_{rad}}, \tag{7.11}$$

where $g(\lambda)$ is the shape function of the radiative emission line (or band), namely the fluorescence wavelength emission spectrum normalized to have a unity integral, that is to satisfy the condition $\int g(\lambda)d\lambda = 1$ (if by a preliminary choice of the $g(\lambda)$ scale, its integral is different than 1, just dividing the preliminary choice by that integral provides the desired normalized spectrum); n is the optical refractive index of the matter; c is the vacuum speed-of-light constant; τ_{rad} is the radiative lifetime of the excited state (level "2" in Fig. 7.12); and $\Delta N \equiv N_2 - N_1$ is the difference between the concentration of atoms in the upper "2" level and atoms in the "1" level. Proving of this important relation is beyond the scope of the present book, and we won't do it here. It is important to notice that τ_{rad} and $g(\lambda)$ in Eq. (7.11) provides a linkage between the stimulated absorption and emission mechanisms. Straightforwardly from Eq. (7.11), light beam amplification is obtained when $N_2 > N_1$ (**population inversion** conditions), and attenuation is obtained when $N_2 < N_1$. Population inversion conditions cannot exist under thermal equilibrium, because according to the Boltzmann distribution, population of a higher-energy state is then smaller than of a lower-energy state (see Eqs. [2.5] and [3.6]). Population inversion can be reached by only violation of thermal equilibrium conditions via excitation of the upper states. In the particular case that we are considering now, it is obtained by applying a forward voltage over a p–n junction, causing injection of minority carriers into the opposite junction sides. The physical process of energy investment aimed at obtaining population inversion is called **pumping**.

A laser is a device that emits light based on the stimulated light emission mechanism. The name **laser** is a phonetic adoption of an acronym describing the physical mechanism: **L**ight **A**mplification by **S**timulated **E**mission of **R**adiation. The fundamental construction of a laser system is described in Figure 7.13. The device contains a pumping system that generates the required population inversion in the amplifying medium, demonstrated in the figure as a circular

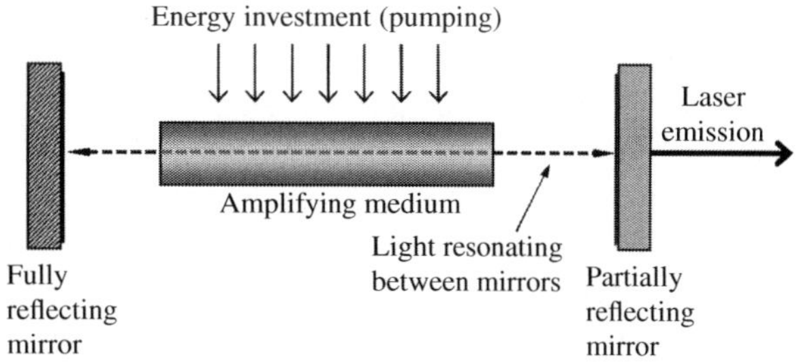

Figure 7.13 Schematic illustration of a laser device construction.

rod. In addition, the laser includes an optical resonator (resonance cavity). The most-simple resonator described in the figure comprises a couple of flat mirrors causing light propagating in the rod axis orientation to echo back and forth between. A spontaneous light emission from an excited state, by chance oriented parallel to the rod axis, will increase its intensity on each crossing through the pumped laser rod. The radiation field energy thus increases on the expense of the energy stored in amplifying medium. One of the mirrors is made partially transmitting, thereby emitting a light beam highly oriented along the rod axis. The emitted beam is quite monochromatic, of a wavelength close to the maximal spectral peak of the spontaneous radiative light emission, and of a common phase ("coherence" property).

7.4.3 Gallium-Arsenide Based Diode Laser

Figure 7.14 demonstrates the fundamental construction of a diode laser based on a gallium arsenide (*GaAs*) semiconductor. The junction is generated by stimulating diffusion of zinc (*Zn*) atoms from the surface into an *n*-type gallium arsenide rod whose electron concentration is $\approx 10^{18}$ cm^{-3}, until achieving zinc (*Zn*) atoms concentration of $\approx 10^{20}$ cm^{-3}. The added zinc atoms convert the zinc in-diffused region into *p*-type. The dimensions of a typical device are length $L \approx 0.2$ mm, width of emitting zone $\Delta x \cong 0.05$ mm, and thickness $w \cong \ell_D \approx 5$ μm (see Section 7.4.1). The electron–hole radiative recombination spectrum in this material is centered at approximately 840 nm wavelength, and its full width at half maximum (FWHM) is $\Delta\lambda_{1/2} \cong 14$ nm. A possible means for creating the optical resonator is by coating the relevant surfaces by dielectric multilayers that form the required optical reflection of the emission wavelengths. Sometimes rather antireflection layers are deposited, and external mirrors are added. These issues are not discussed further in our present book.

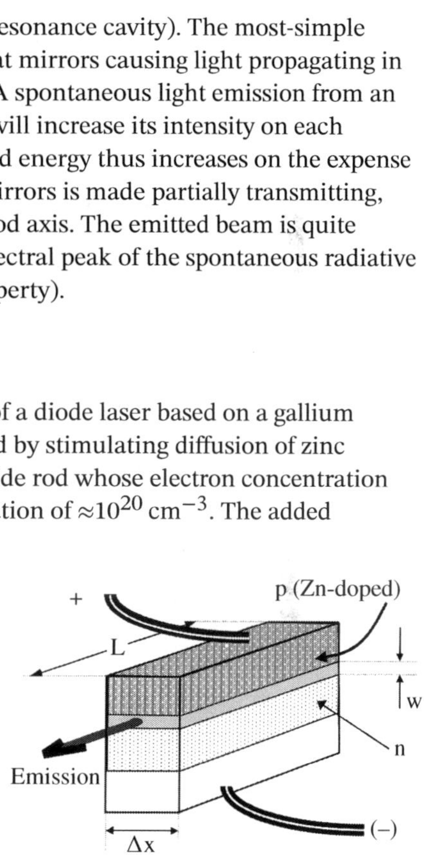

Figure 7.14 A fundamental construction of a diode laser based on a p–n junction of the gallium arsenide (*GaAs*) semiconductor.

For simplicity, we address the light emission as emanating solely from the p-region into which electrons are injected from the n-region. As seen in Figure 7.11, a population inversion is formed for all wavelengths belonging to energies ranging between E_g and $(F_n - F_p)$. In that way, we may express the light gain within the junction as

$$\gamma(\lambda) = g(\lambda)\frac{\lambda^4 N/V}{8\pi n^2 c\tau_{rad}}, \tag{7.12}$$

where N in Eq. (7.12) is the total number of electrons injected into the p-region, $\boldsymbol{V} = wA$ is the light-emitting zone volume, and $A = L \cdot \Delta x$ is the diode cross-sectional area perpendicular to the electrical current orientation (see Fig. 7.13). It is possible to assess the number of injected electrons by the following considerations: under steady-state conditions, the electron recombination rate is equal to their injection rate. The recombination rate is given by $N \cdot (1/\tau_{rad} + 1/\tau_{nr})$, where τ_{nr} is the lifetime of non-radiative recombination. The electron injection rate is given by I/q, where I is the electrical current through the junction. The radiative part N/τ_{rad} of the recombination rate is usually expressed as the product between the current I and the radiative recombination efficiency η. Namely

$$\frac{N}{\tau_{rad}} = \eta \frac{I}{q}.$$

Thus, $N = \eta I \tau_{rad}/q$. For the particular wavelength at the light emission spectral peak λ_{peak}, it is possible to write $g(\lambda_{peak}) \cong 1/\Delta\lambda_{1/2}$, where $\Delta\lambda_{1/2}$ is the full spectral width at half maximum (see Fig. 7.16). Setting all this in Eq. (7.12), we get

$$\gamma(\lambda_{peak}) = \frac{\lambda_{peak}^4 \eta I}{8\pi n^2 qcwL\, \Delta x\, \Delta\lambda_{1/2}} = \eta \frac{\lambda_{peak}^4}{8\pi n^2 qcw\, \Delta\lambda_{1/2}} J, \tag{7.13}$$

where J is the junction current density.

Apparently, light intensity gain in the junction is proportional to the electrical current density. A threshold for the start of laser emission occurs when the gain becomes larger than all losses. There are several loss-causing mechanisms of the light intensity:

(1) Reflection losses;
(2) Diffraction losses;
(3) Light absorption by free electron excitation;
(4) Light absorption by inter band transitions;
(5) Light scattering by structural lattice defects.

The two first mentioned mechanisms will be further discussed in detail.

7.4.3.1 Reflection Loss in the Laser Resonator

A non-fully reflecting mirror in an optical laser resonator causes loss in the radiation field. We denote by R_1 and R_2 the reflections of each of the two laser resonator mirrors (Fig. 7.13). It is customary to represent losses that are actually "lumped" on the mirrors reflecting surfaces as an effective absorption process distributed along the entire length L of the amplifying medium. Thus, reflection losses throughout a back-and-forth propagation of a light beam in the amplifying medium would satisfy the relation

$$\boldsymbol{exp}(-\alpha_R \cdot 2L) = R_1 R_2;$$

or

$$\alpha_R = -\frac{1}{2L} \boldsymbol{ln}(R_1 R_2), \tag{7.14}$$

where α_R is the said medium effective absorption coefficient, providing a loss effect that is equivalent to the mirrors losses. For a typical case, where $R_1 = 100\%$, $R_2 = 90\%$, and $L = 0.2$ mm, we get $\alpha_R \cong 2.6\,\text{cm}^{-1}$.

7.4.3.2 Diffraction Loss in a Laser Resonance Cavity

The amplifying medium in a diode laser is quite thin. It is therefore expected that considerable dispersion of the emitted beam would occur as a result of diffraction. In other words, light emitted in the amplification orientation would disperse by and angle $\Omega = 2\lambda/w$. Namely, only a $\frac{w}{L\Omega} = \frac{w^2}{2\lambda L}$ fraction would remain available for further amplification (see the auxiliary plot in Fig. 7.15). The size of the effective amplification absorption equivalent to the diffraction losses α_{diff} would be

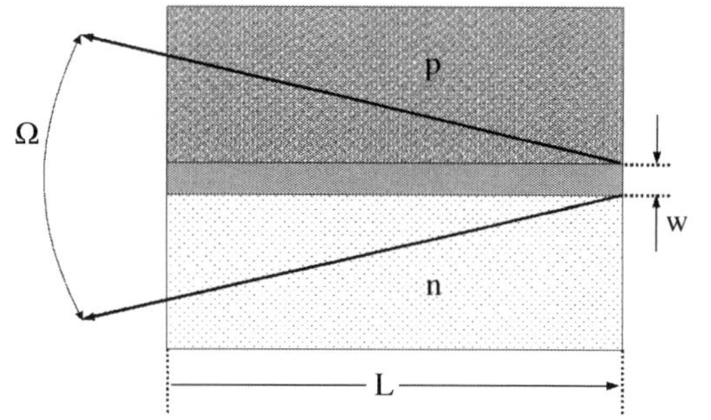

Figure 7.15 Auxiliary drawing illustrating the calculation of suspected losses in laser diodes resulted by diffraction of the emitted light.

$$\alpha_{diff} \cong -\frac{1}{L}\boldsymbol{ln}\left(\frac{w^2}{2\lambda L}\right). \tag{7.15}$$

We now turn to a numerical calculation of expected α_{diff} diffusion loss:

Given:

$$L = 0.2\ \text{mm} = 2 \times 10^{-4}\ \text{m};$$

$$w \cong 5\ \mu\text{m} = 5 \times 10^{-6}\ \text{m};$$

$$\lambda \cong 0.84\ \mu\text{m} = 0.84 \times 10^{-6}\ \text{m}.$$

Setting:

$$\alpha_{diff} \cong -\frac{1}{2 \times 10^{-4}} \ln\left(\frac{25 \times 10^{-12}}{2 \times 0.84 \times 10^{-6} \times 2 \times 10^{-4}}\right) = 1.3 \times 10^4\ \text{m}^{-1} = 130\ \text{cm}^{-1}.$$

The above-obtained losses in a laser resonance cavity are very large; had them been true, they would entirely impair the existence of any lasing. In reality, the diffraction losses are almost entirely balanced by optical reflections from the adjacent n- and p-regions, which exhibit a very high electrical conductivity compared to their junction, which is practically an isolator. The junction thus behaves similarly to a waveguide, comprising an insulating region enclosed between two conducting plates, the p- and n-regions, that reflect light like metallic mirrors. Still, the previously described diffraction occurs fully when the amplified beam exits the junction. Experimental study of diode lasers indicates that in fact, diffraction losses never exceed the order of $\approx 1\ \text{cm}^{-1}$.

When all loss mechanisms are either measured or calculated, they sum to about $\alpha_{loss} \cong 20\ \text{cm}^{-1}$. A **lasing threshold** is defined as the situation where the gain and loss turn equal, namely when

$$\gamma(\lambda_{peak}) = \eta \frac{\lambda_{peak}^4}{8\pi n^2 qcw\, \Delta\lambda_{1/2}} J_{th} = \alpha_{loss},$$

where J_{th} is the current density at the threshold of lasing appearance. Thus, we get

$$J_{th} = \frac{\alpha_{loss} \cdot 8\pi n^2 qcw\, \Delta\lambda_{1/2}}{\eta\, \lambda_{peak}^4}. \tag{7.16}$$

Using Eq. (7.16), we calculate numerically the lasing threshold current using the following data.

Given:

$$\alpha_{loss} = 20\ \text{cm}^{-1} = 2 \times 10^3\ \text{m}^{-1} \qquad w = 5 \times 10^{-6}\ \text{m}$$
$$n = 3.35 \qquad \lambda_{peak} = 0.84 \times 10^{-6}\ \text{m}$$
$$q = 1.6 \times 10^{-19}\ \text{Coulomb} \qquad \Delta\lambda_{1/2} = 14\ \text{nm} = 14 \times 10^{-9}\ \text{m}$$
$$c = 3 \times 10^8\ \text{m/s} \qquad \eta = 1$$

Setting:

$$J_{th} = \frac{2 \times 10^3 \cdot 8 \times 3.14 \times 3.35^2 \times 1.6 \times 10^{-19} \times 3 \times 10^8 \times 5 \times 10^{-6} \times 1.4 \times 10^{-8}}{1 \times 0.84^4 \times 10^{-24}};$$

$$J_{th} = 3.8 \times 10^6\ \frac{\text{A}}{\text{m}^2} = 380\ \frac{\text{A}}{\text{cm}^2}.$$

Experiments verify that such current densities are needed for achieving the laser emissions there exists a fundamental problem concerning the distinction between a regular radiative recombination emission and the laser emission, because the regular radiation emission starts already under quite low currents. The direct way to do it is to measure the light emission spectral shape. The regular radiative recombination emission spectrum is broad, and as already indicated, the emitted photon energy ranges between E_g and $(F_n - F_p)$. (For some approximation purposes, it is assumed to be distributed around the E_g energy). In the specific case of gallium arsenide (*GaAs*), the emission is distributed around the 840 nm wavelength, and its full width at half maximum is 14 nm (see Fig. 7.16). At the very onset of laser emission, the spectral width narrows! Instead of 14 nm, it becomes only about 0.5 nm near the regular spectral radiation peak. It occurs even without any wavelength tuning devices inside the cavity. The reason is that the regular spectral radiation peak also exhibits the wavelength of highest gain. Due to the exponential dependence of the laser emission on the gain, the emission at the highest gain becomes very large compared to its margins (Eq. [7.11]). The said description is demonstrated in Figure 7.16. The light emitted by the diode diverges by approximately 30° vertically and 10° laterally, a direct result of diffraction due to the small dimensions of the emitting region.

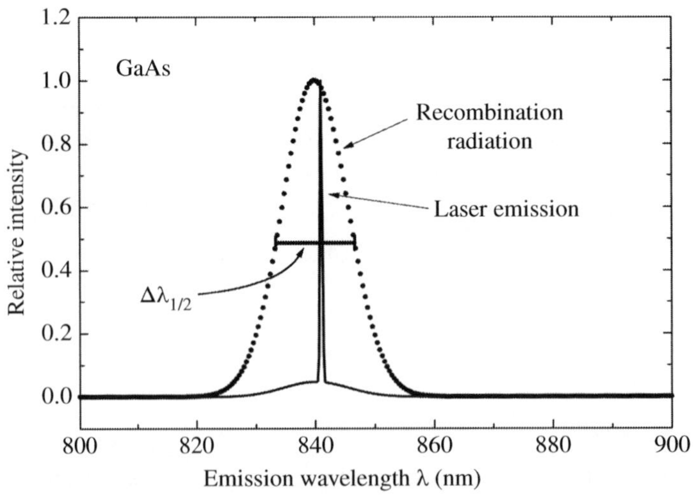

Figure 7.16 Comparison between the spectra of regular radiative recombination emission, and laser emission, in a p–n junction of the gallium arsenide *GaAs* semiconductor. Both spectra are normalized to a 1.0 value at their peaks. The laser emission peak wavelength deviates slightly from that of the regular radiation recombination due to some wavelength dependence of the gain and cavity losses.

The diode lasers technology underwent considerable development during the last decades. Most common diode lasers are based on ternary compounds like gallium aluminum arsenide (*GaAlAs*) and gallium indium arsenide (*GaInAs*). Different content relations between the gallium (*Ga*) and aluminum (*Al*) in the first indicated compound, and between the gallium (*Ga*) and indium (*In*) in the second indicated compound, change the material energy gap E_g, and correspondingly the laser emission wavelength. The said combinations cover spectral ranges between 680 nm on the short wavelengths side and 1100 nm on the long wavelength side. Diode lasers based on indium gallium phosphide (*GaInP*) allow laser emissions over the 630–690 nm spectral range. Use of *IV–VI* elemental binary compounds like lead selenide (*PbSe*) and related ternary combinations like lead strontium selenide (*PbSrSe*) allows far-infrared emissions covering a 2.5–30 μm spectral range.

For quite a number of years, there are already commercial diode arrays consisting of a single row of dozens of individual diodes, or a large number of rows forming two-dimensional arrays of hundreds of diodes in a total area of ~1 cm^2. Such arrays may provide laser radiation power of dozens of Watts continuously (CW), or hundreds Watts during about 100 μs long pulses. Such emissions, when coupled to optical fibers, may be transmitted over long distances.

7.5 Illuminated p–n Junctions

In this section, we analyze the current/voltage characteristics of a p–n junction under external illumination. To facilitate the analysis, we use several simplifying assumptions. First, we neglect possible excitation of charge carriers in the transition range between *p*-type and *n*-type conduction. Such assumption is justified if the junction is very narrow. Then we assume that the illumination occurs only at the n-side of the junction, namely for $x > X_n$ (see Figs. 7.2 and 7.3), forming a uniform charge carrier excitation throughout. The first part of the assumption may be realized by transverse illumination of the n-side; the second part is in principle wrong, because excitation is always larger near the illuminated surface, becoming smaller with increasing distance from the surface. It is, however, possible to approach uniform excitation conditions by using weakly absorbed light, namely of a sufficiently small absorption coefficient, while the crystal thickness is small compared to α^{-1}, where α is the light absorption coefficient. The concentration of charge carriers in the n-region changes under illumination. For illumination durations considerably longer than both excess hole and excess electron lifetimes, these excess carriers' concentrations would be given by

$$p_n^{(ss)} = p_{n0} + G\tau_p^{(n)} = p_{n0}\,\boldsymbol{exp}\left(\Delta F_p^{(n)}/k_BT\right);$$

$$n_n^{(ss)} = n_{n0} + G\tau_n^{(n)} = n_{n0}\,\boldsymbol{exp}\left(\Delta F_n^{(n)}/k_BT\right),$$

where G marks the excess charge carrier generation rate (in m^{-3} s^{-1} units; see also Eq. [6.5] and its related text). The "(n)" superscript relates each quantity to the n-region, and the "(ss)" superscript indicates the steady-state quantity achieved a sufficiently long time after the illumination onset. The carriers' excitation rate is obviously proportional to the light intensity, but we won't analyze that aspect in the present section.

As a result of illumination, and even without application of any external electric field, the crystal is no more under thermal equilibrium. The quasi-Fermi states in the n-region deviate from their former coincidence with the Fermi state prior to illumination (see Section 6.4). The electron quasi-Fermi state F_n rises closer to the conduction band by $\Delta F_n^{(n)}$, and the hole quasi-Fermi state

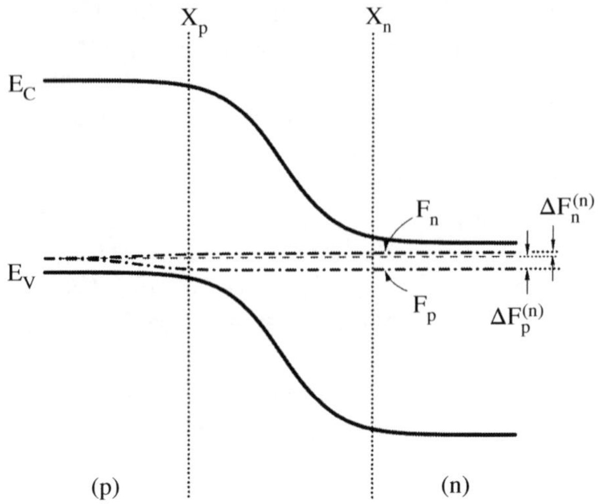

Figure 7.17 Energy band structure of a p–n junction whose n-region is under a steady constant illumination. The Fermi state position without illumination is described by a horizontal straight narrow dashed line.

F_p lowers closer to the valence band by $\Delta F_p^{(n)}$. However, the said quasi-Fermi states join inside the p-region, assuming the value of the Fermi state of before illumination. Figure 7.17 illustrates the above-described energy state scheme.

If all conditions required for calculating the current/voltage characteristics detailed in Section 7.1 are valid for our case, particularly that the applied forward voltages are sufficiently small such that practically the entire voltage develops across the transition region (junction), energy state structure under such voltage V_a appears as illustrated in Figure 7.18.

We continue by developing expressions for the electrical current versus voltage as performed in Section 7.1 after Eq. (7.1), by using similar considerations. Initially, we calculate the hole current density, and for that purpose we express $p_n(X_n)$ as follows:

$$p_n(X_n) = N_V e^{-\frac{H}{k_B T}} = N_V e^{-\frac{H+qV_a-qV_a}{k_B T}} = N_V e^{-\frac{H+qV_a}{k_B T}} \cdot e^{\frac{qV_a}{k_B T}}$$

$$p_n^{(ss)} \cdot e^{\frac{qV_a}{k_B T}} = [p_{n0} \cdot e^{-\frac{\Delta F_p^{(n)}}{k_B T}}]\, e^{\frac{qV_a}{k_B T}}.$$

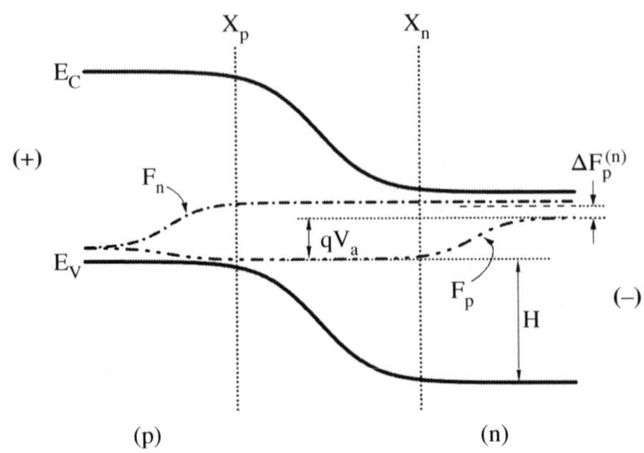

Figure 7.18 Energy band structure in the vicinity of a p–n junction whose n-region is under a steady constant illumination after application of a forward V_a voltage. The Fermi state position that existed at the depth of the n-region before the combined illumination and external voltage application is described by a horizontal straight narrow dashed line.

We continue by following the steps that follow Eq. (7.2), where $p_n^{(ss)} = p_{n0}\boldsymbol{exp}\left(-\Delta F_p^{(n)}/k_BT\right)$ replaces p_{n0} in all equations. The hole current density would be given by

$$J_p = J_p(X_n) = \frac{qD_p}{\sqrt{qD_p\tau_p^{(n)}}}p_n^{(ss)}\,[e^{\frac{qV_a}{k_BT}} - 1].$$

We further assume that the same illumination takes place in the p-region, and we obtain a symmetric expression for the electron current:

$$J_n = J_n(X_p) = \frac{qD_n}{\sqrt{qD_n\tau_n^{(p)}}}n_p^{(ss)}\,[e^{\frac{qV_a}{k_BT}} - 1].$$

The total current would thus be

$$J = J_p + J_n = J_s[e^{\frac{qV_a}{k_BT}} - 1], \text{ where } J_s = \frac{qD_p}{\sqrt{qD_p\tau_p^{(n)}}}p_n^{(ss)} + \frac{qD_n}{\sqrt{qD_n\tau_n^{(p)}}}n_p^{(ss)}. \tag{7.17}$$

The expressions obtained in Eq. (7.17) are quite similar to Eq. (7.4), which describes the dark current density versus voltage (without illumination); the minority carrier concentrations p_{n0} and n_{p0} existing far from the p–n junction under dark conditions are replaced by the steady-state minority carrier concentrations under illumination $p_n^{(ss)}$ and $n_p^{(ss)}$, respectively. These concentrations obviously depend on the light intensity. Figure 7.19 illustrates the obtained room temperature current versus voltage characteristics according to Eq. (7.17), namely for $k_BT = 0.025$ eV and for different excitation G rates.

All current/voltage characteristics exhibit current saturation for a reverse voltage (p-side negative), they all vanish for a zero voltage, and grow steeply under forward voltages (p-side positive). If the dark reverse saturated current is subtracted, the remaining saturated currents under illumination are directly proportional to the electron–hole pairs generation rate G, namely proportional to the incident light intensity. Thus, reverse voltage-biased p–n junctions may and are used as light detectors and light intensity measuring devices. We won't

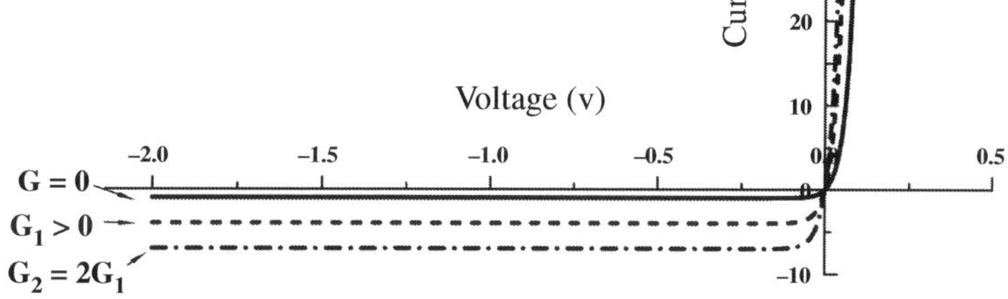

Figure 7.19 Room temperature current density as a function of voltage at in a p–n junction under a constant illumination for different light intensities that induce different rates of electron–hole pairs generation. J_{S0} is the dark saturated reverse current density.

provide more analyses and details here. A reader interested in delving into such issue may refer back to Section 3.9.2, where the responsivity of a photomultiplier is theoretically calculated, and/or to Section 9.2, where both responsivity and specific detectivity are calculated for a mercury cadmium telluride (MCT) far-infrared detector.

In contrast to expectation by Eq. (7.17) and Figure 7.19, under illumination but a zero external voltage, the current density doesn't entirely vanish. A residual current density appears in a reverse direction. It emanates from excess charge carriers generated at the junction potential barrier by either direct light absorption, or as a result of excess electron–hole pairs generated at the p- and n-regions, which then diffuse into the junction. The internal large junction electrical field causes a drift of the carriers, thus causing the said residual current. We won't discuss that effect further.

7.6 Bipolar Junction Transistor

A **transistor** is a semiconductor-designed device exhibiting three external metallic **contacts** to different device regions. It is different from the case of a diode having only two external contacts. Transistors are basically used to achieve amplification of either voltage or current. In the following discussion, we consider a case where the electrical current between two given contacts changes when the current between two other contacts is changed. In Section 7.5, we saw that the reverse current in a p–n junction depends on the minority carrier densities under thermal equilibrium, but it also changes when thermal equilibrium is violated. The example we showed there was the change of minority carrier density by exposure to external illumination. As we shall see, a similar result occurs by injecting minority carriers using a junction under an external forward voltage polarity.

An example of the above case is a p–n–p transistor. That device contains three regions arranged in a series: a p^+-region, an n-region, and a p-region, as displayed in Figure 7.20. The "+" superscript used to indicate the conductivity type of a semiconductor means that it contains a relatively very high concentration of that type carriers. In fact, that construction generates two separate p–n junctions, while the *n*-type region is common for both. Each region is contacted to a metallic conductor. The right-hand side junction is backward biased, saturating its through current. The right-hand side junction is forward biased, causing hole injection into the common n-region. If the n-region is narrow compared to the hole diffusion range, a considerable fraction of the injected holes would reach the right-hand side junction, and increase the through flowing current in a form described in Section 7.5. The p-edge of the reverse-biased junction is called **collector** and marked C. The p-edge of the forward-biased junction is called **emitter** and marked

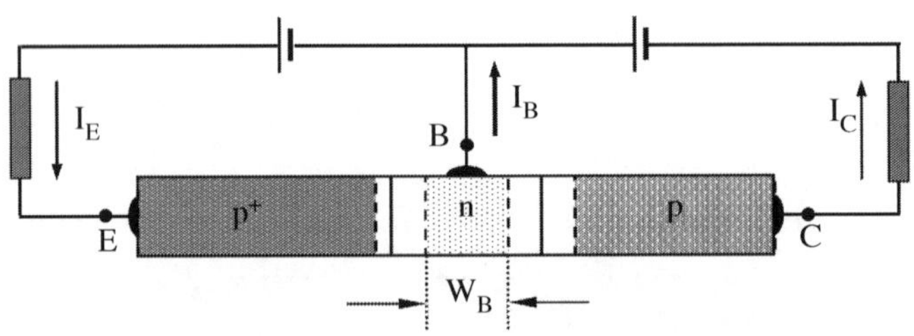

Figure 7.20 Schematic illustration of a p–n–p transistor with contacts construction where the collector junction is backward biased and the emitter junction is forward biased. The junctions are indicated by bright regions.

E. The common n-region is called **base** and marked B. The base width is marked W_B. In the following analysis, we assume it to be constant, independent of the bias voltages and the developing currents, although in fact it exhibits small changes as a function of the voltages developed across the different regions (The voltage dependence of each junction widths is described in Section 7.3.3, particularly Eq. [7.8].)

Under steady-state conditions, the density of injected excess holes from the emitter side into the base region is obtained by solving the excess hole diffusion equation (Eqs. [6.39] and [6.40])

$$\frac{d^2(\delta p_n(x_B))}{dx_B^2} = \frac{\delta p_n(x_B)}{D_p \tau_p}, \tag{7.18}$$

where x_B is the distance from the emitter junction. Boundary conditions for solving the equation are first $\delta p_n(0) = p_{n0}[e^{\frac{qV_{EB}}{k_B T}} - 1]$ (see Eq. [7.2]). We recall that if the base width would have been large compared to the diffusion range ($W_B \gg \sqrt{D_p \tau_p^{(n)}}$), the other boundary condition would have been $\delta p_n(\infty) = 0$, and the equation solution would have been as if an undisturbed diffusion, namely $\delta p_n(x_B) = p_{n0}[e^{\frac{qV_{EB}}{k_B T}} - 1] \cdot e^{-\frac{x_B}{\sqrt{D_p \tau_p^{(n)}}}}$ (see also Eqs. [6.40] and [7.3]). However, in our case, the base region is designed to be rather narrow compared to the diffusion range, which is $W_B \ll \sqrt{D_p \tau_p^{(n)}}$. In addition, its collector side is a reverse-biased junction under current saturation. Therefore, the second boundary condition is $\delta p_n(x_B) = -p_{n0}$ (see last paragraph in Section 7.1).

Instead of solving the diffusion equation accurately, we present an approximate, quite simplistic solution, namely that the concentration of excess charge carriers reduces linearly from its known value on the emitter side till its known value at the collector side. Figure 7.21 displays a model of such linear variation through the base region compared to a case expected for a broad base (a "free diffusion") and also compared to an analytically obtained variation by accurate solution of the diffusion equation (not presented here). The "free diffusion" solution and the accurate solution are given for the $W_B = \sqrt{D_p \tau_p^{(n)}}$ special case. The deviation between the two latter solutions is indeed very small, and it is practically negligible for calculating the emitter and collector currents if $W_B \ll \sqrt{D_p \tau_p^{(n)}}$. An existing difference stems from excess holes recombination while flowing through the base, and it actually determines the base current.

The hole currents through the collector and emitter are the diffusion currents of the excess holes through the relevant junctions. In using the approximate solution for the varying concentrations shown in Figure 7.21, the slope is constant. Thus accordingly, the hole current densities in the emitter and collector are identical and are given by

$$J_{pE} = J_{pC} = -qD_p \frac{d(\delta p_n(x_B))}{dx_B} = \frac{qD_p}{W_B} p_{n0} \cdot e^{\frac{qV_{EB}}{k_B T}}. \tag{7.19}$$

In fact, as seen in Figure 7.21, the true concentrations slope near the emitter junction is a bit larger than the approximate value, and the one near the collector junction is a bit smaller than the approximate one. Therefore, J_{pC} is a bit smaller than J_{pE}. As previously indicated, the difference, to be calculated below, stems from the excess holes recombination while streaming through the base.

The volumetric recombination rate of the excess holes in the base region is given by $U_p(x_B) = \delta p_n(x_B)/\tau_p^{(n)}$ (see Eq. [6.3] and related text). The total recombination rate in the base

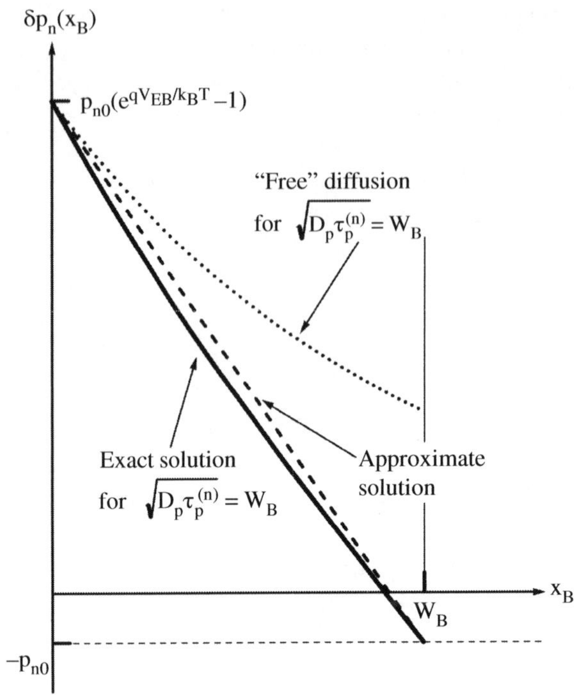

Figure 7.21 Distribution of excess hole density injected into the base region from the emitter side of a p–n–p transistor for a specific case where the base width equals the hole diffusion range: a solid line is an exact solution; a dashed line is an approximate solution. For comparison, a dotted line shows the hole density distribution that would have been obtained had the base width been long compared to the hole diffusion range.

volume is given by $\left(A/\tau_p^{(n)}\right)\int_0^{W_B}\delta p_n(x_B)dx_B$, where A is the device's cross-sectional area (which is the same for both emitter and collector). Multiplying by q provides the difference between the emitter and collector currents, which expresses itself as the base current. Setting $\delta p_n(x_B)$ by the approximate solution, we get

$$I_{pB} = \frac{1}{2}A\left(\frac{W_B^2}{D_p\tau_p^{(n)}}\right)\frac{qD_p}{W_n}p_{n0}\cdot e^{\frac{qV_{EB}}{k_BT}}. \tag{7.20}$$

The meaning of the last result is that as much as the base width is made narrow compared to the diffusion range, the base current becomes smaller. In addition, we would then need to correct the expressions for the emitter and collector currents given in Eq. (7.19). For simplicity, we assume the emitter current to be larger by half of the value given by Eq. (7.20), while the collector current to be smaller by that same half, such that their difference would equal the full I_{pB}. We then get

$$\left.\begin{aligned} I_{pE} &= \frac{qD_pA}{W_B}p_{n0}\cdot e^{\frac{qV_{EB}}{k_BT}}\left[1+\frac{1}{4}\frac{W_B^2}{D_p\tau_p^{(n)}}\right]; \\ I_{pC} &= \frac{qD_pA}{W_B}p_{n0}\cdot e^{\frac{qV_{EB}}{k_BT}}\left[1-\frac{1}{4}\frac{W_B^2}{D_p\tau_p^{(n)}}\right]; \\ I_{pB} &= \frac{qD_pA}{W_B}p_{n0}\cdot e^{\frac{qV_{EB}}{k_BT}}\left[\frac{1}{2}\frac{W_B^2}{D_p\tau_p^{(n)}}\right]. \end{aligned}\right\} \tag{7.21}$$

In principle, there are also additional contributions of electron current. Usually, however, such currents are relatively very small; in the emitter, they are very small because in that p^+-type region, the hole density is very high, and correspondingly, the electron density is several orders of magnitude smaller than the electron density at the n-type base. The electron saturated reverse currents in the collector are thus tiny compared to the hole currents caused by hole injection from the emitter side; they provide the hole currents when $qV_{EB} \gg k_BT$. Thus, the hole currents expression in Eq. (7.21) is in fact an excellent approximation for the total currents. For that reason, the "p" subscript is usually discarded from the currents expressions.

p–n–p transistor devices are usually characterized by ratios between collector-to-emitter and between collector-to-base currents as follows:

$$\left.\begin{aligned} \alpha &\equiv \frac{I_C}{I_E} \cong 1 - \frac{1}{2}\frac{W_B^2}{D_p\tau_p^{(n)}} \qquad \text{for } W_B \ll \sqrt{D_p\tau_p^{(n)}}; \\ \beta &\equiv \frac{I_C}{I_B} \cong \frac{2D_p\tau_p^{(n)}}{W_B^2} \qquad \text{for } W_B \ll \sqrt{D_p\tau_p^{(n)}}. \end{aligned}\right\} \tag{7.22}$$

In high-quality transistors, α is close to 1, and $\beta \gg 1$.

A most common means of characterizing commercial p–n–p transistors is to provide a plot of the collector current I_C as a function of the voltage applied across the entire transistor length V_{CE} in a polarity that makes the collector junction be reverse biased, while the base current I_B is kept constant (namely functioning as the curve parameter). According to Eq. (7.21), it is obvious that setting the size of I_B determines the size of the V_{EB} across the emitter junction. The latter is small compared to the entire voltage across the device, because it is at a forward polarity, and its electrical resistance is thus very small compared to the collector junction resistance. Namely, V_{CE} is only slightly larger than V_{CB}. The reason for describing the characteristics that way is because it serves the most commonly used electrical circuit, where the emitter edge contact is common for two circuits, an arrangement called "common emitter." A schematic description of current/voltage characteristics in a common emitter arrangement is provided in Figure 7.22.

The current/voltage characteristics in Figure 7.22 relate to different base currents, all for a case where the collector-to-emitter currents ratio is $\alpha = 0.945$ and the collector-to-base currents ratio is $\beta = 18$. The latter are obtained for a base width of one third of the hole diffusion range (Eq. [7.22]). The voltages and currents exhibit a negative sign, because the collector junction is reverse biased. All curves, including the one belonging to a zero base current, show a rise of the collector current as a function of an increasing voltage across the entire device. That increase indicates that the collector voltage approaches reverse saturation conditions (namely $|qV_{BE}| \cong |qV_{CE}| \gg k_BT$). We refrain from providing an analytical expression for that part of the curve, because it is fundamentally similar to the one provided in Figure 7.4. In the addressed voltage range ($V_{CE} \lessapprox 2$ V), the saturated currents appear virtually flat. Under much higher voltages, such as dozens volts, the currents exhibit some moderate rise as functions of a rising voltage (not shown in Fig. 7.22). The reason is a narrowing of the base caused by the increasing junction width (Eq. [7.8]) whose role is to increase the α and β quantities (Eq. [7.22]).

The issue of the current/voltage characteristic when the emitter junction is under a reverse polarity is not presented here. We just note that qualitatively, when the two junctions operate under a reverse polarity, the currents are extremely small. An immediate transition from a very high current into a very small one by just flipping the voltage polarity across the emitter junction has many applications in electrical circuits; we won't dwell further on that particular subject.

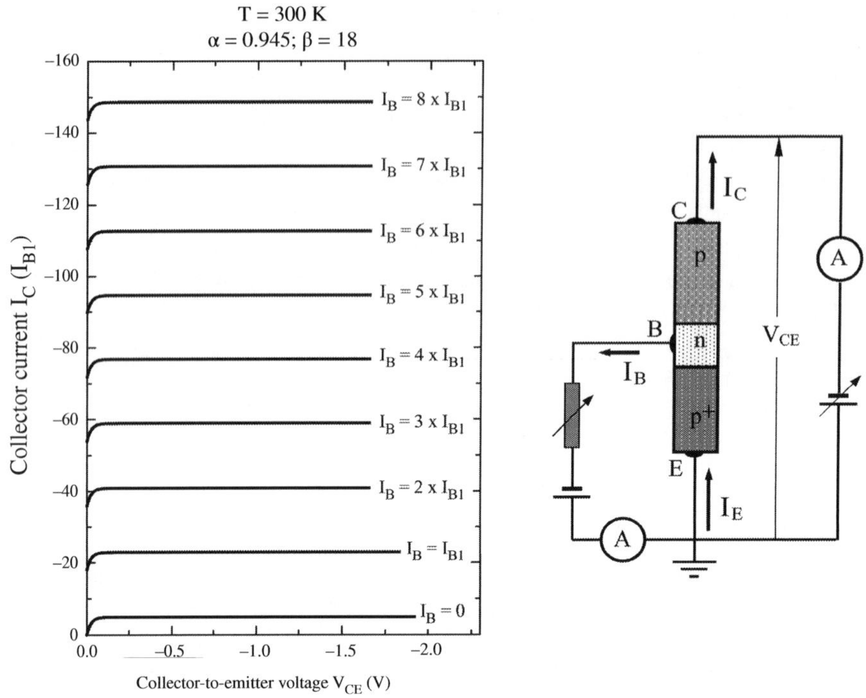

Figure 7.22 Room temperature ($T = 300$ K) current versus voltage characteristics in a p–n–p transistor under a common emitter arrangement for different base currents, where $\alpha = 0.945$ and $\beta = 18$. A plot of an electrical circuit used in measuring such characteristics is presented on the right-hand side.

Junction transistors of n–p–n arrangement exhibit very similar properties as p–n–p transistors, and the technological means of their manufacturing are also quite similar. Considerations on choosing a transistor type in electrical circuits are based on the details of the circuit design, and not on any advantage or disadvantage of any of them.

7.7 Voltage Amplification Circuit

Figure 7.23 presents the standard symbol of a p–n–p bipolar transistor in electrical circuits. The E, B, and C notations are presented to explain its relation to Figure 7.22 but are not usually presented in professional electrical circuits plotting. The arrow orientation is reversed for the n–p–n transistor symbol. Bipolar transistors play important different roles in electrical circuits. A common one that we present functions as a voltage pulses amplifier; at the point denoted as **out**, an amplified voltage signal is obtained relative to a signal fed at a point denoted as **in**. If not otherwise indicated, the voltages are measured relative to ground.

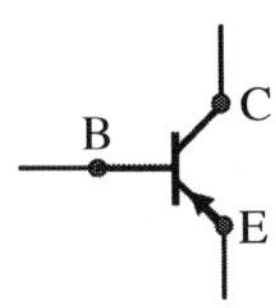

Figure 7.23 Symbol of a bipolar p–n–p junction transistor.

Such a voltage-amplifying circuit is presented in Figure 7.24. The emitter edge is connected to ground through an $R_{in} = 1$ kΩ resistor called **input resistor**. The collector edge is connected to a constant (DC) voltage supply at a reverse polarity relative to the collector junction, through a resistor called **output resistor**, that for demonstration is chosen to be 20 times larger, namely $R_o = 20$ kΩ. The reason for the said resistors' names will be described later. The base edge is connected to a δV_{in} voltage pulses supply connected in series to a regulated constant (DC) voltage supply, at a forward polarity relative to the emitter junction. The latter is smaller than the voltage supplied at the collector edge, whose value is adjusted such that both collector and emitter currents satisfy requirements called **work point**. When a small voltage pulse appears at the base contact, the emitter current changes such that practically the same voltage pulse appears across the input resistor R_{in}. The reason is that the resistance of the emitter junction at its forward polarity is very small compared to R_{in}, and the resistance of the collector junction is very large compared to all other circuit resistances. The input resistance R_{in} practically establishes the load resistance across the entire circuit, and that is the reason for its given name.

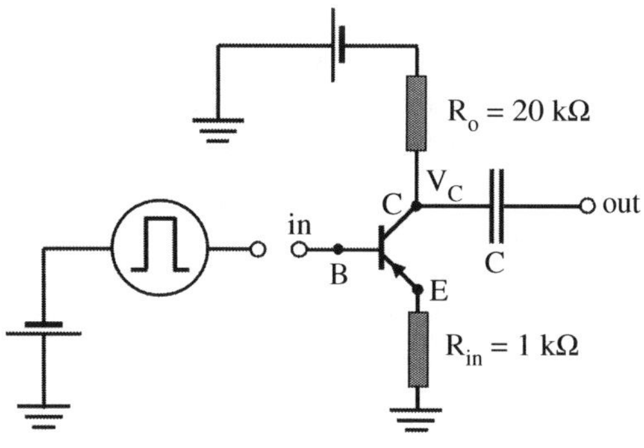

Figure 7.24 Example of an electrical circuit used for amplifying of voltage pulses. For a detailed discussion, see the text.

When a voltage pulse δV_{in} develops at the E point, the current through the input resistor changes by $\delta I = \delta I_E = \delta V_{in}/R_{in}$. The current change through the collector is practically identical. Thus, a voltage change occurs across the output resistor: $\delta V_o \cong \delta I \cdot R_o = \delta V_{in}(R_o/R_{in}) = 20\delta V_{in}$. The orientation of the potential change at point C is opposite to that applied at point B, because the other resistor edge is steadily kept at a constant potential. Thus, the generated output voltage pulses are amplified by an (R_o/R_{in}) factor compared to the input ones, but exhibit an opposite polarity. The R_o resistor also establishes the output circuit impedance, because its resistance is very small compared to the emitter resistance. Because the developed voltage changes are in fact additions to a direct constant voltage supplied at point C (the collector edge electrical contact), that constant component is isolated from the output point "out" by the **C** capacitor.

As in all electronic circuits, the performance of the circuit presented in Figure 7.24 is influenced by "parasitic" capacitances and inductances existing among all circuit parts. Their effects turn quite apparent for very short pulses. It is an important factor in the functioning of electrical circuits in general, but consideration of such effects is beyond the scope of the present book and won't be elaborated upon here.

Exercises for Chapter 7

7.1 Consider a semiconductor whose energy gap is 0.7 eV; at room temperature, the effective conduction band density of states is $2.5 \times 10^{19}\ cm^{-3}$ and of the valence band it is $5.5 \times 10^{19}\ cm^{-3}$. The hole mobility is $1.9 \times 10^3\ cm^2/V \cdot s$, and the electron mobility is $3.9 \times 10^3\ cm^2/V \cdot s$. A p–n junction constructed using that semiconductor has $2 \times 10^{15}\ cm^{-3}$ holes in the p-region and $1.5 \times 10^{16}\ cm^{-3}$ in the n-region. The minority carrier lifetime in the n-region is 150 μs, and in the p-region, it is 80 μs.

(a) Calculate the junction saturated current density under a reverse voltage.
(b) Calculate the junction current density under a forward voltage of 0.2 V.
(c) Given that the junction cross-sectional area is $10^{-2}\ mm^2$, calculate the dynamic junction resistance $R_D \equiv dV/dI$ at a zero voltage limit and at forward voltages of 0.1 V and 0.2 V. What is the effective resistance $R \equiv V/I$ under each of the said voltages?

7.2 (a) Calculate the width of the junction described in Exercise 7.1 without any external voltage; the material relative dielectric constant is $\varepsilon_r = 13$.
(b) An avalanche current occurs in the junction described in Exercise 7.1 at a reverse voltage of 16.5 V. Calculate the mean free path of charge carriers crossing the junction till impact ionization. What is the junction width at the occurrence of avalanche?
(c) Calculate the electrons mean free path using their given mobility and effective conduction band density of Exercise 7.1, and compare it to the mean free path value calculated in 7.2(b) exercise.

7.3 For the purpose of producing a diode laser, a p–n junction was formed using a semiconductor whose energy gap is 0.18 eV and its optical refractive index for light in the infrared region is $n = 4$. The effective conduction band density-of-states is $6 \times 10^{17}\ \text{cm}^{-3}$, and of the valence band, it is $6 \times 10^{18}\ \text{cm}^{-3}$, both at the liquid-nitrogen temperature of 80 K. The hole density on the junction p-side is $8 \times 10^{17}\ \text{cm}^{-3}$, and the electron density on the junction n-side is $2 \times 10^{16}\ \text{cm}^{-3}$. The electron mobility is 80, 000 $\text{cm}^2/\text{V} \cdot \text{s}$, and the hole mobility is 1200 $\text{cm}^2/\text{V} \cdot \text{s}$. The lifetime of excess electrons in the p-region is 100 ns, and their radiative recombination efficiency is 60%. The spectral width of the emitted recombination radiation under a forward voltage polarity across the junction is approximately 125 nm.
(a) Provide an estimate of the central wavelength of the radiative recombination emission when the junction is forward biased.
(b) Calculate the width of the radiation emitting zone of the junction and the radiated beam dispersion angle.
(c) Laser emission by the junction is obtained for current densities that exceed 500 A/cm^2; calculate the total losses in the junction light emission.
(d) Calculate the forward voltage applied across the junction needed to obtain the above-calculated current density.

7.4 The junction described in Exercise 7.1 is illuminated by weakly absorbed light whose absorption coefficient is $\alpha = 0.6\ \text{cm}^{-1}$. The photon flux density is $F = 8 \times 10^{15}\ \text{cm}^{-2}\text{s}^{-1}$. Each photon energy is $h\nu = 0.65$ eV.
(a) Calculate the incidence light intensity on the semiconductor surface in W/cm^2 units.
(b) Given that the light reflectivity at the surface area is $r = 0.45$, calculate the excess hole density in the n-region and the excess electron density in the p-region.
(c) Calculate the junction saturated reverse current density under the illumination (In some of your considerations, you could use the text that follows Fowler's Eq. [3.15].)

7.5 A p–n–p transistor is constructed using the semiconductor presented in Exercise 7.1. What should be the thickness of the n-region to achieve a $\beta = 30$ current amplification factor at the collector? What would then be the ratio α between the collector and emitter currents?

8 Field-Effect Devices

8.1 Space-Charge Layer at a Crystal Surface

An infinitely large and uniform crystal under thermal equilibrium exhibits no space-charge and therefore no bands bending. A crystal of finite dimensions, however, often exhibits space-charge layers near its surfaces, and the bands, which were horizontal inside the crystal, may bend upward or downward close to a surface. There are three different causes for such effects: surface states, proximity of a different material of a different work function, and an external electrostatic field.

8.1.1 Surface States

Chemical bonding of surface atoms are considerably different from those of the inner crystal volume. Simplistically, electrons of a surface atom that would have involved in forming chemical bond in the inner volume remain free to form states that often belong to the forbidden energy gap. In addition, chemical adsorption (chemisorption) or physical adsorption (physisorption) of ambient impurities also involve in formation of such surface states. Occupation of such local "trapping" states causes band bending near the surface. As an example, we consider a specific case of electron trapping states in an n-type semiconductor, which when empty are located below the Fermi state, and the conduction and valence bands are horizontal. Such state will trap electrons and form a negatively charged surface layer. The charge will repel mobile charge carriers (electrons in this specific example). The conduction band will be depleted of electrons near the surface. The conduction and valence bands will be "pushed" upward, together with the trapping level. The degree of upraising depends on the density of trapping levels, but the process will stop when the trapping level crosses the Fermi level, yet no more than about k_BT above it. Such case is displayed schematically in Figure 8.1.

Usually, only a part of the trapping centers become occupied with electrons. The near-surface region turns into a layer of electric dipoles: negative charge trapped in surface states and positive space charge distributed at the near-surface crystal volume. Because the near-surface conduction band is depleted of its charge carriers, that layer is called **depletion layer**. A similar effect causing accumulation of charge carriers occurs when the surface trapping levels are located above the Fermi level when they are filled, and the bands are straight. Then the trapped electrons are thermally excited into the conduction band, and the surface becomes positively charged. Mobile (free) electrons are attracted toward the surface, and the conduction and valence bands are bent downward, along with the trapping level. The process stops when the trapping level crosses the Fermi level downward, till about k_BT below it. That effect is displayed schematically in Figure 8.2. The near-surface region turn into a layer of electric dipoles: positive charge trapped

Conductors, Semiconductors, Insulators, and Crystal-Growth Technology: A Practical Guide,
First Edition. Zeev Burshtein.

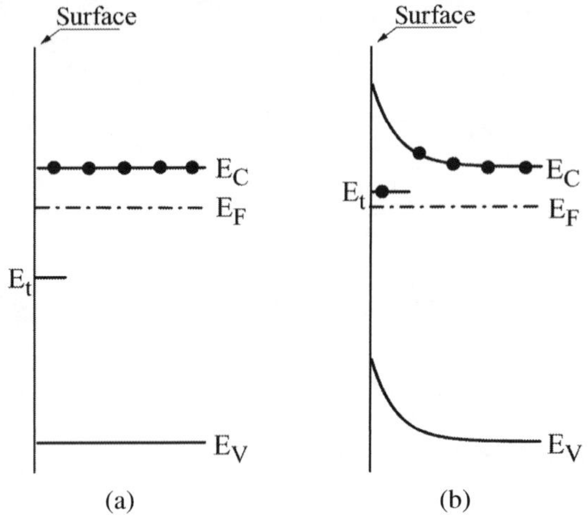

Figure 8.1 (a) Energy band structure of an n-type semiconductor with electron trapping surface states of E_t energy. (b) Band bending caused when the trapping centers actually trap electrons, thus inducing an electron depletion layer.

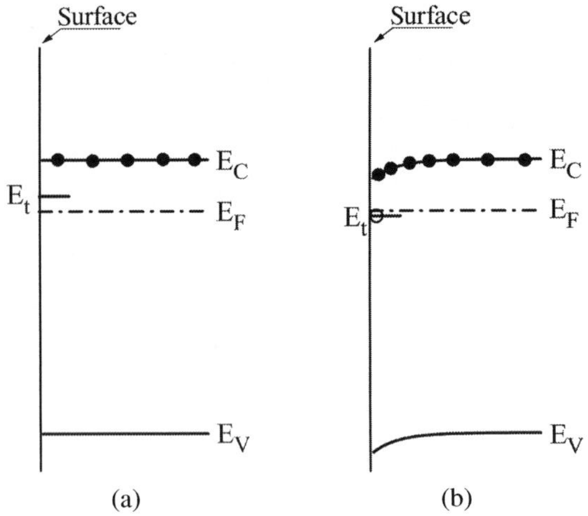

Figure 8.2 (a) Energy band structure of an n-type semiconductor with electron trapping surface states of E_t energy situated above the Fermi level. (b) Band bending caused when the trapping centers actually release trapped electrons into the conduction band, thus inducing an electron accumulation layer.

in surface states, and negative space charge distributed within the crystal inner volume. Because the near-surface conduction band is enriched (accumulated) with charge carriers compared to their inside concentration, that layer is called **accumulation layer**.

When a depletion effect becomes very large, it may bring about a situation where the valence band distance from the Fermi level at the very surface is smaller than the conduction band distance from the Fermi level inside the crystal volume, far from the surface. The conductivity type near the surface reverses and becomes *p*-type, in fact creating a p–n junction. The layer where such change occurs is called **inversion layer**, and it is displayed schematically in Figure 8.3.

It is possible to estimate the effective depth of the layer where such change occurs, by using Poisson's equation (see Eq. [6.25])

$$\frac{d^2V}{dx^2} = \frac{1}{\varepsilon_r \varepsilon_0}\rho(x), \tag{8.1}$$

where V is the potential measured relative to the conduction band potential deep inside the crystal volume, x is the distance from the surface, ε_r is the material dielectric constant, ε_0 is the permittivity of free space, and ρ is the electrical space-charge density. The V potential is sometimes called **potential perturbation**, because it "perturbs" the horizontal linearity of the

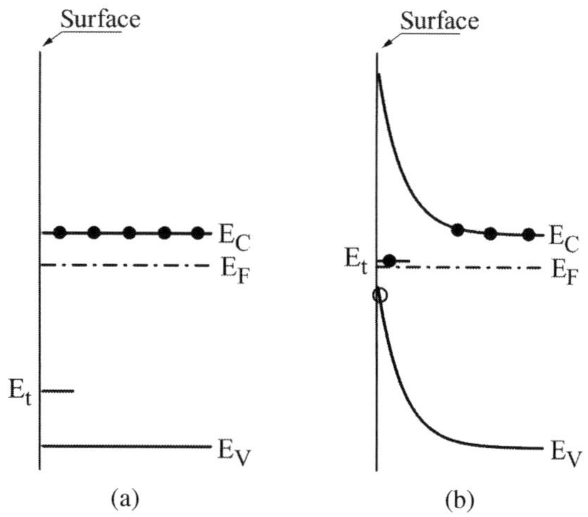

Figure 8.3 (a) Energy band structure of an n-type semiconductor with electron trapping surface states of E_t energy situated below the Fermi level close to the valence band. (b) Band bending caused when the trapping centers actually trap electrons, thus inducing an inversion layer.

inner bands, and not always being a steady property, often stemming from inadvertent adsorption of impurities at the surface. In our following analyses we deal only with cases of n-type crystals, and won't address extreme cases of inversion layers. It would always allow a $p \cong 0$ approximation.

In principle, a distribution of trapping states in the forbidden gap may always exist. The space-charge density is given by

$$\rho(x) = -q\left\{\left[n(x) - n_0\right] + \sum \left[n_D(x) - n_{D0}\right]\right\}, \tag{8.2}$$

where n_D is the density of trapped electrons in a trapping state located inside the crystal, and n_{D0} is their density in the same type center far from the crystal surface; $n(x)$ is the density of free electrons, and n_0 is their density far from the crystal surface. The summing in Eq. (8.2) is over all trapping levels inside the crystal. To simplify our model, we assume that there is only a single type of discrete states inside the crystal of energy E_D above the Fermi state, which lay below, but close to the conduction band, of N_D density. Such assumed levels structure and their related band bending are displayed in Figure 8.4.

The density of free (mobile) electrons occupying the conduction band inside the crystal is given by

$$n(x) = N_C e^{-\frac{E_C - E_F}{k_B T}} = N_C e^{-\frac{E_C(x) - E_C(\infty)}{k_B T}} \cdot e^{-\frac{E_C(\infty) - E_F}{k_B T}} = n_0 e^{\frac{qV(x)}{k_B T}}.$$

In addition, we also have

$$n_D(x) = \frac{N_D}{1 + e^{\frac{E_D - E_F}{k_B T}}} \underset{E_D - E_F \geq 3k_B T}{\cong} n_{D0} \cdot e^{\frac{qV(x)}{k_B T}}. \tag{8.3}$$

Thus, in a case where $E_D - E_F \geq 3k_B T$, the Poisson equation assumes the form

$$\frac{d^2 V(x)}{dx^2} = \frac{q}{\varepsilon_r \varepsilon_0}(n_0 + n_{D0})[e^{\frac{qV(x)}{k_B T}} - 1]. \tag{8.4}$$

It should be noted that when only a single set of traps exists inside the crystal, these traps also function as the donor impurities that form the conductivity type; then $n_0 + n_{D0} = N_D$. Eq. (8.4) is a differential equation, whose solution describes the variation of V as a function of the distance from the crystal surface. Under small perturbations, where always $qV(x) << k_BT$, Poisson's equation assumes the form

$$\frac{d^2V(x)}{dx^2} = \frac{q^2N_D}{\varepsilon_r\varepsilon_0 k_BT}V(x), \tag{8.5}$$

whose solution for the $V(\infty) = 0$ boundary condition (namely when the potential perturbation vanishes far from the surface) is:

$$\left.\begin{aligned} V(x) &= V_S exp(-x/L_D) \\ L_D &\equiv [\varepsilon_r\varepsilon_0 k_BT/q^2N_D]^{1/2} \end{aligned}\right\}, \tag{8.6}$$

where V_S, called **surface potential**, is the size of the potential perturbation at the surface. The solution portrays an exponential decay of the said potential perturbation oriented toward the inner crystal volume. The distance L_D where the perturbation drops by a $1/e$ factor compared to its surface size is called **Debye range**.

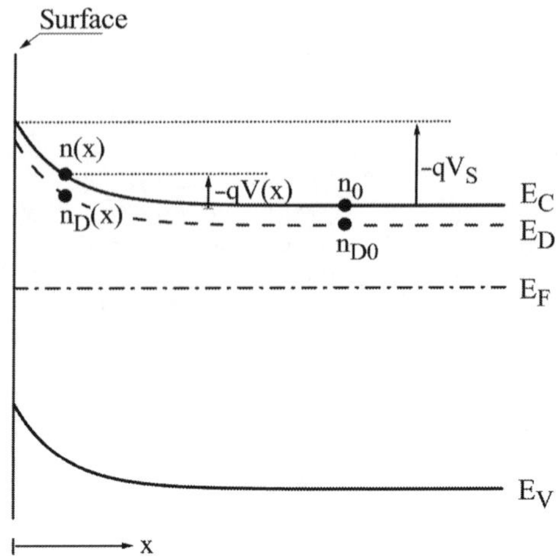

Figure 8.4 Energy band structure of an n-type semiconductor having a shallow electron-trapping level (donor level) inside the crystal, of energy E_D close to the conduction band. The band bending is caused by a negative charge accumulated at the surface.

To get some feeling of the effect order of magnitude, we consider a typical case: an *n*-type crystal having shallow traps, whose free-plus-trapped electrons concentration is 10^{14} cm^{-3}:

Given: $\varepsilon_r = 15$; $\varepsilon_0 = 8.85 \times 10^{-12}$ Coulomb/V·m; $k_BT/q = 0.025$ V;

$q = 1.6 \times 10^{-19}$ Coulomb; $N_D = n_0 + n_{D0} = 10^{14}$ cm^{-3} $= 10^{20}$ m^{-3}.

Setting in Eq. (8.6):

$$L_D = \left(\frac{15 \times 8.85 \times 10^{-12} \times 0.025}{1.6 \times 10^{-19} \times 10^{20}}\right)^{1/2} = 4.55 \times 10^{-7}\ \text{m} = 0.455\ \mu\text{m}.$$

As seen by Eq. (8.6), Debye range is inversely proportional to the square root of the charge-carriers density. Therefore, that effect is quite negligible in metals, for which a calculated Debye range is of an order of a single atom dimensions, or even less.

The size of the surface potential depends on the **surface-charge density** $\boldsymbol{Q_S} = fq\widetilde{N}_t$ of charge trapped at surface trapping states at E_t energy, where $\widetilde{N}_t$ is their density per unit area, and f is the fraction of filled (occupied) states. According to Gauss equation

$$-\boldsymbol{Q_{SC}} = \boldsymbol{Q_S} = \varepsilon_r\varepsilon_0\mathcal{E}_S = -\varepsilon_r\varepsilon_0 \left.\frac{\partial V}{\partial x}\right|_{x=0},$$

where $\boldsymbol{Q_{SC}} \equiv \int_0^\infty \rho(x)dx$, where $\rho(x)$ being the volume density of space charge distributed near the surface inside the crystal, and $\mathcal{E}_S$ is the electric field at the surface. Using Eq. (8.6), we get $|\boldsymbol{Q_{SC}}| = \varepsilon_r\varepsilon_0|V_S|/L_D$, therefore $|V_S| = |\boldsymbol{Q_{SC}}|L_D/\varepsilon_r\varepsilon_0$.

The space-charge region may be attributed with a measure of electrical capacitance $\boldsymbol{C_{SC}}$ and a dynamic electrical capacitance $\widetilde{\boldsymbol{C}}_{SC}$ defined accordingly by

$$\boldsymbol{C_{SC}} \equiv \frac{-\boldsymbol{Q_{SC}}}{V_S}; \quad \widetilde{\boldsymbol{C}}_{SC} \equiv \frac{-\partial \boldsymbol{Q_{SC}}}{\partial V_S}. \tag{8.7}$$

The units of both capacitance quantities are F/m^2. The minus "–" signs appearing in the definitions express the fact that the sign of the electrical space-charge distributed inside the crystal is opposite to the sign of the charge trapped at the crystal surface. The dynamic capacitance concept is especially important when it is possible to perform controlled changes of the surface potential. In Section 8.1.3, we show that the surface potential may be varied by applying an external voltage across the crystal, thus allowing the measurement of the two capacitance types. In the above-described simple model, the capacitance and the dynamic capacitance are equal. Their values are given by

$$C_{SC} = \widetilde{C}_{SC} = \frac{\varepsilon_r \varepsilon_0}{L_D} = q\sqrt{\frac{\varepsilon_r \varepsilon_0 N_D}{k_B T}}. \tag{8.8}$$

In cases of larger band bending, the calculation turns more complex. To shorten the mathematical expressions, we denote by $\mathrm{v} = qV/k_BT$. Poisson Eq. (8.4) then assumes the form

$$\frac{d^2\mathrm{v}}{dx^2} = \frac{1}{L_D{}^2}(e^{\mathrm{v}} - 1). \tag{8.9}$$

To enable solution, we multiply both sides by $\frac{d\mathrm{v}}{dx}$ to get

$$\frac{d\mathrm{v}}{dx} \cdot \frac{d^2\mathrm{v}}{dx^2} = \frac{1}{L_D{}^2} \cdot \frac{d\mathrm{v}}{dx} \cdot (e^{\mathrm{v}} - 1);$$

and by a form changing:

$$\frac{1}{2}\frac{d}{dx}\left[\left(\frac{d\mathrm{v}}{dx}\right)^2\right] = \frac{1}{L_D{}^2} \cdot \frac{d}{dx} \cdot (e^{\mathrm{v}} - \mathrm{v}).$$

Integrating both sides, we get

$$\left(\frac{d\mathrm{v}}{dx}\right)^2 = \frac{2}{L_D{}^2} \cdot \frac{d}{dx} \cdot (e^{\mathrm{v}} - \mathrm{v} - Const).$$

Setting the boundary conditions $\mathrm{v}(\infty) = 0$ and $\left.\frac{d\mathrm{v}}{dx}\right|_{\infty} = 0$, we get that $Const = 1$. We are in fact interested in expressing the space charge distributed inside the crystal, whose sign in opposite to that of the surface charge. Thus

$$Q_{SC} = (-)\varepsilon_r\varepsilon_0\mathcal{E}_S = \varepsilon_r\varepsilon_0 \left.\frac{\partial V}{\partial x}\right|_{x=0} = -sign(\mathrm{v}_S)\frac{\varepsilon_r\varepsilon_0 k_B T}{q} \cdot \frac{\sqrt{2}}{L_D}(e^{\mathrm{v}_S} - \mathrm{v}_S - 1)^{1/2};$$

$$Q_{SC} = -sign(V_S)\frac{\varepsilon_r\varepsilon_0 k_B T}{q} \cdot \frac{\sqrt{2}}{L_D}\left(e^{\frac{qV_S}{k_BT}} - \frac{qV_S}{k_BT} - 1\right)^{1/2}, \tag{8.10}$$

for $-[E_C(\infty) - E_F] < -qV_S < E_g - 2[E_C(\infty) - E_F]$.

Equation (8.10) expresses the surface density of the space charge inside the crystal adjacent to the surface as a function of the surface potential V_S in a *p*-type crystal. As to the surface density of the trapped surface charge, its size is identical, but its sign is opposite.

To delve into and understand the meaning of Eq. (8.10), we consider different boundary cases:

(i) A large perturbation of an inversion layer: $V_S < 0$ and $|qV_S| >> k_BT$.

We may thus write the approximation

$$Q_{SC} \cong \sqrt{2}\frac{\varepsilon_r\varepsilon_0}{L_D} \cdot \frac{k_BT}{q}\left|\frac{qV_S}{k_BT}\right|^{1/2}. \tag{8.11}$$

The dynamic capacitance would be given by

$$\tilde{C}_{SC} \cong \frac{1}{\sqrt{2}} \frac{\varepsilon_r \varepsilon_0}{L_D} \left| \frac{qV_S}{k_B T} \right|^{-1/2}; \tag{8.12}$$

And the surface potential would be given by

$$|V_S| \cong \frac{1}{2} \frac{k_B T}{q} \left(\frac{\varepsilon_r \varepsilon_0}{L_D} \right)^2 \frac{1}{\tilde{C}_{SC}^2} = \frac{\varepsilon_r \varepsilon_0 q N_D}{2} \cdot \frac{1}{\tilde{C}_{SC}^2}. \tag{8.13}$$

In analysis of experimental results, a plot of the surface potential size $|V_S|$ as a function of the inverse squared dynamic capacitance $1/\tilde{C}_{SC}^2$ displays a straight line. Its slope allows extracting the donor density $N_D = n_0 + n_{D0}$. Means of measuring surface potential and dynamic capacitance is presented in Section 8.1.3.

(ii) A large perturbation of an accumulation layer: $V_S > 0$ and $|qV_S| >> k_B T$.

We may thus write the approximation

$$Q_{SC} \cong -\sqrt{2} \frac{\varepsilon_r \varepsilon_0}{L_D} \cdot \frac{k_B T}{q} e^{\frac{qV_S}{2k_B T}}. \tag{8.14}$$

The dynamic capacitance would then be given by

$$\tilde{C}_{SC} \cong \frac{1}{\sqrt{2}} \frac{\varepsilon_r \varepsilon_0}{L_D} e^{\frac{qV_S}{2k_B T}}. \tag{8.15}$$

Thus, the surface potential will be expressed as

$$|V_S| \cong \frac{2k_B T}{q} ln \left(\frac{L_D}{\varepsilon_r \varepsilon_0} \tilde{C}_{SC}^2 \right) = \frac{k_B T}{q} \left[ln \left(\frac{2k_B T}{\varepsilon_r \varepsilon_0 q^2 N_D} \right) + ln \left(\tilde{C}_{SC}^2 \right) \right]. \tag{8.16}$$

In analysis of experimental results, a plot of the surface potential size $|V_S|$ as a function of the natural logarithm of the squared dynamic capacitance $\boldsymbol{ln}\left(\tilde{C}_{SC}^2\right)$ displays a straight line. Its intercept at the ordinate (Y-axis) allows extracting the donor density $N_D = n_0 + n_{D0}$. Means of measuring surface potential and dynamic capacitance is presented in Section 8.1.3.

Figure 8.5 presents a plot of the dynamic capacitance as a function of the surface potential. It is important to notice the transition from a moderate dependence in the carriers' depletion region to an exponential dependence in the accumulation region. That effect is qualitatively understandable, because the depletion is limited to the sum of charge carriers occupying the conduction band when the bands are horizontally flat; on the other hand, there is no limit to the reservoir of charge carriers in an accumulation process. Thus, in the latter case, the dynamic capacitance grows steeply as a function of the surface potential.

In essence, similar effects are obtained in a *p*-type semiconductors, with analogous expressions. We won't provide these expressions here, and the reader will be invited to deal with that problem as an exercise.

8.1.2 Contact Potential

The subject we consider in this section is the semiconductor crystal structure near a metallic contact. When the metal and the semiconductor have different work functions, a potential barrier V_{cp} appears on the semiconductor surface upon the contact formation, without application of any external voltage between the two. That potential barrier is called **contact potential**, and in certain cases, **Schottky barrier**. The barrier occurs when exchange of electrical charge between the metal and the semiconductor is enabled, which brings about a thermodynamic equilibrium between the two (namely having equal Fermi energies). A contact potential formation process, where the semiconductor energy bands are initially horizontally flat, is presented in Figure 8.6. The figure addresses a case where the metal work function $W_{\Phi}^{(m)}$ is larger than the semiconductor work function $W_{\Phi}^{(s)}$. When the two materials are far from each other, and before exchange of charge carriers is enabled, the two-materials system is not under thermal equilibrium; each separate material is characterized by its usual states structure, and no electrical field develops between the two, as displayed in Figure 8.6(a). The absence of thermal equilibrium is emphasized by denoting quasi-Fermi energies in each material, rather than a Fermi energy. The χ quantity appearing in the figure describes the energetic distance between the conduction band bottom E_C and the vacuum free electron energy (vacuum state). That quantity is called **affinity**. Being related to the atomic potential that creates the band structure, the affinity is independent of macroscopic electrical fields across the crystal. When the materials become sufficiently close, some charge carrier exchange is enabled, equalizing the Fermi states, and forming an electrical field between the two.

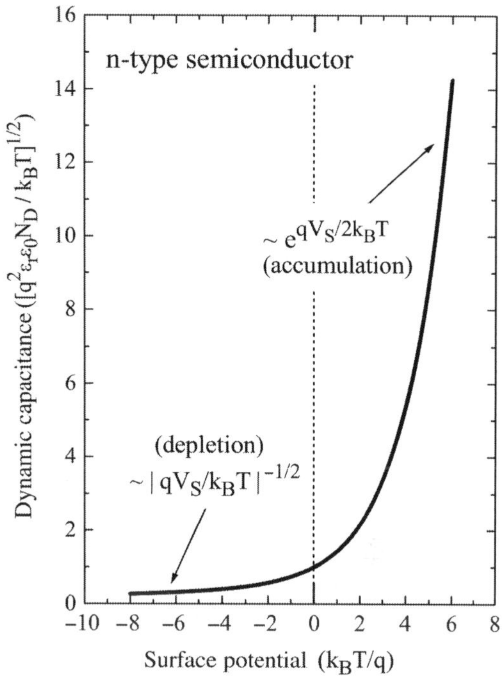

Figure 8.5 Dynamic capacitance of the surface-adjacent space-charge layer inside an n-type semiconductor as a function of the surface potential. A negative surface potential relates to carrier depletion; a positive surface potential relates to carriers accumulation.

When the distance d between the materials is long compared to Debye range L_D, most voltage falls across the separating space, and only a small fraction falls across the space-charge layer inside the crystal (the voltage distribution is actually determined by the ratio between the capacitances of the separating space and of the space-charge layer). Such case is displayed in Figure 8.6(b). Further distance reduction where d is small compared to L_D makes most voltage fall across the space-charge layer. Such case is displayed in Figure 8.6(c). When a physical tight contact is established, the entire voltage $V_{cp} = \left(W_{\Phi}^{(s)} - W_{\Phi}^{(m)}\right)/q$ falls across the semiconductor space-charge layer. In a case of a depletion layer, it is called **Schottky barrier**.

Figure 8.7 displays a case where the metal work function $W_{\Phi}^{(m)}$ is smaller than the semiconductor work function $W_{\Phi}^{(s)}$. When the two materials are far from each other, and no exchange of charge carriers is possible, the two materials system is not under thermal equilibrium; each separate material is characterized by its usual states structure, and no electrical field develops between the two, as displayed in Figure 8.7(a). When the materials move sufficiently close, some charge carrier exchange is enabled, equalizing the Fermi states, and

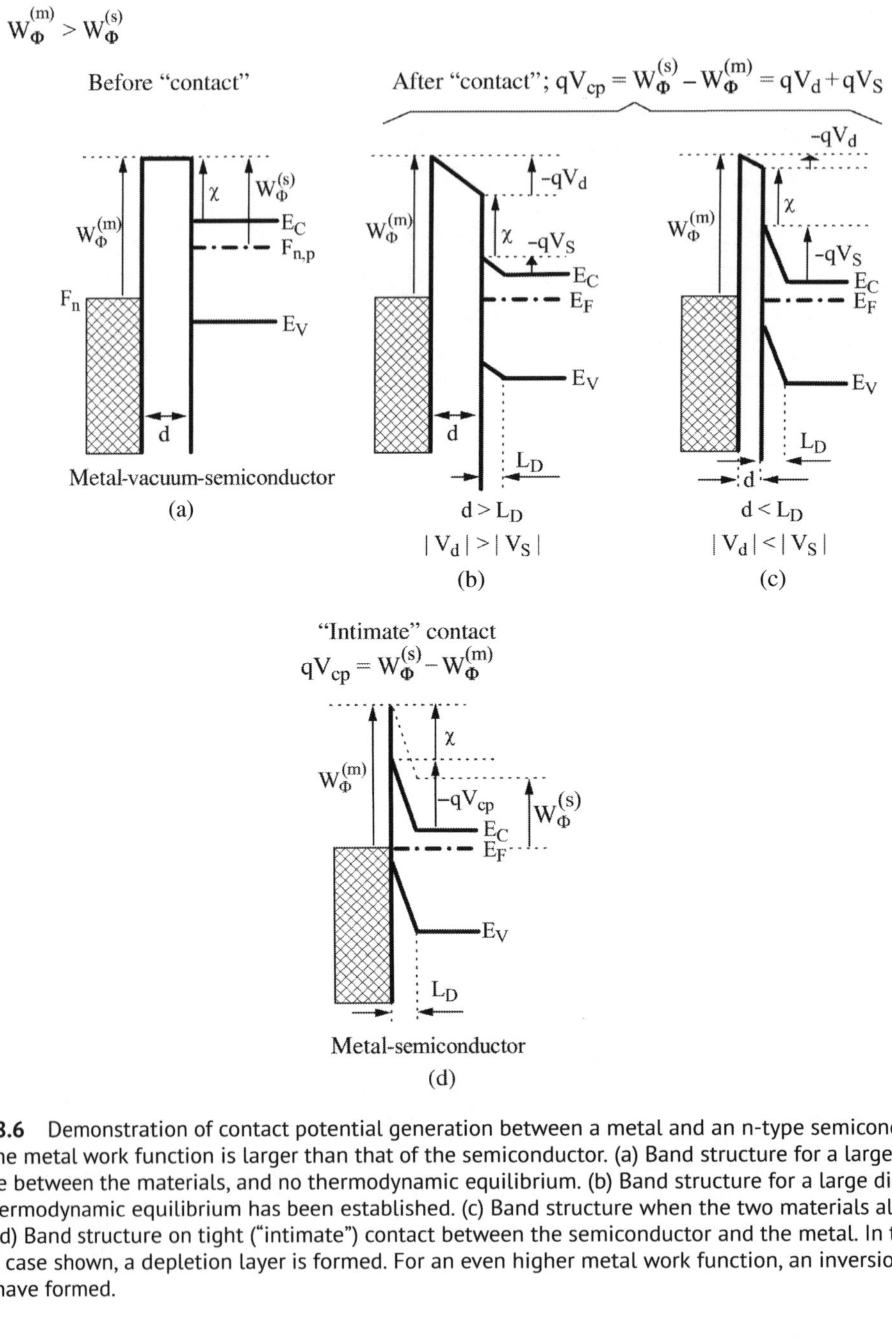

Figure 8.6 Demonstration of contact potential generation between a metal and an n-type semiconductor, when the metal work function is larger than that of the semiconductor. (a) Band structure for a large distance between the materials, and no thermodynamic equilibrium. (b) Band structure for a large distance, after thermodynamic equilibrium has been established. (c) Band structure when the two materials almost touch. (d) Band structure on tight ("intimate") contact between the semiconductor and the metal. In the specific case shown, a depletion layer is formed. For an even higher metal work function, an inversion layer would have formed.

forming an electric field between the two. When the distance d between the materials is still long compared to Debye range L_D, most voltage falls across the separating space, and only a small fraction falls across the space-charge layer inside the crystal. As indicated in analyses of the previous case, here as well the voltage distribution is actually determined by the ratio between capacitance of the separating space and of the space-charge layer. Further distance reduction, where d becomes small compared to L_D, most voltage falls across the space-charge layer, and only the small remainder falls across the empty space between them. Such case is displayed in Figure 8.7(c). When a final tight contact is established, the entire voltage

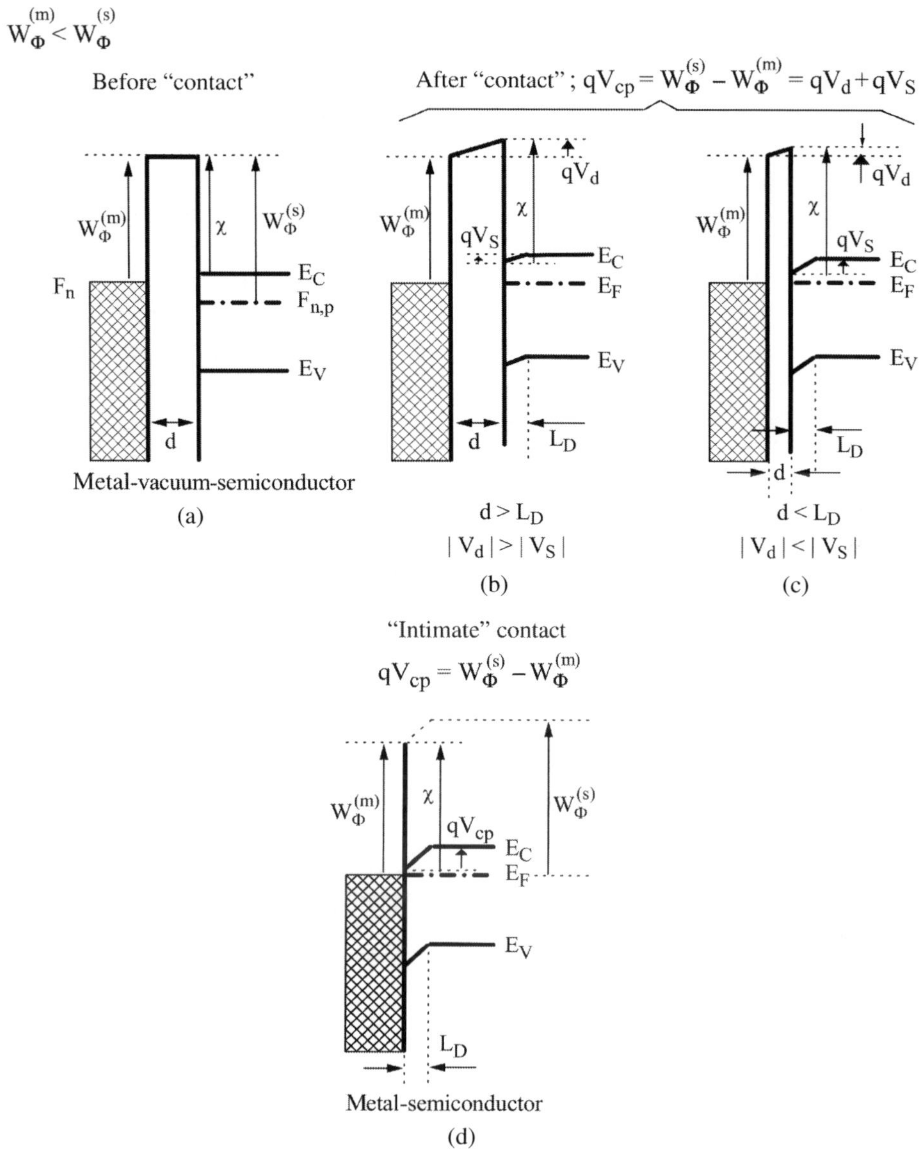

Figure 8.7 Demonstration of contact potential generation between a metal and an n-type semiconductor, when the metal work function is smaller than that of the semiconductor. (a) Band structure for a large distance between the materials, and no thermodynamic equilibrium. (b) Band structure for a large distance, after thermodynamic equilibrium has been established. (c) Band structure when the two materials almost touch. (d) Band structure on tight ("intimate") contact between the semiconductor and the metal. In the specific case shown, an electron accumulation layer is formed.

$V_{cp} = \left(W_\Phi^{(s)} - W_\Phi^{(m)}\right)/q$ falls across the semiconductor space-charge layer. In the present case, it forms an electron accumulation layer.

We won't describe here the different cases where the crystal is a *p*-type semiconductor. They are all analogous to the above-described *n*-type crystal, and the reader is invited to address those cases as an exercise.

In all described cases, the electrical charge distributed in the space-charge region is compensated by an opposite charge residing on the metal contact surface. If the semiconductor exhibits a band bending due to occupied surface states before addition of the metallic contact, the two described effects are summed up.

8.1.3 An External Voltage

Application of an external voltage between the metal and the semiconductor is equivalent to an added difference between the two materials work functions (Section 8.1.2). That voltage is distributed between the semiconductor space-charge region capacitance and the capacitance of the empty space (or of some other insulating medium) existing between the metal and the semiconductor. The distribution of the voltage in the form of an electric field inside the semiconductor is called **field effect**. A straightforward calculation of the capacitances ratio shows that the voltage induced across the space-charge layer $V_S^{(ext)}$ is given by

$$V_S^{(ext)} = V^{(ext)} \frac{1}{1 + (\boldsymbol{C}_S/\boldsymbol{C}_d)}, \tag{8.17}$$

where $\boldsymbol{C}_d = \varepsilon_r \varepsilon_0 / d$ is the capacitance density (capacitance per unit area of the insulating medium of ε_r dielectric constant), separating the semiconductor from the metal. It is readily obvious that when the said medium is very narrow, or under direct contact between the metal and semiconductor, the entire external voltage falls across the space-charge layer.

The contribution of the external voltage to the bands bending is added to the two other factors described in Sections 8.1.2 and 8.1.3; charge carrier surface trapping and a difference between the work functions of the contact metal and the semiconductor. Application of an external electric field and measurement of the space-charge layer capacitance allows determination of the band bending that would have existed without the external voltage application. The most accepted method is application of a constant external voltage added with an alternating (AC) voltage whose amplitude is only several millivolts. The alternating voltage is used for measuring the dynamic capacitance between the contact and the semiconductor, which in fact is the capacitance of the space-charge layer. A typical measuring circuit is described in Figure 8.8. The C_0 is a capacitor of a known value connected in series to the space-charge capacitance. The used alternating voltage is of a very high frequency, such that the C_0 impedance is very small compared to that of the resistor R. Thus, the alternating voltage component plays the role of a capacitors voltage divider. Measuring the alternating voltage across C_0 allows extracting the series space-charge capacitance. To allow an accurate determination of the constant (DC) voltage component across the crystal,

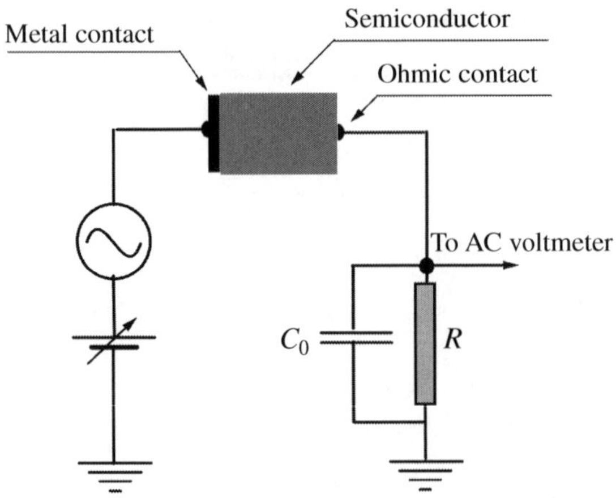

Figure 8.8 An electrical circuit that enables measuring of the space-charge layer capacitance as a function of the surface potential.

the R resistance is made small compared to the crystal resistance. Thus, almost the entire DC applied voltage falls across the space-charge layer. Tuning the DC voltage, and recording the DC voltage at which a steep change in the dynamic space-charge layer capacitance occurs (from a moderate dependence to a steep dependence), identifies the condition where the surface potential is zero (see Fig. 8.5). In other words, it is the condition where bands are horizontally flat till the surface. The external voltage has compensated the band bending caused by the other factors. At room temperature, such situation may be identified at an accuracy of the order of $k_BT/q \cong 25$ mV.

Performing the described measurement requires considerable sophistication, because a voltage application across the semiconductor actually requires two metallic contacts, one that determines the semiconductor surface potential and the other that determines the "metallic contact potential". The first noted must be such that doesn't generate any band bending near the contact. Such contact is called **Ohmic contact**. Achieving that trait requires special means. On the one hand, they include special treatment of the semiconductor surface, like chemical etching and a controlled ambient atmosphere, and on the other hand, they include selecting the metal and the technique of attaching/soldering it onto the crystal surface. The metal selection relates not only to the semiconductor conductivity type, but also to its donor (or acceptor) density, which determines the Fermi energy state position relative to the conduction band, hence also its work function (see Figs. 8.6 and 8.7). The scientific literature includes thousands publications describing methods for obtaining Ohmic contacts to various semiconductors, and much information protected by patents is available from various commercial companies involved in manufacturing of semiconductor devices.

8.2 Field-Effect Transistor (FET)

In Section 7.5, we indicated that a transistor is a semiconductor-designed device exhibiting three external metallic contacts to different device regions. Use of transistors allows obtaining current or voltage amplification. In the device we hereby analyze, the amplification is obtained when a current between two edges changes when an external voltage is applied to an additional third edge. The latter voltage is called **bias voltage**, and the device is called **field-effect transistor** (FET). There are two major kinds of FET: if the bias is applied over a reverse-biased p–n junction, it is called junction-FET (JFET). If the bias is applied across a layer of an insulating oxide material, it is called metal–oxide–semiconductor-FET (MOS-FET). In the following, we delve only on the JFET properties. The characteristics of a MOS-FETs are generally similar, but they are different in some details.

As we already saw, application of a reverse voltage across a junction changes the junction width. That property is used in a JFET to control the width of the device region called **channel**. Figure 8.9 describes schematically the construction of the regions that make up a JFET device having an n-type channel. It also allows understanding the principles of its manufacturing procedure.

The raw material is a plate of a p-type semiconductor, whose hole density is very large and is therefore marked $\boldsymbol{p^+}$. The surface is coated by a material that when doping the semiconductor performs as an electron donor. Diffusion of the dopants layer into the plate is activated by heating to exceed the type compensation density until a considerably thick layer transforms into n-type. That process forms the lower p–n junction between the $\boldsymbol{p^+}$-and the n-regions. A different material is then coated at a center area of the newly formed n-region, which when doping the semiconductor performs as an electron acceptor (namely a hole donor). Diffusion of these

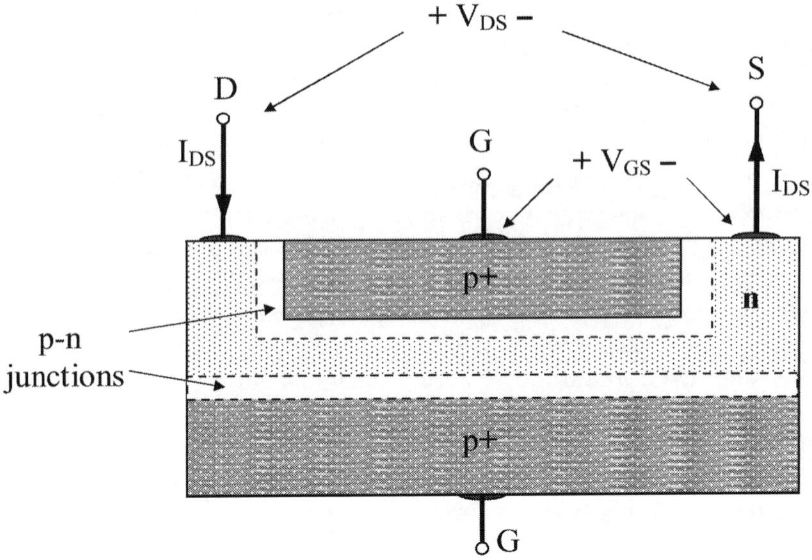

Figure 8.9 Schematic illustration of a JFET construction with an n-type channel. For more details and explanations, see the text.

dopants layer into the mid plate is again activated by heating to exceed the type compensation density, until a part of the newly formed *n*-type region transforms back to become $\boldsymbol{p}^+$-type. That process forms the upper p–n junction between the new $\boldsymbol{p}^+$- and the new n-regions. The $\boldsymbol{p}^+$-regions are connected ("short-circuited") by an outside conductor that also serves as one of the external contacts, marked G and called **gate**. Two edges of the n-region are connected to two external contacts: one, connected to a positive pole of a voltage (or current) supply, is marked D and called **drain**. The second, connected to the negative pole of the same voltage supply, is marked S and called **source**. The $\boldsymbol{p}^+$-regions, which together form the gate, are of a relatively high electrical conductivity compared to all other regions, and therefore always maintain the same potential. The n-region is of a smaller conductivity, and is called **channel**. The junctions are practically insulating, and under a reverse voltage would prevent electrical current flow between the channel and any $\boldsymbol{p}^+$-region.

To perform a mathematical analysis of effects occurring in a JFET, we consider a simpler geometric model, as presented in Figure 8.10. In the model, the different device regions are described as planar tables laying on top of each other. We first provide a qualitative consideration of the effects. The figure describes a special connection between the source S and the gate G, namely $V_{GS} \equiv V_G - V_S = 0$. A current I_{DS} is made to flow between the drain D and the source S. For small currents, the channel behaves like a regular resistor, and I_{DS} is linearly proportional to the V_{DS} voltage applied between the drain and source. The potential along the channel changes linearly, growing from zero at the source till V_{DS} at the drain. The gate and channel behave like a reverse-biased p–n junction. Further growth of the external voltage involves a growth of the potential difference between the channel and gate, causing a junction broadening (see Section 7.3.3 and Eq. [7.8]). The broadening is not uniform, because, as indicated, the voltage between the source and gate varies along the channel. The broadening reduces the channel volume and increases its resistance, thereby reduces the increasing rate of I_{DS} as a function of V_{DS}. On a further increase of V_{DS}, the junctions widths broadening reduces the channel width toward almost entire closure. First to close is the channel at the drain D, a situation called **pinch-off**.

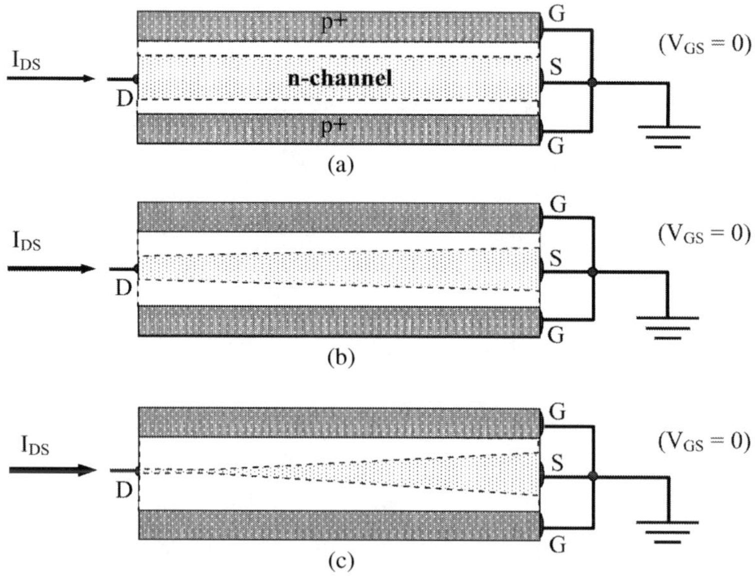

Figure 8.10 A simplified model of an n-channel JFET device construction and demonstration of the channel narrowing as a function of increased current through it. The increasing I_{DS} current between the drain D and source S occurring gradually in cases (a), (b), and (c) is demonstrated by increased thickness of the arrow that indicates its orientation. Additional details and explanations are provided in the text.

The current ceases from growing as a function of V_{DS}, namely it practically saturates. In Figure 8.12 and its following text, we provide a quantitative analysis of the above, qualitatively described current/voltage characteristic.

We turn to discuss qualitatively a different situation, where a reverse voltage V_{GS} is applied between the gate and source as described in Figure 8.11. When this voltage is negative, it establishes a reverse bias across the junctions, thus providing an advanced broadening of the junctions. The channel resistance will start larger, and it would achieve pinch-off condition at smaller V_{DS} voltages compared to that obtained for $V_{GS} = 0$. That way we obtain a set of different current/voltage characteristics, described in Figure 8.12.

We now turn to a quantitative analysis. The JFET in Figure 8.11 is described in detail by the parallel layers model as in Figure 8.10. The distance between the two $\boldsymbol{p}^+$ layers is $2a$, each having a length L and a depth z. The distance from the drain is marked x. The width of each junction – the upper and lower ones – is $W(x)$, and half the channel width is $h(x)$, always satisfying $W(x) + h(x) = a$. Under an I_{DS} current flowing through the channel, an electrical voltage $V_{xS}(x)$ develops along it, whose limits are $V_{xS}(0) = V_{DS}$ and $V_{xS}(L) = 0$.

We consider the special case where the V_{GS} bias voltage between the gate and source is negative, as shown in Figure 8.11. Note that the gate belongs to two $\boldsymbol{p}^+$ regions connected by an electrical wire. The voltage between the gate and the drain is $V_{GD} = V_{GS} - V_{DS}$. The width of each junction near the drain (the $x = 0$ point) is given by (see Eq. [7.8])

$$W(0) = \left(\frac{-2\varepsilon_r \varepsilon_0 V_{GD}}{qN_D} \right)^{1/2},$$

where N_D is the density of donor-type impurities in the channel region. This expression is valid when $N_A >> N_D$; therefore, the concentration of acceptor impurities in the $\boldsymbol{p}^+$-regions has practically no impact on the junction widths. It is also assumed that the junction widths

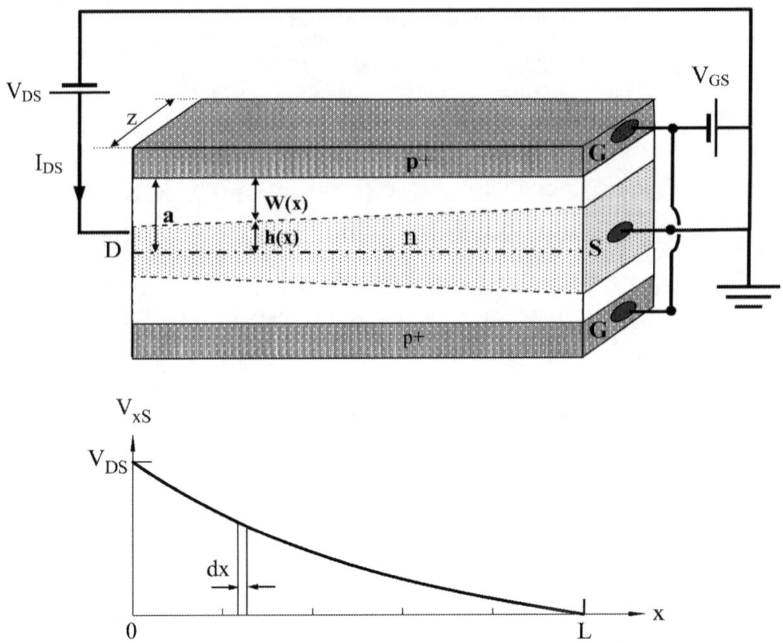

Figure 8.11 Top: the simplified JFET structure of Figure 8.10 under an external non-zero gate–source voltage V_{GS} and a drain-source current I_{DS}. Bottom: schematic illustration of the electrical potential variation along the channel. Details and explanations are provided in the text.

between each $\boldsymbol{p}^+$-region and its adjacent channel are negligibly small under no current and a zero voltage bias. A pinch-off mode occurs when $W(0) = a$. Denoting the voltage between gate and source at pinch-off as V_p, and using the last equation, we get

$$V_p = -\frac{qa^2N_D}{2\varepsilon_r\varepsilon_0}. \tag{8.18}$$

Importantly, if the V_{GS} bias voltage in an n-channel JFET is positive (opposite to the bias polarity displayed at the top of Fig. 8.11), the p–n junctions become forward biased, charge carriers are injected into the n-region, and the former bias effects on the junction width practically disappear. The device properties are thus entirely different. We won't address that case in our present book.

We return to calculate the current/voltage characteristics through the channel. A differential resistance of a dx channel slice (Fig. 8.11) is given by $\rho \cdot dx \cdot 2h(x)$, where ρ is the channel material resistivity. The voltage change across that slice is given by

$$dV_{xS} = -I_{DS}\frac{\rho}{2zh(x)}dx.$$

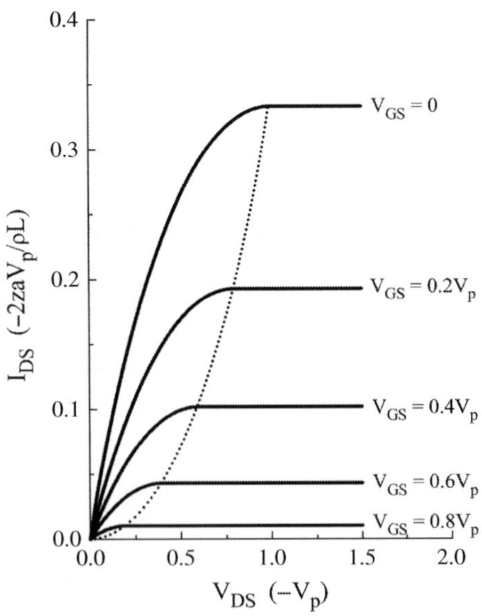

Figure 8.12 Current versus voltage characteristics in a typical JFET device for different V_{GS} bias voltages between the source and gate. The pinch-off voltages and their respectively obtained saturated currents form the dotted curve.

The minus "–" sign on the right-hand side indicates that the V_{xS} voltage becomes smaller on approaching the source point when I_{DS} is positive. Thus, we get

$$I_{DS} = -\frac{2zh(x)}{\rho} \cdot \frac{dV_{xS}}{dx}. \tag{8.19}$$

Next we proceed in calculating the channel width at each point x:

$$h(x) = a - W(x) = a - \left(\frac{-2\varepsilon_r\varepsilon_0 V_{Gx}}{qN_D}\right)^{1/2} = a\left[1 - \left(\frac{V_{GS} - V_{xS}}{V_p}\right)^{1/2}\right].$$

We set it in Eq. (8.19) and get

$$I_{DS} = -\frac{2za}{\rho}\left[1 - \left(\frac{V_{GS} - V_{xS}}{V_p}\right)^{1/2}\right]\frac{dV_{xS}}{dx} = -\frac{2za}{\rho}V_p \cdot \frac{d}{dx}\left[\frac{V_{xS}}{V_p} + \frac{2}{3}\left(\frac{V_{GS} - V_{xS}}{V_p}\right)^{3/2}\right].$$

We next integrate the two sides over x between $x = 0$ and $x = L$, divide by L, and recall that I_{DS} is a constant independent of x, and get:

$$I_{DS} = -\frac{2zaV_p}{\rho L} \cdot \frac{d}{dx}\left[\frac{V_{xs}}{V_p} + \frac{2}{3}\left(\frac{V_{GS} - V_{xS}}{V_p}\right)^{3/2}\right]_0^L;$$

$$I_{DS} = \frac{2za(-V_p)}{\rho L}\left[\frac{-V_{DS}}{V_p} - \frac{2}{3}\left(\frac{V_{GS} - V_{DS}}{V_p}\right)^{3/2} + \frac{2}{3}\left(\frac{V_{GS}}{V_p}\right)^{3/2}\right]. \tag{8.20}$$

At pinch-off conditions, $V_{GS} - V_{DS} = V_{GD} = V_p$, and the current becomes saturated. The saturation current would then be

$$I_{DS}(sat) = \frac{2za(-V_p)}{\rho L}\left[\frac{1}{3} - \left(\frac{V_{GS}}{V_p}\right) + \frac{2}{3}\left(\frac{V_{GS}}{V_p}\right)^{3/2}\right]. \tag{8.21}$$

In the absence of a bias voltage between the gate and source, namely when $V_{GS} = 0$, the saturated current would be

$$I_{DS}(sat) = \frac{2za(-V_p)}{2\rho L}.$$

Equations (8.20) and (8.21) are valid for every $V_{GS} \leq 0$, satisfying $|V_{GS}| \leq |V_p|$, and for every $0 \leq V_{DS} \leq V_{GS} - V_p$. When $V_{DS} \geq V_{GS} - V_p$, the current stabilizes at its saturated value and exhibits only tiny rise as a function of increased voltage between the drain and source. Typical current/voltage characteristics are plotted in Figure 8.12.

8.3 A Source-Follower Circuit

Figure 8.13 presents the standard symbol of an n-channel JFET device in electrical circuits. The D, S, and G notations are presented to explain its relation to Figures 8.10–8.12, but are not usually presented in professional electrical circuits. In the case of a p-channel JFET, the arrow orientation is reversed. In most devices, the dynamic resistance between the gate and source (or drain) is very high, of the order of ~10 GΩ when the gate is biased by a negative polarity voltage relative to both.

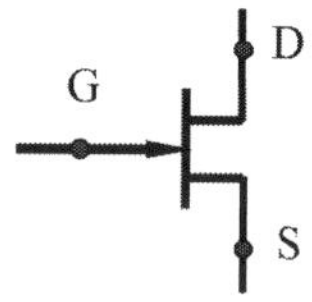

Figure 8.13 Symbol of an n-channel JFET in electrical circuits.

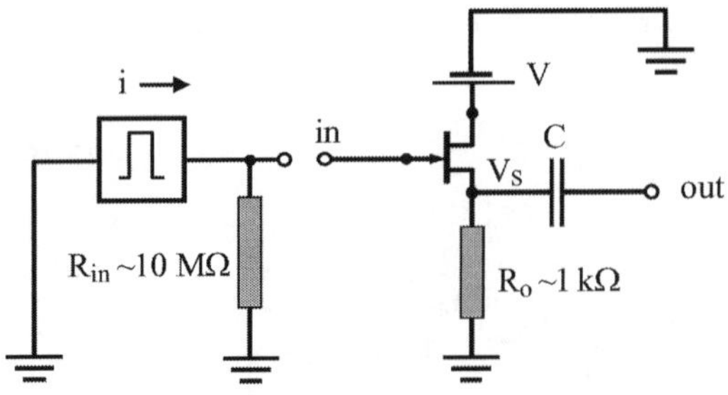

Figure 8.14 Example of a source-follower circuit based on an n-channel JFET. Detailed description and explanations are provided in the text.

A JFET device is used in a variety of electronic circuits. A common circuit presented in Figure 8.14 is called **source-follower circuit**, and is used to reduce by a large factor the "internal impedance" of voltage pulses signals generated by an electrical pulse generator. The supplied voltage pulses are generated by a current pulses supplier connected in series to a very large resistor of the order of ~10 MΩ, and the voltage that develops across it is the measurement goal. When that resistor end is connected to the input end "in" of the impedance conversion circuit, which is the JFET gate connection, the gate resistance establishes the input impedance, and it is sufficiently small to practically diminish the gate potential except during the pulse duration. The drain point is connected to a direct (DC) voltage supplier marked V, and the source point is serially connected to a ~1 kΩ resistor R_o, grounded at its other side. Except for the pulse duration, the gate potential is therefore always negative relative to both drain and source. The voltage V of the supplier is considerably larger than the JFET pinch-off voltage V_p. Thus, the drain-to-source I_{DS} currents always belong to saturation conditions.

To understand how the circuit functions, assume that the drain-to-source I_{DS} current grows gradually starting at zero. Had the source be grounded, the current would rise till the saturation value described in Figure 8.12 for $V_{GS} = 0$. However, along with the current growth, the potential V_S at the edge of the resistor R_o also grows, while the gate potential remains zero. As a result, a negative voltage V_{GS} develops, which narrows the channel, thus increasing its resistance. That effect reduces the current growth. The I_{DS} current would thus grow until balanced by the reduction effect. The balancing point satisfies $\frac{dI_{DS}(sat)}{dV_{GS}} = \frac{dI_{DS}(sat)}{dV_S}$, a condition bearing important conclusions. First, it determines the device work point: $\frac{1}{R_o} = \frac{2za(-V_p)}{\rho L}\left[-\frac{1}{V_p} + \frac{1}{V_p}\left(\frac{V_{GS}}{V_p}\right)^{1/2}\right]$ (the right-hand side is the derivative of Eq. [8.21]). Solving it for V_{GS} as the unknown provides the $V_S = -V_{GS}$ voltage that develops across the R_o resistor. Further, this point exhibits the feature that necessarily $\delta V_S = \delta V_{GS}$. Thus, voltage change appearing on the gate will appear fully at the source. In other words, changes in the source potential follow the changes in the gate potential. That is the reason for the adjective "**source-follower**" granted to such circuit. Notably, the effective impedance of the changes on the source is the R_o resistance while the input impedance is the R_{in} resistance. The functioning of the device may be interpreted by it being in fact a current amplifier, where changes in the gate potential determine the current between the drain and the source. Because the said changes occurring along the R_o resistor are additions to a constant V_S voltage, that constant component is out-filtered from the output "out" point by the large capacitor C.

As in all electronic circuits, the performance of the circuit presented in Figure 8.14 is influenced by "parasitic" capacitances and inductances existing among all circuit parts. Their effects turn quite apparent for extremely short pulses. It is an important factor in the functioning of electrical circuits in general, but consideration of such effects is beyond the scope of the present book and won't be elaborated upon here.

Exercises for Chapter 8

8.1 An n-type semiconductor, contains shallow donor impurities at a concentration of 6.5×10^{14} cm^{-3}; its surface contains electron trapping centers at a concentration of 1.5×10^{9} cm^{-2}. The relative dielectric constant is $\varepsilon_r = 12$. Calculate the characteristic Debye range at room temperature and at −200 °C, the band bending caused at the surface when 20% of the trapping centers are filled with electrons, and the space-charge layer capacitance for each temperature.

What would have been the band bending at each said temperature if the density of trapping centers had been 1.2×10^{11} cm^{-2}, and 60% of them had actually trapped electrons?

8.2 An n-type semiconductor has an energy gap of 0.7 eV and a free electron concentration of 1.5×10^{15} cm^{-3}. The effective conduction band density of states is 2.5×10^{19} cm^{-3}. An electrical contact with gold of $W_{\Phi} = 5.05$ eV work function forms a Schottky barrier of 0.4 eV. Calculate the work function of the semiconductor and its affinity. What should be the work function of a metal contact that would form an electron accumulation layer at the semiconductor surface.

8.3 Provide a drawing and explain in detail the formation of a contact potential between a metal and a p-type semiconductor when the metal work function is smaller than that of the semiconductor and when it is larger than that of the semiconductor. Consult Figures 8.6 and 8.7 and their related text.

8.4 The pinch-off voltage in an n-channel JFET transistor is 8 V. Calculate the channel width if it is known that the density of its free electrons is 1.5×10^{14} cm^{-3} and its dielectric constant is $\varepsilon_r = 11$. It is further given that the electron mobility is 1800 cm^2/V · s, the device width is $z = 0.2$ mm, and its length is $L = 1.3$ mm. Calculate the saturation current when the gate and source are electrically connected (short-circuited).

9 Radiation and Light Detectors

In this chapter, we describe the use of semiconductors to detect electromagnetic radiation in two spectral ranges that are fundamentally quite distant: gamma and X-rays radiation on the short wavelength side, and visible and far infrared on the long wavelength side.

9.1 Gamma and X-rays Radiation Detectors

In this section, we describe the method by which semiconductors may be used to detect gamma and X-rays. The related photon energies range between about 2–3 keV up to several MeV. The name **X-ray** is given to radiation originating by transitions between states of electrons orbiting atomic nuclei. The name **gamma ray** is given to radiation originating by transitions between states of nucleons (protons and neutrons) orbiting and oscillating inside atomic nuclei, which belong to the higher-energy limit in the above-indicated range.

9.1.1 Fundamental Construction of a Gamma and X-rays Detector

Figure 9.1 displays schematically the construction of a gamma and X-rays detector. It comprises a semiconductor crystal in the form of a narrow slab with electrical contacts, each of A area, soldered on its two opposite large surfaces. A large voltage is applied between the contacts. The semiconductor is in fact an insulator, namely it is a semiconductor of a high-energy gap, such that the densities of both free electrons and free holes are practically zero under thermodynamic equilibrium at the measuring temperature. When a high-energy photon (gamma ray or X-ray) impinges on the crystal surface, it penetrates into the crystal and stops by generating a large number of free electron–hole pairs. Under the external electric field, the generated free electrons are collected at the positive electrode and the generated holes are collected at the negative electrode. After both-type carriers are collected, a voltage change δV has developed between the two

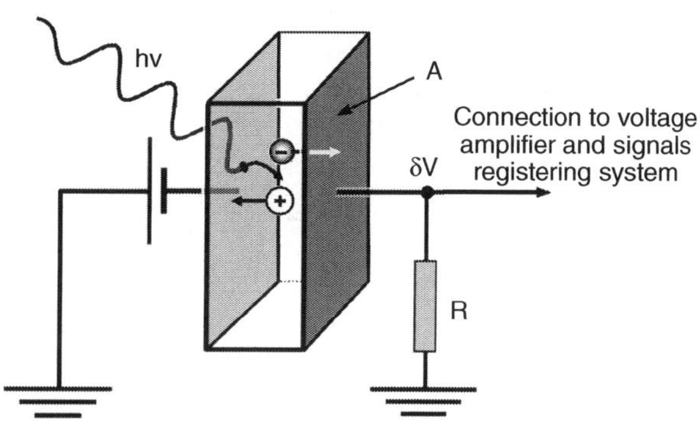

Figure 9.1 Principle structure of an electronic detector of X or gamma radiation. For details see the text.

Conductors, Semiconductors, Insulators, and Crystal-Growth Technology: A Practical Guide, First Edition. Zeev Burshtein.

electrodes. The voltage over the series R resistor grows from zero to δV. That change is recorded by an electronic system. The appearance of the δV "signal" indicates a photon absorption event, and its value expresses the absorbed photon energy. Under proper conditions, the δV signal height is proportional to the absorbed photon energy. Sorting of registered signals makes it possible to describe the energy spectrum of absorbed photons when the detector is placed inside a radiation field.

9.1.2 Mechanisms of Charge Carrier Generation

When a high-energy photon (gamma ray or X-ray) penetrates into the crystal, it may interact with the crystal and even stopped by one of three major processes: electron/positron pair generation, Compton scattering, and a photoelectric effect.

Generation of electron/positron pairs (or in short **pair production**) is a process where the photon energy is absorbed very close to the nucleus of an atom by generating the said pair of particles: an electron and a positron (recall that a positron is a particle whose rest mass and electrical charge are of same size as the electron, but the sign of its charge is opposite, namely it is positive). The threshold for the process occurrence is 1022 keV, which is twice the electron rest mass equivalent energy. The probability for such process occurrence is proportional to the atom squared atomic number: Z^2. The two generated particles usually have a considerable kinetic energy originating from the difference between the photon original energy and the pair rest mass. The pair kinetic energy is then dissipated by generating various excitations inside the matter, like lattice vibrations, but also by generation of free electron/hole pairs, which then contribute to the material conductivity – a matter important for the detection process. When the positron ends up nearly at rest close to an electron, the two annihilate by emitting two photons, 511 keV each, propagating in opposite directions. Generation of two photons rather than just a single one is consistent with the momentum conservation law. Because a detector geometric dimension is finite, each of these photons may pass through and escape the crystal without being reabsorbed. Thus, in a part of the original photon absorption cases, the energy given to the crystal misses 511 keV and, in others, missing 1022 keV. These effects are expressed in the photon energy spectra recorded by the detector.

Compton scattering is an elastic scattering of the incident photon by an unbound (mobile) electron or an electron belonging to an outer shell of the material atoms. On occurrence, the photon changes its propagation orientation and loses some energy, and the electron acquires some kinetic energy that depends on the original photon scattering angle. The probability for the occurrence of Compton scattering by any material atom is proportional to its Z atomic number. The kinetic energy acquired by the electron is then dissipated not only by generating various excitations inside the matter, like lattice vibrations, but also by the generation of free electron/hole pairs contributing to the material conductivity. The scattered photon may pass through and escape the crystal without being reabsorbed. In some cases of original photon scattering, the energy imparted to the crystal lacks some part of the original photon energy. This missing part exhibits a continuous spectrum, because the photon scattering is continuous. This effect is expressed in the energy spectrum of the photons recorded by the detector.

Photoelectric effect is a process where the photon is absorbed by directly exciting a material atom. The probability for occurrence of such absorption by any material atom is proportional to the fifth power of its atomic number, Z^5, and is inversely proportional to the third power of the photon energy. In most cases, an electron belonging to the atom K shell is emitted from the atom with a kinetic energy that equals the difference between the photon energy and its atomic binding energy. The said kinetic energy is dissipated to the surrounding volume by inducing different excitations like lattice vibrations, but also by the generation of free electron/hole pairs contributing to the material conductivity. An electron orbiting in an outer shell of the atom decays into the generated "empty" state at the lower K shell, while emitting an X-ray photon characteristic of the involved atom. In a rather thin detector, or when the event occurs sufficiently close to the detector surface, the emitted photon may pass through and escape the crystal without being reabsorbed. Thus, the energy given to the crystal in a part of the original photon absorption cases is missing X-ray energies that are characteristic of the material constituent atoms. That effect is expressed in the photon energy spectra recorded by the detector.

In view of the above descriptions, an important consideration in choosing a semiconductor material to design an X-ray radiation detector relates to the atomic numbers of its comprising elements. For measuring of photon energies, a high absorption probability is needed. Use of higher Z numbers would allow obtaining absorption while using thin crystals. The absorption probability for each of the described mechanisms obviously depends on the photon energy. In Figure 9.2, we provide a stopping coefficients graph of high-energy photons in a silicon *Si* single crystal. The **stopping coefficient** (or **attenuation coefficient**) concept is similar in essence to the "absorption coefficient" term used for different cases in the near visible optical region. Its role in analytic expressions for X-ray and gamma ray photons is identical when the consideration is limited to the initial absorption events. It is different by the fact that not the entire energy of an interacting photon necessarily remains in the absorbing material.

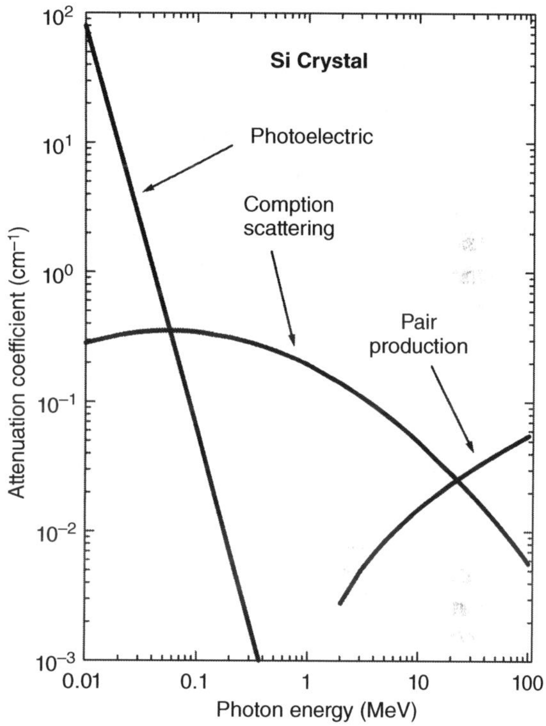

Figure 9.2 Full logarithmic plot of high-energy photon attenuation coefficients as functions of energy in a silicon crystal by photoelectric absorption, Compton scattering, and electron–positron pairs production. For more details, see the text.

9.1.3 Charge Carrier Collection Issues

In order to demonstrate various issues concerning relevant semiconductors properties, we consider a rather extreme case related to detection of low-energy X-rays in the 2–5 keV range. The photoelectric absorption effect dominates in that energy range, and the silicon *Si* attenuation coefficients are of the order of 10^3 cm^{-1} (see Fig. 9.2). Semiconductors whose atoms are of higher atomic Z numbers exhibit even higher attenuation coefficients. Absorption depth of photons exhibiting such energy is quite small, of the order of 10 μm and even less. Generation of charge carrier pairs thus occurs very close to the crystal surface, where enhanced recombination and trapping processes dominate.

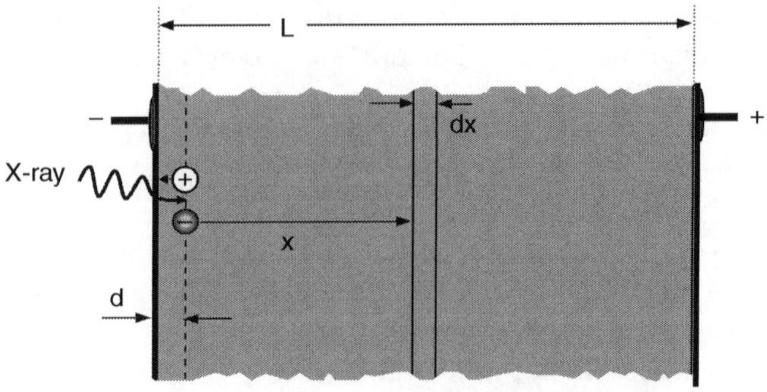

Figure 9.3 Description of charge carrier motion in a slice of an X-ray detector under the action of an external electric field after an X-ray photon is absorbed near the crystal surface contacted to the negative electrode.

Figure 9.3 displays schematically a slice of a radiation detector, indicating the different parameters needed for theoretical analysis of the generated charge carriers, and their collection at the electrodes after absorption of an X-ray photon. We assume that the X-ray is absorbed near one of the electrodes; let it for example be the negative electrode. The absorption occurs at a distance d from the surface, satisfying $d \ll L$, where L is the detector thickness (which is also the distance between the electrodes). Under the external electric field applied between the electrodes, the generated carriers separate, drifting in opposite orientations: the holes are attracted to their near negative electrode, and the electrons are attracted toward the far positive electrode. Because the holes drift over a very short distance, their contribution to the charge induced on the electrodes is negligible compared to the contribution of electrons that drift longer distances into the crystal volume.

We denote by P_S the probability of an electron generated near the crystal surface to escape the near-surface damaged area and be injected into the crystal volume. Its probability to recombine with a hole near the very surface is therefore $1 - P_S$. We assume that P_S is proportional the electron drift velocity toward the opposite electrode $\mu_n\mathcal{E}$, where μ_n is the electron mobility and $\mathcal{E}$ is the external electric field. By analogy, we assume that $(1 - P_S)$, the electron probability to recombine near the surface, is proportional to a quantity of velocity dimensions denoted by s_n, called **surface recombination velocity**. The meaning of the analogy assumption is that

$$\frac{P_S}{\mu_n\mathcal{E}} = \frac{1 - P_S}{s_n}. \tag{9.1}$$

Then we get that

$$P_S = \frac{1}{1 + s_n/\mu_n\mathcal{E}}. \tag{9.2}$$

An electron out-drifted from the surface recombination zone may still be trapped and stopped inside the crystal volume. We denote by τ_n the mean lifetime of a drifting electron. Then the probability that the electron will remain free after crossing a distance x (where $x < L$) is given by $\exp(-x/\mu_n\mathcal{E}\tau_n)$. The $\lambda_n \equiv \mu_n\mathcal{E}\tau_n$ quantity is the mean free path of an electron inside the crystal. The probability $P_B(x)$ for $x < L$ that an electron will get trapped at some distance from the electrode that is shorter than x is given by

$$P_B(x) = 1 - \boldsymbol{exp}(-x/\mu_n\mathcal{E}\tau_n). \tag{9.3}$$

An electron probability to reach an L distance, not getting trapped before, is therefore given by $P_B(L) = 1 - \exp(-L/\mu_n\mathcal{E}\tau_n)$. The electron definitely stops on reaching $x = L$. Therefore, $P_B(x > L) = 0$. The probability $dP_B(x)$ for the electron to get stopped between x and $x + dx$ is given by $dP_B(x) = P_B(x + dx) - P_B(x)$. For $x < L$, we get $dP_B(x) = [dP_B(x)/dx] \cdot dx$; for $x = L$, we get

$dP_B(L) = P_B(L)$. And in summary

$$dP_B(x) = \begin{cases} \frac{dx}{\mu_n \mathcal{E} \tau_n} \boldsymbol{exp}\left(-\frac{x}{\mu_n \mathcal{E} \tau_n}\right) & 0 \le x < L \\ \boldsymbol{exp}\left(-\frac{L}{\mu_n \mathcal{E} \tau_n}\right) & x = L \\ 0 & otherwise \end{cases} . \quad (9.4)$$

A reader acquainted with probability theory would notice that the probability distribution is continuous throughout the entire probability space, except for a single point: $x < L$, where the probability assumes a discrete value.

The two processes, surface recombination and inner crystal trapping, are independent. Therefore, the combined probability that a charge carrier generated near the crystal surface will be injected into the bulk, and then trapped between x and $x + dx$ is given by the product $P_S \cdot dP_B(x)$. In the case of a two parallel electrodes geometry, an electron starting to move at the negative electrode and reaching a distance x from it inside the crystal, induces a charge change between the electrodes given by qx/L, where q is the elementary charge. The expected average charge change for a single generated electron is thus given by

$$q_{\text{av}} = \frac{q}{L}\int_0^L x \cdot P_S \cdot dP_B(x). \quad (9.5)$$

We set the above-obtained expressions for P_S and $dP_B(x)$ and get

$$q_{\text{av}} = \frac{q}{L} \cdot \frac{1}{1 + s_n/\mu_n \mathcal{E}} \int_0^L x \cdot dP_B(x) = \frac{q}{1 + s_n/\mu_n \mathcal{E}} \cdot \frac{\mu_n \mathcal{E} \tau_n}{L}\left[1 - \boldsymbol{exp}\left(-\frac{L}{\mu_n \mathcal{E} \tau_n}\right)\right].$$

The efficiency of charge collection at the positive electrode $\eta \equiv q_{\text{av}}/q = Q/Nq$, where Q is the actually induced charge, and Nq is the total charge that would have reached the positive electrode without surface recombination and bulk trapping, where N electron–hole pairs are generated by an absorbed photon. The charge collection efficiency would thus be

$$\eta = \frac{1}{1 + s_n/\mu_n \mathcal{E}} \cdot \frac{\mu_n \mathcal{E} \tau_n}{L}\left[1 - \boldsymbol{exp}\left(-\frac{L}{\mu_n \mathcal{E} \tau_n}\right)\right]. \quad (9.6)$$

If an external voltage of an opposite polarity is applied, a symmetrical equation is obtained, such that the τ_n, s_n, and μ_n quantities are replaced by τ_p, s_p, and μ_p, respectively.

The charge collection efficiency is a steadily increasing function of the external applied electric field $\mathcal{E}$, and it saturates at a unity ($\eta = 1$) value in the limit of very high fields. We now consider the function behavior in two special cases:

(i) Charge carriers trapping in the inner crystal are negligible compared to surface recombination effects. In such case, $L/\tau_n \ll s_n$, and Eq. (9.6) reduces to the form

$$\eta = \frac{1}{1 + s_n/\mu_n \mathcal{E}} \quad \text{for } L/\tau_n \ll s_n. \quad (9.7)$$

Initially, the charge collection efficiency rises linearly as a function of $\mathcal{E}$, then the rise steadily moderates, and it saturates at a unity ($\eta = 1$) value in the limit of very high fields.

The electric field $\mathcal{E}_{hmc}$, where the charge collection efficiency becomes 0.5 (field of half-maximum collection), is given by $\mathcal{E}_{hmc} = s_n/\mu_n$. That value is independent of the detector thickness L.

(ii) Surface recombination effects are negligible compared to charge carriers trapping in the inner crystal. In that case, $L/\tau_n \gg s_n$, and Eq. (9.6) reduces to a form known as **Hecht equation**:

$$\eta = \frac{\mu_n \mathcal{E} \tau_n}{L}\left[1 - \exp\left(-\frac{L}{\mu_n \mathcal{E} \tau_n}\right)\right]. \tag{9.8}$$

Here as well, the charge collection efficiency rises initially linearly as a function of $\mathcal{E}$, then the rise steadily moderates, and it saturates at a unity ($\eta = 1$) value in the limit of very high fields. The electric field $\mathcal{E}_{hmc}$, where the charge collection efficiency becomes 0.5 (field of half-maximum collection), is given by $\mathcal{E}_{hmc} \cong (0.63/\mu_n\tau_n)L$. Compared to case (i), $\mathcal{E}_{hmc}$ does depend on the detector thickness L.

(iii) A general case. It is possible to show that $\mathcal{E}_{hmc}$ is approximately given by

$$\mathcal{E}_{hmc} \cong \frac{s_n}{\mu_n} + \frac{0.63}{\mu_n\tau_n}L. \tag{9.9}$$

If the voltage polarity is reversed, a symmetric equation is obtained, such that the τ_n, s_n, and μ_n quantities are replaced by τ_p, s_p, and μ_p, respectively.

Figure 9.4 shows a graphic description of the charge collection dependence in an X-ray detector as a function of the collecting electric field. It also indicates the size of the field at which half of the generated charge carriers are collected. Equation (9.9) points out a means of measuring the surface recombination velocity s_n and the mean trapping time (lifetime) τ_n of same-type detectors. A series of detectors should be manufactured using the same semiconductor crystal, same surface treatments and electrode deposition, having, however, different thicknesses at a broad range. The collected charge per single absorbed photon is measured as a function of voltage. The $\mathcal{E}_{hmc}$ is determined for each detector and plotted as a function of the detector thickness L. According to Eq. (9.9), the result displays a linear dependence; the line slope allows

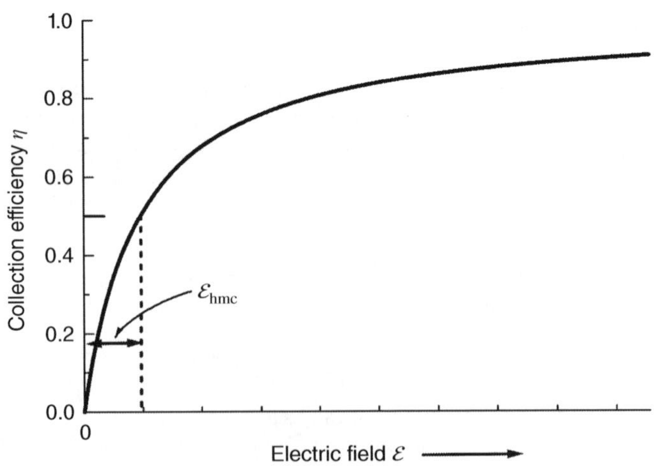

Figure 9.4 Schematic illustration of the charge collection efficiency in an X-ray detector as a function of the applied external electric field between the electrodes.

extraction of the lifetime τ_n, and the intercept on the ordinate (the Y-axis) allows extraction of the recombination velocity s_n.

Notably, a semiconductor carrier lifetime, τ_n specifically, is largely dependent on the material purity and its crystalline perfection. The surface recombination velocity s_n depends strongly on the surface treatment procedures, the surface post-treatment preservation means, and the electrical contacts formations.

To effectively operate a detector requires application of quite high electric fields that would ideally collect all charge carriers generated by an X-ray photon absorption. In reality, there is a practical limit to the field height. Use of voltages exceeding about 2000–3000 V may cause current bursts in the surrounding atmosphere, in the insulating materials used as supports, and sometimes even inside the detector itself. It is thus desirable to achieve maximal charge collection at the lowest possible voltages. For that purpose, $\mathcal{E}_{hmc}$ should be very low. Therefore, lowering the surface recombination velocity s, increasing the carrier mobility μ, and lengthening the carrier trapping times τ, are the goal of device producers. A further goal is to make the detector as thin as possible provided that it is still large compared to the X-ray photon attenuations depth (see Fig. 9.2). In the following, it will appear that other considerations actually imply preference for thicker crystals. A chosen thickness is obviously an optimal compromise between the two contradicting considerations.

To achieve the goal of lowering $\mathcal{E}_{hmc}$, the crystals are grown to a high purity and to a high degree of crystalline perfection, to achieve longest τ_n and τ_p lifetimes. Cutting, polishing, and surface treatment methods are optimized to reduce as much as possible the s_n and s_p charge carrier surface recombination velocities. Exposing the detector to the measured radiation field is chosen such that the photons impinge the electrode for which the crystal mobility–times–lifetime product ($\mu_n\tau_n$ or $\mu_p\tau_p$) is greater. It is important to emphasize that for a radiation field where absorption occurs mostly deep inside the crystal, the surface recombination velocities are not important, but the values of both electron mobility and hole mobility and their lifetimes are very important.

The charge collection efficiency is not the only consideration. The size of a voltage signal δV generated between the electrodes is given by the ratio between the collected charge Q and the detector electrical capacitance C. Therefore, lowering of the detector capacitance would serve to increase the signal. The capacitance is given by

$$C = \frac{\varepsilon_r \varepsilon_0 A}{L}, \tag{9.10}$$

where ε_r is the crystal relative dielectric constant, ε_0 is the permittivity of free space, and A is the electrodes area. Lowering the capacitance is achieved by reducing the electrodes area A and increasing the detector thickness L. Another consideration justifying an increase of the detector thickness is to allow absorption of all impinging photons and minimizing escape of secondary photons generated in the absorption process, as described is Section 9.1.2. A chosen thickness should obviously be an optimal compromise between the contradicting considerations.

9.1.4 Various Technological Considerations

Table 9.1 presents a list of several semiconductors used in the industry to design electronic X-ray and gamma ray radiation detectors, and their properties related to that purpose. Silicon (*Si*) and germanium (*Ge*) exhibit a clear advantage by the high values of their charge carrier mobility, yet a disadvantage by having lower atomic numbers and small energy gaps. Due to the latter property, silicon and germanium detectors could be operated only at very low temperatures (77 K – the

Table 9.1 A list of several semiconductors used in designing of electronic X-ray and gamma ray detectors, and their basic properties.

Material	Chemical symbol	Atomic Z number	Energy gap E_g (eV)	Average energy for e–h pair creation (eV)	$\mu_n \left(\frac{cm^2}{V \cdot s}\right)$	$\mu_p \left(\frac{cm^2}{V \cdot s}\right)$
Silicon	*Si*	14	1.16 (77 K)	3.61	1400	500
Germanium	*Ge*	32	0.74 (77 K)	2.98	3900	1900
Cadmium telluride	*CdTe*	48; 52	1.47 (300 K)	4.43	1000	80
Mercuric iodide	HgI_2	80;53	2.13 (300 K)	4.2	100	4

boiling temperature of liquid nitrogen). Cadmium telluride (*CdTe*) and mercuric iodide (HgI_2) exhibit an obvious advantage by having high atomic numbers of their constituent atoms and higher band gaps, which allow their operation at room temperature. On the other hand, their charge carrier mobilities are considerably smaller. In addition, the carrier lifetimes obtained for cadmium telluride are very short, of the order of 10–200 ns for both electrons and holes. Mercuric iodide exhibits a large asymmetry between electron and hole mobilities. Achievable electron lifetimes are quite long – 100 μs – 1 ms, but achievable hole lifetimes are shorter, only about 1–5 μs. That fact accentuates the asymmetry between electron and hole relevant properties. Advantages of mercuric iodide as a radiation detector are thus limited to radiation fields of 5–50 keV; therefore, the negative electrode should always be oriented to face the radiation field.

An additional technologic consideration relates to the selection of the resistor R connected in series (Fig. 9.1). On the one hand, its resistance should be made sufficiently large to avoid an early discharge to ground of δV before reaching a maximal value. A capacitor discharge is given by the RC product given by $RC = R\frac{\varepsilon_r\varepsilon_0 A}{L}$. The charging duration is approximately given by the transit time of the slowest charge carrier between the electrodes, given by $t_t = L/\mu\mathcal{E}$, where μ is the mobility of the "slow" carrier. For the present consideration, the "slow" carrier is the one for which the product between the mobility and trapping time (lifetime) is smallest. For demonstration, we consider a gamma ray of photon energy considerably greater than about 10 keV, which is absorbed mainly far from the front electrode. In such case, the surface recombination effect may be neglected. When the external electric field is made about 5 times greater than $\mathcal{E}_{hmc}$, most carriers will be collected. To allow sufficiently long signal charging time, we get the condition $R \gg \frac{L\tau}{5\varepsilon_r\varepsilon_0 A}$, where τ is the trapping time of the slow carrier. On the other hand, it is necessary to provide the capacitor sufficient time to discharge of the δV contribution during an average time $t_{\gamma-\gamma}$ between successive absorptions of two gamma photons (the inverse of the photon counting rate). Then we get the condition $R \ll \frac{Lt_{\gamma-\gamma}}{5\varepsilon_r\varepsilon_0 A}$. These two conditions may be satisfied only for a radiation field and detector area such that the average time $t_{\gamma-\gamma}$ between successive absorption of two gamma photons is at least 100–1000 times longer than the trapping time in the detector "slow" charge carrier.

The detector δV signals appearing across the resistor R are amplified and registered by an electronic device called **multichannel analyzer**. The device sorts the signals by their height, and presents a plot of the number of counts obtained per each signal-height segment, called "channel". Each channel belongs to a different energy given to the detector by an absorbed photon. In that way, an image of the radiation field energy spectrum is obtained.

As a conclusion of the present section, in Figure 9.5, we present an example of a radiation spectrum measured using a mercuric iodide HgI_2 detector, and the radiation source is the radioactive ^{241}Am element (the 241 atomic mass americium element isotope of $Z = 95$ atomic number). The americium disintegrates by a nuclear process that involves the emission of a gamma photon of 59.6 keV energy. That energy is expressed in the detector signals δV spectrum by a high peak at a narrow range in the high signal channels (high-energy channels). However, the spectrum also exhibits additional peaks belonging to 27.1 keV and 31.0 keV. They relate to deficiencies of the $K_\alpha = 28.1$ keV and $K_\beta = 32.5$ keV transitions in the iodine atom, whose emitted photons escaped out of the detector volume.

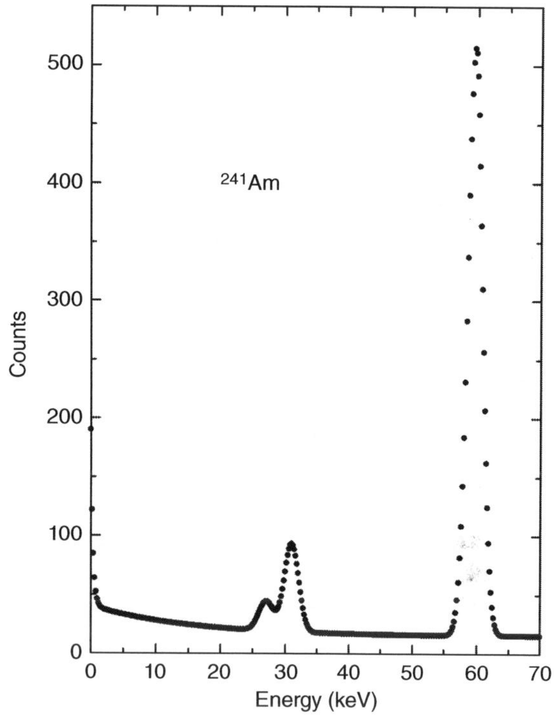

Figure 9.5 An example of a counting versus channel energy spectrum obtained by using a thin mercuric iodide detector exposed to the radiation of ^{241}Am (the artificial radioactive 241 americium element of Z = 95). Registered counts at the very low-energy channels emanate from electronic noise in the detector and in the various amplifiers. For more details, see the text.

9.2 Light Detectors

In this section, we describe the use of semiconductors for the design of light detectors, basically meaning the 0.2–14 μm spectral range; in other words, the ultraviolet in the short wavelengths side, and far infrared on the long wavelengths side. Semiconductors-based light detectors rely on the fact that absorbed light generates free charge carriers in the form of electron–hole pairs. There are three kinds: one kind is **photovoltaic detectors**. Such detector is in fact a p–n junction (Section 7.1). Electron–hole pairs generated near the junction by the absorbed light diffuse into the junction, where they separate into free electrons and free holes due to the very high electric field existing between the p- and n-regions. That separation generates a voltage change signal between the electrodes. Such device doesn't require application of an external electric field. A second kind is **p.i.n** detectors. It is in fact a p–n junction whose transition zone (the "**i**" zone, acronym of the word "insulator") is especially thick. In such structure, most light absorption takes place in the "**i**" region, and the device is under a relatively large voltage at a reverse polarity. The charge carrier couples separate under the electric field and are then collected at the electrodes, forming a voltage change. The light detection is basically similar to the mechanism described for the X-ray and gamma ray detection, yet is different by the fact that the involved photon energies are never considerably larger than the semiconductor energy gap, and any photon absorption generates only a single electron–hole pair. In Section 7.5, we have briefly indicated the possibility of using reverse-biased p–n junctions as light detectors, where absorption in the very junction is negligibly small. The size of such systems response is indeed quite similar, but the response is slow, and it basically expresses the longer lifetime of the excess charge carriers. A third kind are **photo-conducting detectors**. Such detector is based on a uniform semiconductor, whose conductivity changes under light absorption. The detection is based on measuring the change in the specific conductivity. In the present section, we concentrate on discussion of the latter kind of light detectors.

For demonstration we describe in great detail the application of the photoconductivity detection in construction of a mercury cadmium telluride (MCT) (*HgCdTe*) ternary compound detector, where one of its compositions is suitable to detect far-infrared radiation in the 8–14 μm wavelengths range. That spectral range is called "second atmospheric window", because absorption by atmospheric gasses in that particular region is minimal. Radiation detection in that range is important, because it allows "night viewing" of bodies emitting thermal radiation in the vicinity of a room temperature ambiance. An optimal energy gap of a material suitable for that purpose is $E_g \cong 0.09$–0.1 eV. In such low-energy photons, a considerable current in the dark, as well as large thermal noise, occur at room temperature. To reduce these background effects, the detectors are operated at liquid air temperature (77 K).

9.2.1 Light Detection by Photoconductivity

Consider as an example a semiconductor-based detector machined to form a square plane-parallel plate of side length L and thickness d. Electrical contacts are soldered on two opposite sides, and a constant current I is made to flow through. When the detector is exposed to illumination, its resistivity lowers, and simultaneously the voltage between the contacts reduces. The detector signal is the said voltage reduction δV. Figure 9.6 describes the detector geometric shape and the time dependence of the signal when the detector is exposed to a constant illumination for a limited time duration (Section 6.1 and Fig. 6.1). As demonstrated in the figure, the detector exposure duration to light has to be longer than the lifetime τ of excess charge carriers. The latter is a most important detector property. Such detectors are characterized by several specific technological parameters as defined and listed in Table 9.2, to be later discussed in detail.

1) The cutoff wavelength λ_{co} is the wavelength indicating the transition of the detector material from optical transparency to absorption of incidenting light, and correspondingly also the transition from no response for light whose wavelength is longer than λ_{co}, to detection by absorption of light whose wavelength is shorter than λ_{co}. Obviously, the cutoff wavelength represents the semiconductor energy gap E_g by the relation $E_g = hc/\lambda_{co}$, where h is Planck's constant and c is the vacuum speed of light constant.
2) The response time τ is the characteristic time duration required for the δV voltage signal to reach the final value after the illumination onset, and for sufficiently small signals – to decay to zero after the illumination is turned off (see Section 6.1). Practically, response times are often considerably longer than the intrinsic lifetimes of the detector excess charge carriers due to additional response times related to the connected electronic amplifiers.
3) The dark resistance r is the detector resistance in the dark (under no illumination), and its units is Ω. In the special geometric construction presented in Figure 9.6, where the contact lengths are the same as their distance, that resistance is given by $r = \rho/d$, where ρ is the

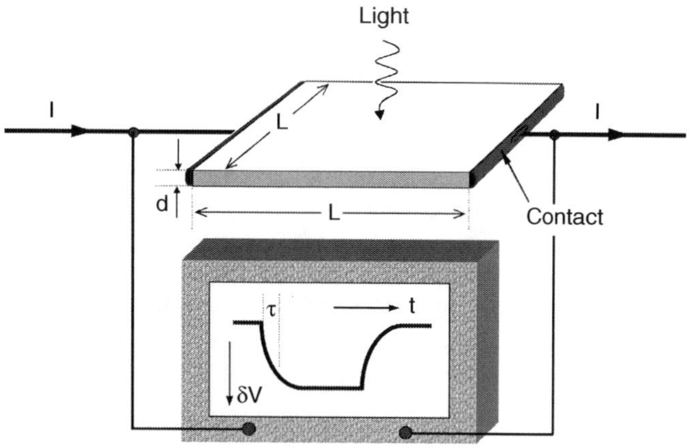

Figure 9.6 The shape and operation principle of a semiconductor light detector. The detector electrical resistance changes under illumination, and simultaneously the voltage between the contacts also changes. The time dependence of that change is described inside a rectangle, as it appears on an oscilloscope screen, when that instrument is used to record the signal.

Table 9.2 A list of characterizing properties of light detectors and their units commonly used units in technology.

Characterizing property	Symbol	Common units
Cutoff wavelength	λ_{co}	μm
Response time	τ	μs
Dark resistance	r	Ω/square
Responsivity	$\mathscr{R}$	V/W
Specific detectivity	D^*	$\text{cmHz}^{1/2}\,\text{W}^{-1}$

semiconductor resistivity. The resistance then depends only on the plate thickness, and is independent of the distance between the contacts. For that reason, the resistance units are often written in the form Ω/square, or Ω/□, as a reminder of the detector plus electrical contacts geometric construction.

4) The responsivity $\mathscr{R}$ is the ratio between the signal height δV and the power P of the light that strikes the detector face:

$$\mathscr{R} \equiv \frac{|\delta V|}{P}. \tag{9.11}$$

Note that this particular ratio appears in the definition of the specific detectivity!

5) The specific detectivity D^* is the "signal-to-noise ratio" corrected by dividing the signal by the incident light power, and by dividing the noise V_n by the square root of the product between the detector area A and the amplifier frequency band Δf. (A discussion concerning the detector noise origin is provided in an example we present shortly). Formally,

$$D^* \equiv \frac{|\delta V|/P}{V_n/(A \cdot \Delta f)^{1/2}}. \tag{9.12}$$

The justification for the above definition is that the signal is always proportional to the power of light striking the detector, and the noise is always proportional to the square root of its area and to the square root of the amplifier frequency response bandwidth. Therefore, the specific detectivity is independent of neither the incident light power nor the detector area and the amplifier frequency response bandwidth, and it characterizes the detector intrinsic properties. In most cases, users are interested in commercial detectors of highest specific detectivity.

9.2.2 Mercury Cadmium Telluride Detectors for the 8–14 μm Range

For demonstration, we delve on the properties of an *HgCdTe* detector at a composition designed to fit the detection of far-infrared radiation in the 8–14 μm wavelength range. We will later see that all characterizing properties listed in Table 9.2 may be calculated on the basis of the material energy gap E_g, the electron mobility μ_n, and their density n_0. The energy gap would be determined by the measured specific gravity of the material; the mobility and carrier density would be obtained via measuring the Hall effect and the specific resistivity (see Sections 3.7.2 and 5.2).

9.2.2.1 Determination of the Cutoff Wavelength

The mercury cadmium telluride semiconductor is a ternary compound whose mercury and cadmium molar fractions are changeable as long as they sum to a unity; in other words, it is a continuous series of compositions $Hg_{1-x}Cd_xTe$, where each exhibits a different energy gap. It may also be regarded as a solid solution of *CdTe* and *HgTe*, namely $Hg_{1-x}Cd_xTe \equiv (HgTe)_{1-x}(CdTe)_x$. The first compound, *CdTe*, is a semiconductor of 1.47 eV energy gap; the second compound, *HgTe*, is a **semimetal**, namely a semiconductor having a negative energy gap of −0.2 eV. The meaning of a "negative energy gap" is that the conduction band bottom is situated below the top of the valence band (by 0.2 eV in the case of mercuric telluride). In the course of cadmium molar content *x* changing from zero to 1, the energy gap changes continuously from −0.2 eV up to +1.47 eV. The dependence of the cadmium molar content exhibits a near-linear dependence. Figure 9.7 displays the said dependence in the $x = 0.1–0.3$ range at the liquid nitrogen temperature. The crystal unit cell size remains quite unchanged through that range. Because the mercury atom is considerably heavier than the cadmium atom (atomic weights 200.6 compared to 112.4, respectively), the compound-specific gravity (density) turns smaller with increasing cadmium content. Accordingly, the compound density is provided as a scale on the top X-axis of the graph. The cutoff wavelength is provided as a scale on the right ordinate (Y-axis) of the graph. The reason for providing these data is to emphasize that it is practically possible to determine the material cutoff wavelength by a simple measurement of its specific gravity.

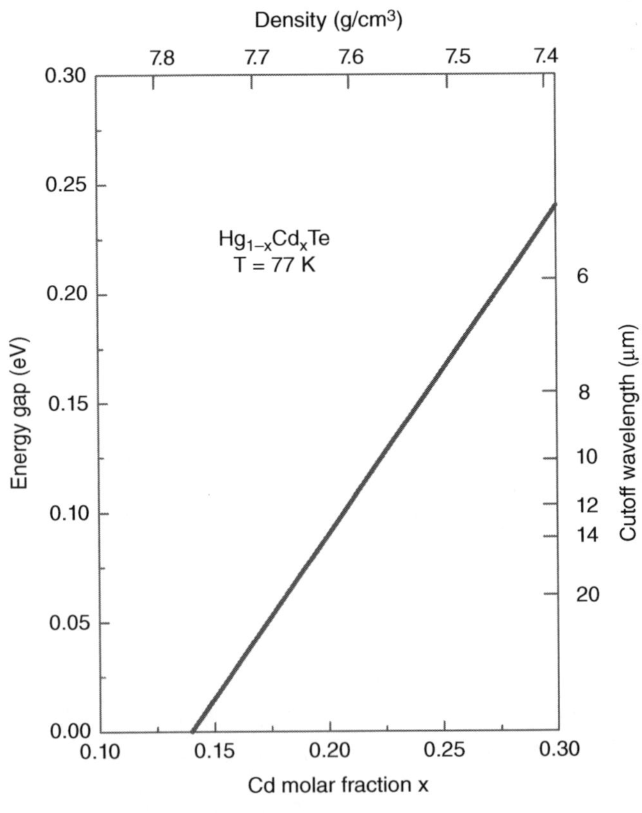

Figure 9.7 The mercuric cadmium telluride energy gap and the optical cutoff wavelength functions of the cadmium molar fraction and material density.

Table 9.3 A list of basic physical properties of the $Hg_{0.8}Cd_{0.2}Te$ semiconductor at a 77 K temperature. m_0 indicates the rest mass of the free electron in vacuum.

Energy gap	$E_g = 0.09\,\text{eV}$
Cutoff wavelength	$\lambda_{co} = 14\,\mu\text{m}$
Electron mobility	$\mu_n = 10^5\,\text{cm}^2/\text{V}\cdot\text{s}$
Hole mobility	$\mu_p = 700\,\text{cm}^2/\text{V}\cdot\text{s}$
Effective electron mass	$m_n = 7\times10^{-3}m_0$
Effective hole mass	$m_p = 0.55m_0$
Relative dielectric constant in the optical range	$\varepsilon_r \cong 14$
Effective conduction band states density	$N_C = 1.9\times10^{15}\,\text{cm}^{-3}$
Effective valence band states density	$N_V = 1.33\times10^{18}\,\text{cm}^{-3}$
Intrinsic concentration of electrons and holes	$n_i = 5.7\times10^{13}\,\text{cm}^{-3}$

By observing Figure 9.7, it appears obvious that the molar fraction of cadmium in MCT should be chosen as $x = 0.2$ to detect light radiation in the 8–14 μm wavelengths range, and never be larger than $x = 0.24$. Table 9.3 provides a list of relevant physical constants of MCT for $x = 0.2$. Except for changes in the energy gap as seen in Figure 9.7, the other properties remain quite similar throughout the entire $0.2 < x < 0.24$ range.

9.2.2.2 Response Time Determination

Under conditions where the capacitance among connection wires is sufficiently small, the measuring system response time will be determined by the intrinsic response time of the detector proper. The response time will thus be identical with the charge carriers' lifetime. That is the case when the detector is sufficiently long such that drift to outside the detector is negligible (see Section 6.5.3.3 case (iii) and also the continuation of the present section). In MCT crystals with an energy gap of about $E_g = 0.1$ eV, which are sufficiently pure and of a high crystalline structure, the decay of excess charge carriers is dominated by an **Auger** process (Section 6.2.3). Theoretically, the lifetime determined by that process is given by

$$\tau \cong S\left(n_i^2/n_0^2\right)(E_g/k_BT)^{3/2}\boldsymbol{exp}(E_g/k_BT), \tag{9.13}$$

where S is a constant of time dimensions, whose size is, among other things, a function of the conduction and valence bands structure. Development of Eq. (9.13) is not within the scope of the present book, and we use it as given. Anyhow, the actual calculation is quite complex and provides only qualitative agreements with experimental results. In an n-type crystal, the free electron density n_0 is determined by the concentration of donor impurities, which is a constant property of each specific crystal. We set in Eq. (9.13) the law of mass action $n_i^2 = n_0p_0 = N_CN_V\,\boldsymbol{exp}(-E_g/k_BT)$ (see Eq. [5.8]), recall that a detector operates at a 77 K temperature, and get $\tau = BE_g^{3/2}n_0^{-2}$, where B is a constant. Experimentally in MCT, $B = 4.45\times10^{25}\,\text{s}\cdot\text{cm}^{-6}\,(\text{eV})^{-3/2}$. We may then design the following technical expression for the carriers' lifetime:

$$\tau[s] \cong 4.45\times10^{25}E_g^{3/2}[\text{eV}]\cdot n_0^{-2}[\text{cm}^{-3}]. \tag{9.14}$$

Notably, the Auger-dominated recombination lifetime is inversely proportional to the squared electron density. It is intuitively understandable, because increase of the electron density increases the probability for an electron to give-up it energy by an Auger process, as described in Section 6.2.3. Electron densities in MCT detectors vary between 5×10^{14} cm^{-3} and 10^{16} cm^{-3}, and the energy gaps vary in the 0.08–0.12 eV range. To demonstrate the relation expressed in Eq. (9.14), in Figure 9.8, we provide full logarithmic plots of the excess electron lifetime as a function of the electron density for several energy gaps. Also note that in a part of the considered electron concentrations, $n_0 \gg N_C$, namely the crystal is then actually degenerate!

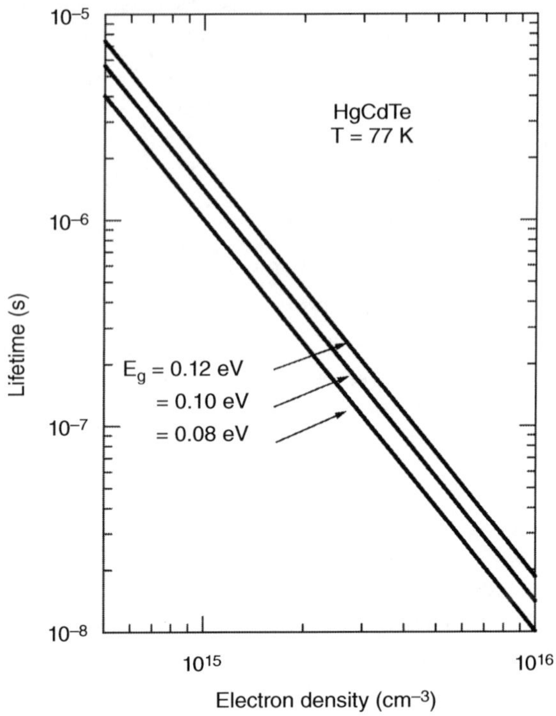

Figure 9.8 Log–log plots of Auger-dominated excess electron lifetime in MCT crystals as functions of the electron density at the 77 K liquid-air temperature for several relevant energy gaps.

Figure 9.8 shows that excess electron lifetimes range between several microseconds at low electron densities and several dozen nanoseconds at high electron densities. To obtain large response signals, we naturally prefer long lifetimes; on the other hand, to achieve a fast responding detector, a short lifetime is preferable. Our following analyses are aimed at finding how to design a detector of highest possible response signals!

Detectors manufactured to have small dimensions and operate by carrying high currents, exhibit a problem by the fact that illumination-created excess electrons drift to outside the crystal into the electrical contacts. The effective response time turns shorter, because it is dominated by the transit time of excess electrons between the detector edges. That time duration, denoted by t_t is given by $t_t = L^2/\mu_p V$ (see Section 6.5.3.6), where μ_p is the hole (minority carrier) mobility, and V is the voltage between the contacts. The voltage obviously depends on the operating current I and the dark resistance.

Other detector properties, namely the specific detectivity D^*, the responsivity $\mathscr{R}$, and the dark resistance r also depend on the detector thickness d, and they grow when the thickness is made smaller. The goal is to have a detector of maximal specific detectivity and responsivity. As we discuss later, there is no advantage in reducing the detector thickness below the excess carrier diffusion range, because the effective lifetime then shortens due to enhanced surface recombination. Our next analysis is thus devoted to calculate the diffusion range of excess charge carriers.

9.2.2.3 Diffusion Range and Optimal Detector Thickness

Let ℓ_D be the diffusion range of excess charge carriers generated in the detector. It is given by

$$\ell_D = (D\tau)^{1/2}, \tag{9.15}$$

where D is the ambipolar diffusivity. Under weak illumination, the ambipolar diffusivity is dominated by the conduction properties of minority carriers, namely holes in our case (see Eq. [6.38] and the related text). To obtain a technical expression for ℓ_D, we set Einstein's relation $D = (k_B T/q)\mu_p$ (Eq. [6.19]), the numerical values of the hole mobility (Table 9.3), the electronic elementary charge, the Boltzmann constant, and the temperature, and get

$$\ell_D[\text{cm}] = 1.55 \times 10^{13} E_g{}^{3/4}[\text{eV}] \cdot n_0{}^{-1}[\text{cm}^{-3}]. \quad (9.16)$$

In Figure 9.9, we present the numerical value of the diffusion range as a function of the electron density for several values of the MCT energy gap. We have previously indicated that the responsivity grows for smaller detector thicknesses; however, when the thickness approaches the size of the diffusion range, excess charge carriers reach the detector surfaces, becoming under the influence of larger recombination processes. The effective lifetime shortens, thus reducing the responsivity. The optimal detector thickness is achieved at approximately three times the diffusion range expressed by Eq. (9.16). Later presented technical equations are based on the assumption that the detector is machined in the form of a square plane-parallel plate of thickness $d_{opt} = 3\ell_D$.

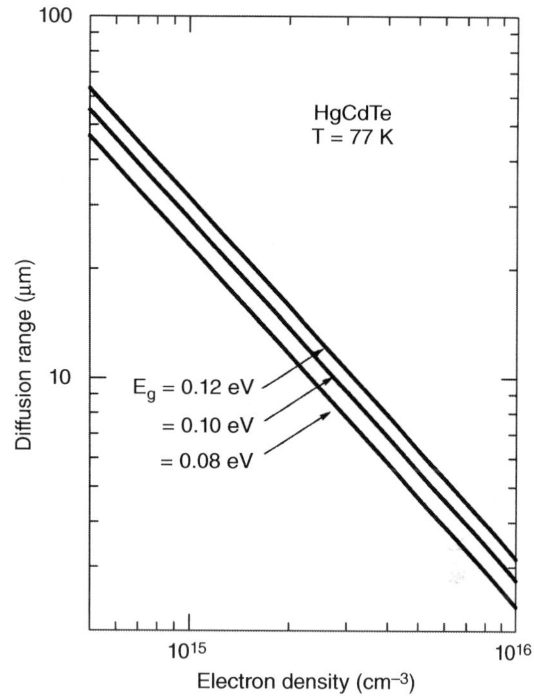

Figure 9.9 Log–log plots of excess carrier diffusion range in n-type MCT crystals as functions of the electron density at the 77 K liquid-air temperature according to Eq. (9.19) for several relevant energy gaps. An optimal detector thickness is chosen as 3 times the diffusion range.

9.2.2.4 Calculation of a Detector Dark Resistance

The detector dark resistance in the particular detector shape shown in Figure 9.6, namely when the length L of each electrical contact is the same as the distance between the contacts, is independent of L; it is given by $r = 1/\sigma d$, where $\sigma = q\mu_n n_0$ is the specific detector conductivity. We neglect the hole contribution to the conductivity, because the crystals are n-type and because the hole mobility is tiny compared to the electron mobility (see Table 9.3). Based on the previous paragraph, we set $d = d_{opt} = 3\ell_D$ and the numerical values of the elementary charge and the electron mobility, and get the following technical expression for the detector dark resistance:

$$r[\Omega/\text{square}] = 1.35 E_g{}^{3/2}[\text{eV}] \quad (9.17)$$

It appears as if the detector dark resistance is independent of the electron density; it turns out so because per each electron concentration of the detector material, the designed detector is assigned a different (optimal) thickness. In the energy gaps range of our present interest, there is only a small change of the dark resistance: the dark resistance of a detector of optimal thickness varies between 8.9 Ω/square for $E_g = 0.08$ eV and 6.6 Ω/square for $E_g = 0.12$ eV.

9.2.2.5 Calculation of a Detector Responsivity

We consider a detector carrying a steady current I, resulting in a steady voltage $V = Ir$ between its contacts, where $r = 1/\sigma d$. The resistance changes under illumination, generating a voltage signal $\delta V = I \cdot \delta r$. If the generated excess charge carriers are evenly distributed throughout the detector, we get $\delta r = -(1/\sigma^2 d)\delta\sigma = -r(\delta\sigma/\sigma) = -r(\delta n/n_0)$. Usually, the generated excess charge carriers are not evenly distributed. However, we would always get that $\delta r = -r(\delta N/N_0)$, where δN is the sum of added charge carriers, and $N_0 = L^2 d n_0$ is the total number of charge carriers inside the detector. We now limit our consideration to a case of a monochromatic light beam of λ wavelength. We denote by F the flux density of photons impinging the detector surface. The units of that quantity are photons per unit area per unit time. Several lifetimes after the start of illumination, the sum of added charge carriers is given by $\delta N = \eta L^2 F\tau$, where η is the quantum efficiency of an electron/hole pair generation by a single photon incidence on the detector surface. We further note the relation $F = \frac{P}{L^2(hc/\lambda)} = \frac{P}{L^2 E_g}\left(\frac{\lambda}{\lambda_{co}}\right)$, where P is the power of the light beam impinging perpendicularly to the detector surface. We set it in the expression for δV and obtain the responsivity expression as

$$\mathcal{R} = \begin{cases} \eta \dfrac{Ir\tau}{E_g L^2 d n_0}\left(\dfrac{\lambda}{\lambda_{co}}\right) = \eta \dfrac{Ir\tau}{E_g N_0}\left(\dfrac{\lambda}{\lambda_{co}}\right) & \text{for } \lambda \le \lambda_{co} \\ 0 & \text{for } \lambda > \lambda_{co} \end{cases} . \tag{9.18}$$

Equation (9.18) is general, valid for any n-type crystal made detector. Setting the above-developed technical expressions for the lifetime, dark resistance, and optimal detector thickness, we get the following technical expression for the MCT detector responsivity:

$$\mathcal{R}[\text{V/W}] = \eta \frac{8.1 \times 10^{30} I[A]}{E_g[\text{eV}] L^2[\text{cm}] n_0{}^2[\text{cm}^{-3}]} \times (\lambda/\lambda_{co}). \tag{9.19}$$

Figure 9.10 presents the numerical responsivity value as a function of the electron density for an MCT square detector whose two side lengths are each $L = 1.0$ mm, and it operates under an $I = 10$ mA current, illumination at the very cutoff wavelength $\lambda = \lambda_{co}$, and the excess charge carrier generation efficiency being $\eta = 100\%$, for several relevant values of the energy gap. Seemingly, it would be possible to increase the responsivity to reach any desired value by just increasing the operating current I (Eqs. [9.18] and [9.19]). However, the need to dissipate the heat thus generated limits the possible

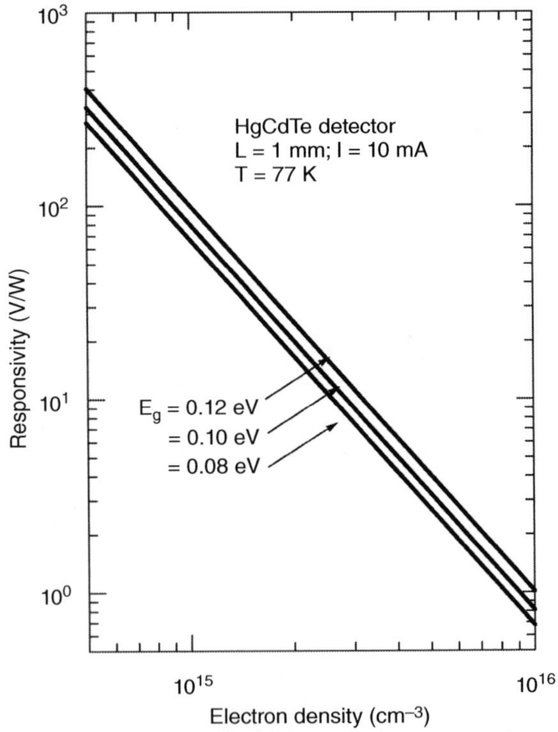

Figure 9.10 Log–log plots of the responsivity as a function of the electron density of a square-shaped MCT detector of optimal thickness and side edge 1.0 mm long operating under a 10 mA current, illumination at the cutoff light frequency, for several relevant energy gaps.

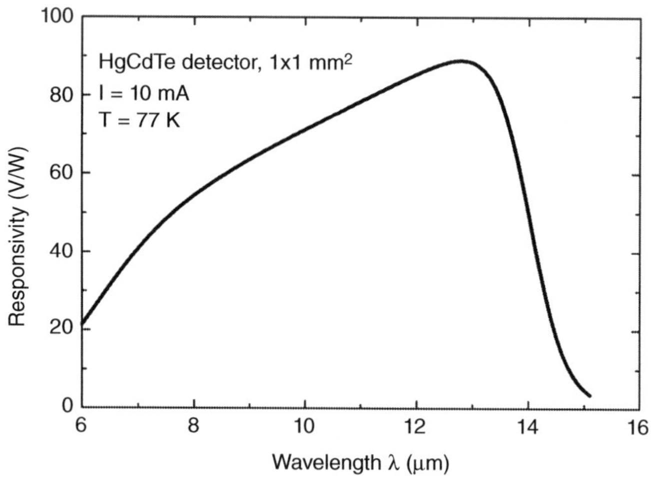

Figure 9.11 Responsivity of a typical commercial MCT detector operating under a 10 mA current as a function of illuminating beam wavelength.

used current; an additional limit relates to the need to limit the drift of excess charge-carriers to outside the detector through the contacts, as already indicated in Section 9.2.2.2.

A further important insight is obtained by considering the responsivity wavelength dependence. Equation (9.17) suggests that the responsivity grows linearly with wavelength till the cutoff wavelength, and then turns zero. In reality, the dropping to zero is not steep, and its width in MCT detectors is about 0.5 μm. The reason for that width is related to the involvement of phonons in the absorption process, but we won't delve into that issue here. In addition, there are a number of other reasons causing the wavelength dependence not to follow an exact linearity below the cutoff wavelength. The most important reasons are wavelength dependence of the optical reflectivity and of the quantum efficiency for the charge carrier generation. Most commercial detectors are sometimes coated with a thin dielectric layer that reduces the detector surface optical reflection (which also serves an important role in preserving the surface electrical properties). Such layers minimize the optical reflection only over a limited range around the cutoff wavelength, and their optical transmission wavelength dependence obviously expresses itself via the detector responsivity versus wavelength dependence. In Figure 9.11, we present a responsivity wavelength dependence of a typical commercial MCT detector.

9.2.2.6 Calculation of a Detector-Specific Detectivity

The specific detectivity is defined as the ratio between the detector responsivity and its corrected electronic noise (Eq. [9.12]). The responsivity calculation is presented in the previous section. As to the electronic noise, we start by emphasizing the way it expresses itself in the signal measurements. In contrast to the smooth form of the signal trace shown in Figure 9.6, Figure 9.12 demonstrates a form of a real trace that includes a sum of the detector response to a square light pulse and the fast noise-originated changes in the voltage V_n. For analysis, in the present section, we assume that the entire noise originates within the detector itself. The calculations follow a quite simplistic model, which despite its simplicity provides an excellent agreement with experimental results.

We first select the amplifier frequency band to optimize the observation of the voltage signal. Determination of the signal size is made when the voltage change between the electrical contacts

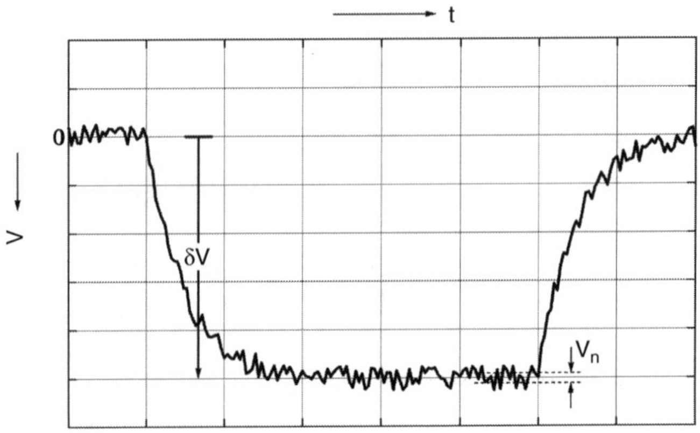

Figure 9.12 An example of the shape of a real voltage signal versus time trace measured between electrical contacts of a detector using an oscilloscope, when the detector surface is illuminated for a short limited time. δV is the voltage signal, and V_n is the average amplitude of voltage changes caused by the electronic noise.

reaches saturation, namely a time duration of $3\tau - 5\tau$ after the illumination onset. For that reason, the electrical amplifiers response must be at least τ or faster to observe the reaching of the signal saturation. They, however, should not be much faster, otherwise they won't form averaging of a high-frequency noise. By that reasoning, the upper limit of the frequency band should be $1/\tau$. On the other hand, the measuring device should be able to trace the signal sufficiently long to observe the signal achieving its saturated value. By that reasoning, the lower limit of the frequency band should be close to zero. We thus conclude that the amplifier frequency band should be $\Delta f \cong 1/\tau$.

We assume that the free electron population in the detector emanate from ionization of $2N_0$ total shallow donor impurities, and that the electrons exhibit the same probability to occupy the conduction band or be trapped by the donor. That assumption can by justified by the quite small value of the MCT effective conduction band states density (Table 9.3), which is of the same order as the free electron density in crystals used for detectors manufacturing. We should further recall that the lifetime τ describes the time required for the system to reach thermal equilibrium after any deviation from such state (Section 6.1). We thus imagine that during a τ time duration, there are $2N_0$ discrete attempts of electrons to choose one out of two options: being free (mobile) in the conduction band, or being trapped (immobile) in a donor impurity. On the average, there will be a total N_0 electrons in a conducting state. By a simple Bernoulli probability distribution,[1] the variance in the sum of free electrons is $\Delta N = \sqrt{\frac{1}{2}N_0}$. That size of deviation is the dominant factor causing the noise in the detector: on increase of carriers number, the voltage between the electrodes lowers and vice versa. The voltage noise will thus be given by $Ir(\Delta N/N_0)$, and the specific detectivity would then be given by

$$D^* = \frac{\eta \dfrac{Ir\tau}{E_g L^2 d n_0}\left(\dfrac{\lambda}{\lambda_{co}}\right)}{Ir(\Delta N/N_0)/L\sqrt{1/\tau}};$$

1 The *Bernoulli* probability distribution is the distribution of N attempts of "success" probability p, and "failure" probability $q = 1 - p$. The "success" mean value is $\overline{N} = pN$, and its variance is $\Delta N = \sqrt{Npq}$.

namely

$$D^* = \eta\frac{(2\tau/dn_0)^{1/2}}{E_g}\left(\frac{\lambda}{\lambda_{co}}\right). \tag{9.20}$$

Equation (9.20) is general, valid for any n-type crystal made detector. Setting the above-developed technical expressions for the lifetime and optimal detector thickness, we get the following technical expression for the MCT detector-specific detectivity when $\eta = 100\%$:

$$D^*[\text{cm}\cdot\text{Hz}^{1/2}\text{W}^{-1}] = 8.56\times10^{24}E_g{}^{-5/8}[\text{eV}]n_0{}^{-1}[\text{cm}^{-3}]\cdot\left(\frac{\lambda}{\lambda_{co}}\right). \tag{9.21}$$

Figure 9.13 provides graphs of the specific detectivity as a function of the electron density in a square-shaped MCT detector under illumination at the cutoff wavelength when the absorption and charge carrier generation efficiencies are 100%, for several relevant energy gaps.

Analyses of the obtained equations and figures yield that in order to design an MCT detector whose responsivity and specific detectivity are highest, it has to be prepared with a lowest possible electron density, a thickness optimized for that density, and signal amplifiers whose response time is the same as the excess charge carriers lifetime. Still, detectors of higher electron densities would be preferable in cases where details of very short light pulses are a goal of the measurement.

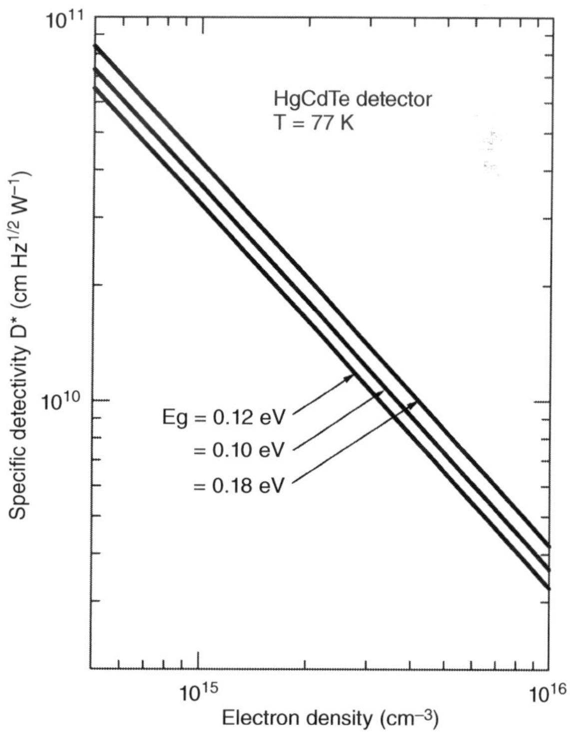

Figure 9.13 Log–log plot of the specific detectivity as a function of the electron density of a square-shaped MCT detector of optimal thickness, for illumination at the cutoff light wavelength λco, for several relevant energy gaps.

Exercises for Chapter 9

9.1 (a) A gamma and X-rays detector is constructed of a 0.8-mm-thick crystal plate whose faces area is 1.35 cm^2. Thin electrical contacts are deposited over the faces, and the voltage applied between them is sufficiently high to collect any generated excess charge carrier at the proper electrode. The crystal relative dielectric constant is 8.8. (i) Calculate the electrical capacitance of the detector. (ii) Calculate the voltage change signal between the electrodes that follows absorption of a gamma photon of 511 keV energy; the absorbed energy conversion constant into electron–hole pairs is 4.4 eV/pair.

(b) The surface recombination velocity of electrons in the crystal is 2.5×10^4 cm/s. What is the collection efficiency, and what would be the voltage change signals for a radiation of 8.6 keV incident on the crystal from the negative electrode side under a collecting voltage of 50 V? The electron trapping time is assumed to be very long. The electron mobility is $\mu_n = 80\,\text{cm}^2/\text{V}\cdot\text{s}$. Calculate the needed collecting voltages to achieve 50% collection and to achieve a 95% collection.

(c) Calculate the charge collection efficiency and the size of voltage change signals under conditions of the previous exercise, except that the average electron trapping time is $\tau_n = 3.0\,\mu\text{s}$. Calculate the approximate voltage for which the collection efficiency becomes 50% and the voltage for which the collection efficiency grows to 95%.

(d) Calculate the collection efficiency and the size of voltage change signals if the external voltage polarity is reversed (namely a 50 V external voltage is applied such that the irradiated electrode is positive). It is given that $\mu_p = 28\,\text{cm}^2/\text{V}\cdot\text{s}$, $\tau_p = 0.8\,\mu\text{s}$, and $s_p = 1.6 \times 10^5$ cm/s.

(e) Draw the detector electrical circuit, and choose the series resistor over which the voltage change signals would be measured, and the collection voltage required for measuring gamma radiation exhibiting photons of energy greater than 100 keV. What would be the maximal possible photon counting rate for such detector.

9.2 (a) The energy gap of a certain semiconductor is 0.22 eV. Calculate the cutoff wavelength for charge carrier generation in that material.

(b) The above-said material is n-type exhibiting an electron mobility of 16,500 $\text{cm}^2/\text{V}\cdot\text{s}$. Calculate the electron density in a case where its specific conductivity is $25\,\Omega^{-1}\,\text{cm}^{-1}$.

(c) The above-said material is machined in the form of a square plane-parallel plate, whose side length is 5.0 mm and its thickness is 0.8 mm, to be used as light detector. Calculate that detector dark resistance.

(d) The lifetime of excess charge carriers in the detector is 1.8 μs. Calculate the detector-specific detectivity, and the responsivity for light at the material cutoff wavelength under a steady current of 50 mA. What would be the responsivity for illumination with light whose wavelength is half the cutoff wavelength?

(e) The hole mobility in the said material is 1300 $\text{cm}^2/\text{V}\cdot\text{s}$. Calculate the transition time of an electron–hole pair generated at the detector center to reach an electrical contact. Consider whether it is required to correct the responsivity calculation performed in (d) exercise. If you decide positively, correct the calculation accordingly.

(f) What would be the signal size generated by a light beam of cutoff wavelength of 0.1 μW power incident on the detector front face? What is the amplitude of the electronic noise generated by the detector when connected to an electronic amplifier whose bandwidth is 200 kHz. How large would be the signal-to-noise ratio?

9.3 (a) The cutoff wavelength of an MCT detector ($Hg_{1-x}Cd_xTe$) is 12.6 μm. Calculate its energy gap in electron-volt units.

(b) The detector response time at the liquid air temperature (77 K) is 1.6 μs. Calculate free (mobile) electron density by using the known expression of the excess electron lifetime on the energy gap and on the free electron density.

(c) The said MCT detector is intended to be shaped in the form of a square plane-parallel plate of 125 μm thickness. What would you recommend as an optimal thickness for the said material, knowing that the hole mobility is 1700 $cm^2/V \cdot s$ and the electron mobility is 85,000 $cm^2/V \cdot s$; what is the detector dark resistance?

(d) The length of the detector side is 2.1 mm. Calculate the detector-specific detectivity and responsivity. What would be the said quantities if the detector thickness was designed according to your recommended value? The operating current in both cases is 10 mA.

(e) Calculate the transit time across the detector plate of an electron–hole pair generated near the positive electrode for the same size of the operating current. Is it needed to correct the calculated values of the specific detectivity and responsivity? Discuss the issue and correct if needed.

(f) What would be the signal size generated by a light beam of 0.1 μW power of cutoff wavelength incident on the detector front face? What is the amplitude of the electronic noise generated by the detector when connected to an electronic amplifier whose bandwidth is 3.0 MHz. How large is the signal-to-noise ratio?

10 Passive Optical Components

10.1 Introduction

In this chapter, we address the possibility that semiconductors, as well as insulating semiconductors, may be used as passive components in optical systems, particularly as prisms, lenses, optical windows, and optical filters and polarizers. A fundamental property for functioning in optical systems relates to the material transparency or opacity to the electromagnetic radiation that belongs to the hosting system field of application. In gross terms, all such materials are transparent to light whose photon energy is smaller than the optical elements' energy gap and opaque to photons of energy larger than the energy gap. There are still additional involved interactions such as the element lattice vibrations, free charge carriers, presence of impurities, and light scattering at lattice imperfections. In the following section, we address specifically the transparency issues in the germanium semiconductor.

10.2 Use of Germanium as a Passive Optical Material

10.2.1 Preamble

Historically, germanium is one of the first studied semiconductors, and it exhibits a variety of technological applications. In the electronics field, it serves as a raw material for production of special transistors. By measuring of electrical conductivity, highly pure germanium is used as a very sensitive thermometer. When added with proper impurities, it is used as a sensitive detector of far infrared radiation, or when being highly pure, it is used for detection of nuclear radiation (see Section 9.1 and Table 9.1). Often added with optical coatings, it is further widely used as a raw material for passive optical components such as prisms, lenses, optical windows, and optical filters for the far infrared radiation in the $2 - 20\,\mu m$ spectral range. In all latter applications, it is most important to reduce the optical absorption coefficient α. As examples, we mention the use of germanium as a front mirror in lasers operating in the infrared: the carbon dioxide (CO_2) gas laser emitting at 10.6 μm, the hydrogen fluoride (HF) gas laser emitting at 2.7 μm, and the deuterium fluoride (DF) gas laser emitting at 3.8 μm. Obviously, front mirror absorption reduces the laser emission power. Furthermore, under high laser emission powers, optical absorption of the front mirror causes its heating, entailing several unwanted effects: the absorption coefficient itself grows; the refractive index changes, becoming inhomogeneous, thus causing deterioration of the laser output coherence and beam quality. The involved temperature gradients may even cause a window fracture. In principle, all optical applications require a high transparency and

Conductors, Semiconductors, Insulators, and Crystal-Growth Technology: A Practical Guide,
First Edition. Zeev Burshtein.

Table 10.1 Basic physical properties of germanium.

No.	Property	Value
1	Atomic number	$Z = 32$
2	Atomic weight	$A = 72.6$
3	Crystal structure: face-centered cubic[a)]	
4	Lattice constant	$a = 5.6575$ Å
5	Density of atoms	$N = 4.42 \times 10^{22}$ cm^{-3}
6	Solid specific gravity (300 K)	$\rho_c = 5.323$ g/cm^3
7	Melt specific gravity (1210 K)	$\rho_\ell = 5.671$ g/cm^3
8	Direct energy gap (300 K)	$E_g^{(d)} = 0.803$ eV
9	Indirect energy gap (300 K)	$E_g^{(i)} = 0.67$ eV
10	Cutoff wavelength	$\lambda_{co} = 1.77$ μm
11	Electron mobility	$\mu_n = 3.9 \times 10^3$ cm^2/V · s
12	Hole mobility	$\mu_p = 1.9 \times 10^3$ cm^2/V · s
13	Intrinsic charge carrier density (300 K)	$n_i = 2.4 \times 10^{13}$ cm^{-3}
14	Specific resistivity (300 K)	$\rho = 46\ \Omega \cdot$ cm
15	Compressibility	$B = 1.3 \times 10^{-12}$ cm^2/dyne
16	Young modulus	$Y = 10 \times 10^{11}$ dyne/cm^2
17	Melting/freezing temperature	$T_f = 937$ °C
18	Boiling temperature	$T_b = 2830$ °C
19	Specific heat (0–100 °C)	$C_p = 0.31$ J/g °C
20	Heat of fusion	$L_m = 3.4 \times 10^4$ J/mole
21	Heat of vaporization (1150 °C)	$L_v = 3.7 \times 10^4$ J/mole
22	Linear expansion coefficient (300 K)	$\alpha = 6.1 \times 10^{-6}$ °C^{-1}
23	Thermal conductivity (300 K)	$K = 0.7$ W/cm °C
24	Relative dielectric constant	$\varepsilon_r = 15.7$
25	Optical refractive index (at 3 μm)	$n = 4.045$
26	Optical refractive index (at 10.6 μm)	$n = 4.0025$
27	Lattice absorption coefficient (at 10.6 μm)	$\alpha_\ell = (0.013 \pm 0.002)$ cm^{-1}

a) A diamond-like structure.

homogeneity of the refractive index. Large efforts are thus invested in minimizing the optical absorption. The production of germanium crystals of minimal absorption at the specific carbon dioxide (CO_2) gas laser wavelength (10.6 μm) is the subject of the following sections.

A summary of germanium physical properties is provided in Table 10.1. The mixed units presented for the different quantities are the ones commonly appearing in the technical literature.

10.2.2 Optical Absorption in Germanium

Germanium is opaque for visible and infrared light up to approximately a 1.8 μm wavelength. In those spectral regions, light is absorbed by exciting electrons from the valence band into the

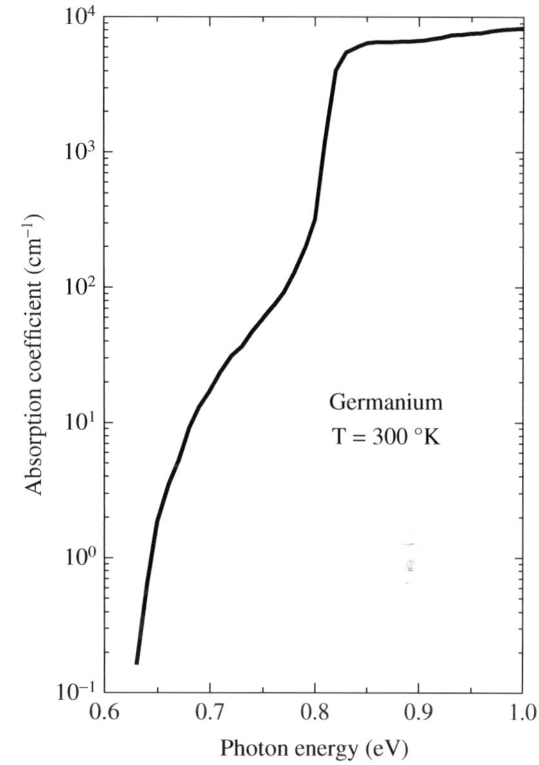

Figure 10.1 A semi-logarithmic plot of the germanium crystal room temperature (300 K) optical absorption coefficient as a function of the photon energy by excitation of electrons from the valence band into the conduction bands.

conduction band (obviously involving the appearance of electrical photoconductivity). The light penetration depth into solid germanium may be as short as 10^{-4} cm. Figure 10.1 provides a semi-logarithmic graph of the light absorption coefficient as a function of the photon energy in the vicinity of the absorption threshold. The shoulder at the lower-energy region relates to the indirect (phonons involved) transition between the valence band and the conduction band. The continued steep rise of absorption expresses the onset of a direct transition into a separate conduction band of a higher energy. Photons belonging to wavelengths longer than about 1.9 μm do not have sufficient energy to induce transition of electrons from the valence band into a conduction band. Therefore, the germanium is basically transparent to wavelengths in that range. However, there are other absorption mechanisms responsible for the fact that the material is not fully transparent in that range. These mechanisms will be discussed below.

10.2.2.1 Lattice Absorption

Lattice light absorption mechanism occurs when the electromagnetic radiation excites lattice vibrations. In ionic crystals, the lattice vibrations are accompanied with electrical dipoles, which oscillate in modes called **optical branches.**[1] Therefore, there exists a large coupling between such motions and electromagnetic radiation, expressing itself by a considerable absorption

1 Lattice oscillations are classified as **acoustic branches** and **optical branches**. A characteristic feature of acoustic branches is that in their long wavelengths range, each unit cell oscillates with almost no distortion of its shape. Optical branches, on the other hand, are characterized by gross shape distortions of each unit cell at their long wavelengths. In short wavelengths, which are of the order of the lattice period, the crystal unit cells distort in both acoustic and optical branches. Light absorption in a crystal may excite lattice waves having the same wavelength as the incident light, namely they are always long wavelengths!

coefficient. In covalent crystals, and germanium is such, dipoles are generated only at a second-order approximation. Therefore, the related optical absorption coefficient is relatively small. A considerable part of absorption in that range involves a simultaneous excitation of more than a single phonon. The optical absorption coefficient in germanium at the 10.6 μm wavelength is $\alpha_\ell = 0.013\,\text{cm}^{-1}$ (see Table 10.1).

10.2.2.2 Impurities Absorption

Light absorption by impurities occurs when the incident electromagnetic radiation excites charge carriers trapped in material impurities into the conduction or valence bands. Recall that impurities which interact that way with the conduction band are addressed as **donors**, and those which interact that way with the valence band are addressed as **acceptors**. The energy levels of donors and acceptors are situated in the forbidden band; donor levels are close to, yet below the conduction band bottom, and acceptor levels are close to, yet above the valence band top. Figure 10.2 provides a schematic illustration of such energy scheme (see also Figs. 5.3 and 5.7).

The absorption coefficient caused by impurities may be generally expressed as $\alpha_{imp} = \sigma_{imp} n_t$, where σ_{imp} is the optical excitation cross section, and n_t is the density of *non-ionized impurities*. Table 5.2 summarizes some experimentally obtained data for several known impurities in germanium – donors and acceptors, by their energy depth E_t, the wavelength at which maximal absorption is obtained λ_{peak}, and the segregation coefficient of the impurity between the solid germanium and its melt upon solidification k. It is worthwhile noting that experimentally, the absorption cross-section peak is obtained at wavelengths that are somewhat shorter than the wavelength corresponding to the absorption threshold (which expresses its energetic depth). The reason relates to the change in the bands' states densities as functions of energy. The given segregation coefficient describes the ratio between the initial impurity density in a germanium melt and the remaining impurity density in the solidifying material at the geometric position where a portion of the melt solidifies. The fact that all segregation coefficients are small compared to a unity is an important property in germanium purification technologies.

In the case of "shallow" impurities satisfying $E_t \ll k_B T$, it is possible to write approximately

$$\frac{n_t}{N_t} \cong \frac{N_t}{N_C}\, \boldsymbol{exp}\left(\frac{E_t}{k_B T}\right), \tag{10.1}$$

where k_B is the Boltzmann constant, T is the absolute temperature, N_t is the total concentration of impurities at E_t depth, and N_C is the effective band states density. As an example, the given

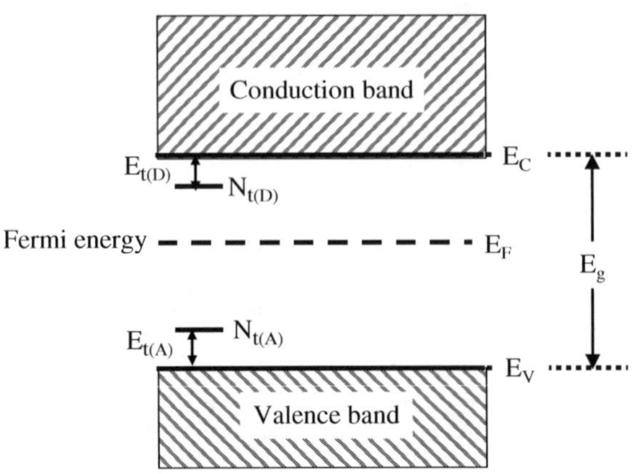

Figure 10.2 Illustration of a semiconductor energy scheme that includes both donor and acceptor states in the forbidden gap.

Table 10.2 Optical absorption cross-sections σ_{imp}, peak absorption wavelengths λ_{peak}, ionization energies E_t, and segregation coefficients k between the solid and melt, for different impurities in germanium.

No.	Impurity	Symbol	Type	σ_{imp} (cm^2)	λ_{peak} (μm)	E_t (eV)	k
1	Phosphorus	P	Donor	1.5×10^{-14}	98	0.0120	8×10^{-2}
2	Arsenic	As	Donor	1.1×10^{-15}	89	0.0127	2×10^{-2}
3	Antimony	Sb	Donor	1.6×10^{-14}	124	0.0096	3×10^{-3}
4	Bismuth	Bi	Donor	1.5×10^{-14}	98	–	4.5×10^{-5}
5	Thallium	Tl	Acceptor	8.7×10^{-15}	95	–	4×10^{-5}
6	Magnesium	Mg	Acceptor	3.9×10^{-16}	10	0.092	2×10^{-6}
7	Copper	Cu	Acceptor	1.0×10^{-15}	23	0.042	1.5×10^{-5}
8	Gold	Au	Acceptor	8×10^{-17}	2	0.05	1.3×10^{-5}

conduction band density in germanium is $N_C = 2.5 \times 10^{19}$ cm^{-3}. Following Eq. (10.1), we have

$$\alpha_{imp} = \sigma_{imp} \left(N_t^2 / N_C \right) \boldsymbol{exp}(E_t / k_B T). \tag{10.2}$$

For a particular example of antimony *Sb* impurity at a concentration $N_t = 10^{15}$ cm^{-3} in germanium, we get $\alpha_{imp} = 9.4 \times 10^{-4}$ cm^{-1} (see the cross-section σ_{imp} datum in Table 10.2).

10.2.2.3 Free (Mobile) Charge Carrier Absorption

The oscillating electrical field component of an incident electromagnetic radiation accelerates the free (mobile) charge carriers in the germanium crystal. The kinetic energy gained by the charge carriers reduces the radiation intensity, or in other words, the process causes absorption of the incident radiation. Subsequently, the accelerated carriers lose their gained kinetic energy by inducing lattice oscillations, a process known as "electron/phonon" interaction. The optical absorption coefficient via that mechanism is theoretically given by

$$\alpha_C = \frac{\lambda^2 q^3}{4\pi^2 \varepsilon_0 c^2 n} \left(\frac{n_0}{m_n^2 \mu_n} + \frac{p_0}{m_p^2 \mu_p} \right) = \sigma_n n_0 + \sigma_p p_0, \tag{10.3}$$

where λ is the incident radiation wavelength, q is the elementary charge size of an electron/proton, c is the speed of light constant, ε_0 is the permittivity of free space, and n is the material (germanium) optical refractive index at the λ wavelength; n_0 and p_0 are the concentrations of free (mobile) electrons and holes, m_n and m_p are their effective masses, and μ_n and μ_p are their mobilities, respectively. We won't provide a proof for Eq. (10.3), and assume it as given.

Equation (10.3) is not fully accurate, because it doesn't take into account often occurring complexities of band structure. It is, however, useful for a qualitative assessment of the cross sections order of magnitude, and their dependence on the incident radiation wavelength. As seen by Eq. (10.3), the thus-defined absorption cross sections σ_n and σ_p of electrons and holes, respectively, would exhibit different values according to their different effective masses and their different mobilities. Experimentally in germanium, the absorption cross section σ_p of holes is approximately 40 times larger than the absorption cross section of electrons, which is $\sigma_n = 1.5 \times 10^{-16}$ cm^2.

For a further analysis of the absorption coefficient in germanium, we recall the "law of mass action" relation $n_0 \cdot p_0 = n_i^2(T)$, where $n_i(T)$ is the carriers' intrinsic concentration, which changes only as a function of the absolute temperature T (see Eq. [5.8] and its following text and also note that at room temperature $n_i(300\ \text{K}) = 2.4 \times 10^{13}\ \text{cm}^{-3}$ – see Table 10.1). We thus obtain

$$\alpha_c = \sigma_n n_0 + \sigma_p n_i^2(T)/n_0. \tag{10.4}$$

Because in germanium $\sigma_p > \sigma_n$, it may be readily concluded that a minimum value for α_c as a function of n_0 is obtained when $n_0 > n_i(T)$, which means that it occurs in an n-type germanium. The free electron concentration for the minimal α_c is $n_0^{min} = n_i(T)(\sigma_p/\sigma_n)^{1/2} \cong 1.5 \times 10^{14}\ \text{cm}^{-3}$, for which the minimal α_c is $\alpha_c^{min} = 2n_i(T)(\sigma_p\sigma_n)^{1/2} = 0.045\ \text{cm}^{-1}$. The electron concentration is obtained in an n-type germanium, whose specific resistivity is $\rho \approx 5 - 15\,\Omega \cdot \text{cm}$. In Figure 10.3, we present a graph showing the room temperature absorption coefficient of germanium for light wavelength of $\lambda = 10.6\ \mu\text{m}$ as a function of the specific resistivity for both n- and p-type materials.

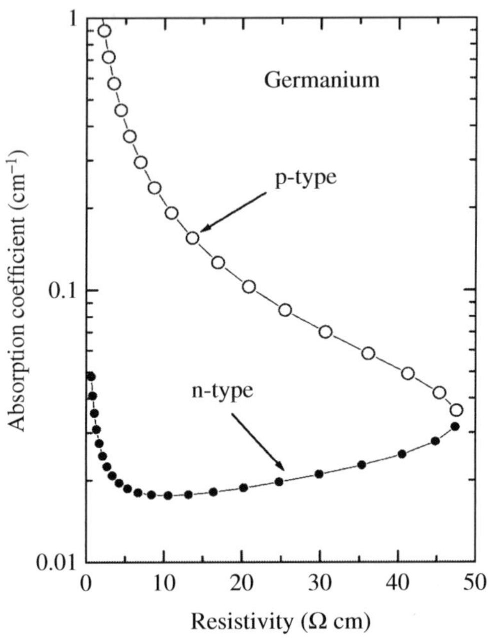

Figure 10.3 Semi-logarithmic plot of the room temperature absorption coefficient of 10.6 μm light in n- and p-type germanium as a function of its specific resistivity for the two cases.

Preparation of germanium crystals having the extreme minimal optical absorption is done by using an initially intrinsic material, namely a most pure material, whose charge carriers' concentrations are dominated by thermal excitation from the valence band into the conduction band. Then, antimony *Sb* impurities are added into the germanium melt, which donate electrons to the conduction band. The reason for choosing antimony over other potential impurities is because its ionization energy is very small while hosted as an impurity in germanium (see Table 10.2). Still, it exhibits a considerable disadvantage by having a very small segregation coefficient (3×10^{-3} – see Table 10.2). Therefore, it would be difficult to subsequently grow from the melt a single crystal having a uniform antimony concentration. The latter effect is qualitatively obvious, considering that those antimony atoms rejected from the solidifying parts actually increase the antimony dopant concentration in the remaining melt. Correspondingly, a continuing growing solid crystal also exhibits a gradual increase of the antimony dopant concentration, hence a corresponding decrease of the grown material resistivity.

The analytic expression describing the impurity density C_S entering a solid matter that gradually solidifies from a melt that contains an initial impurity of a C_0 density, which has a k segregation coefficient, is given by the so-called Pfann's normal freezing equation

$$C_S = kC_0(1 - g)^{k-1}, \tag{10.5}$$

where g is the fraction of the already solidified melt (see Eq. [12.13]). It is based on a realistic assumption that diffusion and material flows inside the melt are sufficient to render uniform the impurity concentration throughout the entire melt at all times.

To moderate the impurity increasing rate, sometimes arsenic (*As*) or phosphorus (*P*) are used as dopants, because their segregation coefficients k are larger than that of antimony (Table 10.2).

10.2.3 Optical Quality Assessment of Grown Germanium Parts

As previously indicated, to properly serve the optical transparency goal, a grown germanium single crystal has to be n-type, with a specific resistivity ranging between 5 and $10\,\Omega\cdot$cm. The grown crystal properties thus have to be assessed compared to its targeted technological properties.

10.2.3.1 Conductivity Type Probing Using a Hot Electrical Contact

In that measurement, use is made of a voltmeter whose two electrically connected metallic wires are placed to contact two close points of a plane-parallel prism cut from the grown crystal. One of the wires is heated up by a couple of dozen degrees Celsius above that of the other one. At the moment of contact forming, the crystal region close to the hot contact assumes an increase of temperature, and the belonging charge carriers diffuse out of it. The hot region thus assumes a space charge, whose sign is the opposite of the charge belonging to the majority charge carriers. This effect expresses itself by a voltage measured by the voltmeter. In an n-type material, the voltmeter indicates a positive charge, and in a p-type material, the voltmeter indicates a negative charge. That simplistic measurement provides an immediate, very reliable indication of the material conductivity type. Figure 10.4 provides a schematic description of the said measuring system.

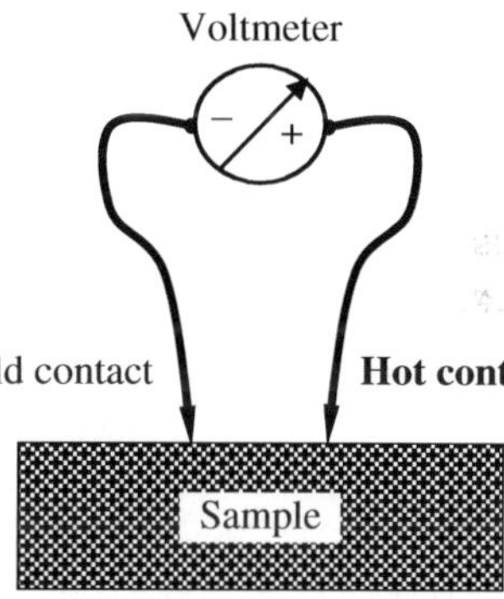

Figure 10.4 Demonstration of the method used for determining the conductivity type of a semiconductor. In an n-type material, electrons close to the hot contact diffuse out of that region, leaving a positive space charge.

10.2.3.2 Four-Point Probe for Specific Resistivity Measurement

In this measurement, four pressure contacts of equal distance arranged in a row are tightened over the sample surface. An electrical current I is driven between the two distant contacts, and the voltage V developed between the two inner contacts is measured. The use of contacts different from those that drive the current is practiced to avoid problems related to contact potentials and voltage drops across the conducting wires, because the related voltmeter is a rather very little current consumer. Figure 10.5 provides a schematic description of the said measuring system.

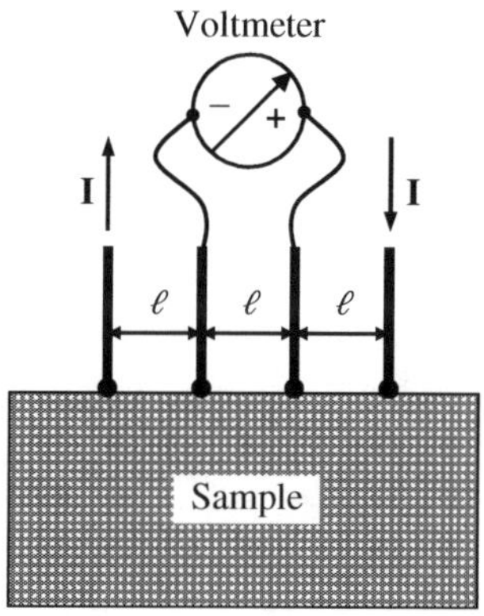

Figure 10.5 A set of electrical connections for measuring specific resistivity using a four-point probe.

Under geometric conditions where all sample dimensions are large compared to the distance ℓ between the electrical contacts, the specific material resistivity is given by

$$\rho = 2\pi\ell\left(\frac{V}{I}\right). \tag{10.6}$$

10.2.3.3 Absorption Coefficient Determination Using Optical Transmission Measurement

As previously indicated, knowing the conductivity type and the electrical resistivity should be sufficient to assess the material absorption coefficient. However, there are quite a number of other effects that could impair the material transparency; mostly the presence of dust particles, especially graphite dust particles emanation from graphite used as raw material of the growth crucible, and sometimes of the heating elements. It is thus quite important to verify the assessed absorption coefficient by a direct measurement. The equipment used for that purpose is a spectrometer, capable of measuring optical transmission over a broad range of the far infrared. The measured quantity is rather the ratio between the intensity transmitted when a beam of the desired wavelength is blocked by a plane-parallel table of the studied material, and the intensity transmitted without the table. That ratio is the optical transmission denoted by T. In our present case, we are interested in the measured transmission at the single wavelength $\lambda = 10.6\,\mu\text{m}$.

The optical transmission through a properly polished plane-parallel plate of thickness d is given by

$$T = \frac{(1-R)^2}{e^{\alpha d} - R^2 e^{-\alpha d}}, \tag{10.7}$$

where α is the absorption coefficient, and the (Fresnel) surface reflection constant R is given by

$$R = \frac{(n-1)^2}{(n+1)^2}, \tag{10.8}$$

where n is the optical refractive index. Under conditions of sufficiently thin plates and sufficiently small absorption coefficient satisfying $\alpha d \ll 1$, we may use the approximation $e^{\pm\alpha d} \cong 1 \pm \alpha d$. Setting it in Eq. (10.8) and solving for α, we get

$$\alpha = \frac{(1-R)^2/T - (1-R)^2}{d(1+R^2)}. \tag{10.9}$$

To further simplify the expression and make it more practical, we denote $T_0 \equiv \dfrac{(1-R)^2}{1-R^2} = \dfrac{1-R}{1+R}$. That quantity expresses the optical transmission of a plate that has the same refractive index but has no absorption ($\alpha \equiv 0$). For germanium, we would get $T_0 = 47.00\%$ at $\lambda = 10.6\,\mu\text{m}$ (obtained by setting the known $n = 4.0025$ in the above expression for T_0; see Table 10.1). An actually measured transmission T is always smaller than T_0 due to all absorption processes. Denoting $\Delta T \equiv T_0 - T$ and setting it in Eq. (10.9), we get

$$\alpha \cong 1.64\,\frac{\Delta T}{d}. \tag{10.10}$$

Equation (10.10) is a simple, very practical expression for calculating the absorption coefficient in germanium sample plates of known thickness.

Exercises for Chapter 10

10.1 The density of antimony atoms as impurities in germanium is $N_t = 10^{15}\,\text{cm}^{-3}$. Calculate and plot a graph of the temperature dependence of their contribution to the absorption coefficient at 10.6 μm between room temperature 27 °C and 100 °C. Consult Eq. (10.2) and its related text, and further recall that the effective conduction band states density is proportional to the 3/2 power of the absolute temperature (see Eq. [5.7]).

10.2 A germanium crystal grows from a germanium melt that contains antimony *Sb* at a density of 5×10^{17} atoms/cm^3. Calculate the density of antimony atoms in the crystal at the regions touching the melt after 10% and after 50% of the melt has solidified.
Also calculate the crystal-specific room temperature resistivity at each said region.
Assuming that all absorption processes related to the 10.6 μm wavelength are due to the free electrons, what will be their contribution to the crystal optical absorption coefficient at $\lambda = 10.6$ μm in each said region?

10.3 A four-point probe device is pressed to touch the surface of a germanium plate whose specific resistivity is $50\,\Omega \cdot \text{cm}$. The distance between the probe contacts is 2 mm. A current of 30 mA is driven through the plate. Calculate the voltage measured by the device voltmeter.

11 History of Crystals Growing and Basic Concepts

11.1 Historic Notes on Crystals Growing

The interest of humans in crystals dates back to very early eras of human history. Humans were attracted to special shapes by which crystals appeared randomly in nature, and to special light refraction that granted some of them special beauty. Mechanical hardness of some, along with rarity and durability, granted them great importance in the minds of people, who used them as jewelry and as a means of payment, an effect that continues even in modern times. The 1669 year may be indicated as the historic time when scientific interest in crystals started. A Danish "naturalist" named Steno discovered that the angles between side planes of mineral quartz crystals are maintained when the crystals are cleaved. By the 1800s, the scientific world was already acquainted with, and understood many optical properties of mineral crystals, and scientists started using them in carrying out of optical studies. In the mid-1900s, sapphire crystals were used in forming bearings for mechanical wheels in watches; the mechanical stiffness of sapphire was effective in preventing wear of the bearings. In the twentieth century, the era between the First and Second World Wars, mineral quartz crystals were used to stabilize frequencies in radio waves equipment. Since then, the technological use of various single crystals has increased at a large rate. Currently, computers based on crystalline silicon (and other materials) became the basis of modern everyday communication equipment.

The growing need of crystalline materials motivated a development of artificial crystal growing industry; in other words, a transition from mineralogy (a specialized geologic branch) toward industry, and a corresponding scientific research and development. Nowadays there are many subjects and technologies related to crystals crowing; some are listed below:

1) *Electronics and microelectronics*: silicon (Si); germanium (Ge); gallium arsenide ($GaAs$); quartz (SiO_2); sapphire (Al_2O_3).
2) *Optics*: calcite ($CaCO_3$); fluorite (MgF_2); quartz; sapphire; germanium; zinc sulfide (ZnS).
3) *Piezoelectricity*: quartz; Rochelle salt ($C_4H_4O_6KNa \cdot 4H_2O$); KDP (KH_2PO_4).
4) *Lasers*: ruby ($Cr:Al_2O_3$); neodymium:YAG ($Nd:Y_3Al_5O_{12}$); titanium:sapphire ($Ti:Al_2O_3$).
5) *Electro-optics and non-linear optics*: $LiNbO_3$; $\beta - BaB_2O_4$; KDP.
6) *Light detectors*: Si; Ge; PbS; $InSb$; mercury cadmium telluride (MCT, $HgCdTe$); LTT ($PbSnTe$).
7) *Electronic gamma and X-ray detectors*: Si; Ge; $CdTe$; HgI_2.
8) *Scintillation gamma and X-ray detectors*: $Tl:NaI$; anthracene ($C_{14}H_{10}$); $CdWO_4$.
9) *High-temperature super alloys*: primarily nickel (Ni)-based metallic alloys.

Conductors, Semiconductors, Insulators, and Crystal-Growth Technology: A Practical Guide,
First Edition. Zeev Burshtein.

11.2 Relevant Scales Related to Crystals

In consideration of any crystal properties, or of different processes involved in its production, it is customary to use three different scales adjusted to the order-of-magnitude of the considered volume.

(i) Macroscopic Scale
That scale involves sizes of several millimeters or more. Such scales are used for describing general external shape of a crystal, temperature gradients while growing, or while functioning as a component in any device, or describing a general distribution of dopants, impurities, or structural faults.

(ii) Microscopic Scale
That scale involves sizes of submillimeter down to a micrometer (μm). It is used to describe intrinsic properties of a crystal, like light scattering, X-ray diffraction, pores, grains and grain boundaries, and distribution of dopants, impurities, or structural faults.

(iii) Atomic Scale
That scale involves sizes between 0.1 μm (1000 Å) and 0.1 nm (1 Å). The lower range of the scale is used to discuss the bonding size among the atoms, and the basic arrangement of the atoms that make up the crystal structure. In the latter limit, it is inappropriate to involve simplistic thermodynamic equilibrium considerations, because the systems involve only very few particles. Such systems exhibit frequent departure from thermodynamic equilibrium.

It is quite often possible to understand physical properties of multi-atomic system as resulting from the behavior of single atoms. In many cases, it is useful to consider properties apparent in the range of one scale in relation to same or other properties under a different scale.

11.3 Definition of a Single Crystal

A **single crystal** (sometimes addressed as a **monocrystal**) is a solid of macroscopic dimensions (at least several millimeters), comprising a **single phase**, whose atoms are arranged orderly according to the same crystallographic structure (see Section 1.1), except perhaps for the presence of structural faults, or the presence of different atoms, whose concentrations doesn't exceed an order of 10^{12} cm^{-3} – 10^{18} cm^{-3} (for comparison, a typical concentration of atoms in a solid is of an order of magnitude of 10^{22} cm^{-3}). The crystalline order is defined by a lattice of a specific symmetry, maintained throughout the entire solid volume.

The reservation regarding the solid uniformity and order allows the concept of "single crystal" to include cases in which there are phenomena of deviation from perfection, as listed below:

1) The presence of atomic **dopants and impurities**. Such atoms may reside in crystalline sites that were previously occupied by "legitimate" atoms; they are then granted the adjective "**substitutional**", or squeezed between them, and then are called "**interstitial**". There are also cases of absence of legitimate atoms; such sites are called "**vacancies**".
2) Defects of crystalline order called "**dislocations**".
3) Defects in crystalline orientation of microscopic regions by less than about 1°, an effect called "**mosaicity**"; such defects occurring in larger than microscopic regions are called "**lineage defects**".

4) Local changes of lattice unit cell size are called "**deformations**"; such changes usually occur only under application of an external mechanical force.

A solid is no more considered as a single crystal under occurrence of the following effect:

1) "**Twinning**": crystalline twins are adjacent regions exhibiting each a different crystallographic orientation. Such twins are usually intertwined throughout the entire solid volume.
2) "**Grains**": These are regions in the solid whose crystalline orientations differ arbitrarily by angles considerably greater than 1°.
3) "**Liquid crystals**": These are solids that exhibit order in only one or two dimensions; where in the other dimension/s, the disorder (mixing) properties are similar to that of a liquid.

Most inorganic solids appear in nature in a crystalline form; however, the range of the order is mostly microscopic, as for example in solid metals. Such form is called "**polycrystalline**". In the microscopic range, the properties of a polycrystalline material are similar to those of a single crystal.

The technological interest in single crystals stems from the fact that their long-range order also casts symmetry of many physical properties; in other words, many physical properties depend on the crystalline orientation. Such effects are called "**anisotropy**". In a polycrystalline material, the physical properties exhibit discontinuity on passing from one crystalline grain to another, whose orientation is different. A grain boundary may also induce a discontinuity in a property for which the single crystal is fully isotropic, because of mechanical forces induced inside a grain due to the surface presence. In a polycrystalline material, averaging of some physical properties occurs along different crystallographic orientations. In order to exploit an orientation-dependent property, it is often necessary to use a single crystal. As an example, consider a crystal physical property whose sign reverses in a reversed orientation; piezoelectricity is such a property (recall that piezoelectricity is the appearance of a voltage across a crystal under application of an external mechanical force). That property may diminish entirely on the macroscopic scale in the polycrystalline phase. Another example of a fundamental change of property in a polycrystalline material versus the single-crystal phase is light scattering off the stressed regions close to the grains boundaries (elasto-optic effect). Grain boundaries are also involved in forming enhanced scattering of charge carrier motion, affecting the material electrical conductivity. They also form favorable locations for concentration of impurities, and channels for their diffusion. It should be also indicated that changes of the material electrical and chemical properties frequently occur at surfaces of macroscopic crystals; they may occur due to the mere discontinuity of the matter existence, due to defects caused by the surfaces preparation procedures, and due to adsorption of ambient impurities. Modern technology is deeply involved in miniaturization of electrical and of other devices, which requires the use of very narrow crystalline layers. In such devices, the surface properties sometimes exhibit a major role on their functioning.

11.4 The Essence of Crystal Growing

Crystal growing is a technological process of having maximal control over the transition from any phase of the matter into the single-crystal phase, whose crystalline order is maintained over macroscopic dimensions.

Most solids found in free nature are polycrystalline, and only very few exhibit a single-crystal structure. This suggests that the growth of single crystals requires special conditions. Indeed, as we shall see later, the industrial process of crystals growing requires a meticulous control over a broad range of physical factors.

Material solidification may start via any aggregated phase. Some crystals may grow directly from the material gaseous phase, for example mercuric iodide (HgI_2) and anthracene ($C_{14}H_{10}$). In other cases, the meant-to-grow material gas is mixed with another inert gas. Under proper conditions, the meant-to-grow material grows epitaxially (layer after layer) on a proper flat substrate. The process is called "vapor phase epitaxy (VPE)". In a more complex option, the meant-to-grow material is still a mixture of not-yet-interacted gasses. It grows epitaxially on a substrate during a chemical reaction of its formation. The process is called "chemical vapor deposition (CVD)".

The starting growth phase may be a liquid. One option is that the meant-to-grow material is dissolved in a proper solvent. Gradual lowering of the solution temperature, or a gradual vaporization of the solvent, induces nucleation and further growth of the solute material in its crystalline form. The process is usually addressed as "solute crystallization" and sometimes as "solute fusion". A quite trivial example is the growth of table salt $NaCl$ out of its aqueous solution. Another option is the direct gradual solidification of the meant-to-grow material from its molten bulk. Some examples are the single-crystal growth of germanium (Ge – see e.g. Section "Optical Quality Assessment of Grown Germanium Parts"), silicon (Si), and neodymium:YAG ($Nd : Y_3Al_5O_{12}$). *Our present book addresses in detail that particular process.*

The starting growth phase may even be solid. One case is conversion of an amorphous or a polycrystalline solid material into a single crystal. An example is the growth of a mercury cadmium telluride ($HgCdTe$) (MCT – see Section 9.2.2). Another case is crystallization of a glass. For example, a quartz glass heated above about 1200 °C starts to grow single-crystal grains. The process is called "devitrification".

It should be noted that all phase changes are similar in the sense that a said change involves a change in the degree of order (namely a change of entropy) while maintaining a constant temperature. That reality has an important meaning in all scales. Thus, finding analogies allows understanding of crystal growing processes.

Each of the above-indicated processes may occur under various conditions; the different crystal growing techniques differ mainly in their technical details aimed at solving specific problems of each material. Each technique exhibits advantages and disadvantages. Most growing processes make use of a single-crystal **seed**, which initially has very small dimensions, but it grows in dimensions while consuming the raw material (often addressed as the **nutrient phase**). In some techniques, no prepared seed is used, and the first spontaneously nucleating grain serves as a seed for the continuing single-crystal growth. In industrial crystals growing, the use of a seed has a great advantage, because it dictates the crystalline orientation for the rest of growth. Spontaneous nucleation, on the other hand, is a very sensitive event, and it is difficult to control with sufficient precision its crystallographic orientation and the uniqueness of its occurrence.

12 Solidification Processes

12.1 Homogeneous Nucleation

On an atomic scale, atoms and molecules comprising a material in its liquid phase are continuously moving, a result of having a kinetic thermal energy. While moving, two or more colliding molecules may bond to form a "solid" cluster (micro-grain). Such bond subsequently breaks due to additional collisions. The said bonded cluster is unstable at any temperature greater than the freezing temperature T_f. In thermodynamic terms, it means that the liquid free energy is larger than that of the solid at temperatures greater than T_f; in other words, energy has to be removed to form a solid. If the temperature rises above T_f, the lifetime of an accidentally formed bonded cluster shortens. As the temperature drops toward T_f, larger solid clusters will appear, preserving their solid nature for longer durations; the process still involves in a continual mode of bonding and breaking. Even if the liquid assumes a temperature that is lower than T_f, no immediate solidification of the entire liquid matter occurs; the reason is the large interface energy, which forms between the solid clusters and the remaining liquid.

For a more analytic description, we consider the energetic state of a solid cluster in its same-composition liquid phase at a temperature T. We denote

F_L – The liquid free energy per unit volume at the T temperature.
F_S – The solid cluster free energy per unit volume at the same T temperature.
$\Delta F_V \equiv F_S - F_L$ – The latent heat of freezing per unit volume.
γ – The surface tension coefficient, which is the free energy per unit area of an interface between a solid cluster and the liquid.

For simplicity, we assume the solid cluster to exhibit a spherical shape of radius r. The difference between the free energies of the same material when it assumes a solid cluster form, compared to a liquid form, is given by

$$\Delta F_N = \Delta F_V \cdot \frac{4}{3}\pi r^3 + 4\pi r^2 \gamma. \quad (12.1)$$

The solidification latent heat per unit volume is negative below the freezing temperature. It is such because a transition from a liquid into a solid involves a reduction of energy (reduction of energy means that the system internal energy lowers). The surface tension coefficient, on the other hand, is positive, and its relative influence is greater when r is smaller. The contributions of a cluster volume and surface area to the difference between the free energies are written in Eq. (12.1). Figure 12.1 demonstrates schematically the two contributions separately, and their summed result.

Conductors, Semiconductors, Insulators, and Crystal-Growth Technology: A Practical Guide,
First Edition. Zeev Burshtein.

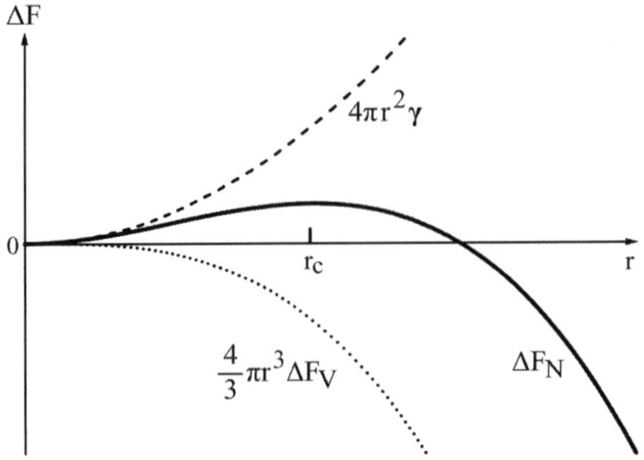

Figure 12.1 Schematic illustration of the volume and surface area free energy contributions in a solid cluster as functions of its r radius. A dashed line – the surface area contribution; a dotted line – the solid volume contribution; A thick continuous line – the sum of the two contributions.

As clearly seen, the cluster volume exhibits a critical radius r_c for which the system internal energy rises when the cluster grows. A cluster, whose radius r is larger than r_c, will continue growing spontaneously while lowering its energy. However, as long as the cluster radius is still smaller than r_c, it has a large probability to reduce its size spontaneously and resume a liquid state. According to the above analysis, a liquid may remain in a liquid state for a long time at a temperature that is lower than the freezing temperature, as long as no solid cluster of a critical radius has formed; such effect is called **"supercooling"** or **"undercooling"**.

A system exhibiting a cluster of r_c radius exhibits maximal energy. The functional derivative of ΔF_N according to r is zero at $r = r_c$. Taking the derivative of the expression in Eq. (12.1), and equating it to zero, yields the relation $r_c = -2\gamma/\Delta F_V$. If the system temperature lowers, the denominator in the expression for r_c increases its size, the critical radius r_c becomes smaller, and the probability for formation of a critical radius increases. Therefore, on cooling a molten material volume below its freezing temperature (supercooling conditions), a solid cluster of critical volume is likely to form, causing rapid solidification of the entire molten material.

12.2 Heterogeneous Nucleation

Heterogeneous nucleation is a volume solidification process out of a liquid or a gaseous nutrient volume on top of a different solid material called **substrate**, which in itself is far from melting or vaporizing conditions. Such technological process is very important in growing of very thin layers of materials used in forming of microelectronic devices. The usability of such devices stems from the fact, that the active part in a microelectronic device is usually narrower than about half a micron (micron ≡ 1 μm). Therefore, in principle, there is no need for growing a thick active crystal for granting it sufficient mechanical stability.

To understand the mechanism of heterogeneous nucleation, we need to consider those solid clusters of the intended for growth material that are attached to the substrate surface. The growth process involves three major steps:

1) Creation of a layer of atoms adsorbed to the substrate surface (condensation process).
2) Creation of a critical two-dimensional cluster comprising the adsorbed atoms, and its further two-dimensional growing.
3) Further growth of the solid layer, which turns into a homogeneous growth process.

In each situation where a solid is in contact with a liquid or a gas, temporary connections are created between the solid surface and the close atoms of the other material. The solid surface atoms produce potential wells, and liquid or gas atoms are temporarily trapped inside these potential wells. The activation energy required to "break free" the trapped atoms is provided by the ambient thermal energy. The depth of the potential well U describes the needed activation energy. The ratio between it and the characteristic k_BT thermal energy of an atomic degree of freedom, where k_B is the Boltzmann constant and T is the absolute temperature, will determine whether the trapping time will be long or short; if that ratio is large, the trapping duration is long; if it is small, the trapping duration is short. The trapped population density is thus temperature dependent. Except for just "breaking free", the adsorbed atoms may also just migrate between the potential wells, and usually, the activation energy for such process is considerably smaller than the one required for the "breaking free" one. These processes obviously depend on the nutrient gas pressure or the nutrient melt components concentrations, according to the involved particular system.

The process may be classified according to three specific relations between the bonding forces:

1) The nutrient bonding energy with the substrate is larger than the bonding energy of the nutrient atoms with their solid clusters. In such case, the initial condensation will occur on the substrate, the atoms will become distributed over the entire substrate area, and the solid will grow in a form of successive layers.
2) The nutrient bonding energy with the substrate is similar to the bonding energy of the nutrient atoms with their solid clusters. There is still a larger probability for the nutrient atoms' condensation on the substrate area than for a critical solid radius formation in the nutrient, thus initiating its homogeneous solid formation. The nutrient atoms condensed over the substrate diffuse across the surface, forming "islands" of bonded atoms. The atoms in each such bonded island form two-dimensional clusters. Here as well, critical size effects are apparent, because of a difference between the island surface area energy and its circumference energy ("circumference tension").
3) The nutrient bonding energy with the substrate is very weak. The situation is not unambiguous. There may still be various geometric forms of two-dimensional clusters on the substrate surface that will still be energetically favored, but usually, heterogeneous solidification will be completely prevented.

It is not at all simple to achieve conditions for proper homogeneous solidification. For example, in crystal growth from a melt, the melt may contain dust particles having melting temperatures that are considerably higher than the hosting melt, for example SiO_2, $NaCl$, or oxides of the melt material itself. In such cases, the heterogeneous nucleation will precede the homogeneous one. In some cases, conditions for heterogeneous nucleation are prepared intentionally. For example, to prevent superheating in the boiling of distilled water heated in glass vessels, it is customary to add special "boiling stones", to play the role of substrates for heterogeneous clustering of water vapor bubbles.

For growing a crystal from its material melt, it is possible to achieve homogeneous solidification by lowering the melt temperature until achieving a solid grain via a homogeneous or heterogeneous cluster formation about a solid particle. For achieving a single crystal, care has to be taken that only a single such solid grain be formed, which then performs as a seed for gradual solidification of the entire melt volume. If such condition is not met, the material will solidify in a polycrystalline form. The only means to overcome this problem, and grow a single crystal, is to use a special seed, as previously indicated. The seed is a single crystal of the same, or of a sufficiently similar material, which is much larger than the critical cluster at the freezing temperature. The seed is arranged to touch the melt; a proper programming of the temperature gradients and of the entire system cooling rate makes it possible to advance the crystal growth around the seed. The use of a selected seed also allows to predetermine the crystallographic orientation of the grown single crystal.

12.3 Layered Growing

Right after formation of a solid cluster inside the nutrient phase, the situation becomes very similar to the case described for the heterogeneous nucleation. It differs by the following aspects:

1) The binding energy among the nutrient phase molecules is similar to the binding energy of the solid phase molecules.
2) The crystalline arrangement of the solid phase molecules is very close to the ideal crystalline arrangement for the solidifying melt molecules, and it establishes a kind of a recipe for the nutrient molecules to organize in a crystalline form. It indeed minimizes the chance for randomness of the crystallographic orientation, which often occurs in heterogeneous solidification.

Figure 12.2 provides a schematic illustration of a possible interface (growth surface) between a fluid nutrient and its solid phase. The figure illustrates identical cubic-shaped solidification regions exhibiting different stabilities and marked by different numbers.

The "1" region indicates a single solidified region over the surface. It bonds to the surface via a single face, and therefore, its boning strength to the surface is quite weak. It can therefore easily migrate over the surface with very small activation energy, or even melt. The "2" region is bonded by two faces; therefore, its bonding strength to the surface is stronger. However, it may still migrate along the step area, or even melt (though less likely). The "3" region, which looks like a step corner, may be described as a single solidified region bonded by three faces to the step corner. The bonding strength for it is thus even stronger; however, it may still disconnect and migrate along the step area, or even melt (though quite unlikely). It is also likely that a "2" region will migrate along its step until trapped at the step corner, thus forming a new step corner. A solidified region that will be trapped inside the "4" region becomes bonded by 4 bonds, thus becoming more

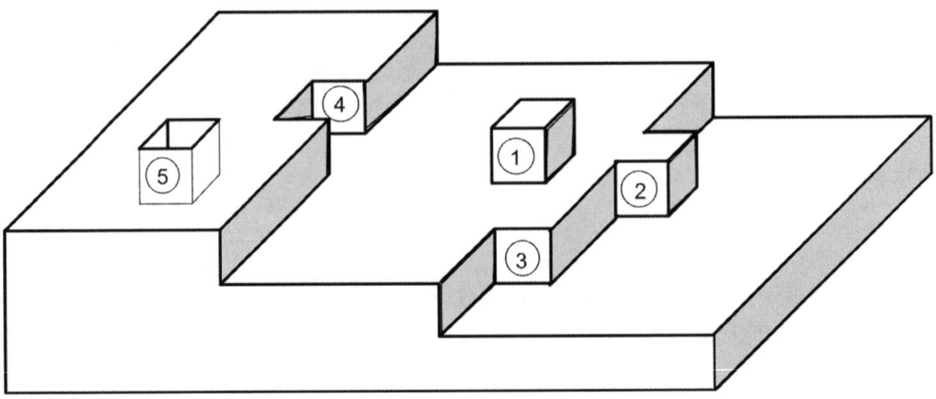

Figure 12.2 A possible characteristic pattern of growth surface condition of a solidifying material.

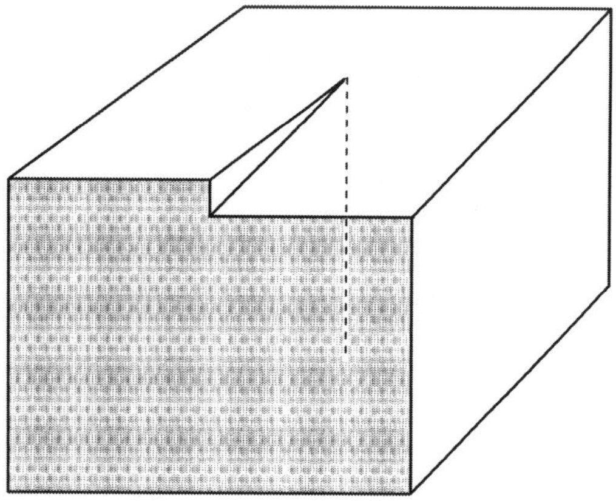

Figure 12.3 Temporary solid surface forms during occurrence of a screw dislocation. The dislocation axis is described by a dashed line.

stable, unlikely to migrate or melt. The "5" region indicates a missing solid region; it provides a bonding possibility to a solidifying region by five faces; as an individual entity, it is most unstable, similar to that of case "1", and it may therefore easily migrate over the surface with very small activation energy. The most stable case is thus a large full face; the most unstable cases are a full large face added with a single solidified region, and a full large face missing a single solidified region. By energy consideration, it is thus obvious that regions "4" and "5" will become filled within a short time. The most probable solidification process is a gradual successive addition of solid layers. Any opening of a new layer with a partial filling exhibits some instability until the partial filling disappears by either melting or by being complemented to become a full layer. The latter process is called **layered growing**. Because the initiation of a new layer is unstable, there is a certain pause in growth between successive layers.

The described pause in the addition process of new layers may be avoided if the layers exhibit a screw dislocation, as described in Figure 12.3.

It turns out that the presence of structural defects may reduce the instability in the process of increasing the volume of the growing solid. The screw dislocation described in Figure 12.3 is a very common structural defect. The screw dislocation results by gliding of a relatively large solid region, where the dislocation axis is identical. As seen in the figure, a linear new step is formed in a part of the growth surface, which makes it energetically preferable for additional growth, because it would bind each solidifying region by two faces. After solidification of many solid units along the step length, actually a new step forms, having a slightly different azimuth angle on the growth surface. The dislocation axis remains unchanged. Eventually, the solid volume growth continues spirally, with an axis that penetrates along a large macroscopic volume of the growing crystal.

12.4 Rough and Smooth Growth Surfaces

The process described in the previous section provides an example that solidification of atoms and molecules over a solid surface depends on the surface characteristic properties. We hereby characterize the growth surface according to the degree of roughness. It is possible to formally define the roughness as the ratio between the surface area measured by a tool of macroscopic accuracy and the surface area measured by a tool of microscopic accuracy. To clarify the issue, consider Figure 12.4. The figure provides cut views perpendicular to the surface area of two

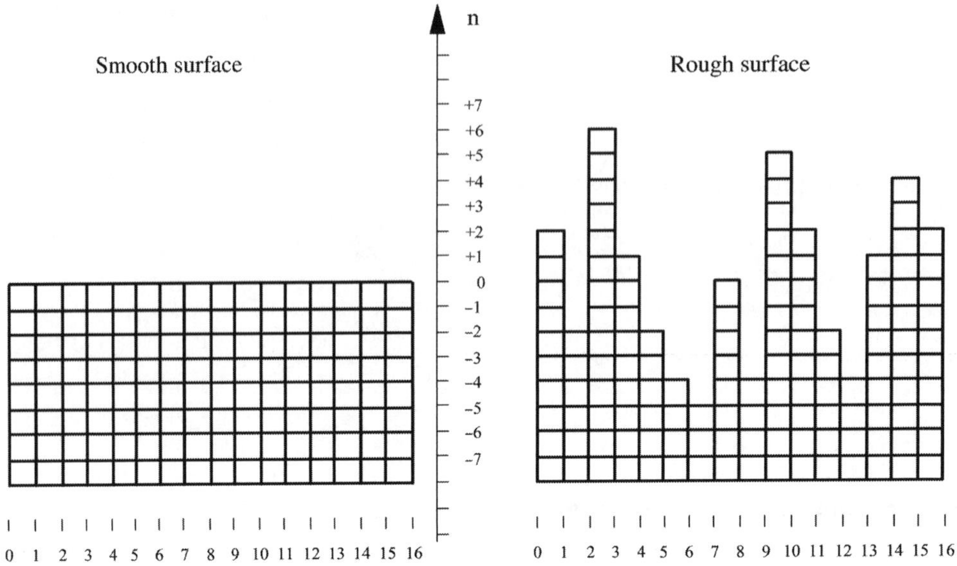

Figure 12.4 Demonstration of a cut view perpendicular to the surface area of a smooth surface (left side), and another cut of a rough surface (right side). The cubes show the solid units, while the liquid (fluid) phase is above. The number of cubes is the same in the two parts of the figure.

solidifying materials; one exhibits a smooth surface and the other exhibits a rough surface. For simplicity, the material molecules are described as cubes. Obviously, the rough surface exhibits a larger area than the smooth surface. The below-presented multilayer model (developed by **Temkin**) provides an example of analyzing the difference between a smooth growth surface and a rough growth surface using a dedicated computer simulation program. The model is of great importance in both the set of assumptions and development, and in the presented conclusions.

12.4.1 Temkin multilayer Model

In this model, we assume that both crystal and melt consist each of collection of layers where each comprise the same number N of molecules. The ordinal number of a layer is denoted by n. An n layer contains N_{nf} fluid molecules and N_{ns} solid molecules, where always $N_{nf} + N_{ns} = N$. For example, a layer entirely solid satisfies $N_{ns} = N$ and $N_{nf} = 0$. The model assumes that the fluid "floats" above the solid; namely, for higher "n" numbers, the fraction of fluid molecules grows. It is further assumed that above any fluid cube, there are only fluid cubes. In other words, there are no solid units that are entirely surrounded by fluid ones.

We denote by C_n the ratio of solid units in the n'th layer, namely $C_n \equiv N_{ns}/N$. Therefore, always $C_{n+1} \leq C_n$, and the boundary conditions are $C_{-\infty} = 1$ and $C_{+\infty} = 0$. The entire system is in thermal contact with a heat reservoir and a volume reservoir, such that the temperature and pressure remain constant throughout the entire process.

The described system has three kinds of bonding between the cubes:

1) A solid–solid bond of binding energy Φ_{ss};
2) A fluid–fluid bond of binding energy Φ_{ff};
3) A solid–fluid bond of binding energy Φ_{sf}.

We define a reference state as a fully smooth surface at $n = 0$. Mathematically it satisfies

$$C_n = 1 \quad \text{for every} \quad -\infty < n \leq 0;$$
$$C_n = 0 \quad \text{for every} \quad 1 \leq n < \infty.$$

The thermodynamic potential function appropriate for constant temperature and pressure is the Gibbs free energy. The difference in the free energy between any state of the system and the reference state is given by $\Delta G = \Delta G_{fs} + \Delta G_{surf}$, where ΔG_{fs} is the change in the Gibbs free energy due to conversion of solid cubes into fluid ones (which is a volume change), and ΔG_{surf} is the change in the Gibbs free energy due to the change in the shape of the interface between the solid and the fluid, namely without any change in the total solid volume. As to a single layer, we may write that

$$\Delta G_{fs} = \frac{\partial G_{fs}}{\partial N_s}\Delta N_s + \frac{\partial G_{fs}}{\partial N_f}\Delta N_f; \tag{12.2}$$

$$\Delta G_{surf} = \Delta E_{surf} - T \cdot \Delta S_{surf}, \tag{12.3}$$

where ΔE_{surf} is the change in the surface energy, and ΔS_{surf} is the change in its entropy. There is no need to consider changes related to the system volume, because by Temkin simplistic model, a solid cube volume is identical to a fluid box volume. We further denote $\partial G_{fs}/\partial N_s \equiv \mu_s$ and $\partial G_{fs}/\partial N_f \equiv \mu_f$.

Consider now a case of a single solid cube turning fluid, namely $\Delta N_f = +1$ and $\Delta N_s = -1$, then $\partial G_{fs} = \mu_f - \mu_s \equiv \Delta\mu$. We further concentrate on the reference case that consists of a smooth interface at $n = 0$. Making it rough involves the turning fluid of cubes in the top layer and of layers below, and simultaneously turning solid of cubes in fluid layers above. Therefore, the change in the Gibbs free energy of the entire solid is given by

$$\Delta G_{fs} = \Delta\mu \cdot N \left\{ \sum_{-\infty}^{0}(1 - C_n) - \sum_{n=1}^{\infty} C_n \right\}. \tag{12.4}$$

We turn to calculate the ΔE_{surf} quantity of Eq. (12.3). We denote by ε the energy needed to generate a "single roughness unit", which means a single bond between a solid box and a fluid box. Using Figure 12.5, we describe a case where one solid cube of a top layer (marked "1") turned fluid, while one fluid tube of a lowest fluid layer turned solid. The total volume of each solid and fluid remains unchanged. The described event produced 8 roughness units, namely 8 new solid–fluid bonds; the process involved the breaking of 4 solid–solid bonds and 4 fluid–fluid bonds, by which, as indicated, 8 solid–fluid bonds were formed. Thus, we have

$$8\varepsilon = 4\Phi_{ss} + 4\Phi_{ff} - 8\Phi_{fs},$$

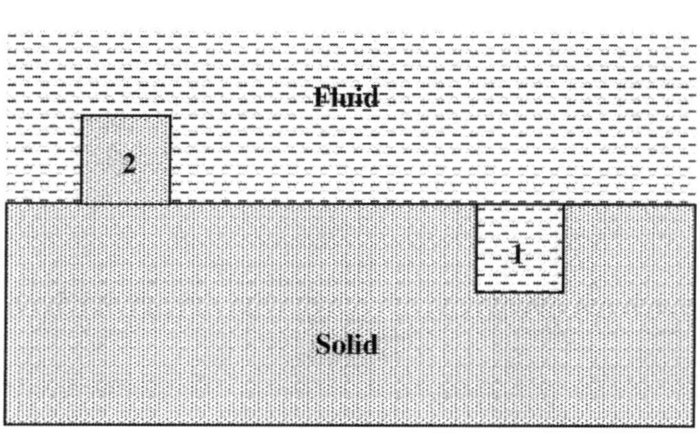

Figure 12.5 Creation of roughness by converting a solid region into fluid (**1**) and a fluid region into solid (**2**).

namely

$$\varepsilon = \frac{1}{2}(\Phi_{ss} + \Phi_{ff}) - \Phi_{fs}. \tag{12.5}$$

The ratio of solid–liquid bonds at the n layer is approximately given by $4C_n(1 - C_n)$ (the ratio of solid cubes is multiplied by the ratio of fluid cubes, then by 4, assuming that each solid cube exhibits 4 solid–liquid bonds; that particular assumption overlooks the possibility of overlapping cases). Under the said approximation, we will have

$$\Delta E_{\text{surf}} = 4\varepsilon N \sum_{n=-\infty}^{\infty} C_n(1 - C_n). \tag{12.6}$$

We turn to calculate the $T \cdot \Delta S_{\text{surf}}$ quantity of Eq. (12.3). We use Boltzmann relation $S = k_B \,\boldsymbol{ln}\, W$, where W is the number of different possible arrangements having the same degree of roughness (note that for a reference surface, $W = \text{Const}$). In a detailed calculation that also uses some simplifying assumption, we get that

$$T \cdot \Delta S_{\text{surf}} = k_B T N \sum_{n=-\infty}^{\infty} (C_n - C_{n+1})\boldsymbol{ln}(C_n - C_{n+1}). \tag{12.7}$$

Summarizing all above-calculated expressions, we get the following expression

$$g \equiv \frac{\Delta G}{N \cdot k_B T} = \beta \left\{ \sum_{n=-\infty}^{0} (1 - C_n) - \sum_{n=1}^{\infty} C_n \right\} + \alpha \sum_{n=-\infty}^{\infty} C_n(1 - C_n)$$
$$- \sum_{n=-\infty}^{\infty} (C_n - C_{n+1})\boldsymbol{ln}(C_n - C_{n+1}), \tag{12.8}$$

where $\alpha \equiv \frac{4\varepsilon}{k_B T}$ and $\beta \equiv \frac{\Delta\mu}{k_B T}$. The α quantity expresses the contribution of the surface roughness to the Gibbs free energy for a constant solid mass, while the β quantity expresses the contribution of the solid mass growth.

A stable state of the solid surface for given values of α and β is obtained by such combinations of C_n where the g quantity (Eq. [12.8]) exhibits a minimum (namely any change in even one of the C_n values causes a rise of g). The finding of the set of such C_n values that provide the searched minimum of g was performed using a numerical computer simulation program. The result analysis shows that the state of the solid growth surface may by classified into two well-defined regions depending on the combination of α and β: one range where the surface state is always smooth; in the other combination, the surface is always rough. On a mathematical (α, β) plane, each region is continuous, and the border between them is very narrow, as shown in Figure 12.6. The area marked "**A**" is where the solid growth layers are always smooth. Here the surface energy α dominates, and the growing solid prefers a minimal surface area, namely a smooth surface. Experimentally, the growth nature is then layered, or exhibits preference of spiral growth around a screw axis as shown in Figure 12.3. A layered growth is difficult to achieve, because such growth requires considerable super cooling. In small α values, on the other hand, the solid volume energy β dominates. The solid growth will be continuous while following the temperature lowering process. The reason is that it is easy to add solid cubes in every location where the temperature drops below T_f: energy gained by the solidification compensates against the growth in roughness.

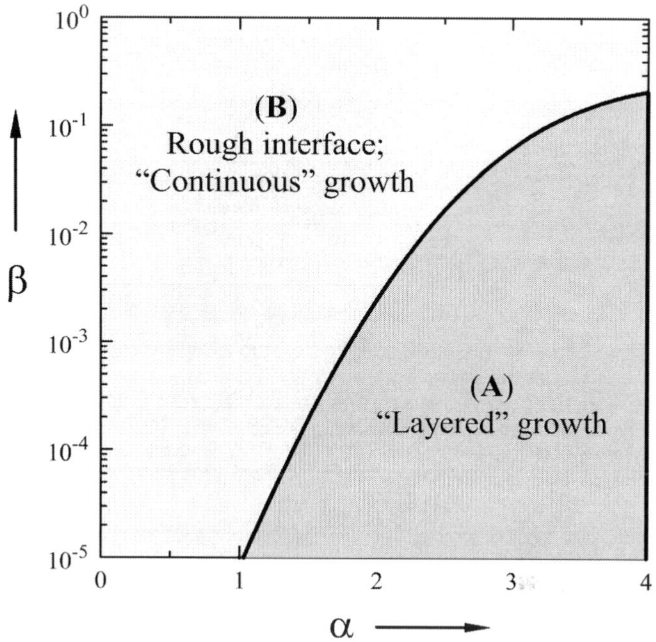

Figure 12.6 An (α,β) parameters plane describing the obtained nature of a solid/fluid interface during a single-crystal growth.

12.4.2 Occurrence of Facets on Grown Crystals

One of the most common features in single crystals, both mineral and industrially grown, is the appearance of facets. The facets are the most planar face characteristic of each crystal according to their relative size, and the angles between them. In fact, that was the property that drew human attention to the existence of a crystalline state. We now have the necessary tools to understand and analyze it.

Consider a solidification process where the freezing temperature geometric isotherm is not planar. Simplistically it is expected that under thermodynamic equilibrium, the solid/melt interfaces would follow the shape of the isotherm. The growing of the crystal may be performed by shifting the isotherm (namely cooling the growth furnace), or by moving the solid part toward the colder furnace zones. In that way, additional solid layers will be added over the already existing solid. There may exist two special cases: when the surface energy is small (α is small) the surface shape will follow the isotherm shape irrespective of its particular shape, as shown in Figure 12.7a. On the other hand, if the surface energy is large (α is large), the solid/melt interface will prefer a planar form, and won't follow precisely the shape of the isotherm, as shown in Figure 12.7b. Still, when a new layer starts to form, its formation is completed very fast. As shown in the figure, there always remains a relatively large volume in a state of super cooling. The value of α depends on the crystallographic plane involved, and obviously on the material. Therefore, the facet occurrence will be accentuated in specific crystallographic orientations, and missing in others. To a large extent, the facet formation expresses the material crystallographic symmetry. A crystal that grows perpendicularly to a plane that favors a facet formation will exhibit a large density of structural defects, such as the screw dislocation (Figure 12.3). For that reason, industrial growing of single crystals favors growth orientations that ensure tracking of the isotherm without overcooling.

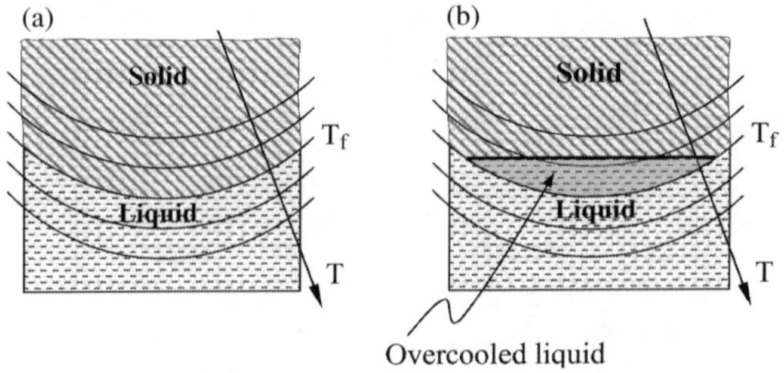

Figure 12.7 Shape of a solidifying front in a matter exhibiting temperature gradients. Isotherms and orientation of the thermal gradients are indicated. (a) Describes a continuous growth with congruence between the freezing temperature isotherm and the solid front, and no liquid overcooling. (b) Describes a layered solidification, where the solidification front forms a facet, while causing formation of an overcooled region in the adjacent liquid.

Notably, considerations related to single crystals growing out of the material melt are equally adaptable to growing out of other nutrient phases. Particularly, growing out of a solution or a gas may involve similar effects of facets formation and oversaturation.

12.5 Solidification Dynamics

The movement of a material solidification front obviously involves a flow of heat. As an example of such case analysis, we consider the situation described in Figure 12.8. The shown system comprises a material contained inside a long "horizontal" tube, where part of it has already solidified, while the rest is still molten. We assume that the heat flow needed to achieve a movement of the solidification front is caused solely by heat flow between the solid and its melt. That assumption establishes a fair approximation at sufficiently low temperatures, where radiation heat loss to the ambient atmosphere may be neglected; when the solidification process is sufficiently fast to justify the former assumption; and when the thermal conduction through the system container walls is sufficiently small compared to the thermal conduction between the addressed solid and its melt. We further assume that no overcooling occurs.

"L" and "S" subscripts will be used to relate any quantity to the liquid or to the solid, respectively. From the heat transfer balance, it is clear that the solid must release both the heat flowing from the liquid and the heat of fusion when the solidification front progresses. Namely, $Q_2 = Q_1 + \rho_S h_f A v$. We set $Q = KAG$ and get

$$K_S A G_S = K_L A G_L + \rho_S h_f A v;$$

whence

$$v = \frac{K_S G_S - K_L G_L}{\rho_S h_f}. \tag{12.9}$$

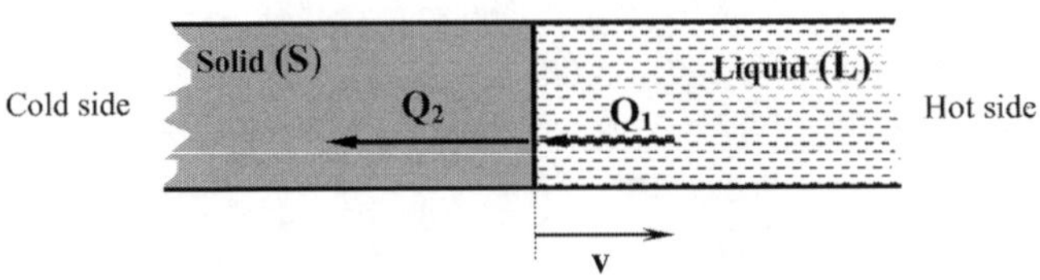

Figure 12.8 A chart of material solidification at a constant velocity and the heat flow during such process.

We denote:

h_f – Solidification latent heat per unit mass in units of J/kg.
ρ – Mass density in units of kg/m^3.
K – Heat conductivity in units of J/s · m · °K.
A – Cross-section area in units of m^2.
G – Temperature gradient ($G \equiv dT/dx$, where T is the temperature and x is a distance oriented from the solid toward the melt) in units of °K/m.
Q – Heat flowing rate in units of J/s.
v – Velocity of the solid surface advancement in units of m/s.

We delve into the physical meaning of the last result by considering several extreme cases:

1) $K_S G_S = K_L G_L$ yielding $v = 0$. That is an equilibrium state, where there is neither solidification nor melting. If the heat conductivities of the melt and solid are identical ($K_S = K_L$), then the temperature gradients from the two sides of the melt/solid interface are also the same.
2) $G_L = 0$; in such case, the solidification velocity assumes a maximal value. The solidification velocity can't become infinitely large, because it is limited by the need to release the solidification latent heat. Obviously, the $G_L = 0$ condition could occur only temporarily, because the emission of the latent heat would change it immediately.
3) $K_L G_L > K_S G_S$; in such case, $v < 0$, which describes a continual melting of the solid. In Figure 12.8, it would mean to reverse the orientation of the plotted arrow, namely the solid/melt interface would move leftward.

There are quite a number of additional processes that affect the heat flow in a real solidification process. Such are

1) Heat transport in the liquid. The rate of heat transport is given by $K_L^{(t)} A G_L$, where $K_L^{(t)}$ is the effective heat conductivity resulting from liquid material convections.
2) Heat radiation from the liquid toward the environment. The rate of heat streaming due to that factor is given by $(\sigma_L A_L {T_L}^4 - \sigma_0 A_L {T_0}^4)$, where σ_L and σ_0 are the radiative heat emission coefficients of the liquid, and the nearby environment, respectively. Their units are J/s · m$^{2°}$K^4; A_L is the outer surface of the liquid, and T_L and T_0 are the absolute temperatures of the liquid and the nearby environment, respectively. The physical law according to which that contribution may be calculated is the Stefan–Boltzmann law (see Eqs. [13.3] and [13.4]).
3) Heat radiation from the solid toward the environment. The rate of heat streaming due to that factor is given by $(\sigma_S A_S {T_S}^4 - \sigma_0 A_S {T_0}^4)$, where σ_S, A_S, and T_S are the radiative heat emission coefficients of the solid, its outer surface area, and absolute temperatures, respectively.
4) Heat transport by the ambient atmosphere. That transport is proportional to the outer surface area of the material in contact with the ambient atmosphere, and depends on that atmosphere composition and the developing gas convection.

Equation (12.9) may be refined by including expressions describing the above-indicated 1–4 processes; they are often quite significant.

Still, Eq. (12.9) demonstrates the importance of temperature gradients for determining single-crystal growth rates. Initiation of a crystal growth involves the generation of temperature gradients in both the solid and the liquid. A system exhibiting low gradients is very unstable, because solid growth rates are rather small, and fluctuations in the temperature gradients will cause gross relative changes in the growth rates. It would cause accentuated effects of overcooling, appearance of facets, and formation of many structural defects. This also complicates the temperature control systems. We should thus strive to design growth systems forming the largest temperature gradients in the liquid. That would also cause large temperature gradients in the solid (otherwise the solid would melt instead of growing). However, large gradients in a solid may cause internal stress or even fracture. Under the above considerations, it is obvious that an optimal arrangement should be tailored for the growth of every specific single crystal.

12.6 Segregation

In this section, we consider a binary molten material mixture, whose "A" and "B" components are soluble in each other by any ratio. A typical **phase diagram** of such case is provided in Figure 12.9. The C concentration expresses the fraction (mass or molar) of "B" in "A." Thus, $C = 0$ case relates to a pure "A" material, and $C = 1$ case relates to a pure "B" material. The T axis is the temperature. The **solidus** line describes the highest temperature at which the materials mixture at the C concentration could be solid (the melting temperature T_m). The **liquidus** line describes the lowest temperature at which the material mixture at the C concentration could be fluid (the freezing temperature T_f). The area between these two lines is gray highlighted; it doesn't represent any material state of thermodynamic equilibrium. It is useful to think of it as cut out and removed from the diagram. It is still important to know that in many practical cases, such combination may exist under a non-thermodynamic equilibrium.

Any process that occurs in a substance requires deviation from thermodynamic equilibrium conditions. However, **a phase diagram is a very important tool in the description of processes, especially those parts of the system that remain close to thermodynamic equilibrium**. Accordingly, Figure 12.9 also provides a description of an occurring process; some liquid of a C_L concentration cools down. On reaching the liquidus line, it can't stay liquid, neither can it skip over the temperature gap required by the diagram to become a solid of the same concentration, because it would require an infinitely fast dissipation of the latent heat. On the other hand, a different skipping along the concentration axis is possible, and this does occur. Tiny amounts of a substance at the C_S concentration are secreted in the liquid without the need to change the temperature. The process is called **segregation** of "B" off "A"; the effect size is characterized by the **segregation coefficient** $k \equiv (C_S/C_L)$. In the process marked in Figure 12.9,

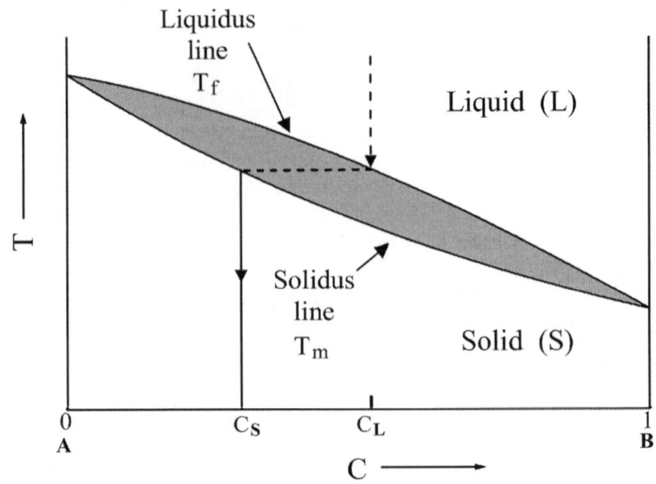

Figure 12.9 Phase diagram of two materials "A" and "B" mixture mutually soluble at any ratio. A process description is added where a solution of a C_L "B" concentration is cooled down, ending as a solid solution of a C_S "B" concentration.

the segregation coefficient is smaller than a unity ($k < 1$), and that property of k is also characterized by an effect where increasing of a "B" material concentration in the mixture reduces the melting temperature. There are other cases, where the segregation coefficient of "A" off "B" is larger than a unity. Such cases express themselves when an increase of the "A" concentration in the mixture increases the melting temperature. Because the liquidus and solidus lines are not linear, the segregation coefficient is not constant, and it depends on the "B" material concentration.

The material mixture whose concentration is larger is called **solvent**; the material mixture whose concentration is smaller is called **solute**.

We delve into further analysis of the described process. Because any amount of formed solid contains less of the "B" component, the remaining liquid is enriched with the "B" component; in other words, C_L grows. It may be illustrated on the phase diagram as a gradual rightward crawling of the liquid concentration along the liquidus line. That part of the process actually distracts the system from the pure thermodynamic equilibrium state. The reason is that the formed solid never exhibits material flows; therefore, the concentration of "B" in the solid will grow gradually. That effect is called **segregation**. Notably, there are systems of different phase diagrams for which the segregation coefficient is a unity (namely $k = 1$). Such cases are described to exhibit a **congruent solidification**; in such cases, the composition of the grown solid is identical to the composition of the nutrient fluid. (In Figure 12.9, congruent solidification occurs trivially for $C = 0$ and $C = 1$.)

We turn to analyze rather a melting process according to the phase diagram of Figure 12.9. A piece of solid material containing a C_S concentration of the "B" component is heated. When the temperature reaches the solidus line, it can neither remain solid nor skip over the temperature gap to become a liquid of the same concentration, because it would require an infinitely fast supply of heat. Instead, the other kind of skipping occurs – in the direction of the concentration axis. Molten material regions at a C_L concentration of "B" are generated, whose amount depend on the amount of the supplied heat. Usually, a molten material density is different than that of the solid. Therefore, great mechanical stress is formed between the solid and the molten region, which eventually causes the solid breaking into large pieces. The heated solid becomes a mixture of a liquid at a C_L concentration, and fragments of solid material of concentrations that are smaller than C_S. Such a state is called **slush**. When the mixture is further heated, the fragments of solid turn smaller, thereby diluting the liquid, whose concentration crawls leftward along the liquidus line; anyhow, the average concentration in the slush remains exactly C_S. At the very moment where the slush temperature reaches the liquidus line at the C_S concentration, the entire material becomes liquid, having a uniform C_S concentration.

The segregation effect emanating from the occurrence of a segregation coefficient different than a unity may as well be described on an atomic scale. Any solidification involves numerous processes of adsorption and escape of the fluid and solid atoms. A crystal growth needs to be carried out very slowly, to allow each atom to settle on the site of the lowest energy; then the best crystal arrangement will be obtained. Different kinds of atoms exhibit different bonding forces with a certain solid. If the minority type of fluid atom has a weaker bonding force to the solid compared to the force of the majority fluid atoms, the adsorbed minority atoms relative concentration will be smaller than their relative concentration in the fluid. Then, we will have $k < 1$. If their bonding force is stronger, we will have $k > 1$.

The process thus described is based on the phase diagram, which is valid under thermodynamic equilibrium conditions. Process analyses may thus be inferred only for very slow processes. Real processes often deviate to some extent from thermodynamic equilibrium, and segregation coefficients would then be described in relation to the actual cases. To refine their definition accordingly, consider Figure 12.10. In that figure, $C_i(L)$ is the solute concentration at the interface

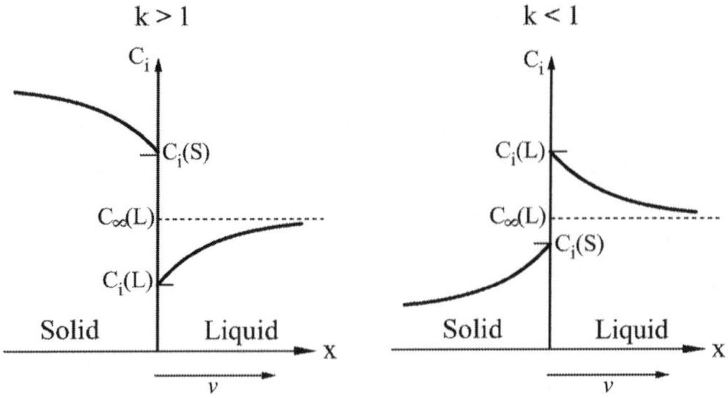

Figure 12.10 Schematic illustration of a solute distribution near the solid/liquid interface during a crystal growth in cases of $k > 1$ and $k < 1$. For a detailed discussion, see the text.

inside the solid, and $C_\infty(L)$ is the solute concentration in the liquid far from the interface, namely in a region exhibiting considerable self-stirring. We thus define two additional segregation coefficients:

$$k_{\text{eff}} \equiv \frac{C_i}{C_\infty(L)}; \quad k_{\text{i}} \equiv \frac{C_i(S)}{C_i(L)}. \tag{12.10}$$

k_i expresses the solute segregation coefficient adjacent to the crystal growth surface, which is influenced by the conditions of the system deviation from exact thermodynamic equilibrium. k_{eff} is the effective segregation coefficient, which serves as the more practical quantity, because it is defined by measurable quantities; the solute concentration is mostly measurable far from the crystal growth interface. Both segregation coefficients defined by Eq. (12.10) depend on the crystal growth velocity, and on the nature of the material stirring in the liquid.

When a solidification is very slow ($v \to 0$), the interface segregation coefficient practically satisfies the relation $k_i = k$, and when the liquid exhibits vigorous stirring (bringing the system closer to thermal equilibrium conditions), the effective segregation coefficient also practically satisfies $k_{\text{eff}} = k$. If the solidification is very fast ($v \to \infty$), escape of a solute atom accidentally bonded to the solid becomes unlikely. Then, the solute concentration in the solid becomes identical to that in the liquid, namely $k_{\text{eff}} = 1$. Occurrence of such process is called **quenching**.

The fact that practical values of k_{eff} depend on conditions of the crystal growth in both the growing crystal and the liquid is the reason why experimental segregation values reported in the scientific literature for the same solid/fluid combinations exhibit considerable variance. The experimental results depend on the growth method, the solidification velocity, the temperature gradients, and even the shape of the material container that affects the material flow in the liquid, which therefore affects the nature of the mixing.

In that respect, it is interesting to indicate that during the 1980s, crystal growing experiments were performed in the outer space under microgravity conditions, thus minimizing liquid flow. It was expected that only diffusion may occur in the liquid, a process that may be easily calculated, and the grown crystals would perhaps exhibit higher quality. To the best of the present author's knowledge, the results did not show a significant change compared to growths made under optimal conditions on Earth.

A connection between the effective segregation coefficient and the crystal growth conditions is expressed in what is known as Burton–Prim–Slichter equation:

$$k_{\text{eff}} = \frac{k_i}{k_i + (1 - k_i)\boldsymbol{exp}\left(-\frac{v\,\delta_D}{D}\right)}, \tag{12.11}$$

where v is the crystal growing velocity, D is the solute diffusion coefficient, and δ_D is the thickness of the boundary layer, whose meaning will be clarified in Section 12.9. As to that quantity, there exists a region where its value is inversely proportional to the crystal growing velocity. By Eq. (12.11), k_{eff} is approximately constant. However, under very slow solidification velocities, or under very fast ones, the boundary layer thickness assumes a constant value; k_{eff} then converges to the values described in the previous discussions at the $v \to 0$ and $v \to \infty$ limits.

On a practical aspect, it is possible to infer the segregation coefficient from the relevant phase diagram of binary materials, such as presented in Figure 12.9. A different possibility is to carry out a solidification process by cooling of a molten material, where the freezing temperature T_{f} is measured for different solute concentrations. The freezing temperature satisfies the well-known cryoscopic relation

$$\frac{dT_{\text{f}}}{\text{d}C} = -\frac{(1-k)RT_{\text{f}}^2}{H_{\text{f}}}, \tag{12.12}$$

where R is the gas constant, C is the solute molar concentration, and H_{f} is the molar freezing latent heat (Eq. [12.12] actually describes the local slope of the liquidus curve). Comparing the results with the cryoscopic relation allows extracting the value of k. Another way is to experimentally solidify the solute-containing material very slowly, and analyze the segregation effect, namely the variation of the solute concentration as a function of the solidified fraction according to Pfann's normal freezing equation. That is the subject of the next section.

12.7 Pfann's Normal Freezing Relation

Assume that a molten material of mass M is initially included in a long bath. The material started to solidify from the (left) side as seen in Figure 12.11, and at a certain moment, a fraction g of the mass has solidified. Without loss of generality, and for demonstration, we assume that the solute segregation coefficient is $k < 1$.

Under such conditions, while the solid fraction grows, the solute concentration in the molten material increases due to the partial rejection from the growing solid. We assume that the added solute dispersion throughout the melt is very fast, such that any added solute to the melt is evenly distributed throughout the entire molten mass. The solute contained in a dg mass that is about to solidify is $dgMC_L$, where C_L is the solute concentration in the melt. The amount of solute in the solid-becoming mass is $kdgMC_L$, and according to our assumption, the melt is enriched by a solute sum of $(1-k)dgMC_L$. That added solute is distributed evenly in the remainder melt; namely

$$dC_L = \frac{dgMC_L(1-k)}{(1-g)M};$$

$$\frac{dC_L}{C_L} = -\frac{k-1}{1-g}dg.$$

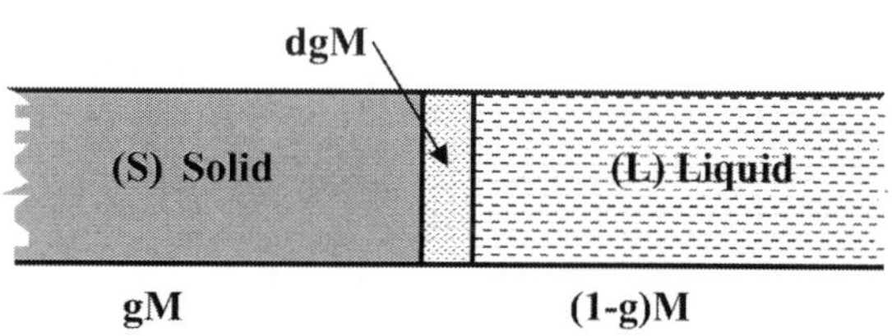

Figure 12.11 Differential representation of a normal freezing.

Integration of both sides yield

$$\ln(C_L/C_0) = \ln(1-g)^{k-1},$$

where $C_0 \equiv C_L(g=0)$, whence we get the expressions known as **Pfann relation**:

$$C_L = C_0(1-g)^{k-1}; \quad C_S = kC_0(1-g)^{k-1}, \tag{12.13}$$

where $C_S(g)$ is the solute concentration in the solid. That equation is based on the assumption that solute dispersion throughout the solid is extremely small, practically negligible. Equation (12.13) is also known as the **normal freezing relation**. It is very practical in describing the solute distribution in cases of very slow crystal growing, especially when the melt is stirred vigorously, for example when the furnace heating is achieved by high-frequency inductive irradiation, an issue to be addressed in more detail in Section 13.4.

As expected, the normal freezing equation predicts that for $k < 1$, the solute concentration in the solid C_S is a steadily increasing function of the solidified fraction g. In a case where $k > 1$, C_S is a steadily decreasing function of g. These are the previously addressed segregation effects. As we shall see in the next section, using that effect allows refining materials off remnant impurities. A complete solidification process exhibits a situation where impurities of $k > 1$ are concentrated in the first solidified material, while impurities of $k < 1$ are concentrated in the last solidified material. If these regions are removed, the remainder solid exhibits a much higher purity state.

12.8 Zone Refining

In the present section, we describe a material purification system where the initial material state is solid, rather than a liquid as described in Section 12.7. The said solid contains a uniform distribution of an impurity of C_0 concentration throughout its entire volume. The system construction is presented in Figure 12.12. A narrow section on the left side of the solid is heated to melt, and the heating is moving very slowly rightward, maintaining a melted zone of width Δ. That way, a slow freezing takes place at the left side of the melted section, and a slow melting takes place at its right side. It is assumed that the melted zone itself mixes well, so that it has a uniform distribution of impurities throughout.

For a mathematical analysis, we consider a material rod of length L, cross-sectional area A, and a melted zone. For simplicity, we assume that the specific gravities ρ of the solid and melt are the same. The impurity concentration in the re-solidified material is marked C_S. Since the refining process started, the melted zone has reached a distance x from the left side. Over a rightward differential motion dx of the heating furnace, the left-side solid acquired more solid mass, containing an amount $C_S\rho A dx$ of the impurity. The amount of impurity added to the melted zone

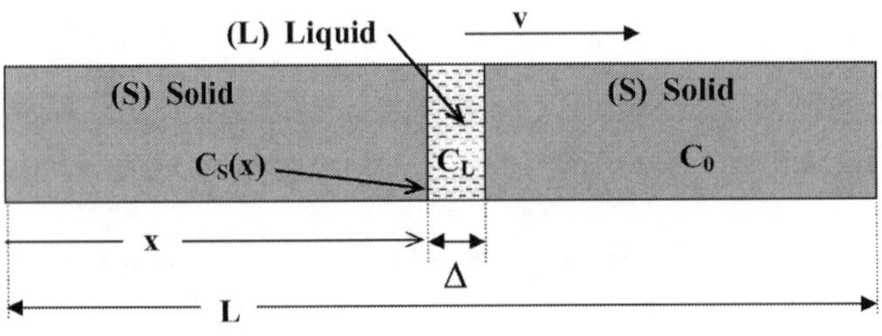

Figure 12.12 Diagram of a zone-refining process.

Δ is $C_0\rho Adx$. The change in impurity content in the melted zone is $(C_0\rho Adx - C_S\rho Adx)$. Thus

$$dC_S = k\frac{C_0\rho Adx - C_S\rho Adx}{\rho A\Delta}.$$

$$dC_S = \frac{k}{\Delta}(C_0 - C_S)dx \Rightarrow \frac{dC_S}{C_0 - C_S} = \frac{k}{\Delta}dx.$$

Integration of the two sides yields

$$\frac{k}{\Delta}x = -\boldsymbol{ln}(C_0 - C_S) + W,$$

where W is an integration constant, whose value is set according to the boundary conditions. Setting the condition $C_0 - C_S(x = 0) = C_0(1 - k)$, we get

$$\frac{k}{\Delta}x = -\boldsymbol{ln}\frac{(C_0 - C_S)}{C_0(1 - k)}.$$

Then, by extracting C_S we get

$$C_S = C_0\left[1 - (1 - k)\boldsymbol{exp}\left(-\frac{k}{\Delta}x\right)\right]. \quad (12.14)$$

Analysis of Eq. (12.14) indicates that at the beginning of the solidification process, when $x = 0$, we get the expected $C_S = kC_0$ result. Further on, when $x \geq 3\Delta/k$, we get $C_S = C_0$. These results may be understood by considering a specific case where $k < 1$. In the beginning of the process, the impurity remaining in the solidifying mass is small according to the segregation coefficient being smaller than a unity. The impurity fraction "supplied" to the melting zone mass is constant; therefore, the impurity concentration in the molten zone gradually grows, until reaching a distance x, where the rate of impurity entering the zone becomes equal to rate of impurity leaving it on its left side, namely a concentration $C_S = C_0$. That state will continue until the molten zone reaches the right end of the system, where the solidification follows a normal freezing process (Eq. [12.13]). A graphical description of impurities' distributions that follow several successive solidification procedures will be presented in Figure 12.15. As noted at the end of Section 12.7, impurities exhibiting a $k > 1$ concentrate at the left side, which was first to solidify, while those exhibiting a $k < 1$ concentrate at the right side, which is last to solidify.

The described zone melting process may be repeated several times from left to right. By the end of each such melting scan, the heating furnace stops moving, cools down, returns to the left side, heated, and begins a new scan. After many scans, the impurities concentrate at the solid ends according to their segregation coefficients and to the width of the molten zone. After finishing a round of zone-refining scans, the solid ends may be cut and removed, and the remained material is prepared for a new round of zone-refining scans.

12.9 Diffusion-Controlled Oriented Solidification

The normal freezing relation (Eq. [12.13]) was obtained by assuming that impurity mixing in the liquid is intense, causing it to achieve a uniform distribution throughout the entire liquid volume. That assumption is never fully valid. There always exists a boundary layer adjacent to the solid surface, where the liquid mixing is restrained. The dispersal of impurities is then dominated by diffusion. The thickness of the boundary layer itself depends on the solidification rate and on the nature of the liquid material flows. Generally, it may be considered as being quite similar to the situation in a zone-refining process: the boundary layer is similar to the melted liquid section, while the remainder liquid, where the concentration of impurities is uniform, is rather similar to the right-side "nutrient solid". The fundamental difference, however, is that the concentration of impurities in the boundary layer is never uniform.

In the following presented model (contrary to an assumed normal freezing case), we assume that material mixing occurrence in the liquid is negligible. Without loss of generality, and for demonstration, we assume that $k < 1$. We denote by C_0 the concentration of impurities while the material is fully molten. The impurity concentration at the first solidified layer is kC_0. Because $k < 1$, the adjacent liquid layer is enriched with impurities, and that concentration will grow steadily during the solid growth, until balanced by a forward diffusion, thus reaching a steady concentration.

The shape of the impurities' distribution concentration in the liquid adjacent to the solidification front is presented in Figure 12.13. Its value C_L in the liquid is larger than C_0 right near the solidification front, and decays toward the C_0 value at growing distances from that front. The factor influencing that decay structure are (i) diffusion in the liquid works to distribute the impurities uniformly throughout the liquid, of a diffusion coefficient D in units of m^2/s. (ii) The solidification that acts as a source for added impurities is adjacent to its front. It is characterized by the front advancing velocity v in units of m/s. The C_a quantity is the added concentration of impurities on the liquid side of the solidification front.

Let x denote the distance of some point in the liquid from a reference position in the solid, and let ξ denote the distance of that point from the solidification point advancing at a velocity v. Their relation is obviously $x = \xi + v\,t$, where t is the time from the start of solidification. In the general case, the concentration of impurities in the liquid is written as $C_L = C_L(x, t)$. We perform a change of variable from x into ξ, and consider the full-time differential of C_L

$$\frac{dC_L}{dt} = \frac{\partial C_L}{\partial \xi} \cdot \frac{\partial \xi}{\partial x} \cdot \frac{dx}{dt} + \frac{\partial C_L}{\partial t}.$$

In our case, $\frac{dx}{dt} = v$ and $\frac{\partial \xi}{\partial x} = 1$. By the second Fick diffusion law, $\frac{\partial C_L}{\partial t} = D\frac{\partial^2 C_L}{\partial \xi^2}$, where D is the diffusion coefficient of impurities in the liquid. Therefore

$$\frac{dC_L}{dt} = v\frac{\partial C_L}{\partial \xi} + D\frac{\partial^2 C_L}{\partial \xi^2}.$$

The solution of that differential equation under a steady state ($dC_L/dt = 0$) and the $C_L(\infty) = C_0$ boundary condition is

$$C_L(\xi) = C_0 + C_a \boldsymbol{exp}\left(-\frac{v}{D}\xi\right).$$

The solution describes an increase of the concentration of liquid impurities by C_a over the average C_0, taking place adjacent to the solidification front. That increase decays exponentially

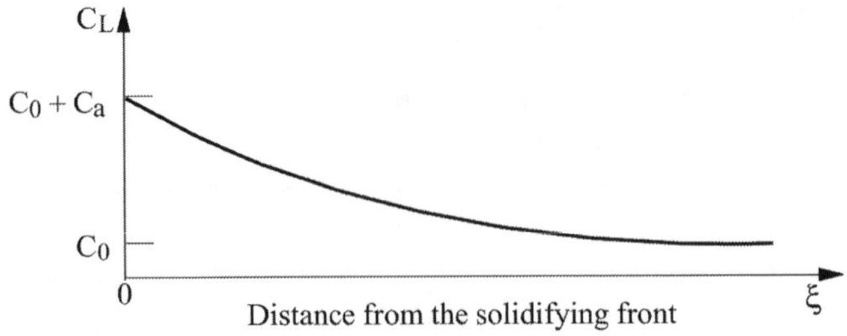

Figure 12.13 Distribution of impurities in the molten material in front of the solidification front, advancing at a velocity v.

with distance from the solidification front, with a characteristic decay range of $\delta_D = D/v$, that expresses the width of the boundary layer. Typical value in solid growth is $v \sim 0.1$ cm/s and diffusion coefficient of impurities $D \sim 10^{-5}$ cm^2/s. Correspondingly, $\delta_D \sim 10^{-4}$ cm. Notably, the boundary layer thickness is expected to be inversely proportional to the solid growth velocity v. In practice, as a result of other effects like the liquid viscosity and development of liquid inner flows mixing, the boundary layer width never becomes larger than a certain limit, obtained under small solid growth velocities.

We now consider in more detail the progressing process before reaching the former-described steady state. By the first moments of solidification, the solidified material will be left with impurities concentration of only $C_S = kC_0$, and the adjacent liquid will just start to accumulate more impurities. On continued solidification, the concentration of liquid impurities increases, accompanied by a simultaneous increase of concentration of impurities inside the solid. The increase continues until the amount of impurities entering the solid becomes equal to that added into the liquid. On reaching a steady state, we would have $C_0 = k(C_0 + C_a)$, whence $C_a = \frac{1-k}{k}C_0$. We set that conclusion in the above expression for $C_L(\xi)$ and get

$$C_L(\xi) = C_0 + \left[1 + \frac{1-k}{k}\right] \boldsymbol{exp}\left(-\frac{v}{D}\xi\right). \tag{12.15}$$

We turn to calculate analytic expressions for the concentration of impurities in the solid at each point before the system reached the steady state. Summarizing our former conclusions:

1) When solidification starts, the concentration of impurities in the solid is kC_0.
2) The concentration of impurities in the solid rises asymptotically toward C_0.
3) By consideration of total impurities conservation, the amount of impurities "rejected" from the solid till steady state is equal to the excess amount of impurities in the boundary layer at the solidification front.

The above-described conditions are presented in Figure 12.14, which describes the gradual initial rise of the concentration of impurities in the solid from the kC_0 value toward the asymptotic C_0 "goal". We assume that the concentration between these two ends displays an *exponential reduction* as a function of x: for a moment, we denote the exponent parameter as α. The sole justification for that assumption is by its mathematical convenience. A fundamental condition is that the increased impurity concentration in the liquid at the solidification front is

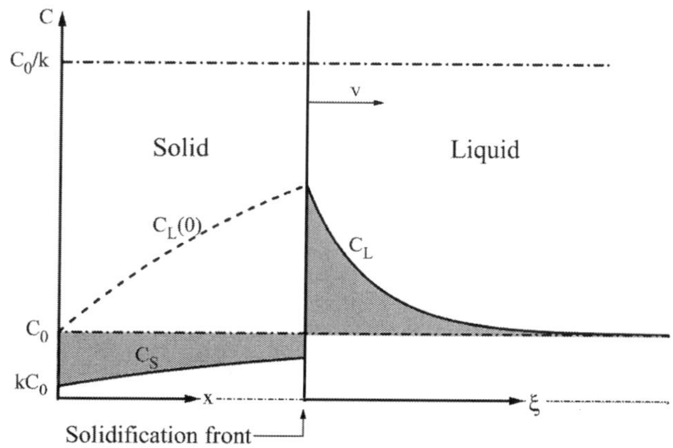

Figure 12.14 Distribution of impurities in the liquid C_L and in the solid C_S at the first part of the solidifying material. $C_L(0)$ shows the concentration of impurities that existed in the liquid adjacent to the front of the solidifying material. The gray-highlighted regions emphasize the sum of impurities "rejected" from the solid (leftward of the solidifying front at $\xi = 0$) and the corresponding sum of impurities added to the liquid (rightward of the solidifying front at $\xi = 0$), respectively. C_0 is the original concentration of impurities in the liquid.

equal to a reduced impurity concentration in the solid at the front left side. Mathematically, the said conditions are

$$(C_0 - C_S) = (1 - k)C_0\boldsymbol{exp}(-\alpha x);$$

$$\int_0^{\infty}(1 - k)C_0e^{-\alpha x}dx = \int_0^{\infty}\frac{(1 - k)}{k}C_0e^{\left(-\frac{v}{D}\xi\right)}d\xi.$$

Performing the integrations, and extracting the α parameter, we get $\alpha = k(v/D)$. Setting in the above first equation, we get

$$C_S(x) = C_0\left[1 - (1 - k)\boldsymbol{exp}\left(-k\frac{v}{D}x\right)\right]. \tag{12.16}$$

As expected, Eq. (12.16) describes an asymptotic rise of C_S toward the constant C_0 value, with a characteristic approaching length D/kv. The concentration of impurities in the solid reaches a C_0 value after several D/kv lengths of the solid/fluid interface advancement.

Notably, the above-described process that is summarized in Figure 12.14 is transient; in most practical systems, it takes place along a fraction of a millimeter of the first solidified material. It also appears at the very left side of the diffusion-dominated solidification curve as shown in Figure 12.15. Obviously, the excess impurities donated to the liquid during the beginning of solidification must enter the solid at the end of the process, when the entire liquid had solidified. This effect starts to be noticed at a distance of approximately $3D/kv$ before the end of solidification

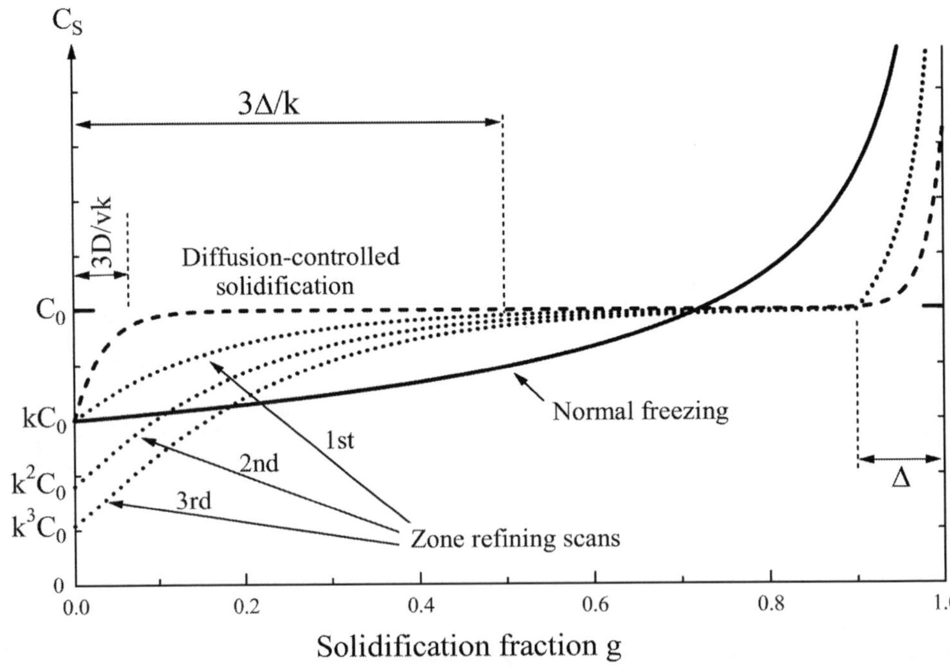

Figure 12.15 Distribution of impurities in the solid C_S during different solidification processes when $k < 1$. The g quantity expresses the solidification fraction in cases of a full fluid steering and a diffusion-controlled solidification, and the fraction of the refined solid material on the last zone-refining scan.

(see Eq. [12.16]), where there is not sufficient liquid to accept the impurities "pushed-off" from the solid. An analytic analysis of the impurities rise in the solid in that range is quite complex and won't be detailed here. A schematic illustration of it appears on right side of Figure 12.15.

12.10 Constitutional Supercooling

The liquid at the opposite side of the solidification front may become supercooled (undercooled) due to two different causes. One is that in order to induce a cause for solidification, the liquid must assume a temperature that is below the equilibrium solid/liquid freezing point T_f. That effect is independent of the presence of impurities, and it is valid for an entirely pure material. The other reason relates to the presence of impurities in the liquid, and it is addressed as **constitutional supercooling**. The concentration of impurities $C_L(\xi)$ in the liquid side of the solidification front is given by Eq. (12.15), and it therefore exhibits a freezing temperature dictated by the liquidus line of a binary solvent/solute system (see Figure12.9). We denote $\frac{dT_f}{dC_L} \equiv m$ and assume for simplicity that m is constant. For demonstration, we assume that the solute presence increases the freezing temperature upon increase of its concentration, namely $m > 0$ and $k > 1$. The solidification temperature T_f (the liquidus line) will thus be given by $T_f = T_0 + mC_L$, where T_0 is the freezing temperature of the pure solvent (which is also equal to its melting temperature). Setting C_L according to Eq. (12.15) for the steady state of the solidification, we get

$$T_f(\xi) = T_0 + mC_0 \left[1 + \frac{1-k}{k} \boldsymbol{exp}\left(-\frac{v}{D}\xi\right)\right].$$

On the other hand, the actual temperature in front of the solidification front is given by

$$T(\xi) = T_0 + m\frac{C_0}{k} + G \cdot \xi,$$

where G is the temperature gradient in the liquid dictated by the heating furnace (note that $T_0 + m\frac{C_0}{k}$ is the solidification front temperature). In Figure 12.16, we provide a schematic illustration of a material freezing temperature T_f dependence on the distance from the solidifying front ξ, and of the actual fluid temperature dependence on ξ, for two different temperature gradients G. As clearly seen, when the temperature gradient is too small (curve T_{ii} in the figure),

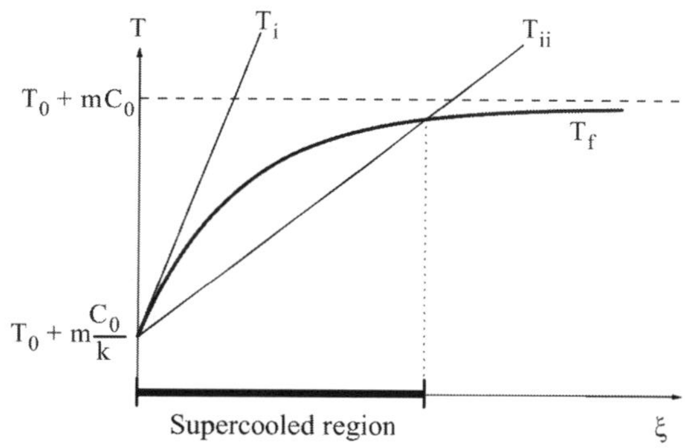

Figure 12.16 Formation of constitutional supercooling: the thermodynamic equilibrium freezing temperature T_f and possible two actual temperatures, T_i and T_{ii}, that may be established in a liquid existing to the liquid (right) side of the solidification front, as functions of the distance from the solidification front ξ. Under small temperature gradients such as for T_{ii}, a supercooled region forms.

quite a considerable region in front of the solidifying front is undercooled (supercooled), namely the actual temperature of that region stays below the freezing temperature T_f. That region might undergo a very fast freezing, involving many structural defects. The special case marked T_i would exhibit no supercooling: it occurs when the slope of the T_f curve starting at $\xi = 0$ and $T_f = T_0 + m\frac{C_0}{k}$ is equal to, or smaller than, the actual temperature gradient at that point. The condition on the temperature gradient G for avoiding a hazard of constitutional supercooling occurrence is $G \geq mC_0\frac{k-1}{k} \cdot \frac{v}{D}$. The common presentation of that important result in most scientific literature is[1]

$$\frac{G}{v} \geq \frac{mC_0}{D} \cdot \frac{k-1}{k}. \tag{12.17}$$

If the fluid exhibits vigorous flows, Eq. (12.17) is quite valid, where D means an effective diffusion coefficient. In growing of single crystals, it is important to avoid the formation of constitutional supercooling. For that purpose, the temperature gradient G in the fluid should be made high, the crystal growing velocity v be made small, and the effective diffusion coefficient be made high by inducing vigorous material mixing in the liquid.

When constitutional supercooling does occur, the liquid in front of the solidification front state is far from thermodynamic equilibrium. This effect often involves a relatively rapid solidification, where a solidifying layer contains a high concentration of impurities. The following solidified layer will exhibit a much smaller concentration of impurities, and so repeatedly over many layers. These layers are addressed as **striations**. The impurities' "rich" layers may be described as having an effective very high segregation coefficient ($k_{eff} \gg 1$), sometimes addressed as **anomalous segregation coefficient**.

An effect of a local high concentration of impurities may occur via other mechanisms. When solidification occurs adjacently to a crystalline natural facet (see Section 12.3), sometimes local fast solidification occurs, namely involving an anomalous segregation coefficient. Such effects are likely to occur in regions of less efficient fluid mixing. Practically, it is difficult to obtain identical mixing throughout the entire fluid; in some regions, the fluid flows are fast, and in others, called **stagnation points**, flows are slow or even practically absent. As an example, we describe a region near the solidification front exhibiting a flow stagnation point (Figure 12.17a). Solidification in that region will be delayed compared to the other regions due to the involved increase in the concentration of impurities. Then, a kind of a dent forms (Figure 12.17b), furthering the flows stagnation – thus deepening the dent (Figure 12.17c). Eventually, the dent may close, forming a kind of an enclave of a liquid containing a high density of impurities. At a later stage of temperature lowering, that liquid enclave solidifies. If it contracts, the enclave forms a bubble or a void. If it expands, the enclave induces a strained vicinity.

Occurrence of bubbles or strained areas is definitely undesirable in the growth of single crystals. To minimize the chances of that possibility, proper measures are taken to overcome the possibility of constitutional supercooling. These measures include maximizing the G/v ratio, meaning increasing as much as possible the temperature gradients G in the growth furnace, and

1 In quite a number of scientific publications, the solidification front velocity v is replaced by the symbol letter R (acronym of the "Rate" word). We refrain from that use to avoid confusion with the gas constant that also employs the R letter.

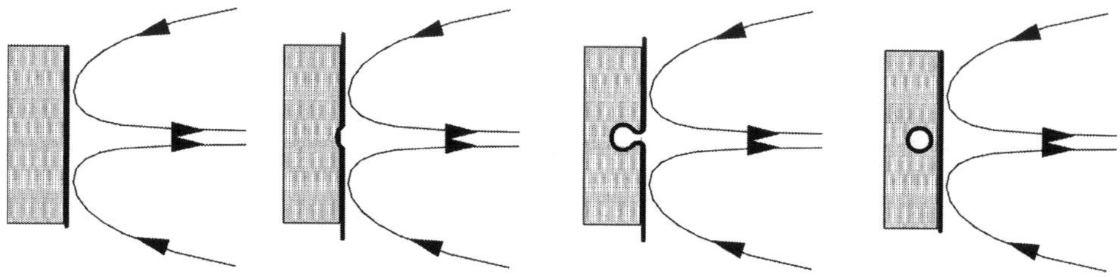

Figure 12.17 Schematic illustration of a bubble formation (a, b, c, and d-cases, successively) inside the solidifying material in a state of constitutional supercooling generated close to a stagnation region of the mixing fluid flows.

minimizing the crystal growth velocity v. It is sometimes possible to increase the effective diffusivity by stimulating vigorous mixing in the liquid.

Notably, there exists an additional mechanism capable of causing bubbles in a growing crystal. If the liquid contains a large amount of dissolved gas, or gas bubbles, it is likely that they will follow the solidifying material, and unite to form gas bubbles inside the solid. In such cases as well, the means for their minimizing is the lowering the solid growth velocity and enhancing vigorous mixing in the liquid.

12.11 Factors Affecting the Segregation Coefficient

In developing the above mathematical expressions, the segregation coefficient k was assumed for simplicity to be a constant quantity. However, already at the initial presentation of the segregation issue in Section 12.6, we presented a general case where the liquidus and solidus lines are not linear (Figure 12.9); therefore, the effective segregation under thermodynamic equilibrium k_0 depends on the solute material density. Furthermore, on departure from strict thermodynamic equilibrium, the effective segregation coefficients may depends on various conditions in the solidifying material. As described in Section 12.6, as the solidification velocity grows, the effective segregation coefficient k_{eff} approaches a unity value 1. As the solidification velocity reduces, it approaches the thermodynamic equilibrium value k_0. The value of the k_i parameter (see Eq. [12.10]) is also not constant, and it depends on the liquid mixing, the solidification velocity, and the effective diffusivity of the solute near the solidification front.

The segregation coefficient value under thermodynamic equilibrium k_0 is constant under a constant solute concentration. Still, the actual value may be changed by changing the solute composition. That means is sometimes used in technology in cases where k_0 is very small compared to a unity "1", due to a single reason or to a combination of the following reasons:

1) The ion size: Impurity ions, whose ionic radius is considerably different from the radius of the original (legitimate) ion residing in the intended crystallographic site for their replacement, sometimes involve shape deformation of the site vicinity due to the "improper dimensions" of the replacing impurity ion. It causes an increase of the free energy in the vicinity of the

replacing ion, usually entailing a segregation constant which is very small compared to a unity. The solubility limit of the ion in the solid is then usually quite small.

2) The ionic charge: Impurity ions, whose ionic electrical charge is different from that of the "legitimate ion" intended to be replaced, would cause an electrical charge imbalance in their close vicinity. Such occurrence also causes shape deformation of the site vicinity due to the "improper electrical charge" of the replacing impurity ion. It causes an increase of the free energy in the vicinity of the replacing ion, usually entailing a segregation constant which is very small compared to a unity. The solubility limit of the ion in the solid is then also quite small.

Having a segregation coefficient much smaller than a unity "1" involves many difficulties in growing crystals exhibiting uniformity in their impurity concentration. That is valid especially if the goal is a single crystal of a large volume. It is possible to change k_0 toward the desired value of 1 by adding an appropriate different solute to the liquid. That method is called **compensation**. There are thus two compensation methods: **size compensation** and **charge compensation**.

12.11.1 Size Compensation

When the reason for k_0 being much smaller than 1 is that the solute atomic ion is larger than the solid original ion in the lattice, it is possible to add a secondary solute, whose atomic ion is smaller than the solid original ion in the lattice. That will make the size of a combined large and small pair of the solute similar to a pair of the original ions in the pure solid. The prospected lattice deformation on entering the growing crystal will be small, and their effective segregation coefficient will become close to 1. Moreover, such an extent increases the solubility of the two types of impurities combined, compared to dissolving each of them separately. It allows obtaining single crystals having a large density of a desired dopant without induced structural defects.

12.11.2 Charge Compensation

When the reason for k_0 being much smaller than 1 is that the solute atomic ion has a different valence than that of the solid original ion in the lattice, it is possible to add a secondary solute, whose atomic ion valence deviates in the opposite direction than the valence deviation of the main solute ion. That will make the size of a combined pair of the two different-type ions valence similar to a pair of the original ions pair in the pure solid. The prospected lattice deformation when they enter the growing crystal will be small, and their effective segregation coefficient will become close to 1. Moreover, such an extent increases the solubility of the two types of impurities combined, compared to dissolving each of them separately. It allows obtaining single crystals having a large density of a desired dopant without induced defects, thus impairing the crystalline quality.

In Table 12.1, we provide a comprehensive list of ionic radiuses of most elements.

Table 12.1 A list of average ionic radiuses of different elements. A given ionic radius is only slightly different in different chemical compositions; in a solid, it also depends slightly on the number of adjacent neighbors.

Z	Element name	Symbol	Valence	Radius (Å)
1	Hydrogen	H	1−	1.36
3	Lithium	Li	1+	0.68
4	Beryllium	Be	2+	0.35
5	Boron	B	3+	0.23
6	Carbon	C	4+	0.16
7	Nitrogen	N	3+	0.16
			5+	0.13
8	Oxygen	O	2−	1.40
			6+	0.10
9	Fluorine	F	1−	1.34
			7+	0.08
11	Sodium	Na	1+	0.97
12	Magnesium	Mg	2+	0.66
13	Aluminum	Al	3+	0.51
14	Silicon	Si	4+	0.42
15	Phosphorus	P	3+	0.44
			5+	0.35
16	Sulphur	S	2−	1.85
			4+	0.37
			6+	0.30
17	Chlorine	Cl	1−	1.81
			5+	0.34
			7+	0.27
19	Potassium	K	1+	1.33
20	Calcium	Ca	2+	0.99
21	Scandium	Sc	3+	0.81
22	Titanium	Ti	2+	0.80
			3+	0.76
			4+	0.62
23	Vanadium	V	2+	0.88
			3+	0.74
			4+	0.63
			5+	0.59
24	Chromium	Cr	3+	0.62
25	Manganese	Mn	2+	0.80
			3+	0.66

Z	Element name	Symbol	Valence	Radius (Å)
			4+	0.60
			7+	0.46
26	Iron	Fe	2+	0.74
			3+	0.64
27	Cobalt	Co	2+	0.72
			3+	0.63
28	Nickel	Ni	2+	0.93
29	Copper	Cu	1+	0.96
			2+	0.72
30	Zinc	Zn	2+	0.74
31	Gallium	Ga	3+	0.62
32	Germanium	Ge	2+	0.73
			4+	0.53
33	Arsenic	As	3+	0.58
			5+	0.46
34	Selenium	Se	2−	1.96
			3+	0.78
			4+	0.50
			6+	0.42
35	Bromine	Br	1−	1.95
			5+	0.47
			7+	0.39
37	Rubidium	Rb	1+	1.47
38	Strontium	Sr	2+	1.12
39	Yttrium	Y	3+	0.92
40	Zirconium	Zr	4+	0.79
41	Niobium	Nb	4+	0.74
			5+	0.69
42	Molybdenum	Mo	4+	0.70
			6+	0.62
43	Technetium	Tc	7+	0.56
44	Ruthenium	Ru	4+	0.67
45	Rhodium	Rh	3+	0.68
46	Palladium	Pd	2+	0.80
			4+	0.65

(Continued)

Table 12.1 (Continued)

Z	Element name	Symbol	Valence	Radius (Å)
47	Silver	Ag	1+	1.26
			2+	0.89
48	Cadmium	Cd	2+	0.97
49	Indium	In	3+	0.81
50	Tin	Sn	2+	0.93
			4+	0.71
51	Antimony	Sb	3+	0.76
			5+	0.62
52	Tellurium	Te	2–	2.18
53	Iodine	I	1–	2.19
			5+	0.62
			7+	0.50
55	Cesium	Cs	1+	1.67
56	Barium	Ba	2+	1.34
57	Lanthanum	La	3+	1.14
58	Cerium	Ce	3+	1.07
			4+	0.94
59	Praseodymium	Pr	3+	1/06
			4+	0.92
60	Neodymium	Nd	3+	1.04
61	Promethium	Pm	3+	1.06
62	Samarium	Sm	3+	1.00
63	Europium	Eu	2+	1.24
			3+	0.98
64	Gadolinium	Gd	3+	0.97
65	Terbium	Tb	3+	0.93
			4+	0.81
66	Dysprosium	Dy	3+	0.92
67	Holmium	Ho	3+	0.91
68	Erbium	Er	3+	0.89
69	Thulium	Tm	3+	0.87
70	Ytterbium	Yb	3+	0.86
71	Lutetium	Lu	3+	0.85
72	Hafnium	Hf	4+	0.78
73	Tantalum	Ta	5+	0.68
74	Tungsten	W	4+	0.70
			6+	0.62

Z	Element name	Symbol	Valence	Radius (Å)
75	Rhenium	Re	4+	0.70
			7+	0.56
76	Osmium	Os	4+	0.66
			6+	0.69
77	Iridium	Ir	4+	0.68
78	Platinum	Pt	2+	0.80
			4+	0.65
79	Gold	Au	1+	1.37
			3+	0.85
80	Mercury	Hg	2+	1.10
81	Thallium	Tl	1+	1.47
			3+	0.95
82	Lead	Pb	2+	1.20
			4+	0.84
83	Bismuth	Bi	3+	0.96
			5+	0.74
84	Polonium	Po	6+	0.67
85	Astatine	At	7+	0.62
87	Francium	Fr	1+	1.8
88	Radium	Ra	2+	1.43
89	Actinium	Ac	3+	1.18
90	Thorium	Th	4+	1.02
91	Protactinium	Pa	3+	1.13
			4+	0.98
			5+	0.89
92	Uranium	U	4+	0.97
			6+	0.80
93	Neptunium	Np	3+	1.10
			4+	0.95
			7+	0.71
94	Plutonium	Pu	3+	1.08
			4+	0.93
95	Americium	Am	3+	1.07
			4+	0.92

Exercises for Chapter 12

12.1 It is given that the critical radius for a solid cluster to induce further solidification of its melt at the freezing temperature is 0.05 mm. The solidification latent heat is 30 cal/g, and the material density is 4.5 g/cm^3. Calculate the surface tension coefficient between the solid and its melt.

12.2 It is known that addition of the Ho_2O_3 compound to a molten YAG ($Y_3Al_5O_{12}$) causes reduction of the freezing temperature by 1 °C per each atomic percent (at. %) of holmium (Ho) ions replacing yttrium (Y) ions in their crystallographic sites. The solidification latent heat of pure YAG is 35 kcal/mole, and its melting temperature is 1950 °C. Calculate the segregation coefficient of the Ho^{3+} ion off YAG.

12.3 The segregation coefficient of a phosphorus (P) impurity off pure germanium is 8×10^{-2}. A melted germanium load that contains 0.1 wt. % phosphorus solidifies very slowly. Calculate the concentration of phosphorus entering germanium after 50% solidification of the load and after 90% solidification of the load.

12.4 Germanium, whose atomic weight is 72.6, is melted inside a very narrow and long horizontal crucible. The germanium heat of fusion is 8.1×10^3 cal/mole, and its specific gravity is 5.32 g/cm^3. The crystal thermal conductivity is 0.7 W/cm°C, and that of the melt is larger by 35%. At a certain moment, a part of the melt has solidified, and the solidification velocity of the solid/melt interface was 10^{-3} cm/s; the temperature gradient in the solid side was 22 °C/cm. Calculate the temperature gradient existing on the melt side of the solid/melt interface.

12.5 Calculate the maximal allowed solidification velocity of a single crystal without constitutional supercooling where the melt-side temperature gradient is 20 °C/cm, and it contains an impurity of 5 at. % whose segregation coefficient is 0.05, the effective diffusivity of impurities in the melt is 10^{-4} cm^2/s, and each atomic percent of it lowers the melting temperature by 12 °C.

13 Furnace Construction Technology

13.1 Preamble

The instrument used for growing crystals at high temperatures is called **furnace**. The growing furnace is a complete set of accessories and devices whose coordinated combination enables the process to be carried out. Each material and growing method requires the construction of a different furnace; however, there are common principles adequate for designing and constructing crystal-growing furnaces. The elements that make up each growing furnace are the crucible that contains the raw material, the immediate surrounding atmosphere, heating system, thermal insulation system, temperature measuring system, and moving systems for the crucible and the growing crystal. Such elements will be described in the present chapter.

13.2 Crucibles

The crucible is the vessel containing the raw material used for the crystal growing. Naturally, the initial consideration in choosing the crucible material is its melting temperature, which needs to be considerably higher than the crystal melting temperature. Another important consideration is its mechanical stability at the working temperature: it should withstand its own weight and the hydrostatic pressure of the melted raw material. Mechanical stability at room temperature is not always a proper indication for use at the intended operation temperature, because quite a number of materials soften upon increase of temperature. Table 13.1 provides a list of popular materials used to function as crucible materials at high temperatures.

A further crucial consideration in choosing a crucible material is the need to have a complete absence of chemical reaction with the material it may contain. For a thorough examination of that issue, referring to the relevant phase diagram provides an important insight. It is quite often seen that oxides tend to form mixtures whose joint melting point is considerably lower than of each separately. A similar effect occurs in alloys of different metals. Therefore, there exists a "practical" consideration, that metallic crucibles are used for melting of oxides, and oxide crucibles are used for melting of metals. As an example, we provide a list of several common examples:

(i) An iridium (Ir) crucible is used to melt alumina (Al_2O_3).
(ii) A platinum (Pt) crucible is used to melt lithium niobate ($LiNbO_3$).
(iii) A graphite (C) crucible is used to melt germanium (Ge).
(iv) A quartz (SiO_2) crucible is used to melt silicon (Si).

Conductors, Semiconductors, Insulators, and Crystal-Growth Technology: A Practical Guide,
First Edition. Zeev Burshtein.

Table 13.1 A list of commonly used crucible materials in high-temperature crystal growing.

Material	Chemical formula	Maximal working temperature (°C)	Comments
Platinum	*Pt*	1770	
Iridium	*Ir*	2450	
Molybdenum	*Mo*	2600	Requires vacuum or oxygen absent ambiance
Tantalum	*Ta*	3000	
Graphite	*C*	3730	Requires vacuum or oxygen absent ambiance
Alumina	Al_2O_3	2050	
Zirconia	ZrO_2	2400	
Quartz glass[a)]	SiO_2	1000	
Boron nitride	*BN*	1700	

a) Often the glass surface is added with a pyrolytic graphite layer. Preparation procedure involves letting the glass stay under a vacuum of 10^{-7} torr for 24 hours at 1100 °C, then exposing it to methane (CH_4) gas at a 200 torr pressure for 1 hour, then cooling to room temperature. Working requires high vacuum, or oxygen-absence conditions.

An interesting example of an unexpected misfit, which emphasizes the need to examine all the aspects related to the material properties and their combinations, is in seeking a crucible material for melting of lead (*Pb*) or of bismuth (*Bi*). Particularly, there is a huge difference in melting temperature between platinum (1770°C) and lead (327°C). Instinctively, one may be tempted to suggest platinum as a proper crucible material for containing melted lead; however, melted lead dissolves platinum! Phase diagrams of the platinum/lead pair clearly reflect that fact.

13.3 Crystal Growth Atmosphere

The furnace atmosphere surrounding the core crystal-growing process influences the process via three important factors: (i) its thermal conductivity; (ii) involvement in chemical reactions among the different furnace components, especially among its hot parts; (iii) evaporation of different furnace parts, particularly the raw material used for the crystal growth; and (iv) designing and monitoring the atmosphere.

(i) Thermal conductivity

A gas thermal conductivity depends not only on the gas type but also on the furnace construction, which allows or prevents the development of gas flows in different parts. If gas flows are absent, the gas thermal conductivity is a function of its pressure, temperature, and atomic mass. An interested reader should refer to specialized books and research publications concentrating on physical properties of gasses. Notably, actual gas pressure measuring in the low-pressure range ("vacuum" range) is based on a direct measurement of their thermal conductivity.

(ii) Chemical reaction

A gas present inside a furnace might react chemically with any part of the growth furnace, especially with its hotter parts. A common means of preventing undesired reaction is to create an ambiance of a noble gas, like argon (*Ar*) or helium (*He*). Sometimes using of a nitrogen (N_2) atmosphere is sufficient. On the other hand, sometimes it is precisely the addition of a certain gas, which creates a reaction that prevents unwanted development in crystal growth. For example, using an atmosphere that contains oxygen (O_2) will prevent decomposition of an oxide raw material; alternatively, an atmosphere containing hydrogen (H_2) would prevent oxidation of various raw materials, especially metals and semiconductors.

(iii) Vaporization

In the very high temperatures existing in different crystal growth furnaces, sometimes different parts of the furnace may undergo some evaporation. That process becomes significant when the said material vapor pressure is higher than the surrounding atmospheric pressure. Sometimes it is possible to minimize the process by increasing the inner atmospheric pressure. In specific cases, vaporization of the raw material occurs, often related to its decomposition. Sometimes addition of a material that contains the volatile component may compensate against the raw material decomposition. For example, in the growth of cadmium telluride (*CdTe*) single crystals, the raw material tends to decompose at its melting temperature (1092°C) with a faster vaporization of the cadmium (*Cd*). The grown crystals would then include an undesired excess of tellurium (*Te*). Addition of a cadmium reservoir near the growing crystal, set at a proper temperature (~880°C), balances the said decomposition.

(iv) Designing and monitoring of the atmosphere

Not always is it possible to grow a single crystal in a regular air ambiance. Air is well known to comprise nitrogen and oxygen that tend to react chemically with various substances. That tendency increases at elevated temperatures. When a growing of some crystal requires the absence of either of these two components, it is possible to fill the furnace growth container with a mixture of other gasses. Except for preventing the undesired reaction, the presence of an inert gas at a sufficiently large pressure might prevent vaporization or chemical disintegration of certain furnace components or of the raw material itself. Sometimes, however, the presence of a proper gas generates a desired chemical reaction that balances the raw material tendency to disintegrate at the used high temperatures. Such need is achieved by closing the inner part of the furnace inside an external container sealed against gas leakage. This is usually achieved by incorporating gaskets into the furnace container connections. A common, often very useful option for preventing undesired chemical reaction, which also contributes to thermal isolation, is to remove the ambient gas entirely by pumping out all gasses, and create high vacuum conditions. In such cases, the container is connected to an out-pumping system adapted to the needed vacuum state. There are simple rotational pumps that allow reaching a vacuum of a residual 10^{-3} torr pressure. A diffusion pump connected in series to the rotational pump allows reducing the residual pressure down to 10^{-6} torr. It is also possible to select the residual pressure to exhibit a preferable composition, by forcing leakage of the desired gases combination from a faucet connected to a tank containing the appropriate gas composition.

13.4 Heating Methods

In experimental studies of crystal growth, a large variety of heating methods are used, from flame heating, electrical current heating, laser beam heating, till concentrated sun rays heating. Each different method obviously exhibits a specific unique advantage. For example, it is possible to grow a crystal inside a sealed container heated with a laser beam through a transparent window. It would allow keeping a large degree of purity and a small-sized heated area.

Industrial crystal growth uses two main methods:

1) Melting using a hydrogen/oxygen flame, a method called "flame fusion". That unique method is known as **Verneuil method** and has been used in growing of sapphire (Al_2O_3) "drop-shaped" single crystals. In our present book, we won't dwell further on that particular method.
2) Electrical resistivity heating; it involves two variations:
   (i) Heating by using resistive electrical wires
   (ii) Heating by coils coupling (inductive heating; also called "RF heating")

(i) Electrical resistivity heating

Resistivity heating is achieved by driving an electrical current through electrically conducting cords. The cord electrical resistance causes the development of a heating power along their lengths. The heating power may be adjusted to receive both a desired temperature and temperature gradients. The cord material may be a metallic alloy, but other options like graphite are quite possible. The conducting material should be adjusted to the special conditions in each furnace. Table 13.2 provides a list of materials used to form heating conductors in crystal-growing furnaces.

Heating elements comprising a metallic wire need to be wound around an insulating support. The issue of electrical insulation becomes severe at very high temperatures, because most insulating materials tend to conduct electrical currents by mechanisms of ionic conduction and/or free charge carrier conduction.

Table 13.2 A list of commonly used electrical conducting materials used in high-temperature crystal growing.

Material name	Chemical formula	Maximal working temperature (°C)	Allowed ambiance
Chromel D	$Ni-Cr$	760	Air
Chromel C	$Ni-Cr$	1100	Air
Nichrome	$Ni-Cr$	1100	Air
Kanthal	$Fe-Al-Cr$	1375	Air
Platinum	Pt	1400	Air
Silicon carbide	SiC	1600	Air
Molybdenum	Mo	1800	Reducing atmosphere
Super Kanthal	$MoSi_2$	1800	Air
Iridium	Ir	1900	Air
Tantalum	Ta	2000	Vacuum
Tungsten	W	2600	Vacuum
Graphite	C	3000	Inert gas
Tungsten	W	3000	Reducing atmosphere
Tungsten	W	3200	Helium

Another problem is that there are differences among the thermal expansion coefficients of different materials. A very common construction in laboratory experiments is a tubular heating element comprising a Kanthal cord wrapped around a quartz glass tube. Often when the temperature increases, the cord's wraps loosen, causing electrical short circuits between adjacent windings. The reason is that the thermal expansion coefficient of Kanthal is considerably larger than the expansion coefficient of the quartz glass. Stabilizing the wraps in their position is best achieved by forming grooves on the tube's outer surface into which the cord is inserted; using of special glues is also an option.

Metallic heating elements also soften at higher temperatures and would collapse if not properly supported. Such problems may be overcome by using graphite or silicon carbide (SiC) for heating. These materials remain rigid throughout broad working temperature ranges.

(ii) Inductive, high-frequency heating

The inductive heating is based on driving of high-frequency (RF; radio frequency) alternating electrical currents through a coil constructed of a long copper pipe. The used frequencies range between about 10 kHz and 450 kHz. The magnetic and electrical fields generated inside the coil volume induce eddy currents in close electrically conducting materials. Any material whose electrical resistivity in not too high (almost insulating) or too small (copper; aluminum) will heat up as a result of the Ohmic resistivity. The Ohmic electrical resistivity of the coil itself is quite small; therefore, its heating by the electrical current is quite small. That heat, or heating caused by the furnace heat-radiating hot parts, is removed by cooling water driven through the pipe itself.

Furnaces using that technology use one of the following three methods of heating:

1) A direct heating of the raw material, an approach that works only if the material is electrically conducting. (if it is insulating, it is possible to disperse inside it several lumps of a conducting material to mediate the heating, provided of course that it won't melt and won't react chemically with the raw material [namely won't contaminate it!]).
2) Heating of the crucible containing the raw material. It works if the crucible itself is electrically conductive.
3) Heating an electrically conducting pipe (usually graphite made), which works as a transducer of the radiated electromagnetic energy into thermal energy. The crucible containing the raw material is positioned inside that pipe.

The main advantage of the inductive heating is the effective avoidance of contaminating the raw material; that is because the coil can be placed outside the crucible zone and be physically isolated from it. It is rather simple to design a vacuum chamber, inside which the crucible and transducer (if needed) be placed, while the coil is outside. In Ohmic heating, on the other hand, the heating element, being the body of highest temperature inside the furnace, might be a source of impurities. The inductive heating also provides a needed possibility to control the temperature profile in the furnace, particularly achieving relatively large temperature gradients if needed.

An important effect related to inductive heating is that the induced eddy currents are limited to the near-surface regions of the transducer, an effect called **skin effect**. As an example, consider a cylindrical body set inside the heating coil, both having the same axis, as shown in Figure 13.1. The eddy currents will be limited to the rim area of the cylinder shell, and vanish exponentially toward the center. The currents' penetration depth δ is defined as the depth at which the currents

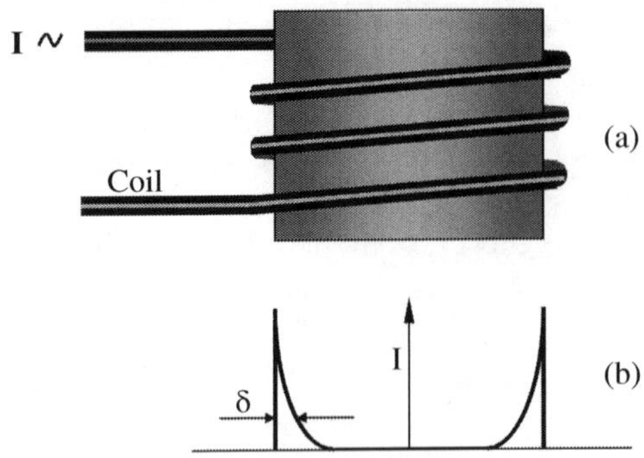

Figure 13.1 (a) A cylindrical electrically conducting material placed under an alternating, high-frequency magnetic field generated by an alternating current in an inductance coil. (b) The induced eddy currents inside the electrically conducting cylinder as a function of the radial position in the conducting cylinder.

drop to $1/e$ of their value on the rim, and they turn smaller for cylinders of higher electrical conductivity and for higher frequency of the electromagnetic field.

Due to the practical absence of heating currents in inner part of the cylinder, it is possible to remove that inner part (namely form a hollow cylinder) with almost no influence on the heating power; the wall thickness of the remaining hollow cylinder is chosen to be about twice or three times the calculated penetration depth δ. Such structure is used when forming a heating transducer. The crucible with its raw material is inserted into the cylindrical hollow. If the crucible itself is aimed at functioning as the heating transducer, its bottom of course is not removed.

13.5 Electrical Insulators

Electrical insulators would always be a part of a furnace, and are used for two purposes: one is to isolate electrical wires on crossing between furnace regions. The other is to support the heating element where needed. Electrical insulators are materials having each a sufficiently large energy gap such that their energy bands don't contain free charge carriers. Still, as the temperature rises, their electrical conductivity rises; it emanates from activation of charge carriers into the energy bands or from activation of ionic conductivity. For that reason, materials that are excellent insulators at room temperature are not always useful as insulators at high temperatures. A partial list of electrical insulators, and limitation for their use at high temperatures, is provided in Table 13.3. Some may be produced at different specific gravities, and be formed into different geometrical shapes.

Table 13.3 A list of commonly used electrically insulating materials used in high-temperature crystal growing and the maximal possible functioning temperatures.

Material name	Chemical formula	Maximal working temperature (°C)	Comments
Quartz glass (fused silica)	SiO_2	1100	Use is possible up to 1350°C, yet with loss of transparency (devitrification). Contact with platinum should be avoided.
Ceramic mixture		1500	Various alumina-based mixtures (Al_2O_3).
Boron nitride	BN	1700	Has a rather high thermal conductivity.
Alumina	Al_2O_3	1900	Clean with no additives.
Beryllia	BeO	2300	Beryllia powder is toxic.
Zirconia	ZrO_2	2300	Has to be stabilized by added Cr_2O_3 or MgO to prevent a phase change from monoclinic into tetragonal at 1170°C. Stabilized ZrO_2 is cubic throughout the entire range.
Magnesia	MgO	2600	Stable only under an oxidizing atmosphere.

13.6 Thermal Insulators

Thermal insulators would always be a part of a furnace and are used to prevent undesired heat dissipation from the crystal growth region, as well as to rather form desired temperature gradients there. The primary considerations in choosing thermally insulating materials are of course their resilience under the expected working temperature, particularly melting point, softening and/or brittleness development, or other kinds of functioning loss. There also exists the need for selecting the proper mechanism of thermal insulation at the working temperature: should it act against thermal conduction, convection, or radiation. Below about 1000°C, the dominating heat dissipation is thermal conduction. In the 1000 $\div$ 1200°C range, convection and radiation are equally dominant in heat dissipation. Above about 1200°C the dominant heat dissipation is radiation. It doesn't mean that above 1200°C, the other mechanisms are negligible; therefore, they also need to be taken care of. In temperatures lower than 1000°C, radiation heat losses are quite negligible. Materials used as electrical insulators are usually also adequate as thermal insulators. An important exception is the boron nitride (BN) material, which is an excellent electrical insulator yet quite a good thermal conductor.

In furnaces operating in a regular air atmosphere, or a different gas at a similar pressure, gas convection is the most efficient mechanism causing heat transfer. In cases where convection is not needed for the crystal growth process, it should be prevented. Without convection, air is an excellent thermal insulator, a property that may be exploited for that purpose. It is usually done by using porous materials, which trap the gas inside their pores. The same materials indicated as good thermal insulators may often be produced as foamed insulating materials, ceramic fabrics, and felts made of ceramic or of quartz fibers. All such materials utilize their own heat insulation properties with the addition of the heat insulation of the gas trapped inside their pores. A most efficient, commercially available insulator is called "zirconia bubbles". They are ball-shaped porous zirconia parts, available at various diameters. They are usually used to fill gaps between hard insulators, thus stopping almost entirely all gas convections.

At very high temperatures, where electromagnetic radiation is the dominant heat loss mechanism, it is possible to use series of metallic radiation screens, which limit the heat loss by reflecting the electromagnetic radiation emitted by the hot zone. That method is especially useful in furnaces operating under vacuum or under a low gas pressure, where heat loss by convection is minimal.

13.7 Contacts Between Materials

On designing a furnace for a specific crystal growth, care should be taken to avoid undesired interaction between different structural materials. Chemical interaction between materials usually rises with increasing temperature. It is thus possible that two different materials, both being mutually inert at room temperature, will interact upon increase of temperature. The required information on any two materials may be obtained by observation of phase diagrams of the two. Some general tendencies are well known. Particularly, almost all oxides, when combined, exhibit lower melting temperature than each separately. Thus, it is impossible to rely on a fact that the melting temperature of a certain oxide is higher than that of another one, that they won't interact chemically at the furnace working temperature. Table 13.4 provides a list of maximal allowed "touching" temperatures of common construction and thermal insulation materials in crystal growth furnaces.

Table 13.4 Maximal possible temperature (°C units) for contact between different materials without experiencing a chemical interaction. A dash "–" indicates cases of yet unknown information.

Material	C	Al_2O_3	BeO	ZrO_2	ThO_2	MgO
C	***	1500	2300	1600	2000	2300
Mo	1100	1900	1900	2000	1900	1900
Ni	1300	1800	1800	1800	1800	1800
Pt	–	1800	–	1700	–	–
Ta	900	1900	1600	1600	1900	1600
Ti	–	1600	1600	1600	1800	1600
W	1300	1900	1800	1600	2200	1800
ZrO_2	–	1700	1900	***	2600	1900
ThO_2	–	1750	2100	2600	***	2200
MgO	–	1800	1800	1900	2200	***

A triple asterisk "***" indicates that same materials trivially don't interact at any temperature.

13.8 Temperature Measurement

Control of a crystal growth requires a continuous accurate measurement of temperature in different positions inside the furnace. The measuring equipment is usually called **thermometer** or addressed by a name that emphasizes the underlying physical effect that allows the temperature determination. All thermometers are based on the measuring of a temperature-dependent physical effect, and on its calibration against common temperature scales (Celsius, Kelvin, or Fahrenheit). The methods used in temperature measurements in crystal growth furnaces are:

1) Melting of known materials
2) Colors
3) Regular (medical-like) thermometers
4) Electrical resistivity of metallic wires
5) Thermistors
6) Thermocouples
7) Pyrometers

In the following, we provide some details and explanation of each of the above-listed methods.

(1) **Temperature measurement by observation of known materials melting**

The melting temperature of a known material provides an accurate calibration point for a temperature scale. Particularly, observation of the raw material state provides the most accurate definition of its melting temperature. Obviously, such approach is not capable of providing a continuous tracing of temperature changes.

(2) **Temperature measurement using colors**

There are materials that change their apparent color when the temperature rises above a characteristic known temperature. In some cases, it was customary to paint a stripe of the material in a selected place inside the furnace, such that observing it from outside allows locating the places where the temperature is higher than a certain value, and the places where the temperature is lower. That method is quite common in metals casting furnaces, for which the required accuracy of the temperature and its gradients is not crucial.

(3) **Temperature measurement using regular (medical-like) thermometers**

The well-known regular (medical-like) thermometers are based on allowing a liquid (mostly mercury) spread into a capillary glass tube. The temperature is calibrated versus the liquid length inside the capillary tube. The use of such thermometers is limited to ranges that may be higher than room temperature by only a couple dozen degrees centigrade. A mercury-based thermometer may function properly up to about 150°C.

(4) **Temperature measurement using resistivity of electrically conducting wires**

Electrically conductive metallic wires change their resistivity as a function of temperature. Usually, the resistivity increases with temperature. It is thus possible to calibrate a wire temperature as a function of its measured resistivity. Table 13.5 provides a list of electrical conductors used as thermometers and their relevant physical properties.

(5) **Temperature measurement using thermistors**

Thermistors are ceramic semiconductors whose thermal resistivity decreases as a function of temperature (in contrast to metallic wires whose resistivity increases with temperature). In certain (though narrow) temperature ranges, their sensitivity to temperature changes is very large compared to that of the metallic wires. There is a large variety of commercial thermistors, each adequate to a different temperature range where its sensitivity is largest.

(6) **Temperature measurement using thermocouples**

The most commonly used thermometer for measuring and monitoring of temperature inside furnaces is a thermocouple. The thermocouple is a contact between two metals that creates an electrical voltage by a **Seebeck effect**. That Seebeck voltage depends on the temperature, an effect used for measuring temperatures up to 2700°C. To achieve best sensitivity over specific temperature ranges, it is possible to choose different combinations of the comprising metals.

The fundamental construction of temperature measurement using a thermocouple is shown in Figure 13.2. Initially, two junctions are made at the edges of two conducting, different metallic wires. One of the wires, marked "**A**" in the figure, is cut in its middle, and the cut edges are connected to a voltmeter. The measured voltage is the difference between the two Seebeck voltages generated at the two junctions. That difference is zero if the junctions reside at the same temperature. If, however, the temperatures are different, the voltmeter will measure a voltage called **thermocouple voltage**. The thermocouple voltage depends on the actual voltages at the

Table 13.5 Conducting wires used in measuring temperature through their measured electrical resistivity.

Material	Chemical symbol	Working temperature range (°C)	Resistivity change factor (% °C^{-1})
Copper	*Cu*	$-200 \div 100$	4×10^{-3}
Nickel	*Ni*	$-150 \div 300$	$5 \div 8 \times 10^{-3}$
Platinum	*Pt*	$-200 \div 1200$	$3 \div 4 \times 10^{-3}$

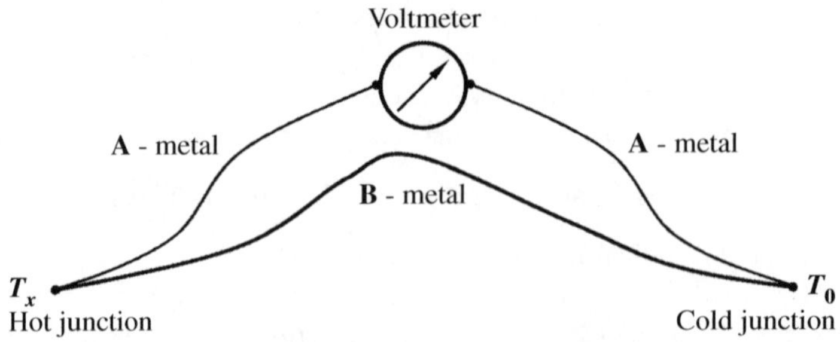

Figure 13.2 A fundamental electrical circuit used for temperature measurement using a thermocouple.

two junctions. It is customary to set one junction at a constant convenient temperature, for example $T_0 = 0°C$ (which is easy to obtain by a mixture of distilled water and ice) and then the thermocouple voltage will depend only on the hot junction temperature marked T_x in Figure 13.2. The T_0 temperature is called **reference temperature**.

Table 13.6 provides a list of common thermocouples used for temperature measurement, including their accepted technical notation in this field and the recommended temperature working ranges. Notably, in the construction shown in Figure 13.2, there are actually two additional junctions – the "**A**" metal edges with the input contacts of the voltmeter. These two contacts should be kept at an identical temperature to avoid undesired bias. Sensitivities of different thermocouples are in the $5 \div 40$ μV/ °C range. Using a proper scaling, the measured voltage can be translated to express the hot junction temperature T_x. There are proper technical

Table 13.6 Thermocouples, commonly used technological names, and working temperature range.

Thermocouple "A" metal – "B" metal	Tech. symbol	Working range (°C)	Comments
Copper – constantan	T	$-250 \div 400$	Wears-out quickly; constantan = *Ni*:*Cu* alloy
Iron – constantan	J	$-200 \div 750$	Wears-out quickly; needs reducing atmosphere
Chromel – constantan	E	$-200 \div 850$	Needs oxidizing atmosphere
Chromel – Alumel[a)]	K	$-200 \div 1350$	Very common; needs oxidizing atmosphere
Platinum – platinum/Rh13%	R	$0 \div 1700$	Expensive; needs oxidizing atmosphere
Platinum – platinum/Rh10%	S	$0 \div 1760$	Expensive; needs oxidizing atmosphere
Platinum/Rh6% – platinum/Rh30%	B	$0 \div 1700$	Expensive; needs oxidizing atmosphere
Iridium – tungsten		$200 \div 2300$	Brittle; requires absence of oxygen
Tungsten – tungsten/Rh26%		$200 \div 2700$	Requires absence of oxygen
Tungsten – tungsten/iridium		$200 \div 2800$	Requires absence of oxygen

a) "Chromel" and "Alumel" are trademarks of different patented metallic alloys.

conversion tables showing the temperature versus measured voltage for different metallic (and metal–alloy) couples. Modern specialized (voltmeter-based) devices often exhibit a direct computer-assisted display of the temperature.

(7) **Temperature measurement using pyrometers**

A pyrometer is a thermometer that measures a material's temperature through the electromagnetic radiation it emits. As we shall see later, the emitted electromagnetic radiation depends mostly not only on the material temperature, but also on some of its other properties to be discussed.

Any material under an external electromagnetic radiation responds to the radiation by a combination of three effects:

1) Incident radiation **absorption**; the absorbed fraction (absorptivity) is denoted by α.
2) Incident radiation **reflection** or scattering; the reflection fraction (reflectivity) is denoted by ρ.
3) Incident radiation **transmission**. The transmission fraction (transmissivity) is denoted by τ.

The obtained values of α, ρ, and τ per each material depend on the radiation wavelength λ. By energy conservation law, the condition $\alpha + \rho + \tau = 1$ always exists for each wavelength component. For any arbitrary spectral intensity distribution $I(\lambda)$, satisfying $I_0 \equiv \int I(\lambda)d\lambda$, it is possible to define average values of absorption, reflection, and transmission according to

$$\overline{\alpha} \equiv \frac{1}{I_0}\int \alpha(\lambda)I(\lambda)d\lambda; \quad \overline{\rho} \equiv \frac{1}{I_0}\int \rho(\lambda)I(\lambda)d\lambda; \quad \overline{\tau} \equiv \frac{1}{I_0}\int \tau(\lambda)I(\lambda)d\lambda,$$

which also satisfy the relation $\overline{\alpha} + \overline{\rho} + \overline{\tau} = 1$. A body that doesn't transmit any radiation, namely satisfying $\overline{\tau} = 0$, is granted the adjective **opaque**. A body that doesn't transmit any radiation nor reflects or scatters it (satisfying both $\overline{\tau} = 0$ and $\overline{\rho} = 0$) is called **black body**, and it satisfies $\overline{\alpha} = 1$ for any distribution of incident electromagnetic radiation.

By basic thermodynamic consideration, any black body must also emit electromagnetic radiation; otherwise, when exposed to an incident radiation field, its temperature would rise to infinity! The emission enables the black body to "dispose" of the energy it has absorbed. That consideration is quite sufficient to suggest that the nature of a black-body emission depends solely on its temperature, and not on any other factor; in other words, there is no relation between the spectrum of the absorbed radiation and that of the emitted radiation. Notably, real bodies are always only approximately black, never entirely black. The best approximation of a black body is rather a small hole in a large container, whose walls are in thermodynamic equilibrium with a heat reservoir at a constant temperature.

A beam entering through the container hole would be reflected and scattered multiple times until mostly absorbed, with practically only zero chance of slipping out, due to the smallness of the hole (Fig. 13.3).

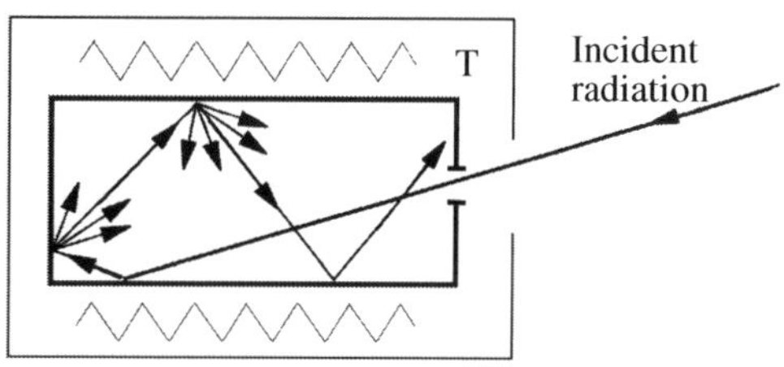

Figure 13.3 Realization of a black body by a small hole in a large container kept inside a furnace at a constant T temperature.

Thermodynamic considerations beyond the scope of the present book allow obtaining an expression of the electromagnetic energy distribution inside the container volume, which maintains a thermodynamic equilibrium with its walls kept at a constant temperature T (**Planck equation**)

$$E_{BB}(\lambda, T)d\lambda dV = \frac{8\pi hc}{\lambda^5} \cdot \frac{1}{\boldsymbol{exp}\left(\frac{hc}{\lambda k_B T}\right)} d\lambda dV, \tag{13.1}$$

where $E_{BB}(\lambda, T)$ is the energy per unit wavelength per unit volume of the container whose walls are at a constant temperature T. The container volume is V, the quantity h is the **Planck constant,** c is the speed of light constant, and k_B is the Boltzmann constant. The electromagnetic radiation emitted from the container hole exhibits the same spectral dependence (namely same dependence on λ) as does $E_{BB}(\lambda, T)$; the distribution of the emitted power density (energy per unit time per unit area of the hole, per unit wavelength) is given by

$$W_{BB}(\lambda, T) = \frac{1}{4} c E_{BB}(\lambda, T). \tag{13.2}$$

Integrating over the entire wavelength range yields the power density of the entire emitted radiation power (energy per unit time per unit hole area) by a uniform 2π spatial angle outside the container

$$\Phi_{BB}(T) = \frac{1}{4} c\sigma T^4, \tag{13.3}$$

where σ is a universal constant: $\sigma = 7.565 \times 10^{-16}\ \mathrm{J \cdot m^{-3}\ K^{-4}}$. The latter expression is called "**Stefan–Boltzmann** law".

We consider now in detail the spectral shape of the black-body radiation. We start by considering the limiting edge cases of very long wavelengths ($hc/\lambda k_B T \ll 1$) and of very short wavelengths ($hc/\lambda k_B T \gg 1$). In these limits, Eq. (13.1) assumes the form

$$E_{BB}(\lambda, T) \cong \begin{cases} \dfrac{8\pi}{\lambda^4} k_B T & hc/\lambda k_B T \ll 1 \\ \dfrac{8\pi hc}{\lambda^5} \boldsymbol{exp}(-hc/\lambda k_B T) & hc/\lambda k_B T \gg 1 \end{cases}.$$

It is readily seen that in the long wavelength range, the radiation intensity becomes smaller as a function of wavelength, while in the short wavelength range, it increases as a function of wavelength. It is thus naturally expected that for a constant temperature, the intensity dependence will reach a maximum at a certain wavelength, that we denote by λ_{peak}. For all wavelengths, the intensity increases as a function of temperature.

Figure 13.4 shows the wavelength dependence of the black-body radiation for different temperatures, which also express the previously indicated properties. An additional important property is the wavelength location of the emission peak. As described in Figure 13.4, the peak wavelength λ_{peak} shifts toward shorter wavelengths as a function of temperature. The said shift follows a simple relation called **Wien's displacement law**

$$\lambda_{peak} \cdot T = C_0 = 2.898 \times 10^{-3}\ \mathrm{m^\circ K}. \tag{13.4}$$

In fact, Wien's displacement law was discovered experimentally several years before the design of Planck equation (Eq. [13.1]), and it is mathematically derivable from it.

To obtain a feeling of the size of that effect, we provide a table that connects the peak emission wavelength to various, temperature-related physical phenomena (Table 13.7).

It would be useful to verify that a black body satisfies the condition $\varepsilon = 1$; a black body is defined as such that absorbs all electromagnetic radiation, namely $\alpha = 1$ (hence also $\overline{\alpha} = 1$).

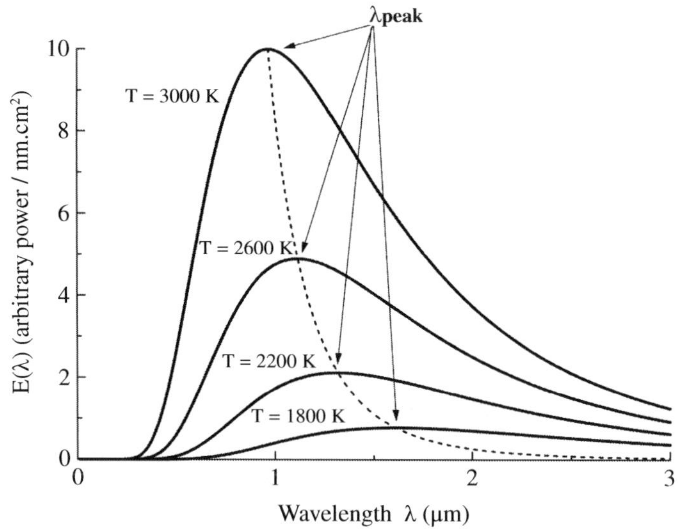

Figure 13.4 Black-body radiation emission spectra at different temperatures. The emission power values $E(\lambda)$ are presented in relative units.

Table 13.7 Black-body emission peak wavelengths at different temperatures.

Temperature	Peak wavelength	Comments
300 °K	~10 μm	Room temperature; namely 27 °C
600 °K	4.8 μm	Lead (*Pb*) melting is at 600 °K
1000 °K	2.9 μm	Aluminum (*Al*) melting is at 933 °K
1500 °K	2.5 μm	Copper (*Cu*) melting is at 1356 °K
2000 °K	1.4 μm	Platinum (*Pt*) melting is at 2045 °K
3000 °K	1.0 μm	Molybdenum (*Mo*) melting is at 2890 °K
4000 °K	0.7 μm	Carbon (*C*) melting is at 3825 °K
6000 °K	0.5 μm	Sun surface temperature is 5800 °K

Under thermodynamic equilibrium between a black body and its vicinity, the radiation it absorbs is equal to the radiation it emits, namely

$$\int \varepsilon \Phi_{BB}(\lambda, T) d\lambda = \int I(\lambda) d\lambda,$$

where $I(\lambda)$ is the spectral distribution of an incident electromagnetic radiation, and it is particularly valid for a black-body radiation $I(\lambda) = \Phi_{BB}(\lambda, T)$. Then necessarily $\varepsilon = 1$. That result means that the extreme radiation absorber (back body) is also the most efficient thermal electromagnetic radiation emitter.

The efficiency of a body thermal radiation emission ε depends not only on its comprising material but also on the form of its surface. The emission efficiency of a rough surface is considerably greater than that of a body whose surface is well polished. The reason is that in rough surfaces, a part of an incident electromagnetic radiation is scattered sideways back and forth, thus increasing their probability of being absorbed. As previously indicated, a body that absorbs well is a better emitter of the thermal radiation. This fact has implications regarding the apparent shape of bodies kept at a high temperature. Scratches and dimples on their surface appear brighter than their smooth surroundings. Thus, such hot bodies are seen as a negative of their appearance at room temperature: when externally illuminated at room temperature, scratches and dimples appear rather darker than their smooth surrounding, because their absorption is larger.

Regular material chunks are usually only partially black; however, their thermal emission is quite similar to that of a black body, yet smaller in size. It can be expressed as following the relation

$$\Phi(\lambda, T) = \varepsilon\Phi_{BB}(\lambda, T). \tag{13.5}$$

The $\varepsilon < 1$ quantity is called **emissivity**. To a small extent, it may depend on the wavelength λ, but that dependence is usually very small. For practical purposes, it is usually assumed to be a characteristic constant of the body, valid for all wavelengths. Table 13.8 provides a partial list of materials and their characteristic emissivity value ε.

The thermal emission properties of materials allow measuring their temperature. It is done using several methods as described below.

(i) **Disappearing filament pyrometer**

Because the radiation emitted power density (power per unit area) of a hot body is a function of its temperature, it is possible to compare a measured temperature to a given temperature by comparing their visible brightness. That comparison is performed by an instrument called "disappearing filament pyrometer". This instrument is a telescope that allows watching the body whose temperature is intended to be measured. The focal plane of the eyepiece is added with an electric current-carrying filament, whose temperature is regulated by a series variable resistor. The filament temperature appears as direct reading on the variable resistor scale. While watching the hot body with the telescope, the variable resistor is dialed until the filament image disappears on the background of the hot body image. The accuracy of such measurement is quite high – only $1 \div 2\,°C$ in the vicinity of $1500\,°C$. A limiting factor of the device sensitivity is that the hot material emissivity has to be taken into account, which is difficult to properly assess (material type and its surface roughness). In addition, any factor that interferes with the emitted radiation such as dust accumulated on the inner side of the furnace windows might cause a change in the measured result.

(ii) **Electronic pyrometer**

That device is also a telescope that collects emitted radiation of the hot object and concentrates it on a photoelectric light detector cell. This cell output depends on the emitted radiation of the

Table 13.8 Experimental emissivity ε values of different materials.

Material	Emissivity ε	Comment	Material	Emissivity ε	Comment
Gold	0.03	500 °C	Soot	0.95	
Nickel	0.15	700 °C	Water	0.96	
Molybdenum	0.13 ÷ 0.26		Asbestos	0.95	
Platinum	0.10	500 °C	Alumina	0.35 ÷ 0.75	
Platinum	0.19	1500 °C	Silver	0.02 ÷ 0.03	
Graphite	0.80		Steels	0.70 ÷ 0.85	
Aluminum	0.04 ÷ 0.07		Tungsten	0.28	2000 °C

hot body, and it is scaled to provide a direct value of the temperature. It has an advantage that the temperature may be recorded continuously, without resorting to human involvement. The limiting factors related to emissivity assessment and absorbing/scattering windows along the observation path are valid for that type of pyrometer as well.

(iii) **Double-color pyrometer**

That type of pyrometer is an extended refinement of the electronic pyrometer, and is aimed at solving the problem of the result dependence on the hot body emissivity ε. It is also a similar telescope, but it splits the entering radiation in two and directs each part into a separate photoelectric detector cell, each passing through an optical filter that transmits a selected different wavelength. Of these two detector cells' readings, the slope of the hot body spectral emission is calculated. That slope depends on the hot body temperature (see Fig. 13.4) but not on ε. The sensitivity of such measurement is more accurate than the previously described ones, and it partially neutralizes the influence of absorbing/scattering windows along the observation path.

13.9 Temperature Control Methods

Crystal growth needs that the temperature at the raw material phase change region (mostly from a liquid into a solid) be either constant or changed in a predetermined controlled manner. In simple furnaces, a previous experimental run may be used to determine the electrical power required for the growth; then the power is changed manually when needed. Such practice is not sufficient for industrial processes, and even in experimental laboratories, it causes many growth failures. Furthermore, it requires a continuous human monitoring during the long times needed for crystal growth. For that reason, automatic temperature monitoring systems were developed, which frees the crystal grower from the need of intervening in most stages of growth.

1) On–off control

An on–off control method is the simplest automatic method, and is sufficient for use in systems where maximal temperature accuracy in not crucial for the crystal growth results. Practically, the method assures that the temperature is kept between certain limits without human intervention. The known domestic system that utilizes such method is a controller based on a **metal couple**, which regulates the water temperature in an electric water boiler. The metal couple comprises two mutually soldered stripes of different metals. Due to a difference between the metal thermal expansion coefficients, the initially straight metal couple bends upon temperature change. That bending is used to connect or disconnect the electrical circuit of the thermal heating element.

Basically, any thermometer could be used as an on–off control through an adequate electrical or electronic switching. The working principle is to disconnect the electrical current to the heating element at the moment where the temperature reached a predetermined high threshold T_{max}. The controlled process system then starts cooling. Because there is always a physical distance between the heating element and the raw material position, the thermometer discovers the temperature lowering after some delay, when the raw material temperature has cooled down to a T_{min} value. Then the thermometer reconnects the electrical current to the heating element. A schematic illustration of the process is provided in Figure 13.5.

It is important to consider several important features of the on–off control. As to the initial "on" heating, the raw material temperature starts to grow quickly at a constant rate. Then the rising rate lowers gradually because of a parallel increase in the heat loss to the surrounding. If heating

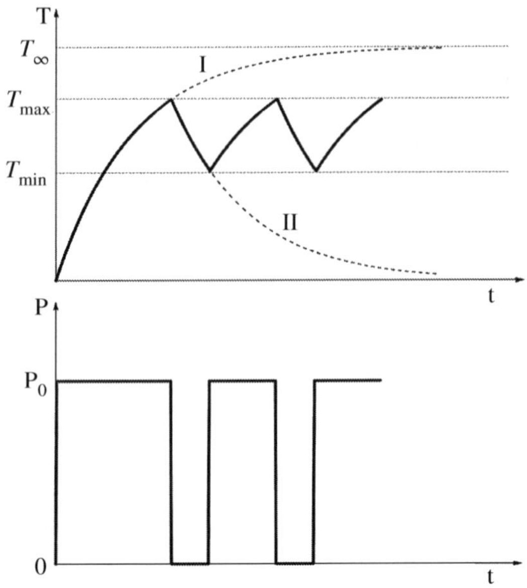

Figure 13.5 On–off control. The lower graph shows the electrical power *P* supplied to the thermal heating element as a function of time t. The upper graph shows the regulated temperature as a function of time t. The dashed curves show prospected curves had the power P_0 continue uninterruptedly (marked I) or had the power remain zero after the first heating period (marked II). The regulated temperature oscillates between T_{max} and T_{min}.

continues, the temperature rise rate becomes zero, and the temperature eventually stabilizes at the T_∞ value (Fig. 13.5). It seems as if such would be ideal for setting a desired temperature. However, if any temperature deviation occurs, it would take a very long time to correct. For that reason, the heating power should be set such that T_∞ is considerably larger than the goal temperature. The natural heat dissipation is also important, because it allows correcting against temperature deviations above T_{max}. The temperature controlling accuracy is defined by the difference $\Delta T \equiv T_{max} - T_{min}$. To reduce ΔT, both temperature detector and heating element should be placed as close as possible to the controlled raw material.

The on–off control method can't overcome temperature variations to the extent needed for most crystal growth systems. Therefore, more sophisticated methods have been developed for that purpose, to be described below.

2) Proportional control

The described shortcomings of the on–off control systems stem from the fact that the system responds by an identical power change for any noticed deviation from the goal temperature, be it large or small. The proportional control method is designed to make the power change proportional to the noticed temperature change: $\Delta T = T - T_0$, where T is the measured temperature and T_0 is the goal temperature. Usually it follows an electronic system that provides an output signal S_p satisfying

$$S_p = -p \cdot \Delta T, \tag{13.6}$$

where p is a positive proportionality factor. The meaning is that for a small ΔT, the control correction will be small, and for a large ΔT, the control correction will be large. The minus "–" sign expresses the principle that the correction establishes a negative feedback to a deviation from the goal of temperature stabilization. Recall that a mass hanging on a spring is also such system, yet it oscillates about the equilibrium position. To limit such oscillations about the T_0 goal temperature, control devices are set to limit the value of S_p such that Eq. (13.6) will function only in a small range ("band"), and above it, S_p remains constant. The range where Eq. (13.6) remains valid is called "proportional band"; in commercial equipment data, it is stated in percents of the goal temperature.

3) Differential control

In order to restrain temperature fluctuations occurring under basic proportional control systems, a correction is added, which is analogous to adding friction in the example of a mass hanging on a spring (such mechanical systems are used in car wheels shock absorbers). That is obtained by adding a correction signal that is proportional to the time derivative of the deviation

$$S_D = -D \cdot \frac{d(\Delta T)}{dt}, \tag{13.7}$$

where D is a positive proportionality factor. The meaning of that correction is that it acts to compensate against quick temperature deviations. Under conditions that are similar to critical damping of free oscillations, the temperature changing rate on returning toward the T_0 goal temperature will be lowered, and won't cross it. The D factor may be tuned to achieve critical damping.

4) Integral control

Combining the two above control principles – "Proportional" and "Differential" would result in obtaining only small oscillations about the goal temperature T_0. Still, if the system exhibits some asymmetry, for example in the heat dissipation in cases of a positive or a negative deviation of temperature, the actual temperature will oscillate around a temperature that deviates from the goal one. To detect this deviation, the actual measured temperature is integrated for a relatively long duration, $t_2 - t_1$. Then, a power correction is given to the heating system with a size that is proportional of the average deviation

$$S_I = -I \cdot \frac{1}{t_2 - t_1} \int_{t_1}^{t_2} \Delta T \ dt, \tag{13.8}$$

where I is an adjustable positive proportionality factor. The described control concept is addressed as "Integral control".

5) PID unified control

Commercial temperature control devices that combine the above three methods are called "PID control systems" acronym of "Proportional–Integral–Differential". In other words, the control signal would be given by

$$S = S_p + S_D + S_I. \tag{13.9}$$

The control is performed by using electrical signals. Therefore, the device is capable of controlling not only a temperature, but also other quantities like the electrical power. Temperature control may be performed using any thermometer whose output is a continuous electrical signal whose size is proportional to the measured temperature (a thermocouple; a pyrometer); controlling the heater power may be realized by measuring the product between current and voltage, and processing an output signal that is proportional to that product. There are also other physical quantities associated with crystal growth that can be controlled. Some will be described in Chapter 14.

All modern temperature control devices are based on signals supplied digitally to proper computer programs.

13.10 Temperature Programming

Most crystal growth processes require programmed changes of the furnace temperature, not just a constant temperature. Such changes are needed mostly in processes carried out at temperatures that are considerably larger than room temperature. It includes the heating duration from room temperature toward the "goal" growth temperature, the crystal growth duration, and often the period of returning down to room temperature. The administered method is the use of a combination of a temperature controller (see Section 13.9) and a temperature programmer. Starting a heating or a cooling program is done as follows: the controller is provided with a goal temperature relating to a particular output voltage of the thermometer; it is called "set point". The programmer, however, provides an additional time-dependent positive or negative voltage to be fed into the controller. When the fed voltage is positive and gradually increases, the controller "is cheated" to believe that the temperature grows, and it acts to gradually lower the temperature at a rate dictated by the fed voltage. If the fed voltage is negative and gradually increases in size, the controller "is cheated" to believe that the temperature lowers, and it acts to gradually increase the temperature at a rate dictated by the fed voltage. In most modern systems, a programmer and a controller are unified into a single electronic device.

13.11 Open-Circuit and Closed-Circuit Control

Control methods in crystal growth systems are often described in terms of "open-circuit" or "closed-circuit". An **open-circuit control** means that the temporal change of a certain process parameter (say the temperature) is predetermined without active control of its actual value (and obviously without option of any feedback). In a **closed-circuit control**, the selected process parameter is continually monitored, and proper feedback is activated to correct deviations. For example, the above-described temperature control of Section 13.9 is a closed-circuit control of the temperature, but not necessarily of the intended process. The most desired control would be a closed-circuit control of the goal process!

13.12 General Behavior of a Temperature-Controlled System

A factor certainly involved in the temperature evolution in a controlled system is the heat capacity of the process hosting furnace, which is added to that of the raw material. Therefore, temperature changes of the raw material are slower than changes in the power of the heating element. That effect might cause "overshooting" of temperature, because a detector, which is often placed near the raw material, will detect the temperature changes at some delay. For that reason, it is desirable that the detector be placed near the heating element, and the furnace region close to the heating element be constructed to have a smallest possible heat capacity. On the other hand, those means are effective to control the heating element temperature rather than the aimed-at process region. That consideration suggests to place the thermometer as close as possible to the raw material crucible. Optimal placing of the thermometer is somewhere between the raw material crucible and the heating element.

13.13 Failures Protection

Crystal growth processes are very sensitive to temperature changes. Therefore, any small fault could cause a complete failure of the process. In the following, we describe some common faults and corresponding means to prevent the growth failure.

1) Electrical power outage for only several seconds may result in a complete failure of a crystal-growing process. The electrical power supply system may be backed up by an emergency electrical power generator or by a sufficiently large array of electric batteries, which start their backup supply within the short time required to save the process. An electrical power outage may also stop the operation of several, often critically needed support systems, like vacuum pumps, cooling water flow pumps, and temperature control systems. Often, if such power outage lasts more than several seconds, it is no more possible to save the crystal growth process. On the other hand, a sudden return of the electrical power to its pre-outage value may harm the support systems. If such occurs, the best option is to stop the growth process. Automatic control systems should include a delay component that stops the heating at a given time after the electrical power outage.
2) Stopping the flow of cooling water will cause overheating of those furnace parts that are not resistant to high temperatures, and therefore may cause permanent damage to the furnace. Therefore, the electrical power supply system of the furnace should include a device that disconnects the heating when the cooling water flow stops. The device that measures the cooling water flow (flow meter) should be installed on the outlet side of the cooling water.
3) If the process occurs under vacuum, or alternatively under a flow of an inert gas, devices that measure their necessary value should be included in the furnace electricity supply system, to disconnect the power when necessary.

It is often desirable that functioning fault of any furnace subsystem will include a warning signal before an ultimate stopping of the heating power is made. Sometimes it is possible to correct the malfunction, thus saving the crystal growth process.

Exercises for Chapter 13

13.1 A fusion furnace is constructed in the form of an empty rectangular box whose inner brick walls are maintained at 1500°C. The furnace has a small side door of 400 cm^2 area, which is used to insert the raw materials.

(a) Calculate the thermal emission peak wavelength when the door opens.

(b) Calculate the power losses when the door is fully open.

14 Crystal Growth Methods

14.1 Preamble

As indicated in Chapter 11, there is a large diversity of methods used for the growth of single crystals. Some methods differ only slightly among them, and they are in fact an evolution of a single basic approach. Others are indeed fundamentally different. Basically, the multitude of methods expresses the need to create optimal conditions for the growth of each selected crystal. The difference between the methods is derived from many factors, among which one can mention both the basic properties of the material of which the crystal is made, such as the melting temperature and the existence of different types of phase changes, and the technological needs it is aimed at serving. In Chapter 11, the growth methods were classified according to the nutrient phase, namely a solid, a liquid, or a gas. By a similar approach, more detailed classifications are made according to various properties such as the composition of the nutrient phase (single element or a compound), use of a solvent or a carrying gas, and their properties. The different growth processes are also classified according to the engineering solution given to technological problems, which in fact are unique for every material and every growth method.

In this chapter, we will describe only a part of the wide variety of existing methods for growing crystals. The considerations influencing the selection of the preferred method for each case will be analyzed, and advantages and disadvantages of each method will be described in detail.

14.2 Choosing the Nutrient Phase

A combination of properties for which a solid nutrient is selected is:

1) The temperature required for activating the crystallization of solid clusters is relatively low.
2) No crystalline phase change occurs between the crystallization temperature and room temperature.
3) The raw material is an element or a stable compound.
4) The vapor pressure at the crystallization temperature is very low, namely no evaporation occurs to any of the comprising components.
5) The surface energy (surface tension coefficient) is large.

A combination of properties for which a liquid (melt) nutrient is selected is:

1) The solidification temperature is low; such property saves many technological difficulties involved in using high temperatures.

Conductors, Semiconductors, Insulators, and Crystal-Growth Technology: A Practical Guide,
First Edition. Zeev Burshtein.

2) No crystalline phase change occurs between the crystallization temperature and room temperature.
3) The raw material is an element or a stable compound, at least up to the melting temperature.
4) The vapor pressure at the crystallization temperature is very low, namely the liquid (melt) chemical composition doesn't change during the crystal growth.
5) The material heat of fusion is relatively small. That feature reduces the probability for the occurrence of facets, crystallographic structural faults, and non-uniform impurity concentrations.
6) The thermal heat conductivity is large; such property is useful to prevent the development of thermal stress in the crystal during cooling to room temperature.

A combination of properties for which a gaseous nutrient is selected is:

1) The material undergoes sublimation and evaporation without chemical decomposition.
2) The vapor pressure is relatively high.
3) The crystal growth temperature is low.
4) The thermal heat conductivity is large; such property is useful to prevent the development of thermal stress in the crystal during cooling to room temperature.
5) No phase change occurs between the growth temperature and room temperature.

The choice of a specific process for the crystal growth is aimed at taking care of specific difficulties emanating from the relative importance of each listed property.

14.3 Phase Diagram-Based Conclusions

One of the most important tools in planning a process of a specific crystal growth, out of a liquid (melt) or out of a different phase, is a phase diagram. In Chapter 12, we presented a specific case of a two-component system (Fig. 12.9). In what follows, we perform a review of four different examples, and discuss in detail some related issues.

(i) The raw material is an element or a stable compound

In such case, there is no need to consult a phase diagram. The only required information concerning the crystal growth is the melting temperature, except in particular cases where the grown crystal undergoes a phase change between the growth temperature and room temperature. If such changes do occur, growing from the melt is not possible, and growing from a solution should be considered as a more favorable option.

(ii) A congruent system

In a **congruent system,** the material melt is a mixture between two components, A and B, at a certain concentration ratio between the two. As long as the melt contains the said two components, the solidifying crystal will contain the same chemically balanced $A_{x_0}B_{1-x_0}$ formula, as given by the vertical line at x_0 in the schematic phase diagram shown in Figure 14.1. If the melt is originally prepared at an x_0 concentration, the composition of the solidified material will be congruent with (same as) that of the melt, which is the source of the name given to that kind of phase diagrams. Congruency is thus a desired condition for the melt composition. However, under practical conditions, even after a very careful preparation of the mixture to be melted, the final accuracy is not absolute, due to factors like weighing errors, minute amounts of material remaining stuck to the measuring tools, and to the tools in which the materials are transferred to the melting crucible. It is also possible that while heating the mixture, one of the components evaporates faster than the other (an effect called "incongruent evaporation"). Therefore, sometimes the beginning of solidification is not always at the desired congruence point. The melt will then be gradually enriched with the component that was originally in excess. Its composition described by the phase diagram will move along the curved side of the diagram, further from the congruence point. That would cause a change in the solidification temperature during the course of the crystal growth. The rate of the change in solidification temperature is not constant, and usually it grows as solidification progresses. That effect is accentuated when the liquidus curve is steeper. Such changes complicate the planning of the growth course, whether it relates to a cooling plot or to moving of the furnace parts. In addition, issues related to constitutional supercooling are expected, as described in detail in Section 12.10.

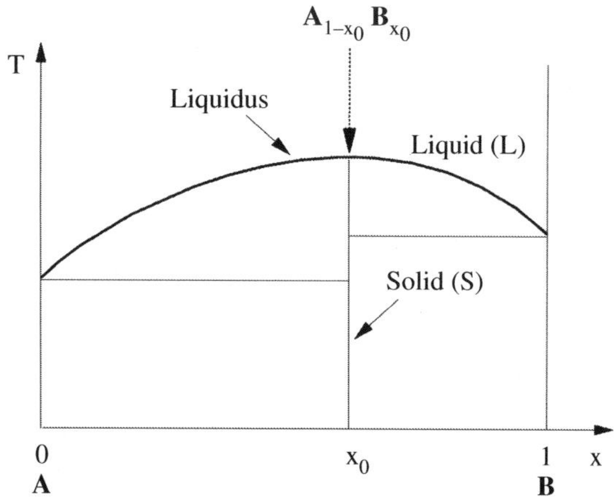

Figure 14.1 A phase diagram of a melt having a single congruent solidification point for a mixture between two materials, **A** and **B**.

An impact of the said complications is that in various single crystal pulling growth methods to be described in Section 14.4, it turns quite difficult to control the geometric shape of the solidifying material. The difficulty in fact grows if the liquidus line in the phase diagram exhibits a large slope. In such cases, it would be better to grow the crystal inside a crucible, such that the crucible shape determines the geometric shape of the grown crystal. If in addition some material evaporation is also significant, it is possible to grow the crystal inside a closed ampoule, say in a method called **Stockbarger method**, as described in Section 14.4.

A system basically defined congruent may still exhibit a more complicated behavior than described in Figure 14.1; namely, it is possible to find other constitution ranges where a solid may equally exist. A schematic phase diagram of such case is presented in Figure 14.2. In some combinations of materials, the possible deviated solid concentration ranges cover several percent in the two-component concentration. If the desired properties of the crystal are tolerable to such deviations, it makes the crystal growth much easier. On the other hand, there are many cases in which changes in the concentrations contradict the required properties. In such cases, precision in preparing the raw material chemical balance is quite critical.

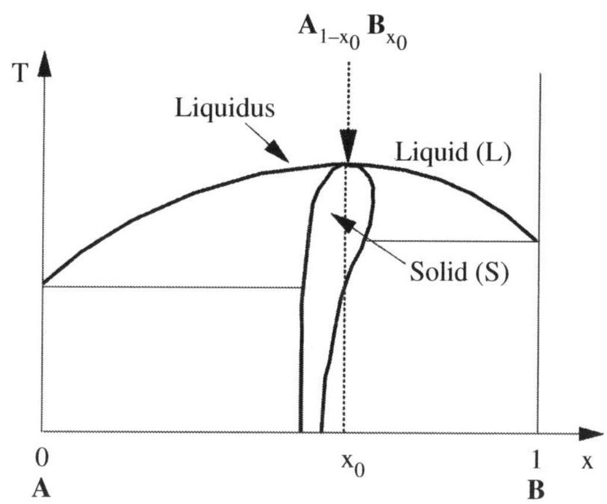

Figure 14.2 A schematic phase diagram of a system where the solid congruent existence range narrows during temperature lowering, and even departs from the chemical composition of the congruent solidification.

Seemingly, if the goal properties of the grown crystal are independent of concentration over some range, it is possible to grow the crystal from a melt in which the chemical balance is not precise. However, in many cases, the very existence range of the solid phase becomes narrower as the temperature drops. The crystal may initially grow perfectly; however, in the course of cooling down to room temperature, separation of different solid phases would occur within the solid material, which will impair the uniformity of its properties. A solution to that problem is to grow the crystal from a melt that is not chemically balanced. The growth should then be performed very slowly, to overcome the expected problem of constitutional supercooling (see Section 12.10). An important example relates to the growth of a lithium niobate ($LiNbO_3$) single crystal.

(iii) An incongruent system

In an incongruent system, the solidifying material chemical composition differs from the melt composition. A known example of such kind is the $Y_3Fe_5O_{12}$ crystal (known as YIG; acronym of yttrium iron garnet). In such cases, it is necessary to grow the crystals by using a non-stoichiometric melt (namely which is not chemically balanced).

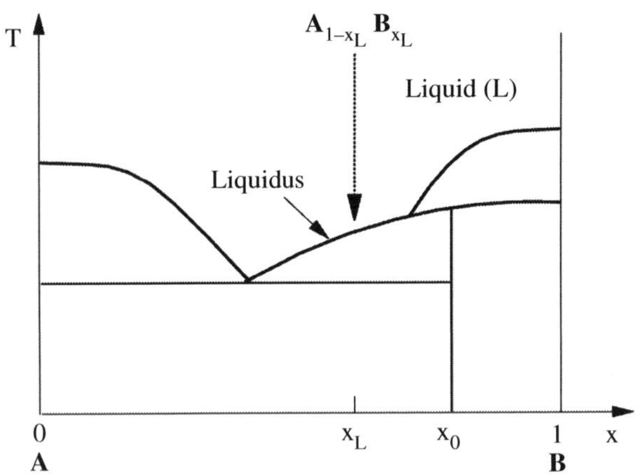

Figure 14.3 A schematic phase diagram of a system that lacks any congruent solidification point. However, it is possible to grow a single crystal of a constant x_0 concentration of **B** from a melt of a different concentration.

A schematic example for such a phase diagram is presented in Figure 14.3. Growing a single crystal from a non-stoichiometric melt involves problems of the crystal quality beyond the issue of a varying chemical composition. During growth, an excess concentration of one of the components develops in the melt adjacent to the solid/melt interface, which is similar to a case of impurity concentration. Solving it requires the generation of intense mixing of the melt, and reducing the growth velocity.

Another situation that dictates growing a crystal from a non-stoichiometric melt occurs when the vapor pressure of one of the components depends strongly on the composition (a case called "non-congruent vaporization"), and compensation against such effect is needed. A well-known example is the growth of gallium arsenide ($GaAs$) crystals. The melt of that particular case is prepared with an excess of gallium (Ga), which reduces the arsenic (As) vapor pressure. Aside, the growing of that particular material also involves a safety problem because of the high toxicity of arsenic.

In a case where the solid undergoes a change of phase structure during cooling to room temperature, it is in principle (and practically) impossible to grow a single crystal at any reasonable quality. However, if the phase diagram liquidus line at some composition belongs to a temperature which is lower than the solid phase change temperature, it is possible to obtain the desired single crystal phase by growing it from a proper non-stoichiometric melt via incongruent solidification.

A different option of growing a crystal at a temperature that is much lower than its melting temperature is to grow it from a solution. Technologically, the two methods are entirely different. For example, the table salt (sodium chloride [$NaCl$]) melts at 850 °C; on the other hand, its growth from an aqueous solution may be done at 25 °C. In principle, growing a single crystal from a solution is in fact an extension of the non-stoichiometric growth principle.

14.4 Single-Crystal Growth from Melt

14.4.1 Growth Inside a Crucible or an Ampoule

(i) **Bridgman growth method**

In the Bridgman method, the crystal is grown inside a long narrow closed ampoule, placed vertically without a seed. The raw material is melted inside the ampoule and positioned inside the furnace in a region of positive temperature gradient (namely the temperature in the lower side is lower than that in the upper side). The furnace designed for that purpose is in the form of a two-region tube. The lower region is stabilized at a T_L temperature that is lower than the raw material melting temperature, and the upper region is stabilized at a T_H temperature that is higher than the raw material melting temperature. The transition between the hot and cold regions should be steep; for that purpose, an annular buffer is added between the regions. The growth is made by moving the ampoule (or crucible) downward, namely from the hot to the cold region. When the melt-containing lower ampoule region reaches the cold region, a solid cluster forms there. It acts as a seed crystal for a continued gradual growth of the entire molten raw material. The solid/melt interface remains steady in a coordinate system attached to the furnace, but it moves upward in a coordinate system attached to the ampoule. Figure 14.4 provides a schematic illustration of the described **Bridgman** growth process.

Features and problems:

1) There is a practical limit on the diameter of the grown crystal, because if it is arranged to be too large, the solid/melt interface becomes concave. That structure restrains the free flows of melt, causing the crystal to exhibit many structural defects (see Section 12.10). Some attempts were made to enhance the flows by rotating the ampoules about their vertical axis. That, however, complicated considerably the furnace construction and was not always useful.

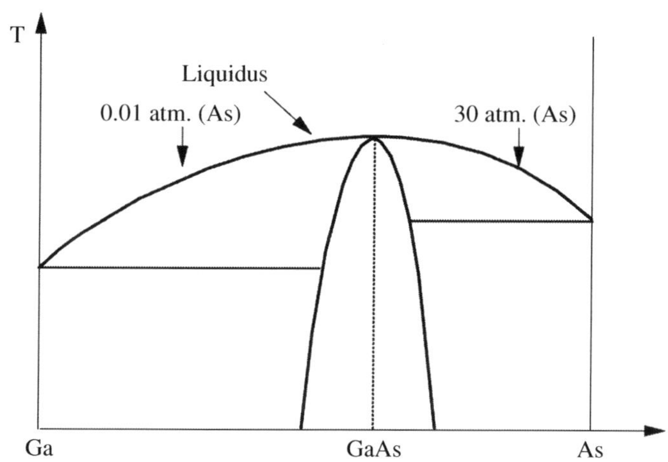

Figure 14.4 A schematic phase diagram of the GaAs two-component system (a similar diagram describes the Ga-P system). The partial vapor pressure of arsenic (As) in the melt is indicated for two different concentrations in the melt.

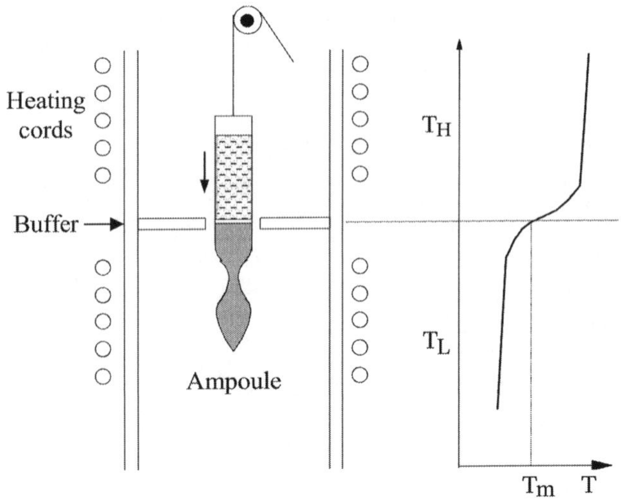

Figure 14.5 A schematic description of the furnace and ampoule structure for a crystal growth by the Bridgman method (left side) and the temperature profile (right side).

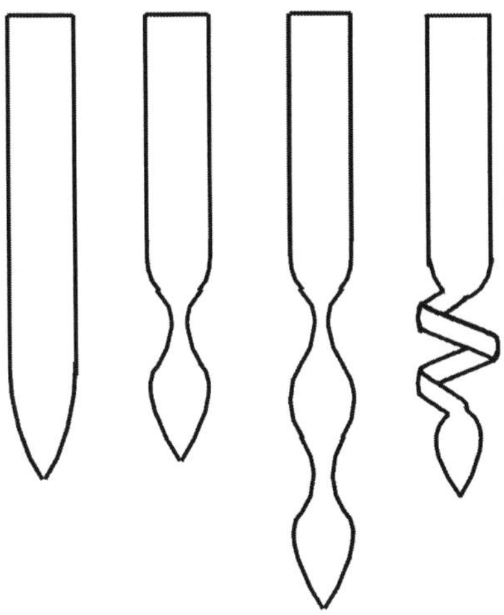

Figure 14.6 Structure of different ampoules used in crystal growing by the Bridgman method.

2) The primary problem related to a Bridgman crystal growth is the probability for a multiplicity of solid clusters creation when the ampoule is lowered to below the melting temperature region, ending with a polycrystalline solid. To reduce such probability, the lowest part of the ampoule is point-shaped, such that the melt volume initially assuming a solidification temperature is very small. Such arrangement is shown in Figure 14.5. Sometimes that measure is not sufficient, and the crystal starts to grow exhibiting several large grains. To cure against such case, a kind of a "spatial filtering" is performed, by adding one or several diameter-narrowing regions along the lower part of the ampoule (see Fig. 14.6). It increases the probability that only one single-crystal orientation will survive when the ampoule broadens, correspondingly dictating the final crystallographic orientation. It is also possible to add brief temperature rises during the initial growth duration. These rises increase the probability that newly formed clusters will melt (see Section 12.1).

3) An additional problem related to the Bridgman growing process is a post-growth stress during cooling to room temperature, especially when the thermal expansion coefficients of the crystal and the ampoule are considerably different. An especially difficult problem occurs for materials of anomalous solidification (expanding during solidification) such as in germanium *Ge*. In some cases, it even causes fracture of the grown crystal. Another problem is adhesion of the crystal to the ampoule walls, which makes it difficult to separate it when the growth is

over. By carefully choosing the material from which the ampoule is made, most of the mentioned problems can be reduced.

4) Quite often, Bridgman growth of certain crystals is used as a preliminary stage for growing by another method. Even if the grown crystal exhibits several grains, it is possible to select a mono-crystalline region that can be used as a seed for growth by another method.

(ii) **Stockbarger growth method**

The Stockbarger growth method is almost entirely similar to Bridgman's but differs by one important detail, which is placing a seed at the ampoule (or crucible) bottom. The seed presence reduces the uncertainty regarding the initial solid cluster formation. Except for that it is subject to all problems occurring in the Bridgman method. The special problems it exhibits are related rather to the seed presence. The process of the raw material melting must include a partial melting of the top part of the seed (obviously not all of it!). Such process requires accuracy in the temperature setting during the heating. In addition, it is imperative that the molten material just above the seed won't penetrate around it and/or under it. Such may occur if there is a too large gap between the seed proper and the "seed housing" region (see Fig. 14.7), or if the melt surface tension coefficient is too small. Such occurrence may result in a growth of a polycrystalline solid as in the case of the basic Bridgman method. Particularly, if the specific gravity of the seed is smaller than the melt's, the melt penetration below it would float the seed upward to the melt surface.

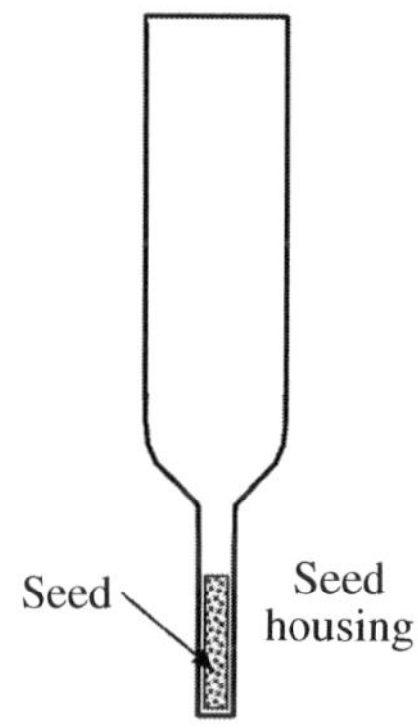

Figure 14.7 Structure of an ampoule used in crystal growth by the Stockbarger method.

(iii) **Schmidt–Viechnicki temperature gradient growth method**

The moving devices used in the Bridgman–Stockbarger method and in the pulling method to be described in Section 14.4.2-**iii** often pose a non-trivial technological complication for the furnace design. These devices need to be very accurate and perform under difficult environmental conditions. Inaccuracy caused by a non-uniform pulling rate, non-axial rotation, or other kinds of vibrations may sometimes cause non-uniform segregation of impurities, and even structural distortions of the grown crystals.

Modern growth systems designed to overcome such problems are often very expensive. The temperature gradient growth method aims at overcoming the need for moving devices being a part of the furnace design. In principle, the system comprises a crucible or an ampoule with a crystalline seed at the bottom, like in the Stockbarger method described above (subsection [**ii**]). To allow such growth process, the temperature gradient in the crucible has to be accurately controlled. Usually, the temperature gradient in the seed region is separately controlled, often by an added separate cooling of the seed housing by flowing of cold helium (*He*) gas through a bore that penetrates through the seed housing (Fig. 14.8).

The growth process includes the melting of the entire raw material inside the crucible, including a small part of the seed. Throughout the entire cooling, the solidification front shifts slowly upward, until the entire melt solidified. The solidification front shape is not planar as simplistically assumed regarding Pfann's normal freezing relation, but the used mathematical relations are equally valid.

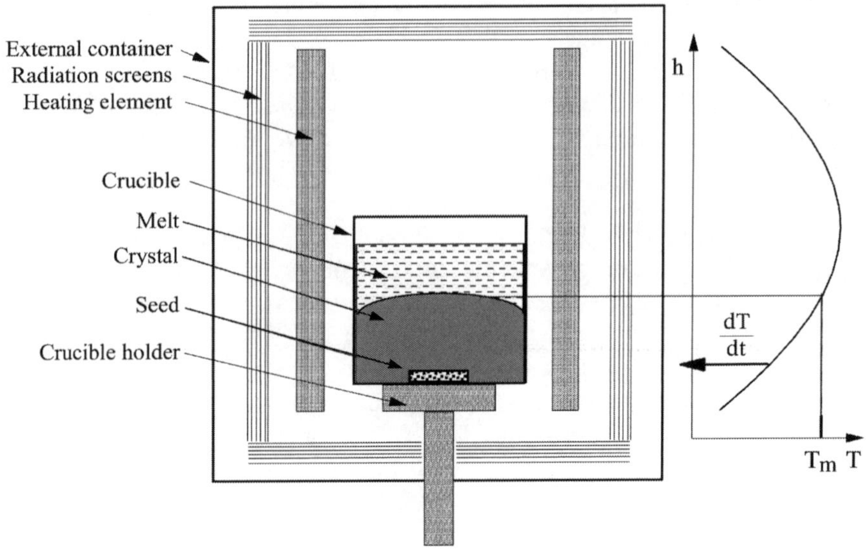

Figure 14.8 Schematic illustration of a system used for crystals growth by the temperature gradient method. The longitudinal temperature gradient prevailing in the furnace is shown on the right side. Over time, the entire temperature profile moves rigidly to the left by a uniform cooling of the entire furnace, at a rate dT/dt.

Features and problems:

1) Due to the relative stability of the system, and especially if heating is performed using a resistive heating element, practically no induced convection develop in the molten raw material. Therefore, the system is best suitable to the growth of uniform pure materials.
2) The solidification takes place inside the crucible; therefore, all implications regarding that fact, as described in connection with the Bridgman and Stockbarger methods, are valid: stress, grain boundaries, and crystal breakage.
3) It is very difficult to measure the seed temperature; therefore, the seed might totally melt at the beginning of the growth process, and the growing solid will assume a polycrystalline structure. That issue is mostly severe in growing of crystals having very high melting temperature in the vicinity of 2000 °C.
4) The cooling rates required to achieve single-crystal growth are very small, often approaching the capability limits of the temperature control equipment.

(iv) **Kyropoulos growth method**

In the Kyropoulos method, a single-crystal seed is mechanically lowered until touching the melt top surface contained in a crucible. The heating element and crucible position are arranged to exhibit a negative temperature gradient (namely, the temperature in the crucible bottom is higher than at the top), and the radial temperature gradient is positive (namely, the temperature at the crucible center is lower than at its circumference). Solidification is performed by a temperature lowering throughout the entire furnace; it starts at the seed and ends at the crucible bottom. The upper parts of the exposed grown crystal often exhibit quite large planar facets, typical of the growth process.

The features and problems are the same as described concerning the Schmidt–Viechnicki temperature gradient growth method (see the above item [**iii**]).

14.4.2 Growth Outside a Crucible or an Ampoule

(i) **Float zone growth method**

The float zone growth method is similar to growth by zone melting, yet without a container (crucible or ampoule) for the raw material. The raw material is prepared as a long cylindrical casting whose axis is vertical. A single-crystal seed is added at the cylinder bottom. The growth process is carried out by zone melting, started at the cylinder bottom and moving upward. The molten zone is sufficiently narrow to avoid spill down (it remains hanging due to its surface tension). An advantage of that process is the absence of intimate contact with any container (a crucible or an ampoule). In that way, contamination by impurities as well as stress problems due to chemical interactions and differences between expansion coefficients, respectively, are minimized. That method would be preferable to grow uniform materials of congruent solidification.

(ii) **Pedestal growth method**

The pedestal growth method is almost identical to the float zone growth method and differs by only a single change: the seed is placed at the top side of the raw material, and the heating scan correspondingly starts at the top side and progresses downward. A schematic illustration is presented in Figure 14.9.

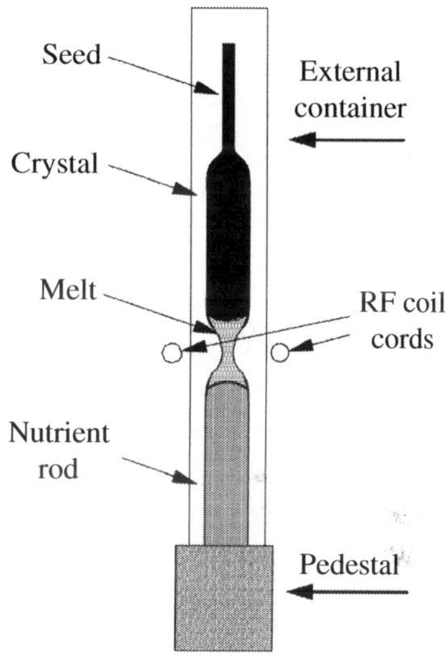

Figure 14.9 Schematic illustration of the "pedestal" crystal growth method.

(iii) **Czochralski pulling-from-melt growth method**

The growth of single crystals by pulling from the melt (Czochralski method) is the most common one in the crystal growth industry. An oriented single-crystal seed is dipped from above into the melt, which is contained in a crucible. The rod-shaped seed is manufactured to have a desired crystallographic orientation, and is oriented perpendicularly to the melt surface. The seed and an attached solidified material are then gradually pulled up to a colder region, usually also being rotated about the seed axis. The purpose of rotation is mainly not only to improve the radial symmetry of the melt thermal profile in the crucible, but also to cause melt flows near the crystal/melt interface, to mix and homogenize the separated impurities that occur for a $k \neq 1$ segregation coefficient case during the crystal growth. The crucible, seed, and seed holder, are contained inside a furnace that creates a negative temperature profile in the upward orientation (namely cold at the top and hotter below). The seed holder plus already grown crystal further support that profile by acting as an upward heat transfer bodies. That feature also helps in maintaining a rather positive radial temperature profile inside the crucible (thus preventing parasitic melt solidification at the crucible inner walls). The furnace power is controlled to maintain a programmed temperature variation at a single specific point of the crucible edge, as measured by a two-metal thermocouple (marked T.C. in Fig. 14.10).

The crucible is usually made of highly pure material that withstands the involved high temperatures and doesn't interact chemically with the raw material melt. The inside furnace atmosphere is often vacuum-pumped, sometimes leaving a constant flow of an inert gas at a minimal pressure (say ~1 torr), or alternatively allowing an inert gas at atmospheric pressure.

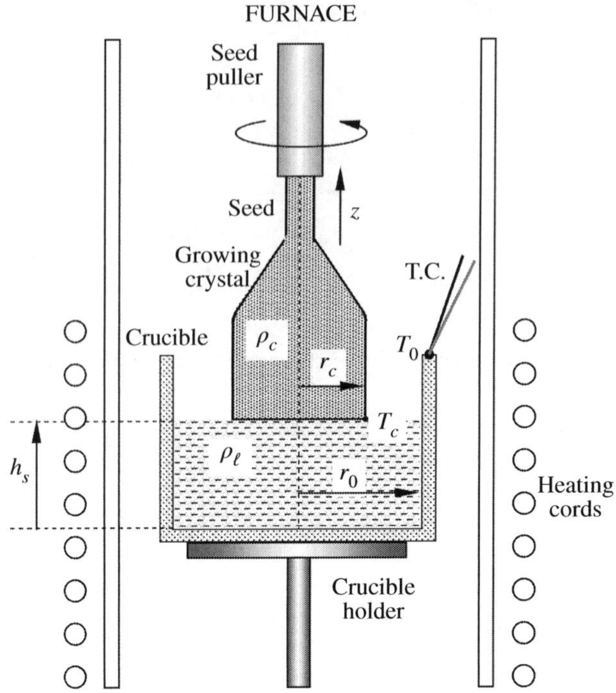

Figure 14.10 Schematic illustration of a single-crystal growth by pulling up from the melt (Czochralski method) used as an auxiliary for calculation of a growth process controlled by a cooling plot.

The basic (quite schematic) construction of a furnace during a crystal-growing process is described in Figure 14.10. We provide here an analysis that shows that it is possible to grow a cylindrically shaped single crystal using a cooling program of the entire furnace. For that goal, we assume a cylindrical symmetry to the heating furnace, including its up-pulling system. According to Figure 14.10, we are interested in growing a crystal whose radius is r_c being pulled out of a cylindrical crucible whose radius is r_0 (obviously satisfying $r_0 > r_c$). The length at which the seed is pulled up is denoted by z, and the melt height above the crucible bottom at some point in time is denoted by h_s. The growth starts at $z = 0$.

Let T_c be the temperature at a point r_c on the circumference of the growing crystal surface that touches the melt top surface. We express T_c as a function of h_s, r_c, and the time t, namely $T_c = T_c(h_s, r_c, t)$. Its full time derivative is then given by

$$\frac{dT_c}{dt} = \frac{\partial T_c}{\partial h_s} \cdot \frac{dh_s}{dt} + \frac{\partial T_c}{\partial r_c} \cdot \frac{dr_c}{dt} + \frac{\partial T_c}{\partial t}. \tag{14.1}$$

Notably, the $\partial T_c/\partial h_s$ quantity is the longitudinal temperature gradient in the crucible, which must be negative, and the $\partial T_c/\partial r_c$ quantity is the radial temperature gradient in the crucible, which must be positive. The former quantity may be continuously measured during the crystal growing, by placing several thermocouples at different heights very close to the crucible side (not shown in Fig. 14.10).

We assume that the furnace power is controlled for setting a desired temperature T_0 at the stable (not moving) crucible. That temperature is thus only time-dependent, namely

$$\frac{dT_0}{dt} = \frac{\partial T_0}{\partial t}.$$

Adding and subtracting the latter equation to the right-hand side of Eq. (14.1), the latter may be rewritten in the form

$$\frac{dT_c}{dt} = \frac{dT_0}{dh_s} + \frac{\partial T_c}{\partial h_s} \cdot \frac{dh_s}{dt} + \frac{\partial T_c}{\partial r_c} \cdot \frac{dr_c}{dt} + \frac{\partial}{\partial t}(T_c - T_0). \tag{14.2}$$

At this stage, we add an assumption whose validity is crucial concerning the ability to monitor the growing crystal radius by a cooling program. We assume that the temperature profile in the crystal growth region of the furnace is "stiff"; namely, any change of T_0 also occurs throughout the entire growth region. It obviously requires that there are no changes in the heat conduction rates throughout the furnace caused by the said temperature change. Under such conditions, the latter partial derivative is zero: $\frac{\partial}{\partial t}(T_c - T_0) = 0$.

While the crystal is being pulled up and grows, the height of the melt surface in the crucible lowers. Equating between the melt mass loss and the crystal mass gain yields the relation

$$\frac{dh_s}{dt} = -\frac{r_c^2}{r_0^2 \frac{\rho_\ell}{\rho_c} - r_c^2} \cdot \frac{dz}{dt}, \tag{14.3}$$

where ρ_ℓ and ρ_c are the specific gravities of the raw material melt and solid, respectively. (In many materials, the ratio of specific gravities satisfies $\rho_\ell/\rho_c > 1$ at the melting point; typically, the melt is several percent denser than the crystal).

For the crystal to grow at the target constant radius r_c, we would further set $dr_c/dt = 0$ in Eq. (14.2). We now consider two special cases concerning the state of the raw material.

(a) Growth using a pure stoichiometric melt

Under such special condition, the solidification front temperature T_c remains the same during the entire growth process, namely $\frac{dT_c}{dt} = 0$. Setting it in Eq. (14.2) along with the above "stiffness" condition on the temperature gradient and Eq. (14.3), we get

$$\frac{dT_0}{dt} = \frac{\partial T_c}{\partial h_s} \cdot \frac{r_c^2}{r_0^2 \frac{\rho_\ell}{\rho_c} - r_c^2} \cdot \frac{dz}{dt}. \tag{14.4}$$

The physical meaning of Eq. (14.4) is very clear: the controlling temperature T_0 has to be lowered at a rate that compensates against the temperature rise that would occur if the melt surface lowers toward hotter regions of the furnace. It may be readily seen that if the up-pulling of a melt uses a narrow seed crystal, the pulled out crystal radius will grow gradually until it reaches the targeted radius r_c, which later on remains unchanged. The equation provides a simple plot to achieve a grown crystal of a constant radius. The involved parameters are the pulling rate dz/dt, the longitudinal temperature gradient $\partial T_c/\partial h_s$ (which is always negative), the target crystal radius r_c, the crucible radius r_0, and the melt-to-solid specific gravity ratio ρ_ℓ/ρ_c. Apart from the solidification temperature, the latter is the only factor relating to the specific material properties. There is still a condition not specifically expressed, which is that the radial temperature gradient must always satisfy $\partial T_c/\partial r_c > 0$, namely that the temperature at the melt circumference (the crucible inner side) is always higher than at the center; otherwise, solidification may be initiated at the crucible inner side.

(b) Growth using a doped or a non-stoichiometric melt

If the raw material is impure (or intentionally doped), or non-stoichiometric, the solidification front temperature T_c changes steadily during the growth process, namely $\frac{dT_c}{dt} \neq 0$. Suppose that the raw material contains a single dopant at a C_0 molar density of a $k < 1$ segregation coefficient. The dopant concentration in the melt then grows, and its solidification temperature reduces, namely $\frac{dT_c}{dC} < 0$. Generally we may write

$$\frac{dT_c}{dt} = \frac{dT_c}{dC} \cdot \frac{dC}{dg} \cdot \frac{dg}{dz} \cdot \frac{dz}{dt}, \tag{14.5}$$

where C is the dopant molar density, and g is the fraction of the solidified raw material. To assess the dT_c/dC quantity, we may use the cryoscopic relation (see Eq. [12.12])

$$\frac{dT_c}{dC} = -\frac{(1-k)RT_c^{\,2}}{H_f}, \tag{14.6}$$

where R is the gas constant, and H_f is the molar solidification latent heat. Assuming the validity of Pfann's normal freezing conditions (Section 12.7 and Eq. [12.3]), we write

$$C(g) = C_0(1-g)^{k-1}, \tag{14.7}$$

whose derivative is

$$\frac{dC}{dg} = C_0(1-k) \cdot (1-g)^{k-2}. \tag{14.8}$$

The pulling rate $\frac{dz}{dt}$ is of course a process parameter that may be determined independently of the specific raw material and crystal properties.

It is interesting to notice that the solidification front temperature T_c would steadily reduce during growth, namely we would always have $\frac{dT_c}{dt} < 0$ for both $k < 1$ and $k > 1$ cases!

Because the crucible radius is constant, the fraction of the solidified material g is given by $g = \frac{h_0 - h_s}{h_0}$, where h_0 is the melt height in the crucible when pulling starts. Thus

$$\frac{dg}{dz} = -\frac{1}{h_0} \cdot \frac{dh_s}{dz} = \frac{1}{h_0} \cdot \frac{r_c^{\,2}}{r_0^{\,2}\frac{\rho_\ell}{\rho_c} - r_c^{\,2}}. \tag{14.9}$$

The last factor in Eq. (14.5) is the pulling rate $\frac{dz}{dt}$, which may be chosen externally.

When all factors appearing on the right-hand side of Eq. (14.5) are given, we can use if for calculating $\frac{dT_c}{dt}$; the cooling plot would then be given by

$$\frac{dT_0}{dt} = \frac{dT_c}{dt} + \frac{\partial T_c}{\partial h_s} \cdot \frac{r_c^{\,2}}{r_0^{\,2}\frac{\rho_\ell}{\rho_c} - r_c^{\,2}} \cdot \frac{dz}{dt}. \tag{14.10}$$

The physical meaning of Eq. (14.10) is clear: The controlling temperature T_0 has to be reduced at a rate that compensates against the temperature rise that would occur if the melt surface lowers toward hotter regions of the furnace, plus the rate of solidification temperature lowering caused by an impurity segregation coefficient $k \neq 1$.

(c) Growth of the single-crystal "shoulders"

The "shoulders" term relates to the growing diameter of the pooled crystal until it assumes the constant diameter. In principle, it is possible to include a desired shoulders shape in the cooling plot.

As an analytic example, we assume a simple growth from a pure stoichiometric melt, then $\frac{dT_c}{dt} = 0$. The cooling plot would then be given by

$$\frac{dT_0}{dt} = \frac{\partial T_c}{\partial h_s} \cdot \frac{r_c^2}{r_0^2 \frac{\rho_\ell}{\rho_c} - r_c^2} \cdot \frac{dz}{dt} + \frac{\partial T_c}{\partial r_c} \cdot \frac{dr_c}{dz} \cdot \frac{dz}{dt}. \tag{14.11}$$

Equation (14.11) may be used as a cooling plot by including a desired shape of $r_c(z)$. It is also reasonable to assume that the longitudinal and radial temperature gradients are identical in size, namely $|\partial T_c/\partial r_c| = |\partial T_c/\partial h_s|$ (the first is positive and the second is negative).

A more practical approach is to adhere to the plot of the intended (large) crystal diameter r_c. The radius will then gradually grow from its size at the first contact with the seed, till reaching the target r_c, which will remain constant until the end of growth. We then use Eq. (14.11) as a differential equation for r_c. Assuming that $\frac{dz}{dt} = const.$, an approximate solution for $r_c(t)$ would be

$$r_c(t) \cong r_c(\infty) \cdot \boldsymbol{tanh}\left\{ -\frac{\frac{\partial T_c}{\partial h_s}}{\frac{\partial T_c}{\partial r_c}} \cdot \frac{r_c^2(\infty)}{r_0^2 \frac{\rho_\ell}{\rho_c} - r_c^2(\infty)} \cdot \frac{\frac{dz}{dt}}{r_c(\infty)} \cdot t \right\}, \tag{14.12}$$

where $r_c(\infty)$ is the target toward which the radius grows asymptotically during the crystal growth.

(d) Other methods of crystal radius controlling in the Czochralski growth

There are quite a number of laboratory and commercial Czochralski crystal-growing system designs, where the growing crystal radius is monitored and controlled.

1) The most direct and simplistic monitoring and controlling the growth radius is by assigning a skilled human controller to watch continually the shape of the growing crystal. If the radius increases or decreases relative to the planned value, the controller manually increases or decreases the heating power, respectively. Such method is quite appropriate in development laboratory growing; in a crystal-growing factory, where many furnaces are simultaneously operating, more automatic control systems are required.
2) Radius control by an automatic measurement of the torsion moment between the rotating crystal and its melt. The underlying physical effect is friction force between the solid solidification front and the melt. The rotating seed holder experiences that friction as a torsion moment, which is proportional to the solidification front area. The seed holder is connected to the pulling and rotating device through a torsion gauge, which feeds an electrical voltage signal proportional to the torque moment into the temperature (or power) control system. A proper feedback assures that the torque moment, hence the interface area, will remain constant.
3) Radius control by measuring the melt height in the crucible. The logic behind it is that while the crystal grows, the melt height in the crucible lowers. Quite a number of techniques were attempted for measuring the melt height, such as measuring the electrical resistance between the melt surface and its bottom; monitoring the deviation of a laser beam reflected from the surface while the melt surface lowers; and a continuous X-ray image processing. Each of these means may be suitable in limited number of cases, of particular crystals.
4) Radius control by measuring the increasing crystal weight rate and correcting the furnace power to maintain the targeted weight growth rate. The seed holder is connected to the pulling

and rotating device through a weight gauge, which feeds an electrical voltage signal proportional to the crystal weight into the temperature (or power) control system. A proper feedback assures that the weight increase rate, hence the crystal radius, will stay constant. In fact, that method is the most common means in the crystal growth industry!

5) There are two critical stages in a Czochralski growth: the initial contact of the seed with the raw material melt might cause a melting of the seed. Therefore, the seed has to be lowered very slowly, and particularly making sure that on touching the melt, a small part of the melt solidifies. Then the furnace temperature needs to be slightly and slowly raised until the added solid melts back. The reason is that a body of quickly solidified material may exhibit a polycrystalline nature, which would then dictate that nature to the continuing solid growth.

The other critical stage is the disconnecting of the grown crystal from the remainder melt at the crucible bottom. If done abruptly, it may cause a thermal shock to the grown crystal, hence its breaking. A recommended procedure is to gradually reduce the radius, thus forming a kind of a narrow tail, and only then performing the abrupt disconnection.

(iv) **Obtaining constant impurities concentration in Czochralski-like growth methods**

An important issue in determining the quality of a single crystal is obtaining the planned dopant (impurity) concentration throughout most of the grown crystal. If the dopant segregation is different than a unity, obtaining a uniform distribution is not trivial (see Sections 12.6–12.11). Assuming that the molten raw material is well stirred, a solution to obtaining a constant concentration in the crystal is to find a way to keep the dopant concentration in the melt at the same concentration it had when the solidification started. That requires providing a proper compensation to the changing concentration in the melt. In the following, we describe such methods related to the Czochralski crystal growth method.

(a) The push–pull method

In the push–pull method, a cylinder of a polycrystalline (cast) material is prepared, having an identical composition with a C_0 dopant concentration as the intended crystal. The raw material melt is prepared to have a C_0/k dopant concentration, where k is the dopant segregation coefficient. While the crystal grows from the melt by pulling up, the cast cylinder is pushed down into the melt at a rate that is identical to that of the crystal growth. It is readily obvious that both the melt and the growing crystal would exhibit the same original dopant concentration throughout the entire crystal growth process (Fig. 14.11).

(b) The two-container method

The two-container method is in principle very similar to the above push–pull method in the sense that it provides the melt with a material that compensates changes in the dopant concentration while the crystal grows. The furnace actually contains two crucibles (containers) inter-connected by a proper pipe at their bottom, thus forming a connected vessels system. A Czochralski-type growth occurs at one of the crucibles (A) containing a melt having a C_0/k dopant concentration. The second crucible (B) contains a melt having a C_0 dopant concentration. While the crystal grows with a $k(C_0/k) = C_0$ dopant concentration, the melt height lowers, inducing an automatic flow of the melt from the B side, which compensates its C_0 dopant concentration loss. A schematic illustration of such system construction is presented in Figure 14.12.

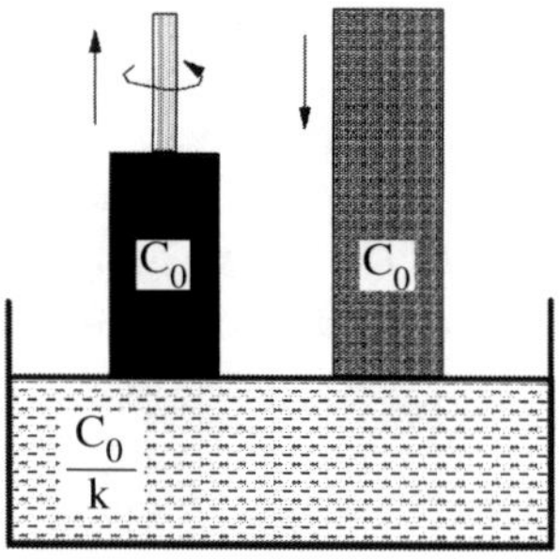

Figure 14.11 Schematic illustration of a push–pull system for receiving a constant impurity concentration in a grown crystal.

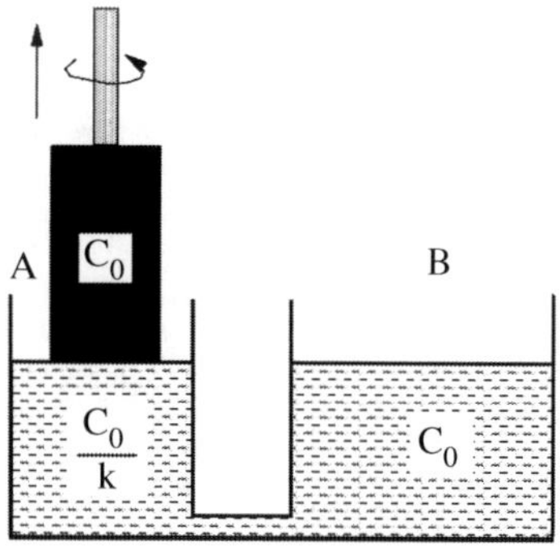

Figure 14.12 Schematic illustration of a two-container system for receiving a constant impurity concentration in a grown crystal.

14.5 Vapor Growing of Single Crystals

Vapor-grown crystals often exhibit higher quality compared to melt-grown crystals, because diffusion in the gas phase is far better than liquid stirring efficiency. Usually, the growth temperature is lower because it is performed below the crystal melting point. It is further possible to exceed or reduce the growth rate by the presence of a carrying inert gas. Each of the methods described below requires special expertise in chemistry and in physical chemistry.

1) Direct vapor growth: Such growth process is performed in a closed container where the raw material is placed at one of its sides while heated to a temperature that is sufficient to induce sublimation or vaporization. The vapor then diffuses toward the other side, which is kept at a temperature that is lower than the melting temperature. There, the material solidifies in the form of a single crystal. Sometimes, a single-crystal seed is used, but often an initial spontaneous solidification is allowed to serve as a seed for further growth. A practical example is the growth of mercuric iodide HgI_2. In that particular case, frequent short and small heating durations are used during the growth to remove “parasitic” spontaneous solidifications appearing on the “legitimate” large, already grown, single-crystal surface.
2) Chemical vapor deposition (**CVD**): A related process involves the production of the raw material by a chemical interaction between two gaseous materials occurring on a substrate inside the growth container. Initially, a different substrate is heated to a temperature that stimulates that interaction, and then the two said gaseous materials are let to enter slowly into the reaction container. The reaction product grows gradually on the intended substrate in the form of a single-crystal layer.
3) Metal–organic chemical vapor deposition (**MOCVD**): That method is a specialized variation of the CVD when deposition of materials having a very low vapor pressure is desired (like some inter-metallic alloys). The used raw materials are organic compounds containing the desired metals that exhibit the desired high vapor pressure. The chemical reaction causes deposition of the metallic part on the substrate, and the organic remainder is pumped out.
4) Molecular beam epitaxy (**MBE**): The principle underlying that method is to deposit a crystalline layer on a selected substrate by using atomic and/or molecular beams incidenting on it. The method allows the growth of crystalline layers at a desired rate, and of a desired layer thickness down to even mono-atomic or mono-molecular layers.

Exercises for Chapter 14

14.1 An indium antimonide (*InSb*) semiconductor single crystal is grown from its melt by the Czochralski method. The ratio between the material melt and the crystal specific gravities (at the 527 °C melting temperature) is 1.14. The growing crucible has an inner diameter of 4.0 cm and is kept under a temperature gradient of −12.5 °C/cm. The crystal is pulled out of the melt at a rate of 2.2 cm/hr, while the furnace temperature is lowered at a rate of $dT_0/dt = -8.3\,°\text{C/hr}$.

(a) Calculate the obtained final diameter of the grown crystal body.

(b) Assuming that the longitudinal and radial temperature gradients are the same, calculate the growth time required for the crystal "shoulders" to reach 85% of the final diameter. (Consult Eq. [14.12] for that purpose.)

(c) The above-described material melt is added with a 5 at. % impurity of a segregation coefficient $k = 0.2$ causing a 40 °C reduction of the freezing temperature. The initial melt height in the crucible is 8.5 cm. Calculate the needed cooling rate for maintaining the same final radius.

15 Examples of Single-Crystal Growth and Mechanical Processing

15.1 Preamble

Conclusions related to the functioning of the end products of single-crystal growth exhibit a decisive influence on the raw material selection and the crystal growth method. Furthermore, post-growth quality assessment of a grown crystal almost always requires mechanical processing like cutting and polishing, and measurement of optical and/or electrical properties (see for example Section 10.2.3). The examples presented in this chapter demonstrate that need. In the examples, emphasis will be placed on the physical understanding of the material properties and its final functioning. The aspects related to the growth procedure will be described only briefly. The last section describes the principles and methods in mechanical processing of crystals into functioning components.

15.2 Growth of Neodymium-YAG (Nd:YAG) for Lasers

15.2.1 Introduction

Section 7.4.2 provides a general description of the physical principles underlying laser operation. The combination of a crystalline matrix with a replacing (impurity) ion (usually of a transition metal or a rare-earth atom) hosting a specific crystalline site is very common in solid-state laser technology. The optical transitions between the ion states allow both the "pumping" process (energy investment by using the ion optical absorption) and the stimulated light emission (resulting from the following ion fluorescence). The major role of the hosting crystalline matrix is to allow the presence of the replacing atom at its ionic state at room temperature. In addition, the ion symmetry reduction due to the specific properties of the hosting site, causes state splitting of the ion, and allows optical transitions that would otherwise be forbidden. Finally, the intimate proximity of the ion to the matrix in its occupying unit cell enables dissipation of excess energy in the form of heat, and the heat dissipation toward the cold environment.

The laser technology employs various, very useful combinations of ions and solid matrices, and research and development in that area is continual and quite intensive. This enables the laser developers to achieve novel combinations of relative advantage. A combination, in fact known since 1964, is of triply ionized neodymium (Nd^{3+}) added as an impurity in the $Y_3Al_5O_{12}$ crystal. The latter is marked with the YAG initials of the material name "yttrium aluminum garnet".

Conductors, Semiconductors, Insulators, and Crystal-Growth Technology: A Practical Guide,
First Edition. Zeev Burshtein.

(The "garnet" term is the name of a family of crystals, all belonging to the same space symmetry). That laser emits a beam at the 1064 nm wavelength, which under proper conditions may be obtained as a continuous beam, or by pulses about 10 ns long. Special arrangements allow obtaining much shorter pulses, only 50 ps long. The original output wavelength can be manipulated to double or triple the frequency, by letting the original beam cross some specific non-linear crystals to obtain pulses of 532 nm or 355 nm, respectively, and also divide the frequency to obtain combinations of longer wavelength. The laser pumping may be performed by different means, like illumination using discharge tubes, or by specific diode lasers (Fig. 15.1).

Nd : *YAG* lasers are used for various technological and industrial purposes, such as medical surgeries, drilling, cutting, soldering of various metals, diamond cutting, and range finding. In Table 15.1, we provide a list of the basic physical properties of the *Nd* : *YAG* crystal.

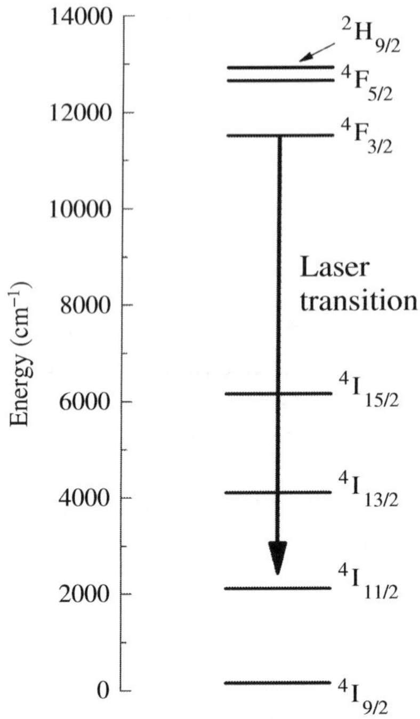

Figure 15.1 Energy levels scheme of Nd^{3+} ions in YAG.

15.2.2 Raw Material Preparation

The simplest approach–preparation of the raw material for the crystal growth is to mix Yttria (Y_2O_3) and alumina (Al_2O_3) powders by a 5 : 3 molar ratio, respectively. Such approach requires great care; in addition to YAG, the quasi-binary $Al_2O_3 - Y_2O_3$ system exhibits the formation of two other compounds; One is $YAlO_3$, which crystallizes in an orthorhombic symmetry of the perovskite family (denoted by the YAP acronym); its melting temperature is 1925 °C. The other compound is $Y_4Al_2O_9$, which crystallizes in a monoclinic symmetry (denoted by the YAM acronym); its melting temperature is 1940 °C. Using a long sintering of yttria and alumina powders at 1940 °C, at the molar ratio appropriate for YAG, traces of YAP and YAM are also formed, most probably due to local non-uniformities in the powders concentrations. These traces not always disappear upon melting, and after a growth of the single crystal, they may be expressed as microscopic light-scattering centers. These effects may be reduced by using especially fine powders, most accurate weighing, and a subsequent intensive mixing before sintering. A detailed phase diagram of the binary $Al_2O_3 - Y_2O_3$ system is presented in Figure 15.2.

The neodymium oxide (Nd_2O_3) dissolves only partially in the YAG crystal. The Nd^{3+} ions replace the Y^{3+} "legitimate" ions in their dodecahedral sites (Wyckoff c-sites of a D_2 symmetry, surrounded by 8 nearest oxygen ions) up to a maximal concentration of 5.0 at. %. (that at. % unit relates to the fraction of replaced Y^{3+} ions, and not to all comprising atoms). Attempts to increase the said replacement fraction results in segregation of undesired phases, mostly alumina (Al_2O_3) and YAP. It may be clearly seen in the phase diagram provided in Figure 15.2. The limit for the possible replacement fraction emanates mainly from the ionic radius difference between them and the yttrium ions: the radius of Nd^{3+} ion in the dodecahedral site is 1.323 Å, while that of Y^{3+}

Table 15.1 Basic physical properties of neodymium:YAG.

No.	Property	Value
1.	Chemical formula of the host crystal	$Y_3Al_5O_{12}$
2.	Molecular weight	$M_w = 593.6$
3.	Crystallographic structure	Body centered cubic
	Space-group #230	$O_h^{10}(I4_1/a\bar{3}2/d)$
4.	Number of formula units in the unit cell	8
5.	Site symmetry of Y^{3+} (and Nd^{3+})	D_2
6.	Lattice constant	$a = 10.005$ Å
7.	Molecular density	$N = 4.62 \times 10^{21}\ \mathrm{cm}^{-3}$
8.	Specific gravity	$\rho = 5.565\ \mathrm{g/cm^3}$
9.	Young modulus	$Y = 29.4 \times 10^3$ Newton/cm^2
10.	Hardness	$H = 8 \div 8.5$ Moh (1215 Knoop)
11.	Stress cracking constant	$R = (1.3 \div 2.6) \times 10^4$ Newton/cm^2
12.	Thermal expansion coefficient	$\alpha = 6.9 \times 10^{-6}\ °\mathrm{C}^{-1}$
13.	Heat capacity	$C_p = 350\ \mathrm{J \cdot Mole^{-1}\ K^{-1}}$
14.	Thermal conductivity	$K = 0.14\ \mathrm{W \cdot cm^{-1}\ K^{-1}}$
15.	Melting (freezing) temperature	$T_m = 1960\ °\mathrm{C}$
16.	T_f change per Nd^{3+} ions concentration	$dT_f/dC = -8\ °\mathrm{C/at.}\ \%\ \mathrm{Nd^{3+}}$
17.	Optical refractive index (at 1.064 μm)	$n = 1.823$
18.	Refractive index change with temperature	$dn/dT = 7.3 \times 10^{-6}\ °\mathrm{C}^{-1}$
19	Relative dielectric constant	$\varepsilon_r = 11.7$
20.	Solubility limit for Nd^{3+} ions	$C_{Nd}^{max} = 5$ at.%
21.	Fluorescence transition in Nd^{3+} ions	$^4F_{3/2} \rightarrow {}^4I_{11/2}$
22.	Fluorescence bandwidth	$\Delta\lambda_{1/2} = 4.5$ Å
23.	Fluorescence transition radiative lifetime	$\tau_{rad} = 550\ \mu\mathrm{s}$
24.	Fluorescence transition effective lifetime	$\tau_{eff} = 230\ \mu\mathrm{s}$

is 1.281 Å. For the same reason, their segregation coefficient is considerably smaller than a unity; its average value over all crystallographic orientation is about $k \cong 0.2$. It is possible to increase the solubility (and simultaneously bring k closer to a unity value) by adding a companion impurity that provides size compensation (see Section 12.11, item 1), like gadolinium Gd^{3+} ions or lutetium Lu^{3+} ions. Usually, such approach is avoided because increasing the neodymium concentration beyond about 1.0 at. % shortens the fluorescence lifetime, hence the laser output efficiency. Commercial $Nd : YAG$ laser rods usually contain neodymium concentrations in the $0.8 \div 1.2$ at. % Nd^{3+} range. Because the segregation coefficient is $k \cong 0.2$, the melt raw material is prepared with a concentration of about 5.0 at. % Nd^{3+}. A partial phase diagram of the "binary" $Y_3Al_5O_{12} - Nd_2O_3$ system is presented in Figure 15.3.

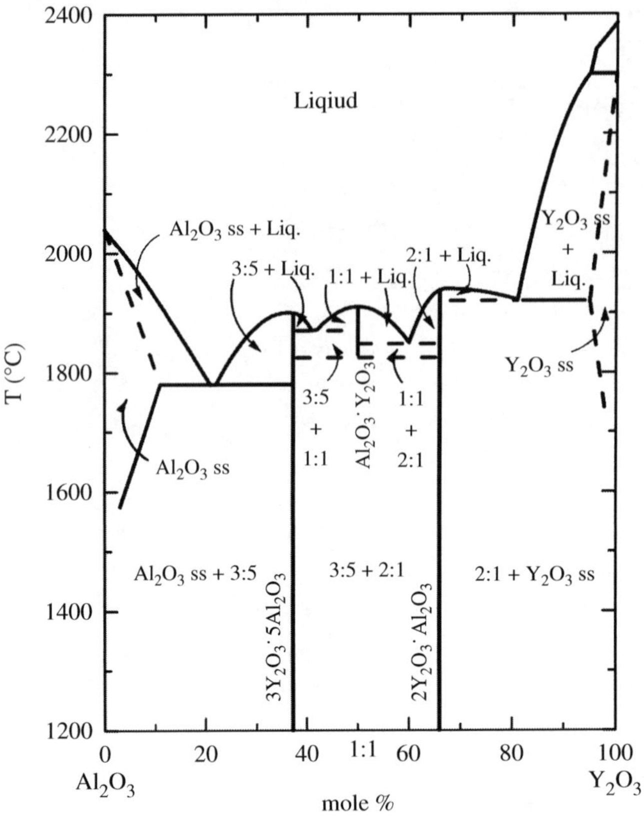

Figure 15.2 Phase diagram of the Al_2O_3-Y_2O_3 binary system. An "ss" note following a chemical formula means "solid-state".

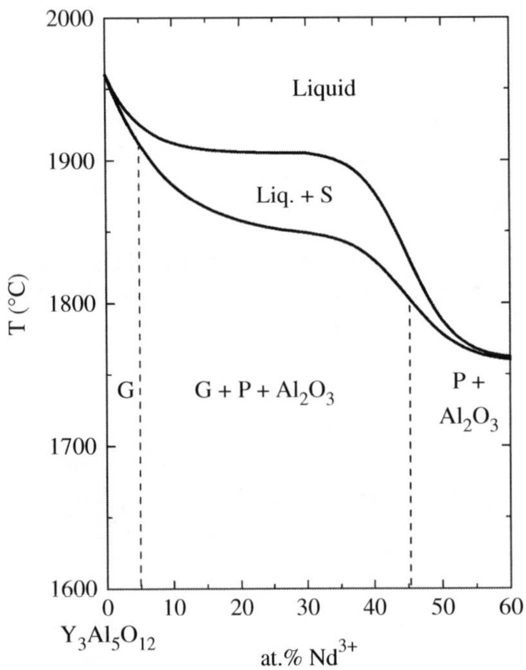

Figure 15.3 Partial phase diagram of the $Y_3Al_5O_{12}$-Nd_2O_3 binary system. A "G" note indicates a garnet structure and a "P" note indicates a perovskite structure.

The raw material preparation includes five stages: drying of each oxide powder, weighing according to required ratios needed for the chemical balance, mixing the powders, compressing, and sintering:

1) Drying of powders: The used powders are Al_2O_3, Y_2O_3, and Nd_2O_3. All are available for purchasing from particular commercial companies, which specialize in that technology, in the form of very fine powders and a high degree of purity. The nominal required purity is 99.999% (whose technological tagging is "5N" – according to the number of "9"s appearing in the indicated purity). As powders, their surface area is very large compared to their volume, and they tend to adsorb water vapor. Dry powders may adsorb up to 1–2% of their weight at a rate of 5×10^{-3} wt. % /min. To assure complete drying, purchased powders are heated in a platinum crucible to a 1000 °C for about 24 hours. Subsequently, the crucible plus powder is transferred into a glass container pumped to a vacuum of 10^{-3} torr to prevent re-adsorption.
2) Weighing: Powder weighing is performed using "analytic" scales, allowing accuracy of 0.1 mg for a load of up to about 1 kg, in a polypropylene container. The weighing is usually made in free air, thus allowing the undesired resumption of water vapor adsorption. It is therefore needed to finish up the weighing within several minutes.
3) Mixing: Weighed amounts of the different powders at ratios appropriate to a chemical balance containing 5.0 at. % Nd^{3+} are filled into a propylene-made container to about a third of its entire volume. The sealed container is then assembled to a special mixer, which performs a three-dimensional motion at a rate of 50 motions per minute for about 24 hours.
4) Compressing: The mixed powder naturally has a very small specific gravity, which disables the full required loading of the growth crucible. Furthermore, heating of powder to a high temperature in an open crucible may cause dispersion of powder particles throughout the entire furnace volume. The powder compression is performed isostatically while inserted in a flexible container made of natural rubber with a wall thickness of about 2.0 mm and a diameter of about 50 mm. The container is tightly bound at its top, and inserted into a high-pressure, fluid containing cell. Following a pressure of about 50, 000 psi, the effective specific gravity of the compressed load rises from about 2.2 g/cm^3 to about 2.8 g/cm^3, which is about half the specific gravity of a perfect YAG crystal.
5) Sintering: The goal of sintering is to increase the specific gravity of the compressed mix, to induce some preliminary chemical reaction between the different components, and not the least is to cause evaporation and burning of organic traces that might have been absorbed to the powder particles during the previous preparation procedures. The sintering is performed inside a platinum crucible heated to 1750 °C in a regular (oxygen containing) ambient atmosphere, and staying under those conditions for about 4 hours. At the end of sintering, the effective specific gravity of the material rises to about 3.5 g/cm^3, approximately 75% of a perfect YAG crystal.

15.2.3 Single-Crystal Seed Preparation

Single-crystal seeds used for growing of large-volume single crystals are produced from earlier grown either YAG or Nd:YAG single crystals. The seed in machined in the form of a cylindrical rod, about 40 mm long and about 2–3 nm diameter, with the rod axis parallel to the [111] crystallographic orientation. The seed is "drilled out" of the single crystal using a hollow drill whose header is impregnated with diamond grains (see also Section 15.5). That mode of production involves some mechanical roughening of the rod cylindrical face, which could impair the single crystalline nature of a later grown crystal. To reduce the surface roughening, the seed is immersed in concentrated phosphoric acid (H_3PO_4) for several minutes at ~300 °C, where the material erosion rate is about 2.5×10^{-4} cm/min. At the end of that process, the seed is washed using distilled water. The obtained surfaces appear smooth as well as transparent, allowing to observe their inside volume in order to select seeds devoid of internal defects.

15.2.4 The Nd:YAG Single-Crystal Growth

The crystal growth is performed using the Czochralski method of pulling from the melt (Section 14.4.2 case [**iii**]). The raw material is melted inside an iridium (Ir) made crucible at a temperature of 1960 °C. The surrounding atmosphere inside the furnace comprises dried nitrogen (N_2) or argon (Ar), with an added ~1.5 vol. % oxygen (O_2), at a flow rate of ~1 liter/min. Reaching of optimal conditions appropriate for starting the growth is a long and tedious process that usually consumes several hours. After melting of the raw material, the seed is lowered slowly toward the melt surface, while rotating at a rate of about 10 revolutions per minute. The seed lower tip is made to just touch the melt surface, and the result is observed. If it causes melting of part of the seed, causing disconnection from the melt, the furnace temperature is lowered by ~0.5 °C. That procedure is repeated until the contact-making causes some melt solidification connected onto the seed lower tip. The furnace temperature is then increased at a rate of ~1 °C/hour until the solidified material disappears, and some minor partial seed melting starts. That is the optimal situation for the growth to start. Importantly, however, if the seed is made of pure YAG, there is no need to wait for its partial melting, because its melting point is higher by about 40 °C from that of the melt (because $dT_f/dC = -8\,°C/at.\%\,Nd^{3+}$, see Table 15.1).
The up-pulling of the seed is at a rate of 0.2 mm/hour while activating the planned diameter control. Usually, the "shoulders" growth is planned to exhibit about a 45 ° angle, then maintaining a constant diameter till reaching the planned length. Then the diameter is reduced down to about 0.5 mm at 45 ° angle (the "tail" region). After the said diameter reduction, the grown crystal is disconnected from the melt by a quick up-pulling of about 2 mm.

15.2.5 Quality Control of a Grown Crystal

For reasons to become quite clear later, the reliable way to assess the quality of a grown Nd:YAG crystal is by measuring the actual functioning of laser rods produced from it. Thus, the crystal growth factory must be also capable to produce laser rods. As a background, in Table 15.2, we provide a typical specifications page for a typical Nd:YAG laser rod. We won't dwell on all indicated subjects. We delve, however, into those topics that cast directly on a grown crystal quality; the related specification items are 1.1, 3.1, 3.3, and 3.4.

(i) Concerning item 1.1. Measurement of the neodymium ions concentration. In the process of rod preparations, edges from their two sides are cut to achieve the specified length. Per each rod, the two cut edges of two opposite sides of each rod are taken for chemical analyses. If the two edges satisfy the specification document requirements, it is most probably true for the rod itself; that is because the impurity distribution usually follows Pfann's normal freezing equation (Eq. [12.13]). For a cases where the impurity segregation coefficient is smaller than 1, the concentration is an increasing function of the solidified fraction.

Table 15.2 Typical specification of commercially used neodymiun:YAG laser rod.

Specifications for a Φ4 × 50 Nd:YAG laser rod
1. *Material* 1.1 $Y_3Al_5O_{12}$ single crystal doped by 1.1 ± 0.1 at. % Nd^{3+}. 1.2 Crystallographic orientation of the rod axis: [111] ± 5 °.
2. *Dimensions* – According to the following drawing: 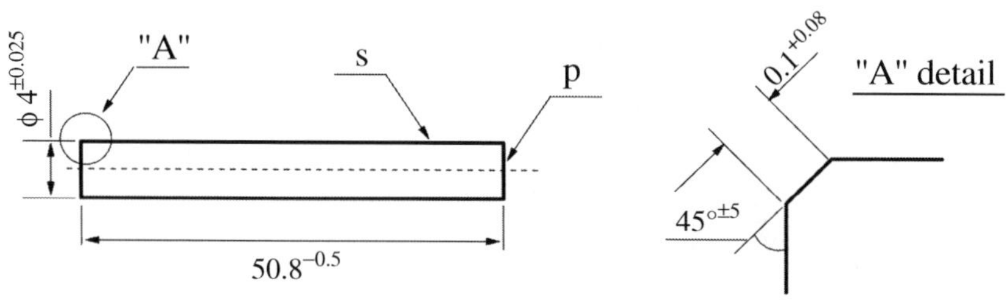
3. *Optical quality* 3.1. The rod is produced using a region outside the crystal core. 3.2. The rod material is uniform, devoid of light scattering centers that would be observed in a darkened room when a HeNe laser beam of 632.8 nm wavelength crosses parallel to its axis. 3.3. Such said beam extinction rate is greater than 23 dB when the rod is placed between crossed polarizers. 3.4. The wavefront distortion is smaller than ½ interference band over the entire rod diameter when measured for a double transition in a Fizeau interferometer at 632.8 nm wavelength.
4. *Finishing* 4.1. The "S" marked cylindrical surface is scrubbed to a Grit-400 degree. 4.2. The "P" marked planar surfaces: 4.2.1. Deviation from perpendicularity of the planar surfaces smaller than 5 arc minutes (5′). 4.2.2. Deviation from parallelism between the planar surfaces smaller than 10 arc seconds (10″). 4.2.3. Deviation from planarity of each planar surface smaller than tenth $\lambda = 632.8$ nm wavelength ($\lambda/10$) across the whole surface. 4.2.4. Optical finishing: scratches and dents on planar surfaces by MIL-O-13830 standard.
5. *Coatings* 5.1. The end rod planar planes anti-reflection coated by a MgF_2 layer to reflect less than 0.25% at normal incidence at $\lambda = 1064$ nm. 5.2. Coating layer resistance according to MIL-C-48497 standard, Sections 3.4.1, 3.4.2, and 3.4.3. 5.3. Resistance to radiation damage up to and above 10^7 light pulses of $\lambda = 1064$ nm wavelength, 10–15 ns long, and 300 MW/cm^2 maximal power.

(ii) Concerning item 3.1: The region from which the rods are produced. Grown Nd:YAG crystals are always very strained along their central region, addressed as "core region". Figure 15.4 provides a photograph of a grown crystal cut perpendicularly to the growth orientation (the [111] crystallographic orientation), by removing both "shoulders" and "tail" regions. The cut surfaces were optically polished to a high planarity. The crystal was placed between crossed optical polarizers and illuminated by white light from the side opposite to the camera. The strained core region expresses itself by changing bright and dark areas over about half the

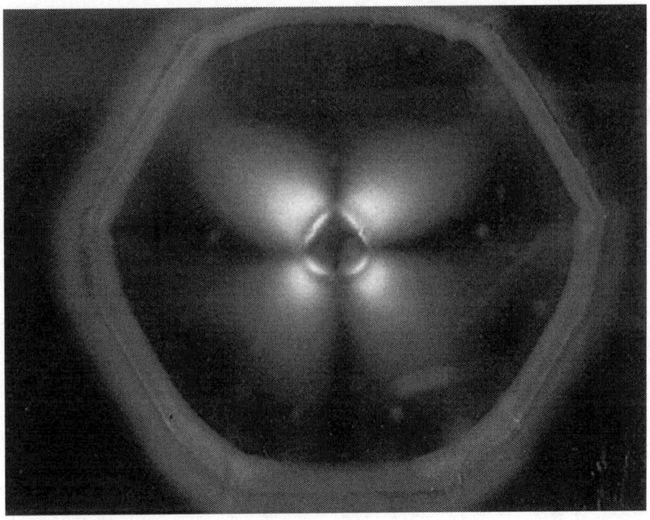

Figure 15.4 Optical color photograph of a of a plane-parallel cylinder of a grown Nd:YAG crystal cut perpendicularly to the growth [111] crystallographic orientation and set between two crossed polarizers. Large parts of the crystal center (core) appear strained (bright). Removing the core by out-drilling usually relaxes most strains, making most remaining parts usable for manufacturing of laser rods (Table 15.2).

radius of the entire crystal. The areas closer to the circumference appear relatively dark and uniform. The reason for the bright area appearance between the crossed polarizers is the presence of strained regions, causing double light refraction, which changes the nature of the crossing linearly polarized light into elliptically polarized light. Also, some color dispersion occurs. (An in-depth account of all optical effects occurring in the strained regions is beyond the scope of the present book). Importantly, at first glance, the shape of the crystal circumference appears hexagonal, having a sixfold rotational symmetry. However, three sides of the apparent "hexagon" are slightly shorter, making it conform to the threefold rotational symmetry about the [111] axis of a cubic crystal.

The existence of a strained core in a grown YAG crystal is unavoidable; however, it may be reduced by reducing the radial temperature gradient in the crucible melt, for example by increasing the rotation rate of the crystal during its upward pulling. In the manufacturing of laser rods, no use is made of the core region. In fact, the first manufacturing stage is to drill out and remove the region. Strain induced by the core then relaxes. A photograph like in Figure 15.4 allows choosing the most unstrained remainder regions as appropriate for laser rod manufacturing. A final confirmation of strain absence in a particular rod is achieved by following specification 3.3 in Table 15.2.

(iii) Concerning item 3.2: Seeking for light scattering centers. The test method is described in detail in the specifications table. If scattering centers do exist in a laser rod, it usually relates to inaccurate preparation of the raw material, such as in the different comprising materials weighing, or a too short mixing of the powders prior to the compressing and sintering stages.

(iv) Concerning item 3.3: Strain measuring in a rod. The measurement is performed in a way similar to the method underlying the measurement describe in the above Section (ii). The intensity of the parallel light beam crossing the rod is measured under two arrangements of the polarizers: $I_{\|}$ when the two polarizers' easy axes are parallel, and I_{+} when the easy axes are crossed. The extinction coefficient (EC) expressed in "decibel" units (dB) is defined as

$$EC \equiv 10\boldsymbol{log}_{10}(I_{\|}/I_{+}). \tag{15.1}$$

The beam polarization entering the rod is determined by the first polarizer. Double refraction caused by strained regions inside the rod contributes an elliptic component to a passing beam polarization state, such that some light will pass through the second, crossed

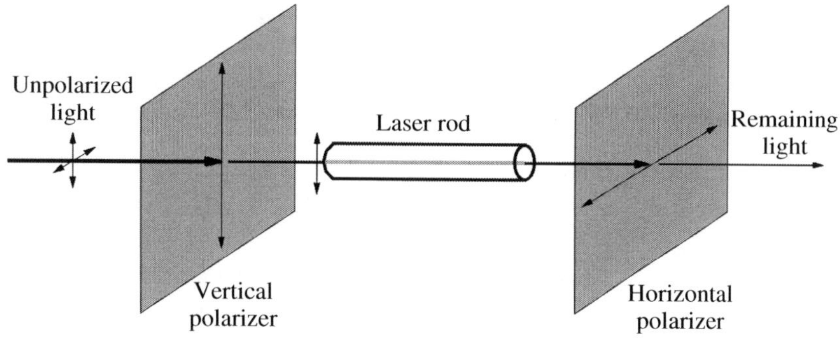

Figure 15.5 Schematic illustration of the elements required for measuring the extinction coefficient of light crossing a laser rod placed between crossed polarizers.

polarizer. A higher EC means less strain in the crystal, namely a better quality. A schematic basic layout of a measuring system is provided in Figure 15.5.

(v) Concerning item 3.4: Measurement of wavefront distortion. The wavefront of light crossing a laser rod may distort as a result of two effects: side planes that are not perfectly planar, and transverse refractive indices that are not perfectly uniform. The latter relates to non-uniformity of the rod material, as well as of strain distribution. The measurement is made using a Fizeau interferometer. The Fizeau interferometer displays the interference between a plane-parallel light beam reflected from a partially transmitting planar mirror and the transmitted beam, which was reflected by a second partially transmitting parallel planar mirror. A slight angular deviation of one of the mirrors from being perfectly parallel to the other generates a series of linear parallel interference intensity lines, which express the gradual change of distance between the two mirrors: the distance between two interference lines expresses a change of $\frac{1}{2}\lambda$ of the optical path. (In terms of reading a topographical map, the said lines describe a $\frac{1}{2}\lambda$ "height difference" between the mirrors). When a laser rod is placed parallel to the beam between the mirrors, the optical length of the part crossing the rod would change. If the refractive index is transversely uniform, it would cause a shifting of the interference lines, but they would remain parallel and linear. If there are transverse changes of the refractive index, the lines would curve. The curving allowed by the specification in Table 15.2 is half the distance between adjacent lines. An illustration of the said description is shown in Figure 15.6.

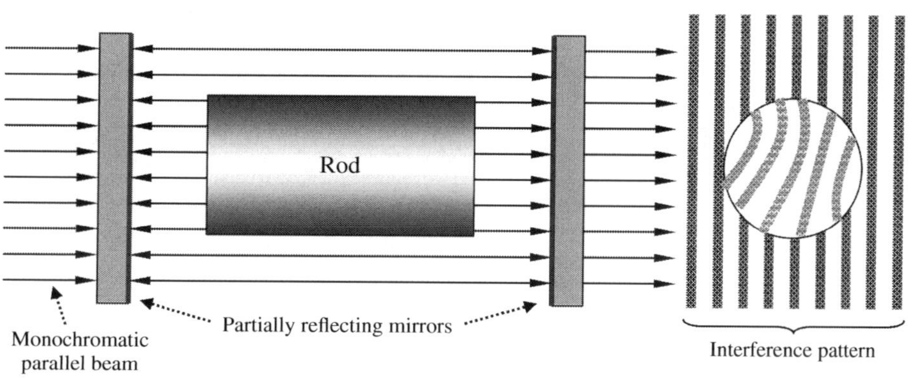

Figure 15.6 Schematic illustration of a Fizeau interferometer, and the interference pattern obtained by introducing a rod that distorts the wavefront by about twice the maximal allowed limit.

15.3 Growth of Zinc Cadmium Telluride Crystals as Substrates and X-ray Detectors

15.3.1 Introduction

The zinc cadmium telluride crystals we address in this section are used, in their possible different compositions, as substrates for MCT far infrared light detectors (see Section 9.2), and electronic detectors of gamma and X-rays (see Section 9.1). The MCT acronym relates to a family of ternary mercury cadmium tellurium compounds $Hg_{1-x}Cd_xTe$. The specific $Hg_{0.8}Cd_{0.2}Te$ compound is used as a very efficient detector for the 8–14 μm far infrared wavelength range. The thickness of such detector rarely exceeds about 50 μm, and therefore, it is often attached to a substrate that provides it with the necessary mechanical stability, and assists in dissipating the heat generated during operation. In cases of thin structures attached to a thick substrate, it is useful to actually grow the thin compound layer by a direct solidification on top of the substrate, a process called "liquid-phase epitaxy" (LPE), and not by gluing. Then it is desirable to have congruence between the crystallographic structures of the substrate and the epitaxed layer. Cadmium telluride ($CdTe$) and MCT in the said composition have the same crystallographic structure. Replacing about 2 at. % of the cadmium Cd^{3+} ions of the $CdTe$ by Zn^{3+} ions renders the two materials the same lattice constant. As to be described later, having such combination enables the illumination the MCT through the substrate, because $CdTe$ is transparent in the 8–14 μm wavelengths range.

A cadmium telluride ($CdTe$) crystal is also suitable to function as a gamma and X-ray detector due to its favorable properties (see Section 9.1): a high room temperature energy gap of 1.47 eV, making it practically an insulator, and the rather large atomic numbers of its constituents – 52 and 48 for tellurium and cadmium, respectively, yielding a high efficiency for gamma and X-ray absorption. Evidently, replacing a fraction of the cadmium atoms by zinc atoms to form a ternary $Cd_{0.8}Zn_{0.2}Te$ crystal increases the energy gap and the mobilities of both holes and electrons.

15.3.2 Structure and Physical Properties of CdTe and CdZnTe Crystals

The stable crystalline phases of cadmium telluride ($CdTe$) and zinc telluride ($ZnTe$) are identical, called "sphalerite" or alternatively "zinc blende". They belong to the #216 space group denoted as $T_d^2(F\bar{4}3m)$. The unit cell in that structure is face-centered cubic (see Section 1.1) and comprises four formula units. Each of the component atoms is surrounded by four atoms of the opposite component, thus forming a tetrahedral T_d point symmetry. The crystal lacks inversion symmetry. Thus, in a slice of the crystal whose faces are perpendicular to the [111] orientation, one face is "metallic" (often called "the A side" – cadmium in the present case) and the opposite face is "non-metallic" (often called "the B side" – tellurium in the present case). These two different sides also express themselves by somewhat different chemical and physical properties.

The following Table 15.3 summarizes the basic properties of cadmium telluride ($CdTe$) and zinc telluride ($ZnTe$). The properties important regarding consideration of forming the "quasi-binary" compound of cadmium telluride and zinc telluride are the different energy gaps and unit cell constants. Due to their identical lattice structure, these two materials may form a "crystalline solid solution" for all molar fractions, namely $Cd_{1-x}Zn_xTe$ for $0 \leq x \leq 1$. The unit cell constants would change nearly linearly as functions of x, between $a = 6.483$ Å for $x = 0$ and $a = 6.1034$ Å for $x = 1$. Similarly, the energy gap would change nearly linearly between $E_g = 1.57$ eV for $x = 0$ and $E_g = 2.15$ eV for $x = 1$.

Table 15.3 Basic physical properties of cadmium telluride CdTe and zinc telluride ZnTe.

No.	Property	*CdTe*	*ZnTe*
1	Molecular weight	$M_w = 240$ g/mole	$M_w = 177.77$ g/mole
2	Crystalline structure	Face-centered cubic	Face-centered cubic
3	Space group	#216; T_d^2 ($F\bar{4}3m$)	#216; T_d^2 ($F\bar{4}3m$)
4	Unit cell constant	$a = 6.483$ Å	$a = 6.1034$ Å
5	Symmetry of occupied sites	T_d	T_d
6	Molecular density	$N = 1.468 \times 10^{22}$ cm^{-3}	$N = 1.759 \times 10^{22}$ cm^{-3}
7	Specific gravity	$\rho = 5.849$ g/cm^3	$\rho = 5.636$ g/cm^3
8	Thermal expansion coefficient	$\alpha = 4.5 \times 10^{-6}$ °C^{-1}	$\alpha = 8.36 \times 10^{-6}$ °C^{-1}
9	Heat capacity	$C_p = 83.35$ J mole^{-1} °K^{-1}	$C_p = 46.86$ J mole^{-1} °K^{-1}
10	Heat conductivity	$K = 0.06$ W cm^{-1} °K^{-1}	$K = 0.18$ W cm^{-1} °K^{-1}
11	Freezing temperature	$T_f = 1092$ °C	$T_f = 1292$ °C
12	Melting vapor pressure	$P_m = 1.5$ atm.	$P_m = 0.8$ atm.
13	Optical transparency range	1–28 μm	0.6–52 μm
14	Energy gap	$E_g = 1.57$ eV	$E_g = 2.15$ eV
15	Electron mobility	$\mu_n \cong 1000$ cm^2/V · s	$\mu_n \cong 500$ cm^2/V · s
16	Hole mobility	$\mu_p \cong 100$ cm^2/V · s	$\mu_p \cong 100$ cm^2/V · s
17	Electron/hole lifetimes	$\tau_{n;\,p} \sim$ 10–200 ns	$\tau_{n;\,p} \sim$ 10–200 ns

To further delve into the crystal growth process, we look at the phase diagram of the *CdTe* compound presented in Figure 15.7. The freezing temperature of the molten $Cd_{1-x}Te_x$ mixture is presented as a function of the tellurium content in the mixture. The balanced $CdTe = Cd_{0.5}Te_{0.5}$ mixture solidifies congruently into a crystalline structure at the 1092 °C temperature, as indicated by the vertical line at $x = 50$ at. % *Te*. The crystalline existence range without precipitation of phases richer by either tellurium or cadmium component is very narrow, about 0.01 at. %. There are additional two points exhibiting eutectic solidification. One is at the tellurium 449 °C freezing temperature (on the excess tellurium side, where pure tellurium *Te* clusters segregate inside the cadmium telluride matrix). The other is at the cadmium 324 °C freezing temperature (on the excess cadmium side, where pure cadmium *Cd* clusters segregate inside the cadmium telluride matrix).

At its melting temperature, the cadmium telluride crystal has quite a high vapor pressure, which is different for each of the separate components. Particularly, the vapor pressure of cadmium is $P(Cd) = 0.7$ atm.; that of tellurium is $P(Te_2) = 5 \times 10^{-3}$ atm. In the melted material, the vapor pressure grows up to 1.5 atm., and the ratio between the contributions of cadmium and tellurium remains the same. The said vapor pressures grow as functions of the melt temperature. The existence of the said two grossly different vapor pressures complicates the crystal growth process, because the preferential volatilization of tellurium off the melt could change its composition, hence also the growing crystal composition. The technological solution involves creation of a high pressure of an inert gas like argon (*Ar*), which slows down the vaporization, and simultaneously creates an intended atmosphere of cadmium vapor that balances the vaporization. That method will be described later in more detail.

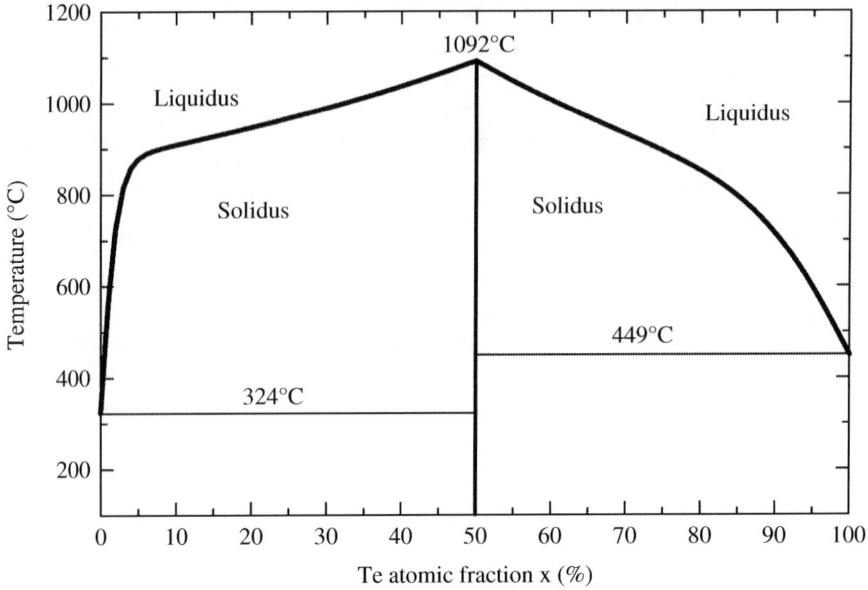

Figure 15.7 Phase diagram in the temperature vs. composition plane of $Cd_{1-x}Te_x$ compound.

An important property of both zinc telluride (*ZnTe*) and cadmium telluride (*CdTe*) is that they are mutually soluble at all relative compositions. The phase diagram expressing that feature is presented in Figure 15.8. It is readily seen that increasing the zinc content in the molten solution increases the freezing temperature, or alternatively said, the segregation constant of zinc in the cadmium telluride is larger than 1. At small concentrations, the segregation coefficient is approximately $k = 1.35$.

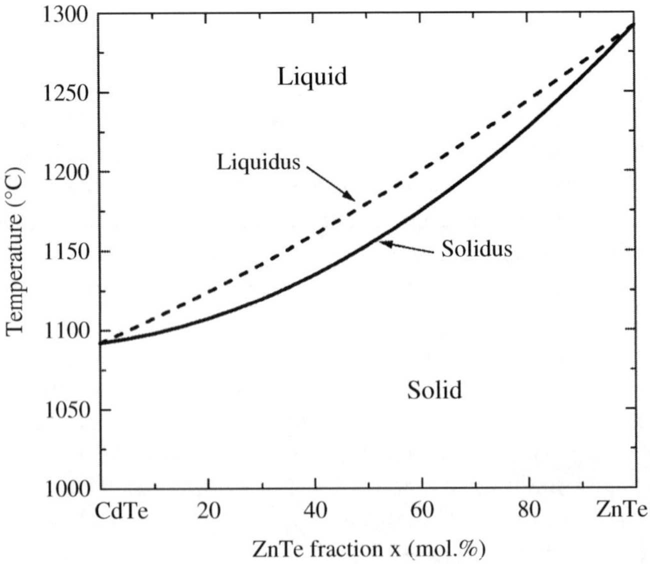

Figure 15.8 Phase diagram in the temperature vs. composition plane of the $Cd_{1-x}Zn_xTe = (CdTe)_{1-x}(ZnTe)_x$ compound.

15.3.3 Growth of Cadmium Zinc Telluride Crystals

A method considered quite appropriate for the growth of $Cd_{1-x}Zn_xTe$ crystals at compositions ranging between $x = 0$ and $x = 0.2$ is in a crucible under a temperature gradient according to Schmidt–Viechnicki (Section 14.4.1 item [iii]). The raw material is melted in a crucible having a solid seed at its bottom inside a furnace maintaining a positive longitudinal temperature gradient and a positive radial temperature gradient. The crystal growth process starts by a slow cooling of the entire furnace after a small part of the seed has melted. That cooling causes a rigid decrease of the temperature profile throughout the entire furnace, especially of the raw material. That way, the solid/melt interface progresses slowly upward until the entire melt solidified. In the specific cases of $Cd_{1-x}Zn_xTe$ crystals growing, the furnace also contains an additional reservoir of zinc (Zn) and cadmium (Cd) melted at a high temperature and further heated to a high temperature in order to produce proper vapor pressures. The cadmium vapor pressure is aimed at balancing the evaporation of cadmium off the raw material melt due to its fundamental high vapor pressure; the zinc vapor pressure is aimed at compensating its diminution from the melt due to the fact that its segregations coefficient k is larger than 1.

Figure 15.9 provides a schematic illustration of the construction of a crucible container in a furnace used for growing cadmium/zinc telluride $Cd_{1-x}Zn_xTe$ single crystals with chemical compositions between $x = 0$ and $x = 0.2$. The container is constructed of especially pure quartz.

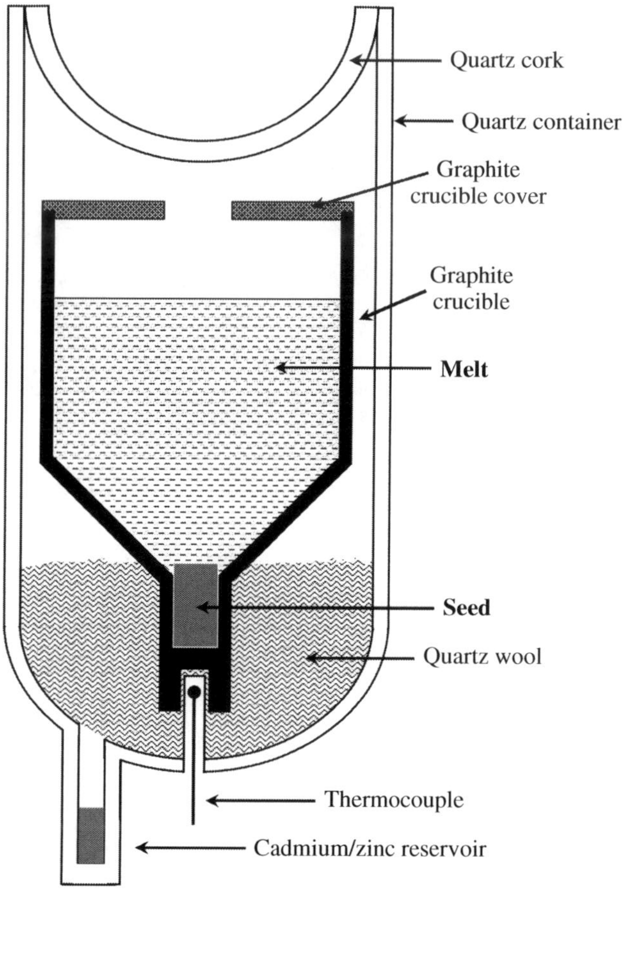

Figure 15.9 Schematic illustration of a crucible container used for growing of cadmium-zinc telluride $Cd_{1-x}Zn_xTe$ single crystals of chemical compositions ranging between x = 0 and x = 0.2.

The crucible is constructed of especially pure graphite. The crucible top has a graphite-made cover, having a central hole whose presence contributes to an increase of the radial temperature gradient inside the crucible. A typical container diameter is 10 cm. Its top is sealed with a quartz cork by welding its circumference with the container top inner circumference. The crucible is shaped with a narrowing bottom, ending with a bore into which the seed would be inserted (seed house). The container close to the crucible bottom and to the seed house contains a quartz wool thermal insulation needed to maintain the required temperature gradients. The seed axis is oriented in the [111] crystallographic orientation, and its chemical composition is the same as the intended composition of the grown crystal. The bottom of the seed house has a narrow bore adjusted for insertion of a narrow quartz tube containing a thermocouple, whose role is to control the furnace temperature. Other thermocouples not shown in the figure are attached to several positions of the container. Their role is to verify the required temperature gradients in the crucible, its rigidity throughout the growth, and especially monitor and control the temperature of the cadmium/zinc reservoir. The quartz container is sealed at room temperature while containing an argon (Ar) gas at a pressure of 200 torr.

The longitudinal temperature gradients found to be most appropriate are 8 °C/cm in the seed house and 3 °C/cm in the melt. The radial temperature gradients are smaller, only 0.5–1.0 °C/cm. The cooling rates are about 0.5–1.0 °C/hour.

The ratio between the zinc and cadmium concentrations in the compensation reservoir reflects their intended ratio for the grown crystal. The reservoir suitable temperature is about 230 °C lower than the growing crystal freezing temperature. For example, in growing a pure $CdTe$ crystal without zinc ($x = 0$ case), the initial, cadmium containing reservoir temperature would be about 860 °C. In growing a $Cd_{0.8}Zn_{0.2}Te$ crystal ($x = 0.2$ case), whose freezing temperature is 1100 °C, the initial, cadmium/zinc mixture-containing reservoir temperature would be about 880 °C. Notably, over a typical growth process, the reservoir temperature reduces by a total of about 20 °C.

15.3.4 Quality Control of Grown Cadmium Zinc Telluride Crystals

(i) $Cd_{0.96}Zn_{0.04}Te$ crystals as substrates for MCT crystals

Obtaining of favorable conditions for making a substrate suitable for MCT crystals (mercury cadmium telluride crystals $Cd_{0.8}Hg_{0.2}Te$ used as infrared light detectors in the 8–14 μm range; see Section 9.2.2) requires a match between the lattice unit-cell constant values of the two materials. The unit cell constant of the said MCT crystal is a = 6.467 Å, and the same is required to be the unit cell constant of the grown $Cd_{0.96}Zn_{0.04}Te$ crystal. The measurement is performed by analysis of monoenergetic X-ray diffraction off the grown crystal powder obtained by shredding a piece of it. The often used X-ray source is a copper anode tube. The copper $K_{\alpha 1}$ emission wavelength is 1.5406 Å (photon energy 8.048 keV). Substrates are prepared as parallel plates oriented perpendicularly to the [111] crystallographic orientation, but other orientations are also possible. The preferred deposition of the MCT is the LPE (liquid phase epitaxy), but other methods are also possible such as CVD (chemical vapor deposition) and MBE (molecular beam epitaxy); see Section 14.5. We won't elaborate here further on these technologies.

(ii) $Cd_{0.8}Zn_{0.2}Te$ crystals as gamma and X-ray detectors

The fundamental factors related to the operation of insulating semiconductors as gamma and X-ray detectors is addressed in detail in Section 9.1. $Cd_{0.8}Zn_{0.2}Te$ single crystals are one important

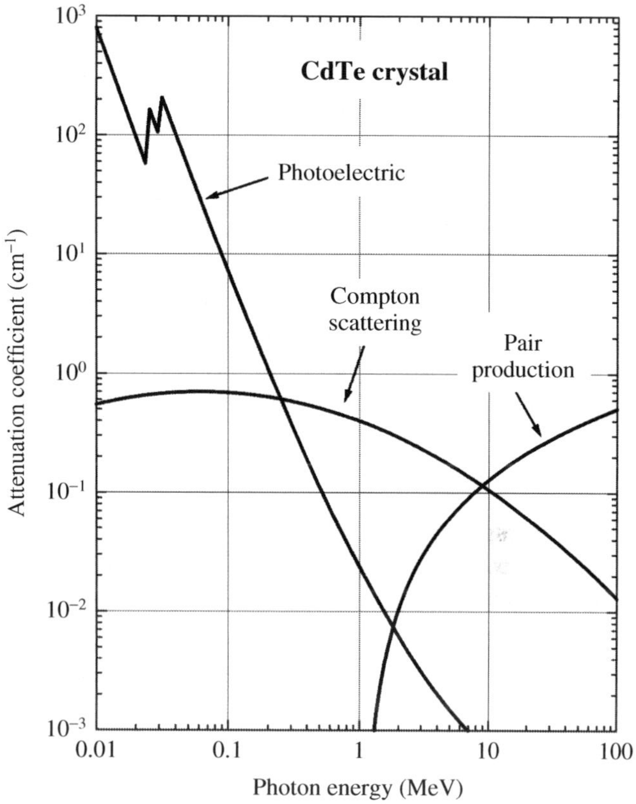

Figure 15.10 Full logarithmic plot of high-energy photon attenuation coefficients as functions of energy in a cadmium telluride crystal by photoelectric absorption, by Compton scattering, and by electron–positron pairs production. For more details, see the text.

example of that important technological field. Figure 15.10 provides a spectrum of the stopping factors of photons belonging to the said spectral range in a cadmium telluride (*CdTe*) crystal; the one belonging to $Cd_{0.8}Zn_{0.2}Te$ crystals is very similar, and as a detector, it has an advantage of having larger hole and electron mobilities. Up to photon energies of about 200 keV, the dominant stopping process is the photoelectric effect. Between about 300 keV and 7–8 MeV, stopping is dominated by electron/positron pairs production. The primary use of cadmium telluride and cadmium zinc telluride detectors for X-ray spectroscopy is up to energies of about 200 keV.

The most important property needed for efficient operation of an electronic detector used for gamma and X-rays spectroscopy is the mean free path of charge carriers generated by the absorbed photons. In order to collect all generated charge carriers at the electrodes, the mean free paths of the two type carriers – holes and electrons – need to be long compared to the crystal thickness, which is of the order of 1–4 mm. These mean free paths, ℓ_p and ℓ_n for holes and electrons, respectively, also depend on the external electrical field $\mathcal{E}$ applied across the detector, and are given by

$$\ell_p = \mu_p \tau_p \mathcal{E}; \quad \ell_n = \mu_n \tau_n \mathcal{E}, \tag{15.2}$$

where μ_p and μ_n are the hole and electron mobilities, respectively, and τ_p and τ_n are their lifetimes, respectively. The "bottle neck" for achieving maximal carriers' collection is the carriers having the shorter mean path. In most materials, they are the holes. Therefore, assessment of a potential use of a cut-and-polished crystalline plate to serve as a high-energy photons detector is measuring their holes $\mu_p\tau_p$ product.

The measurement is performed using an electrical circuit similar to the one shown in Figure 9.1. The positive electrode comprises a very thin metal (~ 1000 Å). That electrode is exposed to a very short light pulse, whose photon energy is larger than the detector energy band gap (wavelength shorter than about 730 nm). It is possible to use frequency-doubled neodymium–YAG laser pulses ($\lambda = 532$ nm) that are only several nanoseconds long, or nitrogen N_2 laser pulses ($\lambda = 337.1$ nm) that are even shorter, only about 1 ns long. Most photons penetrate through the said very thin positive electrode and are fully absorbed inside the crystal, very close to the illuminated electrode (1–2 μm), which is very small compared to the typical detector plate thickness of $L \cong 1$–4 mm. Under the external electrical field $\mathcal{E} = V/L$, a transient current develops, inducing a transient voltage $\delta V(t)$ across the series resistor R. That resistor is chosen to have a sufficiently small resistance, such that discharge duration RC of electrical charge accumulated at the detector electrodes is small compared to the holes transit time $t_t = L^2/\mu_p V$ between the electrodes. The detector capacitance is given by $C = \varepsilon_r\varepsilon_0 A/L$, where A is the detector surface area, ε_r is its relative dielectric constant, and ε_0 is the permittivity of free space. Observation of the voltage signal trace $\delta V(t)$ recorded by an oscilloscope reflects the temporal current dependence $I(t) = \delta V(t)/R$.

During a light-pulse absorption, a large number of excess charge carriers, holes, and electrons are generated in the absorbing thin layer. The electrons drift toward the adjacent contact, and their contribution to the current ends within a negligibly short time duration. The holes are injected into the detector bulk, forming a narrow front drifting toward the negative electrode, thus forming an electrical current, until reaching the negative electrode after a t_t transit time. While drifting, the number of contributing holes, and the electrical current, decay exponentially with a common decay time τ_p. Upon reaching the negative electrode, the current ends abruptly. Analysis of the current time-trace allows calculating both the hole mobility μ_p and the hole lifetime τ_p. A typical shape of a current time-trace is presented in Figure 15.11.

Notably, a simple reversing of the external voltage polarity allows as well determining the electron mobility μ_n and their trapping lifetime τ_n. Room temperature-measured values of all said parameters in grown $Cd_{0.8}Zn_{0.2}Te$ single crystals are $\mu_n = 1350\ \mathrm{cm^2/V \cdot s}$; $\mu_p = 120\ \mathrm{cm^2/V \cdot s}$; $\tau_n \cong 100$ ns; $\tau_p \cong 200$ ns.

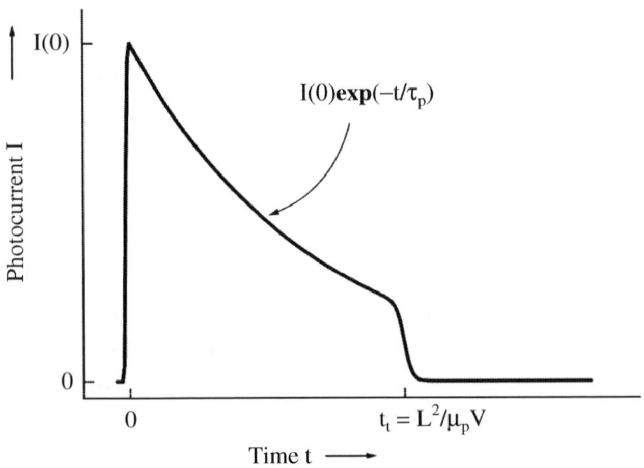

Figure 15.11 Time dependence of the current that follows a short cluster of excess holes generated at the positive side of an external voltage-biased insulating semiconductor. Holes trapping out of the drifting excess hole cluster occurs during drifting, until the cluster, reduced in concentration, exits the semiconductor.

15.4 Crystal Processing

15.4.1 Preamble

Processing of single-crystal components is an important aspect among the large number of subjects related to single-crystal growth technologies. The processing goal is to have the grown material assume the appropriate shape, physical dimensions, surface quality, and proper crystallographic orientation. Processing includes cutting, abrading, grinding, polishing, and surface brushing up. In this chapter, we provide a systematic review of each of these processes.

15.4.2 Crystal Cutting

(i) **Cleaving**

Quite a number of crystals cleave easily parallel to characteristic crystallographic planes. It occurs when the molecules bonding perpendicularly to these planes are considerably smaller than the molecules bonding strengths within the planes. After orienting the crystal, the crystal is simply placed on a table; a sharp knife is positioned on the crystal, oriented parallel to the cleavage plane. A light and quick hammer blow on top of the knife usually causes the crystal to cleave. The revealed surfaces are usually perfectly planar except for "staircase" patterns of a semi-macroscopic scale. The hammer blow strength and the accuracy of the knife orientation are different for different planes in different crystals. In hard crystals, the hammer blow should usually be harder, and very fast, and it is often sufficient to use a sharp hard metal pin, only roughly oriented parallel to the intended plane. In soft crystals, the used knife should be very sharp, moved slowly parallel to the intended plane, and very accurately oriented parallel to the cleavage plane. The cleaving often assumes a shape similar to a "book opening", and cleaved crystal parts assume some plastic distortions due to planes sliding.

The cleaving method is of course preferable only for those applications requiring surface planes oriented parallel to the natural cleaving planes.

(ii) **Cutting using a wire saw**

That cutting method is based on stretched wire moving along. The trivial means to achieve this situation is to weld the two ends of the wire, wrap the wire around two pulleys, stretch the wire, and rotate the pulleys – see Figure 15.12. The cutting is performed by lifting the crystal gradually toward the moving wire. Cutting is achieved as the wire removes material from the crystal along its contact with the crystal. There are three basic methods used to cause the said material

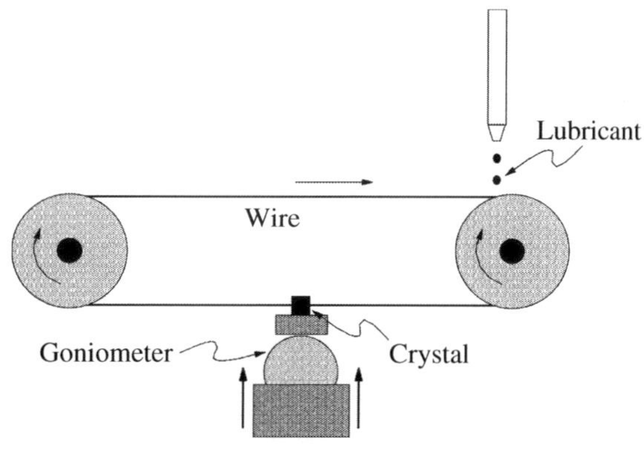

Figure 15.12 Schematic illustration of crystal cutting using a wire saw. The crystal is held on top of a goniometer, allowing a fine orientation, and lifted toward a stretched wire wrapped around two rotating pulleys. Dripping of a lubricating liquid with the abrading powder assists in cutting the crystal.

removal: mechanical abrasion (the abrading material is embedded in the wire or flowed as suspension in a cooling liquid), dissolving using a proper solvent, and chemical corrosion.

There are four materials participating in the mechanical abrasion: the wire, the crystal, the abrading material, and the lubricating fluid. The abrading material is the hardest, in the form of fine-grained powder, whose grains size is usually chosen between about 50 μm and 0.1 μm. Used materials are powders of diamond, alumina Al_2O_3, and silicon carbide SiC. The material used to form the wire must be softer than the crystal material; it is very common to use copper wires, but other kind wires are also possible, depending on the cut crystal. The lubricating fluid has to be sufficiently viscous so as not to drop off the wire too much while it moves, and necessarily chemically inert with respect to both wire and cut crystal. Very common lubricants are mixtures of water, alcohol, and inorganic oil. A suspension of the lubricating liquid with the abrading powder continuously wets the moving wire. Rubbing between the wet wire and the cut crystal sticks the abrading powder into the wire (which is the softer one). The crystal cutting is actually performed by the moving powder particles that are stuck in the wire.

Cutting by dissolution is done without any abrading powder, but with a lubricant that wets the wire and dissolves the crystal but doesn't interact chemically with neither. The wire motion over the crystal removes some material by dissolving it. The solution is driven by the wire and is washed away by a new lubricant during the continued movement of the wire.

Cutting by chemical corrosion is quite similar to cutting by dissolution. The lubricant wets the wire but doesn't interact with it chemically, but it reacts chemically with the crystal, forming a compound that dissolves in the lubricant. The corroded compound in the lubricant is driven away by the moving wire, and washed away by a new lubricant during the continued movement of the wire.

The advantage of all wire cutting stems from the flexibility of the cutting wire, which limits the occurrence of strains and fractures on the cut crystal surface. A disadvantage is that cut surfaces are not always sufficiently planar. Cutting using a disk saw yields better planarity of cut surfaces.

(iii) **Cutting using a disk saw**

That cutting method is based on a thin metallic (usually brass or steel made) circular disk, rotating about its spindle. The near-circumference zone is headed with a soft metal implanted with diamond powder particles or with particles of a different hard material. Cutting is done by moving the crystal toward the rotating disk, allowing removal of material off it by the hard powder particles implanted on the disk heading. A lubricating liquid is flown over the cut zone in order to wash the cut region from straying chips of the crystal and of the cutting disk (see Fig. 15.13). To assure that the disk doesn't bend while cutting (which might yield distorted cut planes), the thickness of the disk should never be less than about a tenth of the cutting length (that ratio naturally depends on the stiffness level of the crystal; for softer crystals, it is possible to use thinner disks).

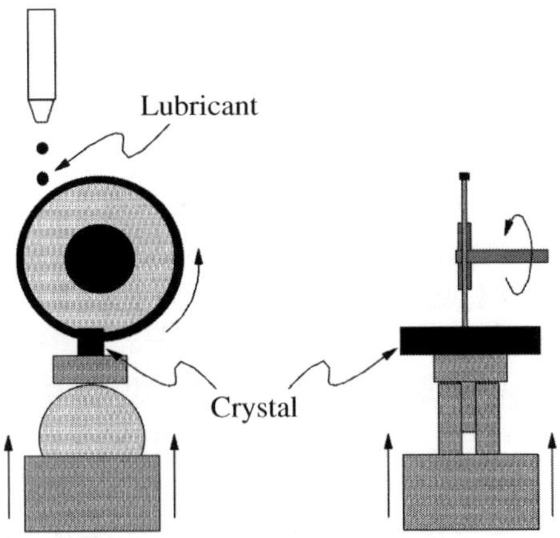

Figure 15.13 Schematic illustration of crystal cutting using a disk saw. The crystal is held on top of a goniometer allowing a fine orientation, and lifted toward a rotating disk whose heading is added with a soft metal implanted with diamond powder particles. Dripping of a lubricating liquid assists in washing off straying chips of the crystal and of the cutting disk.

Cutting by a disk saw, and in fact also by a wire saw, causes some loss of the grown crystal material. An important modification of the disk saw is an annular-structured disk where the cutting side is the annular (inner) side, which is headed with a soft metal implanted with hard material powder particles. The disk circumference is out-stretched by a drum, which also grants it with the rotational motion (see Fig. 15.14). The out-stretching prevents the disk from bending while cutting, which allows using disks of reduced thickness of only a hundredth of the cutting length. That arrangement reduces the loss of the grown crystal material.

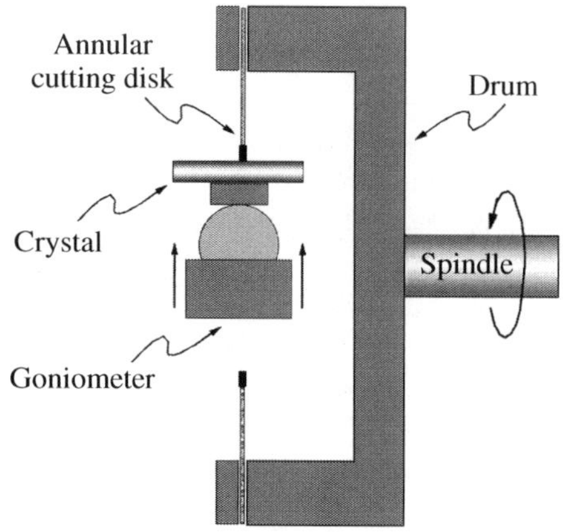

Figure 15.14 Schematic illustration of crystal cutting using an annular disk saw. Stretching the disk radially outward by the holding drum makes it more stable, allows using thinner disks, thus saving of grown crystals material.

(iv) **Drilling and processing of cylindrical rods**

For various purposes, and specifically for producing of laser rods (see Section 15.2.4 and Table 15.2), grown single crystals need to be processed to obtain cylindrically shaped rods. The first stage of obtaining a cylinder is drilling the grown crystal using a diamond-impregnated hollow drill. Such drill comprises a metallic, usually steel-made tube, whose lower end contains a header of a relatively soft metal in which diamond grains or other hard powder grains are embedded, as described for the disk saw in the previous section. The hollow drill is attached to a special drilling machine with a hollow holder through which a lubricating/cooling liquid is flown (Fig. 15.15). The drilling is performed by lowering the spinning drill toward the grown crystal, which is harnessed to the machine tray and oriented in the required crystal direction. The rod remaining inside the inner drill volume has a cylindrical shape, whose diameter is only slightly smaller than the drill inner diameter. The taken out rod forms the raw material for the next stage of processing, which aims at obtaining the accurate planned rod diameter, and of the roughness degree of its cylindrical shell.

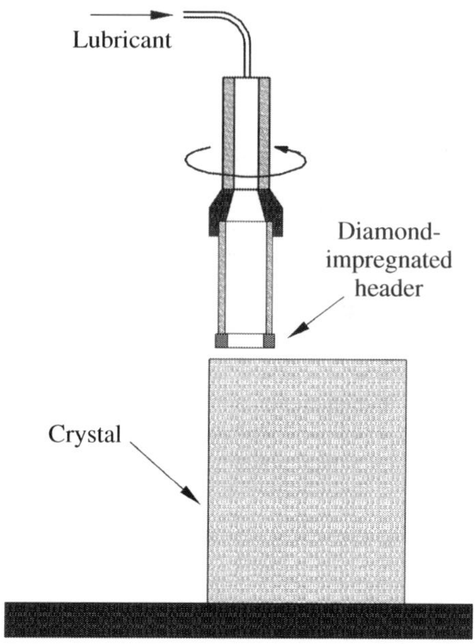

Figure 15.15 Schematic illustration of producing a cylindrical rod from a crystal by drilling with a hollow drill.

The said processing stage is performed by center-less grinding. A grinding wheel rotating at a high speed about its center is lowered toward the rod, while the rod itself is forced to rotate about its center by a slowly rotating "moving wheel" rubbing against the rod cylindrical shell (Fig. 15.16). The grinding wheel is in a sense similar to a cutting wheel, but it is much thicker, typically several millimeters. The rubbing between the rod and the grinding wheel gradually removes material from the rod cylindrical shell, thus reducing its diameter. Forming a tiny angle between the stopper edge and the grinding wheel axis induces a gradual movement of the rod in the orientation of the wheel axis. This measure allows grinding of the entire rod length even when the rod is considerably longer than the width of the grinding wheel. The gradual reduction of the rod diameter while moving parallel to the wheel axis means that the rod axis changes slightly during the process; it is in contrast to a usual machining using a lathe, where the axis of the processed material is fixed. That is the reason why the process described in Figure 15.16 is granted the "center-less" adjective.

While grinding, a lubrication/cooling liquid is flown over the rod in order to wash it from straying chips of the crystalline rod and of the grinding wheel. The center-less grinding allows obtaining improved accuracy for the rod diameter compared to lathing. Control of the degree of roughness of the rod cylindrical shell is obtained by selecting the degree of fineness of the abrasive powder embedded in the head of the grinding wheel.

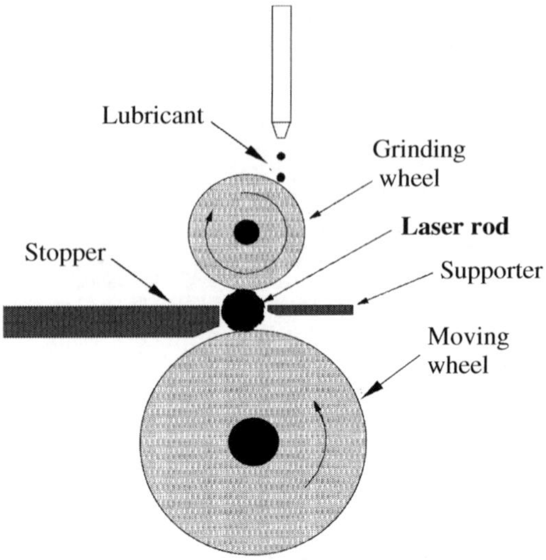

Figure 15.16 Schematic illustration of the grinding procedure in a "center-less" grinder. A grinding wheel rotating at high speed is lowered toward a cylindrical rod forced to rotate around its axis by a moving wheel rubbing against its cylindrical shell. One horizontal table adjacent to the cylinder serves as a stop to determine the cylinder position; a second horizontal table supports against its falling in opposite direction.

15.4.3 Crystals Polishing and Brushing Up

The previously described cutting and processing techniques very often yield quite rough surfaces. In most applications, especially optical ones, some surfaces need to be highly planar and smooth, such that their remaining roughness would be smaller than any wavelength in the visible region (0.4–0.7 μm). A polishing process usually includes three stages of different roughness degree: grinding, polishing, and brushing up. In all stages, the subject surface is made to move over a planar surface (often in the form of a circular wheel – see Fig. 15.17), which grinds/polishes the surface. There are four factors playing a role in the process: the crystal, the grinding/polishing wheel, the grinding powder, and the lubricant. The grinding powder has to be the hardest; the grinding/polishing wheel has to be the softest. When the crystal moves over the wheel (while the wheel also rotates slowly about its spindle), the grinding/polishing powder particles stick into the wheel, and these particles actually perform the grinding. The accuracy of the obtained crystal face planarity depends on the wheel surface planarity, the shape of the crystal holder (which is grinded together with the crystal), and the form of moving the crystal holder (plus crystal) over the wheel surface.

The grinding stage is usually performed using hard material "coarse" powder particles whose

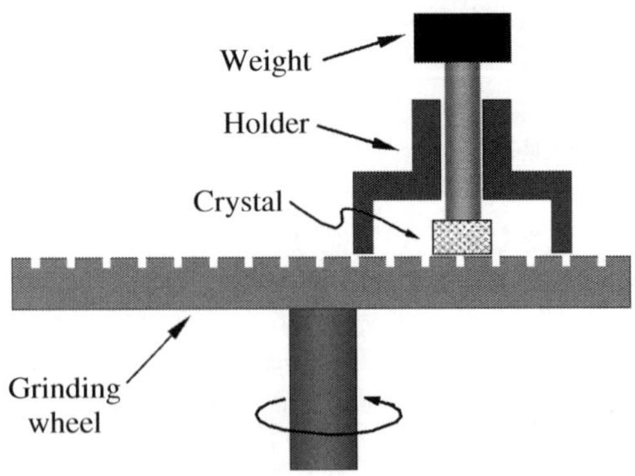

Figure 15.17 Schematic illustration of crystal surface polishing using a grinding wheel. The grooves role is to drain and store the lubricant during the polishing.

size is no more than about 25 μm. Grinding is often also used to reduce the element dimensions to close to the final required ones. Then, the processing is done using several other wheels, in which the powder particles are successively smaller. It is important to wash the crystalline element carefully off all remaining powder particles on each change of wheel, because any remaining coarser particle would cause the appearance of scratches on the polished surface. Using powder particles of about 5 μm dimensions usually yields near optical quality to the polished surface. Using powder particles smaller than about 1 μm, usually yields excellent optical quality to the polished surface. These numbers provide a very rough approximation, and the actually obtained quality depends on all relevant factors. The top grinding wheel surface material (the polishing material) is usually made of a soft metal – copper *Cu*, tin *Sn*, and aluminum *Al* alloys, or even various kinds of tar. Sometimes, use is made of various types of fabric or felt, stretched or glued on top of the grinding wheel. The latter cases allow obtaining very high degrees of surface brushing up, yet the overall surface planarity is harmed. The reason is that the shape of the grinding wheel surface doesn't remain perfectly planar due to some dragging of the fabrics by the processed element.

An important special case is the polishing (and brushing up) of the planar faces of solid-state laser rods (the bases that are perpendicular to the rod axis – see Table 15.2 item 4.2). To carry out that polishing and brushing up process, a special metallic holder is used in the form of a large cylinder whose height matches the required rod length (Fig. 15.18). The holder contains several through-holes drilled perpendicularly to the surfaces. They are aimed at holding the needed laser rods, and prepared to meet the required dimensions and tolerances, of which Table 15.2 is a typical example. The rods are inserted into the holes, and their ends are glued to the holder top and bottom, using a glue whose hardening doesn't induce stress, and it softens at a relatively low temperature. Polishing the rods is performed by actually polishing the two sides of the combined holder plus rods, until reaching the required parallelism.

The described polishing and brushing up process is mechanical, often causing breakage and contamination of a near-surface layer of even several micrometers thick. Removal of such layer is important for semiconductors aimed at usage in electrical circuits. That is done by chemical

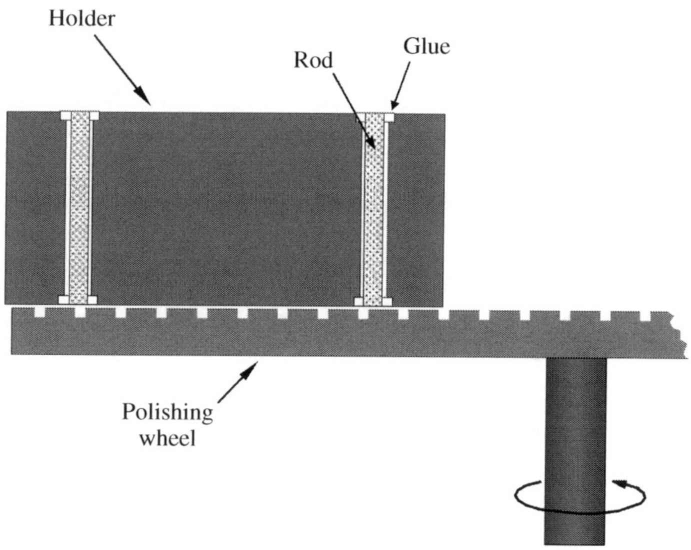

Figure 15.18 Schematic illustration of a laser rod holder while being polished on a polishing wheel. For details, see the text.

erosion and washing (at a cost of slight planarity reduction). Sometimes passivation against chemical reaction with impurities is needed, especially air-contained ones. That may be achieved by surface covering using a special passivation layer that doesn't interfere with the element functioning. It is usually done by evaporating the layer inside a high-vacuum chamber.

Exercises for Chapter 15

15.1 A charge of 500 g raw material is prepared using powders of yttria Y_2O_3, alumina Al_2O_3, and neodymia Nd_2O_3 for growing of $Nd^{3+}: YAG$ single crystals. The density of neodymium ions in the melt that replaces a part of the yttrium ions should be 4.5 at. %. Calculate the weight of each type powder included.

15.2 A plate of a cadmium zinc telluride $Cd_{0.8}Zn_{0.2}Te$ single crystal of 2.3 mm thickness and 2.6 cm^2 surface area is intended to be used as gamma and X-ray detector. As a quality control, the electron and hole mobilities are calculated based on their measured transit times through the plate using short light pulses of $\lambda = 337.1$ nm wavelength. The expected results are $\mu_n = 1350\,cm^2/V\cdot s$ and $\mu_p = 120\,cm^2/V\cdot s$. Draw the electrical circuit used for that purpose, calculate the expected transit times of holes and electrons under an external voltage of 550 V, and adjust the value of the series resistor that would allow a clear measurement of the transit times. It is given that the relative dielectric constant of the material is 11.6. (It is recommended to also consult Section 9.1.1.)

Appendix A Greek Alphabet and Phonetic Names*

Lower case	Capital	Name	Lower case	Capital	Name
α	A	Alfa	ν	N	Nee
β	B	betta	ξ	Ξ	Ksee
γ	Γ	Gamma	ο	O	Ommikron
δ	Δ	Delta	π	Π	Pee
ε	E	Epsilon	ρ	P	Ro
ζ	Z	Zetta	σ	Σ	Sigma
η	H	Etta	τ	T	Tau
θ	Θ	Tetta	υ	Υ	Yppsilon
ι	I	Yotta	φ	Φ	Fee
κ	K	Kappa	χ	X	Khee
λ	Λ	Lammda	ψ	Ψ	Psee
μ	M	Mee	ω	Ω	Ommega

*The indicated (phonetic) names are aimed at pointing out the proper pronunciation by using the English alphabet symbols as pronounced by most North-American English speakers. Capital Greek letters in gray-lighted boxes are not considered in mathematics as Greek, but rather as Roman letters, due to their shape similarity.

Conductors, Semiconductors, Insulators, and Crystal-Growth Technology: A Practical Guide, First Edition. Zeev Burshtein.

Appendix B Table of Physical Constants

Name	Symbol, and size in MKS units
Vacuum speed of light	$c = 2.9979 \times 10^{8}$ m/s
Elementary charge on electron/proton	$q = 1.6022 \times 10^{-19}$ Coulomb
Planck constant	$h = 6.6262 \times 10^{-34}$ J · s
Reduced Planck constant	$\hbar \equiv h/2\pi = 1.0546 \times 10^{-34}$ J · s
Electron rest mass	$m = 9.1096 \times 10^{-31}$ kg
Proton rest mass	$M_p = 1.673 \times 10^{-27}$ kg
Avogadro number	$\mathrm{N_A} = 6.0222 \times 10^{23}$ mole^{-1}
Boltzmann constant	$k_B = 1.3806 \times 10^{-23}$ J/**deg**
Permittivity of free space	$\varepsilon_0 = 8.85 \times 10^{-12}$ Coulomb/V · m
Permeability of free space	$\mu_0 = 4\pi \times 10^{-7}$ Weber/A · m

Conductors, Semiconductors, Insulators, and Crystal-Growth Technology: A Practical Guide,
First Edition. Zeev Burshtein.

Appendix C Literature References for Further Reading

Books

1 F. Seitz, *The modern theory of Solids*, McGraw-Hill, 1940.
2 M. Hansen, *Constitution of Binary Alloys* (Metallurgy and Metallurgical Engineering Series). McGraw-Hill Book Company, 1958.
3 R. H. Bube, *Photoconductivity of Solids*, Wiley, 1960.
4 A. Rose, *Concepts in photoconductivity and Allied Problems*, Interscience, 1963.
5 A. Many, Y. Goldstein, and N. B. Grover, *Semiconductor Surfaces*, North Holland, 1963.
6 W. G. Pfann, *Zone Melting*, Wiley. 1966.
7 R. H. Tredgold, *Space-charge conduction in Solids*, Elsevier, 1966.
8 R. K. Willardson and A. C. Beer, editors, *Semiconductors and Semimetals*, (book series) Academic Press, 1966–1975.
9 W. D. Kingery, Introduction to Ceramics, Wiley, 1967.
10 M. A. Lampert, and P. Mark, *Current Injection in Solids*, Academic press, 1970.
11 R. A. Laudise, *The Growth of Single Crystals*, Prentice Hall, 1970.
12 C. Kittel, *Introduction to Solid-state Physics*, 5th edition, Wiley, 1971.
13 B. G. Streetman, *Solid State Electronic Devices*, Prentice-Hall, 1972.
14 J. C. Brice, *The Growth of Crystals From Liquids*. North Holland, 1973.
15 C. H. L. Goodman, *Crystal Growth, Theory & Technology*, Plenum, Vol. I, 1974; Vol. II, 1978.
16 E. Kaldis and H. J. Scheel, eds. *1976 Crystal Growth & Materials* (Current Topics in Material Science Vol. II), North Holland, 1977.
17 W. Bardsley, D. T. J. Hurle and J. B. Mullin, eds. *Crystal growth, a tutorial approach*, North Holland, 1979.
18 F. Rosenberger, Fundamentals of Crystal Growth, Springer, 1981.
19 A. W. Vere, Crystal growth, Plenum, 1987.
20 K. Debertin and R. G. Helmer, *Gamma and X-ray Spectrometry with Semiconductor Devices*, North Holland, 1988.
21 Peter Y. Yu and Manuel Cardona, *Fundamentals of Semiconductors*. 3rd edition, Springer 2001.

Conductors, Semiconductors, Insulators, and Crystal-Growth Technology: A Practical Guide, First Edition. Zeev Burshtein.

Index

Conductors, Semiconductors, Insulators, and Crystal-Growth Technology: A Practical Guide, First Edition. Zeev Burshtein.

Happy Snacking, Don't Die!

Foraging & Vegan Recipes with Black Forager

Alexis Nikole Nelson

with Jeanelle Hayner Olson

Photography by JB Douglas

SIMON ELEMENT

NEW YORK AMSTERDAM/ANTWERP LONDON
TORONTO SYDNEY/MELBOURNE NEW DELHI

SIMON ELEMENT

An Imprint of Simon & Schuster, LLC
1230 Avenue of the Americas
New York, NY 10020

Photography by Alexis Nikole Nelson (pages 33, 60, 59, 100, 150)

Photography by Jesse Akozbek (pages 93, 163, 164, 183, 219, 237, 328, 245)

Camille George, Wild Glory Bakery (pages 251, 259)

Pottery and Ceramic Prop Credits:
Waldo Pottery (page 97)
Mud Witch Ceramics (page 135)
Pink Sand Ceramics (page 231)
Chuck Strasshofer III (page 251)
Ceramics Made By J (page 189)

First Simon Element hardcover edition September 2026

SIMON ELEMENT and colophon are registered trademarks of Simon & Schuster, LLC

Simon & Schuster strongly believes in freedom of expression and stands against censorship in all its forms. For more information, visit BooksBelong.com.

For information about special discounts for bulk purchases, please contact Simon & Schuster Special Sales at 1-866-506-1949 or business@simonandschuster.com.

The Simon & Schuster Speakers Bureau can bring authors to your live event. For more information or to book an event, contact the Simon & Schuster Speakers Bureau at 1-866-248-3049 or visit our website at www.simonspeakers.com.

Interior design by Jen Wang

Manufactured in China

10 9 8 7 6 5 4 3 2 1

Library of Congress Control Number: 2025946010

ISBN 978-1-6680-0254-4
ISBN 978-1-6680-0255-1 (ebook)

To my mother, her mother, and this planet
(who in a way is everyone's mother)

ENTS

INTRODUCTION

I don't know how many people have noticed this, but I haven't changed my Instagram profile picture since 2018. It's a bit nondescript, and while I'm technically in it, you can't see my face. What you do see is my shadow on pavement, holding a tote bag with something leafy spilling out of it. What you don't see is that it's like a zillion degrees outside, with 350 percent humidity, and I've just decided to start a silly little IG page for foraging.

It's midsummer in Columbus, and like most Ohio summers, it is *hot*. Temperatures are in the nineties and the humidity makes walking feel like swimming in a hot tub with all your clothes on. Even so, my boss, Stacie, and I far preferred to be outside than stuck in the office, and we always conducted our weekly one-on-one meeting outside, usually while taking a walk. (It was also a great excuse for me to share edible plants and plant knowledge with Stacie on company time, let's be honest.) At the time, I was the head of organic social media, and Stacie oversaw all communications at our company.

We went to my favorite patch of mugwort, which by this time was taller than both of us—and neither of us are short. We also gathered a few late burdock stalks and threw them in our totes with our work laptops. We snacked on some wild cherries we found as we walked. Not unexpectedly, our conversation shifted from deliverables to our shared love of wild foods. I took a deep breath and told her that I was thinking of starting a social media page just to keep track of all my weird, foraged concoctions and creations, separate from my personal page.

I was nervous to bring this up because Stacie already ran a very successful cocktail Instagram, often using foraged and wild elements. So I didn't want her to feel like I was stepping on her toes, but I also really wanted her advice. Stacie had always been a master of branding, so unsurprisingly, her first order of business was coming up with a handle for my page. Her handle was @gardencocktails (now @schoolhousehomestead), which so succinctly and perfectly summed up her content. I, on the other hand, had been scrambling for a handle that felt simple, but interesting; honest, but exciting. As we talked, Stacie coached me to figure out the heart of what this page meant to me. She asked me what I wanted people to know the second they looked at my content. I wanted them to know, even without seeing me, that a Black woman was taking these photos and writing about these plants.

"Maybe something like . . . Black Forager?" I winced as I said the name out loud.

"I like Black Forager!" she immediately replied.

Stacie was also brutally honest when she didn't like something, so for her to respond this way meant a lot.

"Yeah, I just feel like I don't see anyone who . . . looks like me, you know?" Stacie understood. She loved that I was doing this as a way to talk about plants and celebrate my heritage while doing so.

I smiled and turned away from the bright sun for a second. I saw our shadows, standing tall, holding bags with all manner of plants, and just felt really good in that moment. I snapped a photo on my phone to capture it.

Later that very same day, I created @blackforager. My profile picture is the one from that day: my shadow in the summer sun, holding my tote, frilly mugwort peeking out of the top.

ON HERITAGE

My mom's family has been kicking it in North America since the 1600s. My great-(times, like, nine) grandfather was brought here as an indentured servant, lucky enough to be employed by someone who was willing to let him buy his full freedom. A racist and classist privilege, but a privilege nonetheless.

He bought his growing family a tract of New England farmland and they tried their best in a new place. (Soon, European farming practices would fully clear-cut much of the forest and render the soil underneath too sandy and devoid of nutrients to be arable. But they didn't know this at the time. Okay, let's stay focused!)

My father's family wasn't so lucky. They were brought to the Southern states nearly one hundred years later, and I barely know anything about them. According to the same United States that was just starting to accept my matrilineal Northern ancestors, the folks in my patrilineal line, Black and Indigenous, weren't citizens. Weren't people. Weren't worth considering. As such, something I'm sure of is that foraging wild greens, either by them or by other members of their plantation community, helped them survive.

Come the 1870 census, my father's family, like thousands of other Black families, showed up as if by magic in public record with a name. (Can you imagine an alien trying to guess what happened between the 1860 and 1870 census?! Where did all these Black folks come from?) Our family, the Nelsons, were sharecroppers. And while they didn't benefit directly from the crops they coaxed out of the soil, they began to secure at least the tiniest bag. (For my pre-1990s crowd, or anyone reading this book more than ten years in the future, to "secure the bag" is to make money!) This proved to be enough to grow a family and create a legacy that my own family would one day help me discover.

Cut to my own childhood: Mom's in the garden, Dad's in the kitchen. That was the way it was when I was wee bitty.

If you wanted time with my mom, you got your hands dirty. You memorized the names of the plants in the garden, which ones were tasty and which ones would give you a heart attack, which ones she planted and which ones just showed up on their own. You learned that plants don't have to be there "on purpose" to be wonderful and useful.

If you wanted time with my dad, you washed up and got your hands clean. You learned which flavors played well together and which cooking methods brought out their best elements. You learned not to touch hot pans and how to focus on the dance among hands and knives and greens.

Since my ancestors gave so much of themselves to the work of cultivating relationships with the soil, the plants, and the signs of the seasons, some by choice and others not, it seems inevitable that I would feel most myself when I do the same. I come from corn and poke and cotton and wild garlic and so many others. And a lot of what soothes the little empty space in my heart and mind, the space where I'd otherwise put the names and stories of my unknown ancestors, is placing some of the plants we had in common there instead.

People always ask how I became who I am, but when I look at my lineage, I ask how I could have become anything else.

ON COMMUNITY

"I love your sunflower fence!" I called to the two figures working in their yard. I had never considered something so whimsical: a fence of pure sunshine and dopamine!

I had been on one of my wanders, something I'd been doing lately to clear my head. After a long week of work, I'd just start walking until I decided to come home. This trip accidentally took me five miles from home, lost in my own sad thoughts. Suddenly, a row of seven-foot flowers stopped me dead in my tracks.

The two figures waved me in, and for a split second, stepping through the wall of flowers made me feel like a little kid in my mom's garden again.

The mystery people, with their magic house and magic yard, were Tyler and John. They were artists and gardeners, and by the end of our conversation they'd sent me home with a huge pot of anise hyssop, a native plant that tastes of sweet black licorice and summer heat. But before I left, John mentioned the word "foraging" and it was like the sleeper agent in me was shaken awake! I knew instantly it was a word for all the times I'd gathered wood sorrel or beach plums or honeysuckle flowers, but I suddenly had a *word* for it.

The next day, I went to the library and checked out as many foraging books as I could carry home in my tote bag.

Reflecting on the kindness and knowledge from so many people has me tearing up like a goober as I write this, and all I can hope is that I'm one of those people for you. You're already cooler than me for actually reading intros, so you deserve it!

But I'd also like to list all the people who gave me the gifts of their plants and fungi and food prowess, in book or video or conversation, and thank them from the bottom of my big silly heart, because they're the reason I could even dream of doing this book.

In no particular order, my biggest thanks to:

Linda Black Elk	**Alan Bergo**
Luke Black Elk	**Gabrielle Cerberville**
Sam Thayer	**Val Rossman**
John Kallas	**Hank Shaw**
Marie Viljoen	**Stacie Grissom**
Ellen Zachos	**Tim Clemens**
Pascal Baudar	**Blair Miller**
Lisa M. Rose	**Jeremy Umansky**
Adam Haritan	**Scott Stimpson**

And all of my Montessori teachers for letting a five-year-old be "best friends" with a tree, and not making me feel weird about it.

And YOU for your care and time and the money you paid for this book, and for your faith in me.

I'm positive I could keep adding to this list for hours, but hey, we've got a book to read!

WHY I FORAGE

- Easy and free source of nutritionally dense foods. A lot of what's in the grocery store is bred to last a long time and not offend any tastebuds (so it's on the mild side). We've also lost a lot of minerals in the soil over the last century. All of this means the veggies and fruits in the store don't necessarily have the same micronutrient content they used to. Foraging can help fill in those gaps!
- Fun flavors to play with in the kitchen that you typically can't find in grocery stores. Things like garlic mustard and milkweed flowers are wonderful and unique. You can't find anything like them at the store. (At least not right now . . . maybe one of you will change that!) For my foodie friends, this means a whole new world of playing with textures, flavors, and colors on your plate.
- It's meditative to turn off your "plant blindness" and tune in to your surroundings. There have been some great studies coming out of Japan, another chronically stressed and online population, on the benefits of forest bathing, or taking time to tune in to the plants, birdsong, and smells of the space you're walking through. It lowers your blood pressure and helps with stress management. I used to roll my eyes when my mom would tell me to go for a walk after a rough day, but as always, she was right!
- As a Filthy Vegan, a funny phrase I came up with to turn the "annoying vegan" trope on its head, foraging opens up my culinary world to so many additional options, especially in the realm of proteins (since there are so many amazing fungi out here in these forest streets).
- It's a great way to connect with where you live. In this age where renting is more prevalent in city spaces than owning, we don't always get the chance to form a bond with our neighbors and neighborhoods anymore. Taking the time to get to know your plant neighbors helps even the most urban neighborhood (like mine) seem full of nature and whimsy, even if you have to look close to find it. I love a good dose of whimsy, and honestly, I think we all need more of it! I'm trying to live my best Beatrix Potter painting life.
- We all need to touch some grass, to be frank. I'm the last one who has the right to call anyone "chronically online," but as we spend more time on our phones and less with each other, it becomes easier to feel angry and disconnected. But even when we feel that way, we're still all connected to one another and the natural spaces that surround us, and foraging is a great way to remember that!

Speaking of FILTHY VEGANS

Maybe you already knew this when you bought this book, maybe not, but I am vegan. Surprise!

So, in my recipes, you'll see vegan butter and plant milks of various kinds because that's how I cook in my kitchen. But, and I cannot stress this enough, *I'm not your mom!* (Please imagine me singing that, not yelling it.) You can replace these with whatever cooking fat or fatty spread or milk that may be chilling in your fridge. If a recipe looks tasty to you and you want to serve it alongside a preferred protein, it's cool! I'm not here to dictate everything you make in your home. Foraging is supposed to help you cut back on consumption, so I would never ask you to go and buy new things when what you have will work perfectly.

FORAGING TOOLS

While I feel like the spirit of foraging calls us to do more with less, there are some tools I recommend for field and kitchen to help you maximize your finds and stay safe!

FOR THE FIELD

- **A multi-tool** like a Swiss Army knife or something similar. Great for snipping, sawing, and tweezing out little splinters.
- **A garden knife** or Hori Hori. Kind of like a trowel, but more menacing! Great for getting up tubers or cutting through roots.
- **Bug spray!** I'm terrified of ticks! After getting a case of tick-induced Bell's palsy, which made half my face numb and tingly for eight months, *I DO NOT MESS AROUND.* I use sprays with permethrin to treat my hiking clothes before hitting the woods. Make sure to treat your clothes outside and away from kids and pets!
- **Boots**, because you never know when a trail, even a paved one, is gonna be slippery or muddy. These don't need to be ultra-technical waterproof hiking boots, just something with support and traction that you don't mind getting messy in.
- **A bag or satchel for your harvest!** I like bringing fabric bags so any mushrooms I gather can spread their spores while I walk, and my greens don't get too hot and wilty. Finally, a use for all the free tote bags you've ever received, which if you're like me, number in, like, the hundreds.
- **A way to track your findings.** I usually just drop pins on Google Maps on my phone, but lots of folks like to keep a physical notebook to write down when and where they find their goodies. (Bonus, you can sketch out the things you find or just doodle in the sun.)
- **A water bottle.** Obviously! Make sure to hydrate before, during, and after your frolicking.
- **Sun protection.** Another obvious one! Whether you prefer sunscreen, a wide hat, long sleeves and pants, or all of the above, your body will thank you.
- **A little hand sanitizer** in case you run into poison ivy/oak/sumac. The alcohol will help break down the oils before they react with your skin!

FOR THE KITCHEN

- **Jars. so many jars.** Your kitchen will look like an old-timey apothecary (isn't that kind of what you wanted?), and jars are so easy to find en masse at charity shops if you don't feel like paying a pretty premium online or at the grocery store. You can also save glass jars from things like pasta sauce, pickles, and sauerkraut and just scrub off the labels. All sizes are useful!
- **A tiny food processor.** People still clown my tiny old food processor from college, but it's so handy when you've gathered enough to make a sauce or a small dish, but not necessarily enough to warrant using a big blender.
- **A tall pot.** Excellent for making jellies (which tend to foam so much more than you expect while they cook!) and for hot water bath–canning anything nice and acidic for long-term storage on the shelf, since the jars need to be under at least an inch of boiling water to be canned safely!
- **A chalk marker**, for writing the date and the contents of your jars right on the jar or lid! Washes off super easily.
- **Your freezer**, great for making fruits and blanched greens and sautéed fungi stand the test of time. Winter—you will weep with gratitude.
- **A food dehydrator.** Okay, okay, while this isn't a must-have, it's seriously a game-changer for extending the life and use of your greens, flowers, and fungi! Dries everything to a crisp so it can easily be stored in all those jars you're hoarding.

Places Where YOU SHOULD NOT FORAGE

- **Golf courses.** *SO MANY HERBICIDES*, you guys. And so much fertilizer! Please do not put the greens that somehow survived here into your body, I'm begging you.
- **Superfund sites.** The name sounds so fun, doesn't it? But they're areas where industrial waste is stored for . . . ever? Lots of times these sites haven't been properly capped (especially if it's an old site) and many of them have been converted into park spaces. You can go on the EPA's website and look up Superfunds and recent spills in your area. But don't let the recent spills map ruin your life, they're not always things that are hazardous, but they do still have to be reported.
- **Along railroad tracks.** I have to remind myself of this one every year because the sumacs near the tracks always look so *lush*, but railroad companies have to do a lot of herbicide spraying to keep the tracks clear. I'd give tracks a solid twenty-foot radius.
- **Next to old painted houses.** I have been told that lead paint chips taste sweet. (It's why unsupervised kids used to sometimes eat them in the 1970s and '80s. Great!) Alas, and as you can probably guess, lead is a tummy no-no, and can persist in the soil for decades. *Decades!*
- **Le dog pee corridor.** I pass on anything growing within a foot of the sidewalk! I know it won't kill you, but it's just kinda icky to think about.
- Places that have (or seem like they could have) been treated with **glyphosate** within the last two months. Glyphosate is a popular herbicide, often sold as weed killer like Roundup, and when it's fresh it's awful for the body. There are plenty of studies on exposure to wet glyphosate upping the chances of developing cancer and Parkinson's, so if you see irregular yellowing and wilting on a series of plants, leave things in that area alone for at least two months. Now, while it's dangerous fresh, the microbes in our soil break it down! It usually takes around thirty days (it has a soil half-life of fourteen days), but I want all of you to be supersafe, so we're gonna call it a clean sixty. That also means if someone treated your yard before you moved in, but hasn't recently, those plants could be fair game as long as your yard doesn't check any of the other boxes above.

For my sweet city mice, a note about urban areas: I think some folks have a sense that cities are less ideal, or downright dangerous places to forage. On some level, I understand the feeling.

Sure, you've got dog pee and, well, other pee, and construction dust and vehicle exhaust and smog and public spaces that are maintained by, in many cases, increasingly tenuous funding. You've also got bleaker stuff like heavy metals (not the music genre!) and things like PAHs (polycyclic aromatic hydrocarbons). Amid all this, it might be easy to assume, based on the much higher volume of humans and businesses per square foot and their many impacts on the environment, that urban areas would have more of those. Right? Actually, not really.

See, the good news/bad news (jk, it's mostly bad news) is that heavy metals and PAHs are *everywhere* now. Even industrial farms—yes, the ones that grow most of our vegetables, including the ones at fancy grocery stores—have them, thanks to things like irrigation tubing that's been in use for the last twenty years or so. So as much as we like to envision rural, and even just non-urban, environments as somehow more pure, or clean, or less impacted by practices that have been actively harming the earth for decades, those places are often just as impacted as urban areas. So awareness of the kinds of sites I mention above is what will really help you discern the best places for you to forage.

Additionally, some plants, and parts of plants, uptake things like heavy metals more than others. So while I would eat a hackberry in Central Park (and I have!), I would not eat, like, a burdock root from Central Park. Know what I mean? Mushrooms are another great example. Mushrooms are known bioremediators (i.e., they help clean the soil after we've . . . well, ruined it) and picking one off a tree means the tree has done a good bit of filtering first, so the mushroom has not had to remediate quite so much. So for me, mushrooms that are growing off of trees in large cities feel safer than ones growing out of the ground. But between you and me, I have been even less careful: Once I brought home an entire suitcase of Brooklyn graveyard maitake! You don't have to do that! I guess what I'm saying is that foraging in urban areas is really up to you and what you're comfortable with. And thankfully, many cities have a lot of accessible green spaces on their outskirts, so it's sometimes as easy as a little half-day or day trip to get to what might be an easier/more comfortable spot to forage.

If you're a city dweller or just curious, Marie Viljoen is an incredible resource on urban-specific foraging. She's been living and foraging in New York City for decades, and her books (especially *Forage, Harvest, Feast*) show the bounty she finds and eats in the greater metro-NYC area. And her social media is absolutely worth a look!

THE NOTICING GAME

A lot of what I do when I travel to forage with groups of people is lead plant walks. And I start all of my plant walk classes with this exercise to help our silly lil' brains go from "Okay, hi, hello, I need to look at my phone every five minutes" to "Wowie wow, is that a baby jewelweed plant?"

I call it the Noticing Game. There are no losers, only winners, because everyone's brain feels better at the end. You can even do the Noticing Game when you're having a rough day and not even planning to go out and forage. It'll make your mind a little calmer and clearer, that's the Loud Plant Lady guarantee.

1. Step outside.
2. Close your eyes.
3. Say three things you hear around you. It can be out loud, but it doesn't have to be! (Some of us just like being the Weird Neighbor.)
4. Open your eyes.
5. Look around and find a plant, any plant! It can be as big as a tree, or as small as the little sprout coming up in the sidewalk.
6. Walk on up to the plant.
7. Say three things you notice about the plant. (Again, out loud or not.) It can be the shape of its leaves, the color of its flowers, its height. Anything you can see is fair game.
8. Close your eyes again.
9. Say another three things you hear.
10. Open your eyes and see if you can come up with another three things you notice about your new plant buddy. You might find you're noticing totally new things!
11. By now you will have activated a clearer, calmer, plant-noticing part of your noggin. You're cleared to go a-questing for some tasty treats!

How to Find What's in Season in Your Area

Now, this book is (gently! kinda!) organized by season. Meaning these will match up for folks living in similar growing zones (I'm in 6b, if you were wondering), but won't necessarily match for everyone.

If you live farther south, some summer finds become spring finds. Chanterelle season solidly starts in July where I live, but starts in mid-May in Mississippi! If you live farther north, your juneberries might be more like July berries, or your field garlic may not pop up until it's solidly spring.

Here are some tips to zero in on when things are popping (Al Roker voice) *in your neck of the woods.*

- **On your phone:** A very easy method is to look up a particular species on iNaturalist. It's an app used to track populations of plants, fungi, animals, algae, sometimes even lil' guys like diatoms or bacteria, all over the world. I love it because you can use it to contribute to citizen science programs, and scientists and specialists can go through your uploads and confirm what you've found! If you type in the name of what plant or mushroom you're looking for, not only can you see where people have found it in your area, you can see the date that the picture they took was taken, which can give you a hint about how far along that specimen was at that time. A good note is people tend to notice plants more when they are flowering or fruiting (because plant blindness), which is great because that's usually what we're gathering.
- **On the ground:** The old-fashioned way is to go out to spaces where you'd expect to see a plant at any given time during the year, and see where it is in its growth cycle. This can sometimes be depressing, as you may miss the flowers or fruit or tender growth, but you can always take notes for the following year! Set a reminder on your phone so you don't forget! I do a little bit of both methods, especially now that I often go after plants that people aren't regularly uploading to iNat!

WEATHER & FORAGING

- For early spring plants, **ground temperature** is going to be what sets off their growth. So, if you have unseasonably warm weather at the tail end of winter, it's not abnormal to see some early spring plants in late February or early March.
- **Wildfires and controlled burns** can jumpstart the growth of morels and also small berry shrubs. They're also great for fire-resistant trees like oaks, because they cut back on competition for resources around the trunks of the trees, allowing them to be healthier, and have bigger acorn crops.
- **Rain**, of course, plays a huge part in plant growth. Too much rain can make berries lacking in flavor, even some of the very sweet ones like juneberries. There's nothing you can really do about it, just something to be aware of! It might lead to you wanting to add a smidge more sugar to recipes than you otherwise would have.

  Rain also often leads to an influx of fungi growth, since they need warm and wet conditions to fruit (aka make mushrooms). And the fruiting bodies are the part we eat! So, a great time to go looking for chicken of the woods, for example, would be after an early summer rainstorm. Give it about twenty-four hours so they have time to fruit—mushrooms are famously fast growers!
- **The first frost** kills off a *lot* of leafy plants, which is why I often descend into an absolute preservation frenzy the week leading up to our first expected one. So, if there are any tender greens you want to save for the winter (a lot of spring greens will have a second round if your fall season is long enough), then *GO GO GO GO!*
- Don't be afraid to go out and forage **during the winter**! Lots of plants, like our edible conifers, still have so much to offer, and fungi like enoki and winter oysters don't give a hoot about freezing temps! It's also fun to get to know outdoor spaces at every point of the year. Enjoy how much light makes it to the forest floor when there are no leaves on the trees. Enjoy how quiet it is when there's nothing rustling in the breeze. It's actually really peaceful!

Okay, I've said everything I need to say! (That's a lie! I could go on forever!) But truly, I've given you my every tip and trick. Time to get out there.

Happy Snacking, and Don't Die!

S
P

MAGNOLI

THE MAGNOLIA IS an ancient yet universally recognizable genus of trees and shrubs. Magnolias finished evolving even before bees did, hence their large blossoms and sturdy pollinator bits, originally meant for beetles and other hefty insects. What I find so curious is that nearly everyone recognizes magnolias. We love them as ornamentals, they're extraordinarily resilient to city life, and lots of places on this lovely lil' planet have their own native species. (Shout out to the southern and sweetbay magnolias!)

But no one knows they're edible!

The petals range from ginger-reminiscent to purely floral to a little citrusy and herbaceous, and are such an abundant source of flavor for that gorgeous two weeks each year when the magnolias proudly march our sleepy butts into spring. Even the leaves can be used to impart flavor! When I found out that sweetbay magnolias have their name because their leaves can stand in to provide a mildly sweeter flavor than a regular bay leaf, my mind was blown. I learned while dining at Kajitsu in NYC about hoba miso, a traditional Japanese dish cooked on dried and soaked magnolia leaves placed on charcoal, where that magnolia leaf adds a layer of sweetish, gingery flavor to the dish.

I grew up with a big Magnolia × soulangeana tree in my front yard, and in late winter I'd sometimes snap off a fuzzy flower bud, cup it to my nose, and inhale deeply while dreaming of the warm days soon to come. Little did I know back then that those buds can be used to make tea and flavor rice.

You didn't ask, but if there was one thing I had to smell, like, forever, it'd be magnolia blossoms (with black cherry flowers as a very close second).

WHAT TO LOOK FOR:

- Magnolia **flowers** are bowl- or star-shaped and can come in white, pink, yellow, green, or purple. Southern magnolia flowers are large, cream-colored, and especially fragrant. Asian magnolia species like star and saucer will bloom before leaves emerge, covering the tree in large, sweet smelling, tulip-like blooms (hence why they're often called tulip trees, a nickname they share with tulip poplars, even though their blooms don't really look alike!). American species like southern and sweetbay produce leaves, or "leaf out," first, and flower after. All flowers have sturdy central reproductive bits, which should be discarded when cooking as they are tough and a little bitter!
- Magnolia **buds** are fuzzy, grayish, and point upward. These lil' babies persist through the winter and are fragrant when broken. If collecting buds for tea, mid- to late-winter is best. Too early and it won't be fragrant, too late and you may as well wait for them to bloom! Be sure not to take too many; every bud taken is one less flower!
- Magnolia **leaves** are long, glossy, and oval-shaped. Southern magnolias also have a rust-colored down on the underside.

PICKLED MAGNOLIA FLOWERS

These are perfect for adding to sandwiches, salads, sushi, and so much more! You can also dehydrate the pickled petals to use as a seasoning in much the same way you'd use dried ginger, or for anything that needs a lil' punch.

MAKES ABOUT 2 CUPS OF PICKLED PETALS AND 2 CUPS OF FLAVORED VINEGAR

About 6 cups packed magnolia petals, cleaned of any debris or inner bits

2 cups rice vinegar

½ cup granulated sugar

2 tablespoons sea salt

Pack the magnolia petals into a clean, quart-sized jar, pressing down on the petals as needed to make more room. You can fit more than you think! And remember the petals will shrink in the brine. Set aside.

In a medium saucepan over medium-high heat, combine the vinegar, sugar, and salt and bring to a simmer, stirring to dissolve the salt and sugar completely. Remove the pan from the heat.

Carefully pour the hot brine over the petals and stir briefly to ensure all the petals make contact with the brine. The petals will float and that's normal! They'll settle down soon. Set aside to cool completely.

Seal the jar and store in the refrigerator. These will keep for about 1 month, unopened, and about another week once opened.

MAGNOLIA BUD TEA

MAKES 20 CUPS OF TEA

20 unopened magnolia buds, cleaned of any dried outer leaves

Boiling water

Preheat the oven to its lowest heat, ideally around 150°F and no more than 200°F. (Many ovens have a "keep warm" function that can be adjusted to a lower heat.)

On a rimmed baking sheet, spread the magnolia buds in an even layer. Dehydrate the buds in the warm oven until fully dried and fragrant (they'll smell a bit gingery), about 4 hours. Set aside and let cool completely.

Drop 1 dried magnolia bud in a cup of just-boiled water and let steep for 5 to 10 minutes, depending on your taste. If you're lucky, the bud may bloom in the water!

Store the dried, cooled buds in a clean, airtight jar for up to 3 months.

THICCC MAGNOLIA SYRUP & MAGNOLIA SNAP COOKIES

You can use this luscious syrup for flavoring cookies, cakes, and even adding some flair to savory dishes! The more scented the magnolia, the stronger the flavor, so leave the less-scented babes on the trees to look pretty and harvest the ones you can smell from a block away. Great to use in place of molasses when you'd like to add a floral, gingery kick!

Thiccc Magnolia Syrup

MAKES ABOUT 2 CUPS

1½ cups granulated sugar

4 cups loosely packed, fragrant magnolia petals, cleaned of any debris or inner bits

In a medium saucepan over medium heat, combine the sugar with 1 cup of water. As the mixture warms, stir until the sugar dissolves. Add the magnolia petals—they'll turn brown, which, while sad, is a normal side effect of them oxidizing. Cook, stirring constantly, until the petals are small and brown, about 3 minutes. Remove the pan from the heat. Set aside for about 10 minutes to let the petals continue steeping.

Strain the syrup into a clean jar, pressing gently on the petals to release any excess syrup. (If you're going ahead to make the Magnolia Snap Cookies on the next page, hang on to those sweet, syrupy petals! You'll use them in the cookies.)

Cover and store in the fridge for up to 2 weeks, or in the freezer for up to 6 months.

Magnolia Snap Cookies

If you've come here fresh from making the Thiccc Magnolia Syrup, great! You've probably got those delectable little syrupy petals sitting around, ready to go. If not, no problem, the cookies are just as good without them. However, if you've got magnolia petals hanging around that haven't yet been processed, may I recommend freezing some? Sounds funny, but you can grate them into the dough here (like, a tablespoon) for an extra floral-gingery kick.

This recipe makes enough cookies to give some away to friends and neighbors as a "thank goodness it's spring" gift and still have plenty left for yourself!

MAKES ABOUT 3 DOZEN COOKIES

1½ cups packed light brown sugar

1 cup (2 sticks) unsalted vegan butter, room temperature

1 tablespoon vanilla extract or Nocino (page 199)

¾ cup Thiccc Magnolia Syrup (page 39)

3¼ cups all-purpose flour

1 tablespoon baking soda

Frozen petals from the magnolia syrup, grated, for rolling into the cookies (optional)

2 teaspoons ground cinnamon

½ teaspoon sea salt

Granulated sugar, for rolling

Preheat the oven to 350°F. Line two baking sheets with parchment paper and set aside.

In a large bowl, add the brown sugar, butter, and vanilla. Use a hand mixer on medium speed to beat until thoroughly combined, about 1 minute. Add the Magnolia Syrup and continue beating on low speed (to avoid syrup splashes!) until combined.

In a medium bowl, whisk together the flour, baking soda, grated frozen magnolia petals (if using), cinnamon, and salt. Add the flour mixture to the brown sugar–butter mixture and mix with a silicone spatula or wooden spoon to combine. The dough might be a bit dry but should stick together easily.

Scoop the dough into roughly 2-tablespoon-sized portions and roll the dough between your hands to form a ball. If you'd like to fill the cookies with sweet, jammy magnolia petals, form the dough around a pinch of the grated petals and then proceed to shape it into a ball. Set the dough balls on the prepared baking sheets and continue with the remaining dough, and magnolia petals, if using.

Add a few tablespoons of sugar to a shallow bowl and roll the dough balls in the sugar until they're fully coated. Return the sugary dough balls to the prepared baking sheets, spacing them about 2 inches apart—these babies like to spread out!

Bake for 12 to 15 minutes, until the cookies are set around the edges but still soft in the middle. (They'll set up more as they cool.) Let the cookies cool on the baking sheet for about 5 minutes, then transfer them to a rack to cool completely. Store in an airtight container at room temperature for up to 1 week.

DANDELI

OH, DANDELIONS. You either love them, or hate them with the white-hot intensity of a thousand suns.

We know that humans have been eating them for as far back in history as we can translate, but within the last hundred years, especially in the United States, eating dandelions has been heavily associated with the Great Depression, poverty, and immigrant families.

However, anyone who thinks these guys are all bitter greens and lawn domination needs to think again! Dandelions have so much to offer, including being edible from the tippy top of the flowers to the tippy bottom of their roots. In fact, I personally feel dandelion flowers are tragically overlooked as a snack, a hair accessory, and a general day brightener! With everyone's talk about dandelions' bitter greens, you'd never think that the flowers taste delightfully warm, sunny, and sweet—almost like honey!

And I know what you're saying: "Alexis, you're a biased party." And you are not wrong! But I'll tell you a little story.

Not too long ago while I was harvesting hawthorn berries on a corporate campus (that I will not name, thank you), a Very Important Man in a suit came walking toward me to ask what I was doing. Needless to say, I was pretty nervous, and had a feeling I was about to get in trouble.

But then I was surprised! When I told him I was foraging, he immediately launched into a story from his childhood in which his friend's mom used to make a dandelion wine that she'd occasionally let him and his friends sneak as teenagers.

He said he hadn't had any since, and that he hadn't thought about it in ages, but suddenly he was really excited at the prospect of maybe finding out where to get or make some. The memory of the flavor of the flowers had stuck with him for literal decades. Not only that, but I think the dandelions kind of got me out of trouble because after telling me that story, he smiled, gave *me* his name, and said if anybody gave me trouble to tell them he said it was okay.

Dandelions: Good for your body, good for your soul, good for not getting in trouble!

If they aren't in bloom—the easiest time to recognize them—look for a basal rosette of the telltale toothed leaves. If it's early spring, you'll maybe see little underdeveloped flower buds hiding in the center (which make incredible stand-ins for capers, by the way, but alas, only so much can fit in one book!).

WHAT TO LOOK FOR:

- The **flowers** themselves are a sunshiny yellow, with small yellow hairs interspersed among the long rectangular petals. They smell faintly of honey! In early spring, you can often see a cluster of small flower buds nestled into the center of that rosette, which shows how many flowers that particular plant will make this year. Stems are hollow (and can actually be used as drinking straws). Flowers then turn into those telltale, wish-making poofballs of seeds. A fun experiment to do with children is to submerge one of those poofballs into a cup of water, and watch it emerge fully dry! They're water-resistant to make sure their seeds don't get bogged down every time it rains.
- Dandelions' deeply serrated **leaves** are the inspiration for their name, from the French *dent de lion*, or "tooth of the lion." The leaves start from one central point, starting out as a lovely rosette, and unfurling as the plant grows. When broken, the leaves ooze a small bit of milky sap that has a distinctly sweet fragrance.
- The **roots** often go deceptively deep into the ground. The roots are coffee-brown, and also leak sap when broken. They tend to stay in one big central taproot, unless you have very rocky soil like me, then you get all sorts of funky shapes. Sometimes they look like lil' people!
- **Lookalike alert!** Dandelions will never bolt into one large upright stem, which is a great way to tell the difference between them and their cousins, the wild lettuces! Wild lettuce *is* edible, but much more bitter, in my opinion.

DANDELION ROOT COFFEE

When I was a kid, I remember being told that dandelion roots could be used as a coffee replacement in times of financial duress. (Looking at you, Great Depression.) But as I grew older and became mildly obsessed with coffee, I was skeptical because often what people say tastes like coffee just tastes like sad, bitter water. I know I might get flack from the coffee purists, but of all the substitutes out there, roasted dandelion root coffee is truly the best. It boasts the roasty, toasty notes of the coffee you love and adds a bit of mocha flavor to the cup. For the biggest roots (and, therefore, the biggest bang for your energetic buck when digging these up), try to nab them before the dandelion plant flowers.

SERVES 2

30 dandelion roots, scrubbed, thin ends trimmed, and sliced into ½- to 1-inch chunks

Boiling water

Preheat the oven to 400°F. Line a baking sheet with parchment paper.

Spread out the dandelion root pieces on the prepared baking sheet and roast until browned and completely dry, about 30 minutes. Set aside.

For a lighter-tasting coffee, steep 2 tablespoons of the dried root in 1 cup of hot water for 10 to 15 minutes. For a stronger flavor, grind the dried root first, then proceed as above. Strain and enjoy.

Store any dried dandelion root in an airtight jar for up to 6 months.

DANDELION FLOWER FRITTERS

If you're looking to introduce children and/or picky eaters (shout-out to my adult picky eaters!) to eating dandelions, this is the way! In a recipe that celebrates their sunny floral flavor and adds everyone's favorite elements—crunch and fried things—you're given a delightful mouth celebration of springtime that even buttered-noodle eaters will love.

SERVES 6

¾ cup unsweetened nondairy milk (I like soy in this recipe)

1 Starch Egg (page 177)

1 tablespoon maple syrup

½ cup all-purpose flour

½ cup cornmeal

2 teaspoons sea salt, plus more for sprinkling

½ teaspoon ground cayenne

Neutral oil, such as sunflower or grapeseed, for frying

About 60 fresh dandelion flowers

In a large bowl, whisk together the milk, Starch Egg, and maple syrup until combined. Add the flour, cornmeal, salt, and cayenne right on top. Use your whisk to gently combine the dry ingredients, then whisk everything together until smooth. (This will save you the fuss of dirtying an extra mixing bowl. These fritters are not that serious!)

Heat a few inches of the oil to 325°F in a medium, high-sided skillet or saucepan over medium-high heat. Set a rack over a baking sheet and place it near your frying station.

Working in batches, pinch the stem end of the dandelion flowers and dip them briefly into the batter, swirling them around a bit to coat the petals. Shake off any excess batter and slowly, carefully drop the battered flowers into the oil. Avoid crowding the pan so the fritters cook evenly. Cook until you see a ring of golden brown around the outside of the flower, 1 to 2 minutes. Flip and cook for 1 more minute. If you're finding the fritters are browning too quickly, you can reduce the heat slightly.

Use a slotted spoon to remove the fritters and transfer them to the rack. While they're still hot, sprinkle with a bit of sea salt, then let cool. Repeat with the remaining dandelions and batter. Remove the pan from the heat.

These are great on their own or served with ketchup, or with the Garlic Mustard Aioli (page 64) if you want to get fancy!

MY BEST DANDELION GREENS

This recipe is meant to put to bed every bit of dandelion green slander I see on the internet (and it's a lot of slander). Put away all cries of "It's too bitter!" and enjoy the ample free greens that are likely available in your backyard and neighborhood. Blanching them first is the key to removing that bitterness. It's a real game changer that also unlocks flavors for so many other initially bitter greens and vegetables, like mustard greens! Garlic mustard greens! Dame's rocket! And even other stuff from the grocery store, like escarole! And broccoli rabe! The flavors here are sort of Asian-leaning but you can swap the soy and maple syrup for extra-virgin olive oil, garlic, and a squeeze of lemon. Or go super simple and just season with salt and pepper.

SERVES 2

Sea salt, for blanching

6 cups packed dandelion greens, roughly chopped

3 tablespoons avocado or canola oil

2 teaspoons soy sauce

2 teaspoons maple syrup

2 teaspoons pickling liquid from Pickled Magnolia Flowers (page 36) or rice vinegar

Sesame seeds, for serving

Bring a large pot of salted water to a boil over medium-high heat. Add the chopped greens and cook until just tender but still bright in color, 1 to 2 minutes. Drain immediately and rinse under cold water. Squeeze out as much excess water as possible and set aside.

Heat the oil in a medium skillet over medium-high heat. Add the greens and stir briefly to coat them in the hot oil. Add the soy sauce, maple syrup, and pickling liquid and cook, stirring frequently, until the liquid has reduced slightly and become a bit saucy, about 5 minutes. Remove the pan from the heat.

Sprinkle the sesame seeds over the greens and serve immediately.

VIOLETS

PURPLE IS SAID to be a color of royalty, and violets really do deserve a crown. (A cute, tiny one!) Violets are a member of the viola family and related to pansies, which are also edible. In both cases, though, don't go munching at your local nursery because you don't always know what they've been treated with. The ones in your garden are fair game!

The purple color in violets is due to anthocyanins—yes, the same guys that make butterfly pea flowers do that cool blue-to-pink color-changing trick when exposed to an acid! For what it's worth, they also change to teal when exposed to a base, but it's hard to find things that are basic and taste good to drink. But I guess if you like the taste of baking soda, I won't stop you!

Violets like to grow in large patches, so gathering enough flowers in spring to make a bottle of color-changing syrup or lemonade is easy peasy, and a whole lot cheaper than buying color-changing tea online. And I don't know if the syrup actually tastes like artificial grape flavor, or if the purple is just really good at fooling my feeble, nostalgia-addled lizard brain, but it's delicious either way!

Did you know that the leaves of the violet plant are also edible? Okay, sure, they don't change color, but not all edible things have to be magic and exciting. Sometimes your girl just wants a salad or a cozy green soup. (And, yes, there's a soup in this chapter!)

I frequently get asked if yellow or white violets will also make color-changing syrup. Since they're not purple, it means they don't have that magical anthocyanin, which means, sadly, they don't do the color-changing thing! You can, however, still candy them (page 53) for beauty-licious food and drink accessorizing. Different regions have different species that vary in flavor, but as long as it's purple, it'll still do the thing!

WHAT TO LOOK FOR:

- The delicate **flowers** emerge shortly after the leaves, and while they are usually deep purple, you can also find white, yellow, pink, and combinations of all three, and all are edible. Much like their (also edible) cousins the pansies, violet flowers consist of five petals, with the top center petal often being bigger than the rest. If you're lucky enough to live where *Viola odorata* grows, it has a distinct grape candy smell, but all violets have more flavor than their fragrance would lead you to believe.
- The forest green, heart-shaped **leaves** of violets are often a declaration that spring is here to stay! They tend to grow in dense clumps, as they spread by rhizomes (horizontal underground stems), and don't have much of a smell other than "green" when broken.

CANDIED VIOLETS

If you ever woke up and said to yourself, "Wow, I wish I lived in Victorian England!" 1) No you don't, tuberculosis is crazy! And 2) You can pretend you're in Victorian England by making and decorating with these candied violets. They are traditionally made with egg whites, but I've given you a vegan alternative so everyone can enjoy them! And really, the flavor is no different. These look so dang cute on top of loaf cakes, cupcakes, or muffins. You can also place them on top of ice cubes to make your summer drinks look ever-so-fancy.

MAKES 20

20 violet flowers

½ cup aquafaba (the liquid from cooked chickpeas, canned is fine!)

1 cup granulated sugar

Preheat the oven to its lowest heat, ideally around 150°F and no more than 200°F. (Many ovens have a "keep warm" function that can be adjusted to a lower heat.) Line a baking sheet with parchment paper.

Lay the violets on the parchment, spacing them evenly. Use a pastry brush or even your finger to apply a thin layer of the aquafaba to the surface of each violet.

Place the sugar in a small bowl. Using a small pair of tongs or your fingers, carefully dip the painted side of each violet in the sugar and return it to the baking sheet.

Bake the violets until crisp and dry, rotating the baking sheet occasionally to help the violets dry evenly, anywhere from 2 to 4 hours, depending on your oven.

Store the candied violets in an airtight container for up to 1 week.

COLOR-CHANGING VIOLET SYRUP & VIOLET LEMONADE

Whoever said chemistry is boring has never done this science experiment! The pigment that gives many violets their signature deep-purple color is pH-reactive, which means the pigment will change colors when mixed with an acid or a base. If you're lucky (or unlucky) like me, and your water is very hard, you get to enjoy the color of basic violets, which is a deep blue. This can also be achieved with baking soda, but that tastes yucky and we're not doing that. The real magic happens when the pigment is introduced to a nice acidic juice, like lemon! It takes the color from a rich, deep purple to a Barbie hot pink. I think the children call this "yassifying." Very fun to serve to friends! Everyone will gasp in wonder when you pour the violet syrup into a pitcher of lemon juice and water. You're welcome for the ego boost—you deserve it!

Violet Syrup

MAKES ABOUT 3 CUPS

2 cups loosely packed fresh violet flowers

2 cups just-boiled water

2 cups granulated sugar

Add the violets to a clean, quart-sized jar. Carefully pour the hot water over the flowers and set aside to cool to room temperature. Seal the jar and refrigerate overnight.

Strain the pigmented liquid into a medium saucepan. Use the back of a wooden spoon or spatula to press gently on the flowers to release any remaining liquid and pigment.

Set the pan over medium-low heat and add the sugar. Gently heat the mixture, stirring frequently, until the sugar is nearly dissolved, 3 to 5 minutes. Remove the pan from the heat and continue stirring to help the residual heat dissolve the remaining sugar. (This will help prevent overheating the liquid and losing any flavor.)

Transfer the syrup to a clean, quart-sized jar. Let the syrup cool completely, seal the jar, and store in the refrigerator for up to 1 month.

RECIPE CONTINUES

Violet Lemonade

MAKES 6 CUPS

2 cups cold water

2 cups fresh lemon juice

2 cups Violet Syrup

Ice, as needed

In a pitcher, combine the cold water and lemon juice. Now, gather your guests around—a very important step!—and with a flourish, pour in the Violet Syrup and watch the color turn to a gorgeous shade of pink. Bask in adoration. Add ice to the lemonade and serve immediately.

VIOLET LEAF SOUP

I concocted this soup when my partner was quite literally bedridden with Covid. In addition to having a butt-ton of vitamin C, violet leaves get delightfully slippy when cooked (*mucilaginous* for the people who like gross words! Sorry!), and can be very soothing to the throat. It ended up being so good that we kept the recipe in rotation in our house and I think you might too. If you're presenting this to a picky or skeptical eater, you can tell them to think of it as a smooth cream of broccoli soup. And if you've got any extra greens or ramp leaves lying around, they're great in here!

SERVES 6

2 tablespoons extra-virgin olive oil

1 small yellow onion, chopped, or 1 cup chopped field garlic

6 cups packed violet leaves (a few rogue violet flowers are okay!), roughly chopped

Sea salt and freshly ground black pepper

4 cups vegetable stock

½ cup unsweetened nondairy cream, like coconut or almond

Heat the oil in a heavy-bottomed soup pot over medium-high heat. Add the onion and cook, stirring frequently, until translucent and fragrant, 5 to 7 minutes. (If you're using field garlic, cook until fragrant and bright green, about 3 minutes.) Add the violet leaves and season generously with salt and pepper. Toss to coat the greens in the oil and onions and cook, stirring, until the leaves are just wilted, about 5 minutes.

Add the stock and bring the mixture to a simmer. Reduce the heat to low and simmer until the greens are very tender and the soup is thickened slightly, about 20 minutes. Remove the pot from the heat.

Carefully transfer the soup to a high-speed blender and puree until completely smooth. (Depending on the size of your blender, you may want to do this in batches. An immersion blender is great here too.)

Return the pureed soup to the pot, stir in the cream, and adjust the seasoning as needed.

Divide the soup among the bowls and serve.

MORELS

FIRST, A CONFESSION: I used to think morels were vastly overrated. Then I ate my first morel!

Morels are everything you want in a mushroom and nothing you don't. Perfectly savory, packed with earthy umami flavor, they don't get slimy when cooked—what else could you ask for in a tiny clump of protein escaping from the earth?

If you're on the eastern side of the United States like me, we have a fairly distinct morel season in the spring and then we don't see them again until the next year. If you're on the West coast, you have a morel season too, but depending on the moisture-temperature balance in your microclimate, you might have a much longer season than we do out east. You likely also get burn morels! That is, morels that will grow in soil that's been recently disturbed by fires. (The weather where I live in Ohio is already wild, but I gotta confess, burn morels have gotten me thinking more than once about moving west.)

Now, since this is the first fungus we're dealing with in this book, I want to clear the air. Plucking versus cutting mushrooms of any type does not matter as they are simply fruiting bodies attached to long pieces of mycelium, and if I catch any of you arguing about it on the internet, I'm not gonna do anything but I will be silently disappointed! Pluck or cut it, you do you.

Part of what makes the morel so magical is that, unlike oysters and shiitakes, we have not yet figured out a way to grow them ourselves on demand. In other words, the conditions morels need in order to fruit are not replicable by us silly humans, and I love that. Honestly, I hope we never do figure it out so we've got a solid ten months a year to yearn for them.

Now, I recommend this for any foraged mushroom: Make sure you give these guys a hearty swish around in some vinegar- or salt-spiked water, then let them dry out thoroughly, or dry-fry out the extra moisture before cooking them however you prefer. This is especially important for morels, which are hollow on the inside and therefore ideal homes for lil' spiders and other teeny forest fauna.

As for where to find morels, there's a lot of advice from a lot of people and a lot of it differs. Helpful, right? Some folks say they grow on the edges of wooded areas—is that true, or are we just more likely to *see* them at the edge of a wooded area? Other people swear morels like to congregate at the base of certain types of trees, such as oaks, elms, ash, aspens, apple trees, and most other hardwoods. And as I mentioned, disturbed ground and burn sites are also primo morel-spotting spaces!

Morels have a pretty distinctive look, so much so that it's nearly impossible to mistake them for anything else, except maybe gyromitra (see below). Which means although they might be elusive, they're a great mushroom for a beginning forager.

WHAT TO LOOK FOR:

- With their pitted, usually oblong **caps**, morels are one of the most recognizable fungi! They can vary in color from cream to gray to brown and are hollow inside (which makes them so nice for stuffing!).
- Morels have a cream-colored **stem** (or **stipe**, in fungi-speak), and are also always hollow all the way through when sliced.
- **Lookalike alert!** Slicing through the mushroom to check that it's hollow is a great way to make sure you don't end up with a morel's main lookalike, gyromitra, which have solid stems and more of a wrinkled brain look to their caps. Many gyromitra are also edible after processing, but the cooking process requires a bit more finesse, so it's good to make sure you can tell the difference.

STUFFED MORELS

You can stuff these bad boys with any nut of your choosing—I adore acorns but I get that acorns and morels are not at all in season at the same time in half the country, and not everyone has a giant acorn storage unit like . . . some people (it's me! I'm some people)—so we're using walnuts as the guiding light here. This recipe is for the times when you feel like a woodland Scrooge McDuck, swimming in a pool of morel abundance, and you want to show off your haul a little bit, but in a low-key way. This is a forager's quiet luxury dish.

SERVES 6

½ cup walnut halves

8 fresh sage or mugwort leaves

1 garlic clove or ¼ cup roughly chopped field garlic

2 tablespoons extra-virgin olive oil, plus more for drizzling

Sea salt and freshly ground black pepper

½ cup panko bread crumbs

¼ cup grated vegan Parmesan cheese or nutritional yeast

12 large morel mushrooms, cleaned and halved lengthwise

Preheat the oven to 400°F. Line a baking sheet with parchment paper and set aside.

Add the walnuts to a small, dry skillet and set over medium heat. Shaking the pan frequently, toast the walnuts until just fragrant, about 5 minutes. Remove the pan from the heat.

In a food processor, add the toasted walnuts, sage, garlic, and olive oil, and season generously with salt and pepper. Pulse just a few times, until the mixture resembles coarse crumbles. Transfer the mixture to a medium bowl, add the bread crumbs and Parmesan, and stir to combine. If the mixture looks a little dry, add another tablespoon of olive oil and stir again.

Set the morel halves on the prepared baking sheet, drizzle with olive oil, and season with salt and pepper. Bake until the edges of the morels sizzle slightly, 8 to 10 minutes.

Carefully spoon a few teaspoons of the stuffing mixture on each warm mushroom. Return the morels to the oven and bake until the stuffing is just golden, another 10 minutes.

Drizzle the stuffed mushrooms with olive oil and serve immediately.

MY PLAIN & SIMPLE MORELS

Morels don't need a lot to shine! This recipe is super easy and lets them take their rightful place center stage. Know how sometimes porcini mushrooms can taste a little bit porky? Morels' flavor can pull a little bit beefy sometimes, which makes this dish a substantial showstopper!

SERVES 2

2 tablespoons unsalted vegan butter, plus more for serving

1 cup morel mushrooms, cleaned and halved

1 teaspoon fresh lemon juice

Sea salt

Field garlic, fresh chives, and/or parsley, chopped, for serving (optional)

Melt the butter in a large skillet over medium heat. Add the morels and cook, stirring occasionally, until they have softened and started to brown along the edges, 7 to 10 minutes. Add the lemon juice and cook for another minute. Remove the pan from the heat.

Salt the morels to taste and add a bit more butter if desired. Toss with the field garlic, chives, and/or parsley, if using, and serve immediately.

MOREL POLENTA FRIES
with Garlic Mustard Aioli

Unlike almost every reality show couple, polenta and mushrooms really are a match made in heaven! This recipe works with any mushroom, so if you, like so many of us (me included!) went on a morel hunt and came back with one, three, or zero morels but everything else growing in the woods, this recipe is still here for you! The dipping sauce is so easy, it's just the Garlic Mustard Pesto (page 85) and some vegan mayo! You can sub out any pesto or condiment and whiz it up into that mayo for a dreamy dip with these crunchy beauties.

MAKES ABOUT 20 FRIES

FOR THE MOREL POLENTA FRIES

Nonstick cooking spray, for the pan

3 tablespoons unsalted vegan butter

4 cups chopped fresh morels

Sea salt and freshly ground black pepper

1 large garlic clove, chopped, or 2 tablespoons chopped fresh field garlic

2¼ cups vegetable broth

1½ cups unsweetened nondairy milk

1½ cups dry polenta

2 teaspoons nutritional yeast

Extra-virgin olive oil, for brushing

FOR THE GARLIC MUSTARD AIOLI

1 cup vegan mayonnaise

3 tablespoons Garlic Mustard Pesto (page 85), or store-bought pesto

MAKE THE FRIES: Spray a 9 x 13-inch baking sheet or straight-sided baking pan with cooking spray and set aside.

Melt 2 tablespoons of the butter in a large nonstick skillet over medium-high heat. Add the morels and toss in the butter. Cook, stirring frequently, until the morels release their liquid, about 5 minutes. Season generously with salt and pepper and reduce the heat to medium-low. Cook, stirring occasionally, until the morels are browned and a bit crispy on the edges, about 10 more minutes. Add the garlic in the final minute of cooking and stir through, cooking until fragrant. Remove the pan from the heat and set aside.

In a medium saucepan over medium-high heat, add the broth and milk and bring to a simmer. Stream in the polenta and whisk vigorously to eliminate any lumps. Reduce the heat to medium-low and continue whisking and cooking for 5 minutes. Reduce the heat to low, cover the pan, and cook for another 5 minutes. Taste the polenta; it may be fully softened at this point. If not, whisk again, cover, and cook for another 5 minutes. Remove the pan from the heat.

Stir in the remaining tablespoon of butter and season generously with salt and pepper. Fold in the morels and garlic and taste once more for seasoning. Pour the polenta in the prepared pan and smooth into an even layer. Cover and chill for at least 2 hours or overnight.

Preheat the oven to 425°F. Line a baking sheet with parchment paper and set a rack on top. Set aside.

Flip the pan with the polenta onto a cutting board. Cut the cooled, solid polenta in half widthwise, then into about 10 rows of thick fries—you'll end up with about 20 pieces.

Place the polenta fries on the rack, leaving about ½ inch of space between the fries. Brush each fry generously on all sides with olive oil. Roast for about 20 minutes, until the edges begin to crisp. Flip the fries and return to the oven for another 15 to 20 minutes, until crispy all over. Set aside to cool slightly.

MAKE THE AIOLI: In a small bowl, whisk together the mayo and pesto until well combined. Serve the fries with the aioli alongside for dipping.

JAPANESE & GIANT KNOTWEED

FEWER PLANTS STRIKE more fear in the hearts of environmentalists and gardeners than Japanese and giant knotweed. Boasting the ability to both withstand most mild herbicides and grow multiple inches a day in the spring, these guys are edible, thank goodness, or this would be a much more disastrous story.

Originally brought to North America as a cute ornamental plant, knotweed is among many other such plants that are now a scourge upon humanity, such as, I don't know, maybe a little guy you've heard of called giant hogweed? Or Bradford pears (page 208)?? Anyway, knotweed is so dang good at the reproduction game that cuttings can take root and start new colonies. Cool, but also terrifying in, like, a zombie plant way? Which is why I always remind folks that when you harvest knotweed, you need to eat all of it, or boil or burn what you don't eat, and don't drop any bits on your walk home!

Rhubarb fans, rejoice: Japanese and giant knotweed both taste incredibly bright and lemony due to their oxalic acid content. And, because of this acidity, knotweed lends itself very well to both savory and sweet preparations.

As far as how they look, these adventurous members of the buckwheat family look similar to bamboo, thanks to their visible joints and hollow stems. They are green with signature small red dots and highlights, especially around the joints. An easy way to scout them for the coming season is to look for the dried canes from the previous year's growth. That's also a good indicator that the area hasn't been sprayed with heavy pesticides by your town, city, municipality, or whoever's taking care of that space.

E D

My test for whether a piece of knotweed is tender enough to cook is to see what happens when you try snapping it off the plant. If it's tender, it'll snap off with an incredibly satisfying knock or *pop* sound. If you have to struggle with it or wrench it, it's too fibrous! I mean, you should still pull it out and burn it, it's just too tough for your teeth bones.

WHAT TO LOOK FOR:

- Knotweed **stems** start as tiny pink shoots just peeking out of the ground, but soon become green with small red spots and blushing along their telltale joints. I often find them along river and creek beds, but they'll honestly grow anywhere they're given opportunity. So, like, *please* don't drop any while you're harvesting!
- Knotweed plants have heart-shaped **leaves**, red markings, and water-filled empty cavities—basically, the thing that goes *POP* when you snap a stem. Knotweed can reach rather tall heights, and folks often confuse them with bamboo, but their snapping stems and leaves make them pretty distinctive.

KNOTWEED GRANITA

What this dish lacks in looks, it more than makes up for in flavor—a phrase that I have maybe said about myself! This knotweed granita is a great way to use a lot of an invasive plant. Once when I served this at a dinner, guests were understandably very apprehensive about this dish sitting in front of them; it does look for all the world like a dish of frozen swamp water. But! At the end of the night, everyone asked for the recipe. Now, if you'd rather not have to cajole your dining companions into eating this as is, you can strain the puree through a fine-mesh sieve before freezing it. This will yield a very beautiful pink juice that people will not believe came from such an annoying plant.

SERVES 6

1½ cups chopped tender Japanese knotweed stems and leaves

1 cup granulated sugar

Pinch of Acorn Starch (page 176) or cornstarch

In a medium saucepan over medium-high heat, combine 2 cups of water, the chopped knotweed, and the sugar and bring to a simmer. Reduce the heat to medium-low and cook until the knotweed is tender and the resulting syrup is tart, about 10 minutes. Remove the pan from the heat and stir in the Acorn Starch.

Carefully transfer the mixture to a blender and blend until completely smooth.

Pour the blended mixture (which will absolutely look like Shrek's swamp) into any baking dish that will fit in your freezer. Carefully place the dish in the freezer and freeze for 1 hour. Use a fork to scrape and break up the ice crystals, return the dish to the freezer, and repeat until the mixture is no longer slushy.

Transfer the mixture to freezer-safe containers and save for a cool, sweet, tart treat during the dog days of summer.

LEMONY KNOTWEED POPPYSEED SCONES

Knotweed's mucilaginous qualities (I'm sorry! But that's the word!) make these scones extra moist and fluffy. We absolutely lost our minds the first time we tasted these! Like, the idea sounded great in theory but we figured we'd need a few rounds of testing to nail it. The scones were so good on the first try that we had to make a concerted effort to save some for my partner, who not only raved about them but finished them for breakfast the next morning! These taste like a classic, tender, lemon poppyseed scone, but they harness the natural twang and sweetness of knotweed for a flavor that's familiar yet totally unique. Make sure you're harvesting the tender knotweed stems and leaves—the stem should snap audibly when you pick it. We devoured these before we could even get around to making a glaze, but I've included an easy recipe for it to make these even more special.

MAKES ABOUT 10

FOR THE SCONES

2 cups roughly chopped tender Japanese knotweed stems and leaves

¾ cup granulated sugar

Grated zest of 1 lemon

2 cups all-purpose flour

4 teaspoons baking powder

¾ cup (1½ sticks) cold unsalted vegan butter, cut into ½-inch cubes

½ cup unsweetened soy milk

2 tablespoons poppyseeds

FOR THE GLAZE (OPTIONAL)

1 cup powdered sugar

1 tablespoon unsweetened soy milk

½ teaspoon vanilla extract

Candied Violets (page 53), lilacs, or forsythia flowers, or additional poppyseeds, for topping (optional)

MAKE THE SCONES: Line a baking sheet with parchment paper and set aside.

In a medium saucepan over medium-high heat, add the knotweed and 2 teaspoons of water. Cover and let the knotweed steam, stirring occasionally, until it wilts and is a vibrant green color, 3 to 5 minutes. Remove the pan from the heat and let cool completely.

In a large bowl, add the sugar and lemon zest and use your fingers to rub them together to release those citrus oils. Add the flour and baking powder and whisk to combine. Add the butter and toss it gently in the dry mixture to coat. Cut or rub the butter into the dry mixture so it's uniformly clumpy. (This will make sense when you see it!) Set aside.

In a cheesecloth or nut milk bag set over a small bowl, place the cooled cooked knotweed. Tie off or twist the bag closed and gently squeeze over the bowl to strain and extract the liquid from the knotweed—you'll need just ½ cup for this batch. Discard the remaining cooked knotweed.

In a medium bowl, combine ½ cup of the knotweed liquid with the milk and poppyseeds.

RECIPE CONTINUES

Add the knotweed-milk mixture to the flour mixture and use a spatula to gently fold and combine, checking for any unincorporated dry spots. If you find your mixture is too wet, add a dusting of flour. If it's too dry, dribble in a teaspoon of the knotweed liquid. Cover the dough and chill for at least 30 minutes and up to overnight.

Preheat the oven to 400°F.

On the prepared baking sheet, dollop ¼-cup–sized scoops of the dough, spaced about 3 inches apart. (If you run out of room on the baking sheet, save the remaining dough to bake off after the first round of scones.)

Bake until the scones are golden and fragrant, and a toothpick comes out clean, 10 to 15 minutes. Let the scones rest on the pan, undisturbed, for 5 minutes.

MAKE THE GLAZE (OPTIONAL): In a small bowl, whisk together the powdered sugar, milk, and vanilla until smooth. Drizzle all over the scones.

Top the scones with edible flowers (if using), and serve warm or at room temperature.

KNOTWEED DIP

Knotweed sure can do a lot of things well. Sometimes she's sour, sometimes she's sweet, sometimes she's savory—take that, Sour Patch Kids! This is one of my favorite ways to take advantage of the fact that cooked knotweed gets a little, well, slippy; something I hear a lot of my followers rightfully complain about. This recipe turns that textural bug into a feature by making a dip that's smooth and dreamy, ready for all your chips and veg! Here, we're using only stalks since the leaves can get a bit stringy.

SERVES 4

¼ cup neutral oil, such as sunflower or grapeseed

2½ cups chopped tender knotweed stalks

1 tablespoon sea salt or Field Garlic Salt (page 221)

Pinch of garlic powder or 1 teaspoon chopped fresh field garlic

Chips, crusty bread, or crudités, for dipping

Heat the oil in a large skillet over medium heat. Add the knotweed stalks and cook, stirring frequently, until they're very tender but before they begin to brown, about 7 minutes. Remove the pan from the heat.

Stir in the salt and garlic powder and use a fork to mix and mash the stalks into the oil to form a loose dip consistency.

Serve warm with chips, bread, or crudités.

MUGWOR

MUGWORT IS THE NAME used to refer to several numbers of the *Artemisia* genus, which grow globally across the northern hemisphere. In this book I'm usually referring to our sweet baby *Artemisia vulgaris,* which feels like a disrespectful species name because ain't nothin' vulgar about her. Now, a more famous cousin of hers that you may be more used to seeing in the kitchen is your run-of-the-mill garden sage, and because of how closely they're related, their aromatics are extremely similar and they can be used more or less interchangeably in recipes with just a few notes. In the Mugwort Green Tea recipe (page 77) I go over the process of what, in China, is called the "kill green," which is necessary if you are drying your mugwort to keep it from getting distastefully bitter like the sneaky minx she is.

These recipes can also be used with most native *Artemisia* species, though make sure that you are not, say, going and harvesting a bunch of white sage in a place you're not supposed to. That's rude!

In many East Asian and South Asian countries, mugwort and its cousins are used as a green and an herb, whereas in many European countries it has historically been used to flavor alcoholic beverages, especially before the popularity of hops as a beer flavoring. Many of us are likely familiar with the most famous artemisia-based spirit, absinthe—which, no, will not make you hallucinate, contrary to what the critically acclaimed film *Moulin Rouge!* may lead you to believe.

Also, I will note that especially in the last two hundred years of Western herbalism, mugwort has often been maligned due to some folks worrying the thujone in it could be abortive in high quantities. However, I'm here to tell you that no sane person would eat or drink enough mugwort to be dealing with thujone issues so please, please, please, with a wild cherry on top, do not make me wrong.

Now, if you don't mind, I'm going to go make myself a cup of mugwort green tea and watch *Moulin Rouge!* for the thirty-seventh time.

WHAT TO LOOK FOR:

- She's known as the "Queen of Changes" due to her **leaves** changing shape as she grows! When small, the leaves are lacy and somewhat fernlike, with a silver underside. As the plant gets close to flowering a couple of months later, the leaves become long and lancelike in shape. Mugwort grows anywhere people have disturbed the ground, like at the edges of yards, fields, homes, etc.
- **Lookalike alert!** Some confuse mugwort with ragweed or Spanish needles, but when broken, mugwort sports a signature sage smell (which makes sense, as it's in the same genus as the sages!).

MUGWORT GREEN TEA

My internet friend Shell (@wild_food_around_the_world) showed me this method from her native China. We're treating the leaves in a manner similar to how green tea leaves are treated to combat excess bitterness and develop sweet, sagey flavor. As far as what constitutes a tender top, if it's too tough to snap off with just your hand, it's too tough for this! Use a few dried leaves when you want green tea but maybe don't want the caffeine that comes with it!

MAKES ABOUT 1 CUP OF DRIED LEAVES

2 cups loosely packed tender mugwort tops or leaves, rinsed in cold water

Place a steamer basket in a medium saucepan. Add about 1 inch of water, place the pan over medium-high heat, and bring the water to a simmer. Place the mugwort leaves in the basket and steam for 2 minutes. Remove the pan from the heat and remove the basket from the pan.

Working with one leaf or frond at a time, gently but firmly flatten each leaf between your hands and place it in a large skillet. Repeat until there are enough leaves to form a single layer in the pan. Set the skillet over the lowest heat possible (this is where that back burner comes in handy!) and carefully watch as the leaves begin to dry. Use a finger or a wooden chopstick to maneuver the leaves around the pan until they're evenly dried. Transfer the dried leaves to a rack to cool.

Repeat with the remaining mugwort leaves, flattening and drying in batches as needed. Let the leaves sit on the rack overnight to finish cooling.

Transfer the dried leaves to a glass jar and seal tightly. Store in a cool, dry place for up to 6 months. For tea, use 2 to 4 leaves per cup of just-boiled water.

MUGWORT MATCHA MUG CAKE

This is a light, herbal-y cousin to those genius microwave mug cakes all over the internet. It's cozy, it's sweet, it's honestly a perfect complement to movie night on the couch. It does call for the dried mugwort leaves from Mugwort Green Tea (page 77), but fear not, you can also just use store-bought matcha in a pinch! How to make Mugwort Matcha? Put your mugwort green tea leaves in a spice grinder, set to fine, and away you go!

SERVES 1

3 tablespoons all-purpose flour

2 tablespoons powdered sugar

1½ teaspoons ground Mugwort Green Tea (page 77)

Pinch of sea salt

Pinch of baking soda

2 tablespoons unsweetened nondairy milk

1 tablespoon avocado or sunflower oil

In a microwave-safe mug, use a fork to whisk together the flour, powdered sugar, Mugwort Green Tea, salt, and baking soda. Add the milk and oil and mix again until smooth.

Microwave for 1 minute, let cool slightly, and enjoy warm. This is lovely on its own, but a little scoop of ice cream never hurt anyone either.

POTATO MUGWORT PANCAKES

Potatoes and sage are a love story as old as cooking itself. But here, I argue that potatoes and mugwort should be a chapter in that story too! Mugwort's sweet, sagey aromatic qualities, along with its underlying herbaceous, green, and even pleasantly bitter flavor, perfectly round out a fatty, crispy potato pancake. This is a great way to use a non-native plant that often enjoys creeping into spaces where it was not invited; you may . . . invite it in (heyo!!) after trying this dish.

MAKES 8 PANCAKES

2 medium russet potatoes, peeled and shredded on the large holes of a box grater

1 cup packed fresh mugwort leaves, chopped

¼ cup all-purpose flour

1 teaspoon sea salt

½ teaspoon freshly ground black pepper

Neutral oil, such as sunflower or grapeseed, for the pan

Flaky salt, for serving (optional)

Vegan butter or nondairy sour cream, for serving (optional)

Place the shredded potatoes in a clean kitchen towel and gather up the corners of the towel. Twist the towel tightly to squeeze out the excess water from the potatoes.

Transfer the potatoes to a large bowl and add the chopped mugwort leaves, flour, salt, and pepper and mix well to combine.

Heat a thin layer of oil in a large skillet over medium-high heat. Scoop the potato-mugwort mixture into a baseball-sized mound and place it in the hot oil. Let this cook for 1 minute, reduce the heat to medium, and use a spatula to flatten the mound into a pancake. Cook until deeply golden, 4 to 5 minutes, and flip. Cook for another 4 to 5 minutes. Transfer the pancake to a rack and sprinkle with flaky salt, if using. Repeat with the remaining potato-mugwort mixture. Remove the pan from the heat.

Cut the pancakes into wedges. Serve with vegan butter or sour cream if you like.

GARLIC M

GARLIC MUSTARD WAS introduced to North America by Western European settlers who used it as both food and medicine. In the United Kingdom, garlic mustard goes by the very adorable name of Jack-by-the-hedge. Cute, right? At least more imaginative than garlic mustard. If I could rename it but make it accurate for North America, it'd be Jack-by-the-oh-my-God-it's-everywhere.

That's right, this member of the brassica family, cousin to all-stars such as Brussels sprouts and kale, is incredibly invasive in the eastern half of North America. And its cousin, black mustard, is very invasive on the western half of the continent. An invasive, garlicky species for everyone!

The silver lining is, once you learn to recognize it, you'll see it everywhere. So, you really can go crazy with it, don't hold back!

But the big ol' gray cloud surrounding that silver lining is this: Not only is garlic mustard excellent at producing hundreds of seeds per plant, and not only do those seeds have an extremely high germination rate, but its roots are allelopathic, which means they release chemicals into the ground that suppress their competitors. Basically, they're real jerks to other plants.

So, I implore you, when you come across it, pull it up! All of it! Even if you don't plan on eating it! And please do be careful when you're walking by while it's in flower or seeding, lest you accidentally track some of those abhorrent seeds into other green spaces.

The other silver lining when it comes to garlic mustard, and the reason it's got its own section in this book, is that it's very tasty. In my opinion, the greens definitely belong in the bitter greens category, and I do think they have more going on than, say, a dandelion leaf, but I don't think they deserve to be made outcasts by people who shun everything that doesn't taste like arugula or spinach.

USTARD

I do recommend that you try a little bite of garlic mustard greens raw, if only to gauge your bitterness tolerance and see if you prefer them blanched before you use them. (And, yes, to feel like a cute little bunny rabbit, munching away on a wild green!)

WHAT TO LOOK FOR:

- Garlic mustard **leaves** can be hard to spot in young plants, as they blend in so well with their surroundings. In the plant's first year, it appears as a low-lying rosette of circular leaves with deep-set veins. Some folks say the leaves look similar to those of violets, but the depth of the veins and the rosette growing pattern are a great way to quickly tell the difference. Also, when the stem is broken it smells strongly of garlic, hence its name!
- In its second year, the rosette bolts, the leaves on the bolting stem become more pointed, and it sports **white flowers** with four petals in true brassica fashion. These flowers become lengthy seedpods filled with hundreds of tiny seeds, which helps them spread so prolifically.

GARLIC MUSTARD PESTO

A fresh and feisty pesto for a feisty plant! And by *feisty*, I mean an invasive meanie that would like to take over so many of our woodlands. So, make this pesto often! You're helping the ecosystem by doing so. Pestos are a super gateway method to use greens that you might be unfamiliar with, and you're welcome to swap out a portion (or all!) of the garlic mustard leaves for whatever other green you have lying around, like chickweed or carrot tops.

MAKES ABOUT ½ CUP

2 cups packed garlic mustard leaves

¼ cup chopped field garlic greens or 1 garlic clove

¼ cup walnut halves and pieces

1 tablespoon nutritional yeast

Sea salt and freshly ground black pepper

½ cup extra-virgin olive oil

Prepare an ice bath and set it next to the stove.

Bring a medium pot of water to a boil over medium-high heat. Add the garlic mustard leaves and cook just until they turn a vibrant green color, about 1 minute. Remove the pot from the heat.

Quickly transfer the blanched greens to the ice bath and let sit for 2 minutes. Drain thoroughly and use your hands to squeeze as much excess water from the greens as you can.

In a food processor, add the blanched greens, field garlic, walnuts, nutritional yeast, and salt and pepper to taste. Pulse until just combined. With the processor running, stream in the olive oil and process until smooth and combined, about 1 minute.

Transfer the pesto to a jar, seal, and store in the refrigerator for up to 1 week.

GARLIC MUSTARD SCISSOR-CUT NOODLES

This recipe was originally going to be dumplings, each one filled and shaped by hand. Cute! Beautiful! But maybe not a project everyone wants to embark on. Instead, we turned to a method from northern China that results in a simple but delicious noodle with the texture of a chewy dumpling. Now you get the childlike joy of just snipping pieces of dough directly into boiling water! We've doubled down on the garlic mustard here: The leaves are in both the dough and the sauce. Bonus, if you've got tender new garlic mustard flowers, they're a beautiful garnish! Who knew that invasive species control could be so elegant? A note about the garlic mustard plant itself: It changes as it "ages." In early spring, the whole plant is fair game—this is a great time to harvest the root if you'd like to use it. (Scrub it well; a toothbrush works great!) As the season goes on, the stem becomes woody, so just use the leaves and flowers if it's late spring.

SERVES 2 TO 4

FOR THE NOODLES

1½ cups all-purpose flour

½ cup finely chopped garlic mustard leaves

1 small garlic mustard root, scrubbed and finely chopped (optional)

Sea salt

FOR THE SAUCE

1 tablespoon extra-virgin olive oil, plus more for serving

3 garlic cloves, thinly sliced

3 cups packed garlic mustard leaves, roughly chopped

Sea salt and freshly ground black pepper

Grated zest of 1 lemon

¼ cup walnuts, finely chopped

Garlic mustard blossoms, for garnish (optional)

MAKE THE NOODLE DOUGH: In a medium bowl, combine the flour, chopped garlic mustard, garlic mustard root (if using), a pinch of salt, and ½ cup of water and mix to combine. Turn out the dough on a work surface and knead it gently just until it's smooth and a little springy, about 3 minutes. Return the dough to the bowl, cover, and set aside for 15 minutes.

MAKE THE SAUCE: In a large skillet over medium heat, heat the oil. Add the garlic and cook until fragrant, just about 1 minute. Add the chopped garlic mustard, a pinch of salt, and a few grinds of pepper. Cook, stirring frequently, until the greens have mostly wilted down, about 3 minutes. Remove the pan from the heat.

Cook the noodles: Fill a large pot with water and a big pinch of salt. Bring the water to a boil over medium-high heat.

Return the pan with the olive oil and wilted greens to medium-low heat.

Holding the ball of dough in your non-dominant hand and a pair of kitchen scissors in your dominant hand, snip off small

RECIPE CONTINUES

pieces of the dough directly into the boiling water. It helps to rotate the ball slightly as you go. Once all the dough is cut and the noodles are all floating to the top of the water, remove the pot from the heat.

Use a slotted spoon or spider to transfer the cooked noodles directly into the pan with the wilted greens. The starchy water clinging to the noodles will help form a sauce with the wilted greens and garlic. Shake the pan gently to help coat the noodles, and use a spatula or wooden spoon to stir and flip them. If the noodles seem dry, add a splash more cooking water. Add the lemon zest and toss to combine. Taste and add salt and a few cracks of pepper as needed. Remove the pan from the heat.

Divide the noodles among bowls and top with a drizzle of olive oil, the chopped walnuts, and garlic mustard blossoms (if using).

STUFFED GARLIC MUSTARD LEAVES

Sometimes you eat a dolma and think longingly, "Y'know, I wish this leaf brought a little more flavor to the table." Worry not! These brassica leaves bring a hefty garlicky punch, immediately tempered by a soft and tasty rice filling, all wrapped up in a very cute and tiny package. It's a perfect little snack or appetizer. Have I convinced you to eat all of the garlic mustard yet? Because there is so much out there! I like to warm the walnuts here in a pan just until they're toasty and fragrant, but it's not mandatory.

SERVES 4 TO 6

Sea salt and freshly ground black pepper

About 40 fresh garlic mustard leaves

Juice of ½ lemon

¼ cup walnut halves

1 cup hot cooked rice (white or brown is fine)

1 tablespoon unsalted vegan butter, melted

Extra-virgin olive oil, for drizzling

Bring a medium pot of salted water to a boil over medium-high heat. Add the garlic mustard leaves and cook until just wilted, about 1 minute. Remove the pot from the heat.

Transfer the leaves to a colander and drain well. Squeeze the lemon juice over the leaves and use your hands to toss and coat the leaves in the lemon juice. Set aside.

In a food processor, add the walnuts and pulse until they resemble a coarse meal, with the largest pieces about the same size as a grain of cooked rice.

In a medium bowl, combine the rice, ground walnuts, and butter, and season generously with salt and pepper.

Working with one garlic mustard leaf at a time, flatten the leaf and place the stem end closest to you. Place a 1-teaspoon-sized portion of the filling in the center of the leaf. Fold the outer edges of the leaf over the filling, followed by the bottom edge of the leaf. Continue rolling from the bottom up to make a tiny packet. Transfer the stuffed garlic mustard leaf to a plate and repeat with the remaining ingredients.

Drizzle the stuffed garlic mustard leaves with the olive oil and serve.

U
S
M

WOOD EAR MUSHROOMS

I LOVE WHEN A PLANT'S common name makes sense. Wood ear mushrooms look like ears and they grow on wood. Refreshingly simple! But did you know that in Japan, one of the names for wood ears translates to "wood jellyfish"? Still makes a lot of sense, but is a lot more playful and I love it.

Now, members of the *Auricularia* genus have been eaten readily for thousands of years in China, Japan, and Korea, but are a more recent addition to the American and European culinary foodways. We were late to the party!

For many Western palates, I find that adding something wiggly and cartilaginous (sorry, but that is absolutely the word!) to a dish can sometimes be surprising, because it's not a texture that frequently shows up in our cuisine. But please give our girl a chance! She holds on to flavors like a dream, and the melt-in-your-mouth cartilaginous-ness is such a welcome addition to noodle dishes, wilted green salads, or—wait for it—gummy candy!

When it comes to wood ears, my recipes tend to pull inspiration from their treatment in Eastern Asian cuisine, so if you're on the prowl for more recipes because you've decided you can't get enough of this cartilaginous mushroom, I'd start there. How many more times can I say "cartilaginous"? Cartilaginous! It's like exposure therapy! I'm getting you comfortable with the idea of a food that feels like cartilage. You're welcome?

If you are stumped on where to find our good wiggly buddy, they are a wood-rotting species that, at least in Ohio, I typically find on dead or dying box elders near creeks and streams.

WHAT TO LOOK FOR:

- Wood ears **grow** directly from dead or dying wood. When you find one, you'll usually find lots! Don't be afraid to peel back some bark when you see a few; you'll likely uncover a few more that way.
- In terms of **appearance**, the top side can range from a dark-brown to a tan color, and the underside is generally a slightly frosted white. Since they're so, well, squishy, they don't necessarily stand up with a distinct stipe and cap the way other mushrooms do, so it's likely you'll see that top side first when you do find them.
- Wood ears are *very* wobbly to the **touch** when fresh. They'll dry to a crisp when there's not enough rain, but do not fear! That's just a survival tactic, and they'll rehydrate with the next spring or summer shower.

RAMEN BOWL with Wood Ears

One of the tried-and-true ways to serve wood ears is in a dish like this, showcasing their sort of chewy, wiggly texture. My partner, who is very anti-mushroom in all things, actually loved ramen with wood ears when it was served to us at a Japanese restaurant in NYC. Granted, he took the first bite because he didn't want to be rude, but then he finished the whole dang thing because he loved it! Other than in the woods, the place you're most likely to encounter fresh wood ears is probably in an Asian grocery store or international market. If you can't find any fresh, those same places will sell them dried, and it's very easy to rehydrate them! The broth here is on the lighter, summery side, but you can give it a more fall/winter vibe by using the maitake mushroom base from the Cream of Mushroom Soup (page 188). This recipe does use a bit of tahini to make the broth slightly creamy and a little richer, but the soup is still great if you want to omit it!

SERVES 2

1 tablespoon neutral oil, such as sunflower or grapeseed

4 scallions, sliced, white and green parts separated

1 cup sliced fresh wood ear mushrooms

3 garlic cloves, chopped

1 tablespoon grated fresh ginger

4 cups vegetable broth or ¼ cup Wild Summer Greens Stock (page 143) mixed with 4 cups water

2 (3-ounce) blocks dry ramen noodles

2 tablespoons tahini (optional)

2 teaspoons soy sauce, plus more

1 teaspoon mirin, plus more

Chili oil, for serving (optional)

In a medium soup pot over medium heat, heat the oil. Add the white parts of the scallion and cook, stirring occasionally, until translucent, about 3 minutes. Add the mushrooms, garlic, and ginger and cook, stirring frequently, until fragrant, about 2 minutes. Add the broth, increase the heat to high, and bring to a simmer. Add the noodles and cook according to the package instructions, usually just a few minutes. Remove the pot from the heat.

Stir in the tahini (if using), soy sauce, and mirin. Taste and adjust the seasonings, adding more soy sauce for salty punch and more mirin for brightness.

Divide the soup between bowls, top with the reserved scallion greens, and drizzle with chili oil (if using).

WOOD EAR SOUR CANDY

Listen, this one's for the weirdos. And that's coming from a bona fide weirdo (it's me)! Not many people look at a mushroom and say, "I wanna candy that!" But us weirdos, we're trailblazers. So, if you too have witnessed the gelatinous consistency of fresh wood ears and said, "Hmmm . . . candy does that!" this one's for you! (Alright, you don't actually have to be weird to make these because they're really good and deceptively easy.)

MAKES 20 CANDIES

2 cups granulated sugar

20 fresh wood ear mushrooms, cleaned

Maple sugar or more granulated sugar, for coating

In a medium saucepan over medium-high heat, combine the sugar with 2 cups of water. Stir well to combine and bring the mixture to a low boil. Reduce the heat to low and simmer, stirring occasionally, until the sugar has completely dissolved, about 5 minutes. Remove the pan from the heat.

Add the mushrooms to the syrup and stir to coat. Cover the pan and let the mushrooms sit in the syrup overnight, or at least 6 hours.

Preheat the oven to its lowest heat, ideally around 150°F and no more than 200°F. (Many ovens have a "keep warm" function that can be adjusted to a lower heat.) Line a baking sheet with parchment paper.

Using a fork or a pair of tongs, remove the mushrooms from the syrup, shaking off any excess syrup. Place the mushrooms on the baking sheet, spaced evenly apart. Dehydrate in the oven until the mushrooms feel sticky to the touch, about 2 hours, checking every 30 minutes or so. Set aside to cool slightly.

Fill a small bowl about halfway with maple sugar. Roll and lightly toss the sticky candied mushrooms in the sugar and transfer to a plate.

Serve immediately or store in an airtight container in a cool, dark place for up to 1 week.

JUNEBER

JUNEBERRY. SERVICEBERRY. SHADBUSH. Sarvis. Sugarplum. Wild plum. Saskatoon berry. Chuckley pear. *Amelanchier*. One of the general rules of ethnobotany is that if a plant has a metric butt-ton of names, it is, and has probably long been, important to a lot of different people!

I was raised calling them *juneberries* because in Ohio, the berries emerge, drumroll please, in June! The name *shadbush* comes from the fact that at least in North America, shad (that's a fish, don't worry I had to look it up too) historically begin their yearly run through creeks, streams, and rivers at the same time that the shadbush blooms. It also is one of the earliest-blooming plants, typically flowering between late February and early May, depending on what growing zone you live in. There is a legend that says the name *serviceberry* came from the fact that the ground will have thawed enough to resume funeral services around when the flowers bloom. Although people haven't found significant proof that it's the real reason for this name, it's still a pretty metal thing to think about.

Because they're one of the earliest plants to flower, juneberries are also one of the earliest plants to fruit! (Okay, sometimes they're beat out by wild strawberries, but this ain't wild strawberries' chapter.) They are also extraordinarily common in city agriculture because cities, especially in northern parts of the United States or in Canada, love plants that add visual interest earlier in the year when things are typically drab and gray. What that means for you, my friend, is that there is a ton of delicious berries likely hiding under your nose! The best juneberries taste like a combination of sweet apples and blueberries, with a distinct almondy crunch from the edible seeds in the center. It's like if the best fruits and marzipan got married and had a very well-adjusted child!

RIES

Flavor does tend to vary a great deal from tree to tree, so taste around, find your favorites, and you can plan to harvest the most from those trees each year.

A final fun fact: Juneberries have been and continue to be a popular fruit to use in pemmican, the Native American food that many lovingly call the OG energy bar!

WHAT TO LOOK FOR:

- Juneberries are often one of the first plants to **flower** each spring, or even late winter (a great time to scout them before spring becomes overwhelming!). Flowers are small and white, with five petals on each bloom, and grow in clusters near the ends of branches. Closed flower buds are a grayish color and slightly pubescent, which is science-people for "hairy," not "angsty and listening to a lot of angry music and forgetting to put on deodorant."
- The **berries** are crowned and grow in bunches, usually three to six to a bunch. They don't all ripen at the same time, so you can visit the same tree over and over again during their all-too-short season.
- If you're scouting juneberry **trees** when they're not blooming or fruiting, you'll know them by their smooth, light-gray bark, often loved by lichens. The trees themselves are small, typically maxing out in cities around nine feet tall. Leaves are ovate (or oval-shaped!), with a light serration, and turn a brilliant red-orange in the fall.

JUNEBERRY LEAF TEA

Maybe you discovered a local juneberry tree (yay!) but just after the fruiting period (naw!). Don't worry, the tree still has many gifts for us! Using the traditional fermentation method that many leaves in the rose family go through for tea (like raspberry leaves), this recipe is a deceptively easy process and the result is herbaceous, sweet, summer in a cup.

MAKES ABOUT 1 CUP OF DRIED LEAVES

3 cups lightly packed fresh, green juneberry leaves

Working with one leaf at a time, gently but firmly press or roll the leaf between your hands. This will bruise the leaf slightly and help it to release its compounds and ferment. Add the leaf to a clean jar and repeat with the remaining leaves.

Seal the jar and store it in a dark place for about 24 hours to lightly ferment.

Preheat the oven to its lowest heat, ideally around 150°F and no more than 200°F. (Many ovens have a "keep warm" function that can be adjusted to a lower heat.)

Place the leaves on a baking sheet in an even layer. Dehydrate the leaves in the oven until completely dry and crisp, anywhere from 2 to 6 hours, depending on your oven and the leaves you've gathered. Set aside the pan to let the dried leaves cool completely.

Transfer the dried leaves to a jar, seal, and store in a dark place for up to 6 months. Use 3 to 4 leaves for each cup of hot tea, and 6 to 8 leaves for iced tea.

JUNEBERRY HAND PIES

The only thing that makes pie better is making it portable. These pocket pies are what I use to introduce new people to the world of foraged food. For example, this is one of the few recipes my very food-cautious father gets excited about! No one will think foraging is full of only vegetal flavors after this. I like my pie filling thiccc and jammy, due to fond memories of gas station mini pies with my mom, but you do you! Vegan butter, even straight from the fridge, can tend to soften up quicker than regular butter, so I like to freeze mine for 5 minutes or so before using.

MAKES ABOUT 10 PIES

FOR THE JAM

3 cups juneberries

1 cup granulated sugar

Juice of 1 lemon

1 to 2 tablespoons Acorn Starch (page 176) or cornstarch, depending on desired thickness

¼ teaspoon ground cinnamon

FOR THE PASTRY DOUGH

4 cups all-purpose flour, plus more for dusting

¼ cup granulated sugar, plus more for sprinkling

¾ teaspoon sea salt

1 cup (2 sticks) very cold unsalted vegan butter, cut into ½-inch cubes

1 cup ice water, plus more as needed

Unsweetened nondairy milk, for brushing

MAKE THE JAM: Combine the juneberries and sugar in a medium saucepan over medium-high heat. Bring the mixture to a boil and reduce the heat to medium. Cook, stirring frequently to help the berries and sugar break down, until the mixture reaches a deep, uniform color and a spoon dragged across the bottom of the pan leaves a slight trail, 7 to 10 minutes. Remove the pan from the heat.

Whisk in the lemon juice, Acorn Starch, and cinnamon. Set aside to let cool completely.

MAKE THE PASTRY DOUGH: In a large bowl, whisk together the flour, sugar, and salt. Add the cold cubed butter and toss it in the dry mixture to coat the cubes. Use your fingers to flatten the butter pieces into little shards, tossing the mixture occasionally to keep the butter coated, until all the pieces are flattened. Add about ½ cup of the ice water and use your fingers to gently toss the mixture—this helps introduce the water to the mix without overworking and heating up the butter. Add more water a tablespoon or two at a time until the dough is still a bit crumbly but holds together when pressed. You likely won't need more than 1 cup of ice water, but if your environment is very dry, you can add more water, just a splash or two at a time.

Turn the dough out onto a clean kitchen towel. Gather up the corners and sides of the towel to catch any stray bits of dough and twist the towel to help compress the dough into a little

RECIPE CONTINUES

pouch shape. Set the pouch on the counter and gently press it into a thick disc. Chill the dough for 30 minutes.

ASSEMBLE THE HAND PIES: Preheat the oven to 375°F. Line two baking sheets with parchment paper and set aside.

Divide the dough into 12 equal pieces and gently shape each piece into a ball. Set the dough balls aside under a kitchen towel to keep them from drying out.

Dust the counter with flour and roll a dough ball into about a 6-inch round. Spoon 2 heaping tablespoons of the juneberry jam in the center of the round and fold the dough over to make a half-moon shape. Press out any excess air pockets, working from the center out. Crimp the dough to seal the edges in any style you like—a fork works just fine! Set the finished hand pie on the prepared baking sheet and repeat with the remaining dough and jam.

Brush each hand pie with a bit of milk and sprinkle with a little sugar. Bake for 20 minutes, or until the pies are fragrant and the edges are golden brown. Let cool for about 10 minutes before serving.

JUNEBERRY OAT SQUARES

For the last few years, I've had a tradition of stashing a gallon bag of juneberries in the freezer so that on the winter solstice, I can get a taste of warmer days. These bars are simultaneously cozy, making them great for winter, and also light, fresh, and a reminder of warmer days to come. If you close your eyes, it's almost like it's June again. These are the weird rituals I have to do so I don't go absolutely crazy in deepest, darkest winter. Maybe you're the same way!

MAKES 9 BARS

FOR THE CRUST

1 cup all-purpose flour

1 cup old-fashioned oats

½ cup packed light brown sugar

1 teaspoon baking powder

Pinch of sea salt

Pinch of dulse seaweed flakes, optional

¾ cup (1½ sticks) unsalted vegan butter, softened

1½ teaspoons Nocino (page 199) or vanilla extract

FOR THE FILLING

2 cups fresh or frozen juneberries

3 tablespoons granulated sugar

2 tablespoons fresh lemon juice

2 teaspoons Acorn Starch (page 176) or cornstarch

FOR THE TOPPING

1 cup old-fashioned oats

¼ cup packed light brown sugar

3 tablespoons unsalted vegan butter, melted

2 tablespoons finely chopped walnuts

Pinch of ground allspice

MAKE THE CRUST: Preheat the oven to 350°F.

In a large bowl, whisk together the flour, oats, brown sugar, baking powder, salt, and dulse flakes, if using. Add the butter and Nocino and use a fork or your hands to work the butter into the mixture until it reaches a crumbly consistency.

Press the mixture evenly into the base of a 9 x 9-inch baking dish.

MAKE THE FILLING: In a medium bowl, toss together the juneberries, sugar, lemon juice, and Acorn Starch. Spread the mixture in an even layer over the crust in the baking dish.

MAKE THE TOPPING: In a medium bowl, add the oats, brown sugar, butter, walnuts, and allspice, working the mixture together with a fork until combined. Sprinkle the mixture evenly over the juneberries.

Bake for 30 to 40 minutes, until the topping is golden and the juneberries bubble slightly. Transfer the baking dish to a rack and let the bars cool completely before cutting.

MILKWEE

MILKWEED FLOWERS OFFER us one of those quintessential smells of summer. Many of us probably weren't even aware that they were responsible for the sweet aroma, a bit reminiscent of vanilla, wafting through the summer breezes, dancing around us as we enjoyed not being in school for a few months.

And while many people can recognize different milkweeds, many are surprised to find out that quite a few are edible and have been eaten traditionally here in North America for thousands of years! The most notable edible species are common milkweed, showy milkweed, and swamp milkweed. I've also heard that poke milkweed is edible, but it's a bit rarer where I live, whereas the other three are frequently planted in parks and public and residential gardens.

Now, maybe you're thinking, "But Alexis, what about the monarch butterflies?" A completely valid question! First, monarch caterpillars enjoy munching on milkweed leaves, which we are not using for any of these recipes. (And if you're a more advanced forager and plan to use milkweed shoots or leaves in a recipe, I implore you to learn what a monarch egg looks like so you can check your foliage thoroughly before consumption!!)

Second, while care for our dwindling monarch numbers is top of mind, I believe that telling more people that milkweed is edible and delicious will encourage folks to plant different milkweeds for their region, giving monarchs a better string of plants to harvest from and lay eggs on.

And third, I should also note that foragers are not the reason for the dwindling monarch numbers. Farms, extensively spraying all milkweed that shows up on their property and in hedgerows, are much more responsible.

I've heard many people's concerns about how much milkweed needs to be cooked. Fun story for any of us familiar with the seminal literary foraging classic *Stalking the Wild Asparagus*: It is believed by many modern-day ethnobotanists that our good buddy Euell Gibbons was actually snacking on dogbane, not milkweed. If the name doesn't say it all (seriously, who would want to drive dogs away!), dogbane is poisonous, milkweed is not. That said, they do look very similar when they are little shoots, which is part of the reason why I didn't include any shoot recipes in this book!

So, get out there, plant some milkweed, and feed the monarchs and yourself!

WHAT TO LOOK FOR:

- Fragrant milkweed **flower** clusters are globes made up of hundreds of small flowers in varying shades of pink and cream. Each flower looks like a little shooting star, with five pointed petals at the end of a long flower stem.
- The **seedpods** (best harvested when 1.5 inches or smaller) look a little alien. They're solid green with lumps, and are shaped like a teardrop. Winter is a great time to ID large stands of milkweed, as the dry seedpods often persist until the next growing season, and milkweed is a perennial so it'll come back in the same spot!
- **Lookalike alert!** Milkweed plants start out in spring as cute, chunky lil' shoots. The shoots will have small, oval leaves hugging close to the stem, growing straight up. The leaves and stem will be downy, which is a great way to make sure you're looking at common milkweed, not hemp dogbane (which is poisonous). Thankfully, a short while into the growing season, dogbane branches out wildly, differentiating it from milkweed's singular stalk from which all leaves, flower clusters, and seedpods emerge!

FRIED MILKWEED BUDS

If you need to wrap your mind around what this will look like, think of when chicken wings get veganized and then restaurants feel like they need to rename them using some weird, hard-partying misspelling of chicken wings, like chik'n wyngz or something. This is a milkweed wyng. Or at least, it's sort of drummy-shaped like a wing (sorry, wyng) but the interior is soft and creamy while the outside is crisp and crunchy. Excellent with any dipping sauce (I love a vegan ranch), or just your favorite hot sauce.

SERVES 4

About 2 cups fresh, young milkweed buds (1 to 2 inches in length)

2 cups unsweetened nondairy milk (I like soy milk for this)

2 cups fine cornmeal, plus more as needed

1 teaspoon sea salt, plus more for sprinkling

½ teaspoon paprika

½ teaspoon garlic powder

¼ teaspoon freshly ground black pepper

Neutral oil, such as sunflower or grapeseed, for frying

Dipping sauce of choice or hot sauce, for serving

Fill a large bowl with ice water and set aside.

Fill a large pot with water and bring to a boil over high heat. Add the milkweed buds and blanch for 2 minutes. Remove the pot from the heat.

Using tongs or a slotted spoon, transfer the milkweed buds to the prepared ice bath. Let sit for 5 minutes, until thoroughly cooled. Drain the milkweed buds.

In a large bowl, combine the blanched buds and the milk. Let the buds soak in the milk for 1 to 2 hours at room temperature.

In a low, shallow bowl, add the cornmeal, salt, paprika, garlic powder, and pepper and whisk with a fork to combine. Set aside.

Add about 2 inches of oil to a heavy, high-sided frying pan. Place the pan over medium-high heat and heat to 350°F. Line a baking sheet with paper towels and set it near your frying station.

Using a fork, remove a milkweed bud from the milk and roll it around in the cornmeal mixture, coating it thoroughly. Shake off any excess cornmeal and carefully place the battered bud into the hot oil. Fry until crisp and golden, about 2 minutes per side. Using tongs or a slotted spoon, transfer the fried milkweed buds to the baking sheet and sprinkle with salt. Repeat, working in batches, until all the milkweed buds are fried. Remove the pan from the heat.

Serve immediately with your dippin' sauce of choice or a few shakes of hot sauce.

MILKWEED NECTAR

Congratulations, you've found the most Fairy Princess recipe in this book. Not only does this let you imbibe the sweet scent of summertime flowers, it's also hot pink when it's done! It truly does taste the way common milkweed flowers smell, and if you've smelled milkweed flowers, you know this is the highest of compliments.

MAKES ABOUT 3 CUPS

2 cups cold filtered or distilled water (tap water can kill the yeast on the flowers)

2 cups fresh, whole milkweed flowers

2 cups granulated sugar

Juice of ½ lemon

In a large pitcher, combine the water, milkweed flowers, sugar, and lemon juice. Cover loosely with a clean kitchen towel (just to keep any debris or bugs from falling in) and let sit at room temperature for at least 48 hours, stirring a few times a day.

Strain and bottle the nectar based on how you'd like to use it (see below). Seal the nectar in an airtight bottle or jar and store in the refrigerator for up to 2 weeks.

If you strain and bottle the nectar just after the first 48 hours, the result will taste like an infused lemonade. Super fun for summer parties!

If you strain and bottle between 48 and 72 hours, the wild yeast will take off and you'll end up with a fizzy milkweed nectar–soda concoction. Also lovely for gatherings.

If you strain and bottle after 72 hours, you'll have a milkweed vinegar ready for fun shrubs and cooking.

STUFFED MILKWEED PODS

You're gonna want to make sure you gather these guys when they're the size of your thumb, or smaller. Any bigger than that and they'll be too tough; those lovely silks inside will be brown, not white; and you won't be able to harness their silky, creamy qualities. (If you do get any pods with brown silks inside, remove and discard them and sub a bit of vegan cheese, like mozzarella, into your stuffing mixture, and hold back a bit on the nutritional yeast.) Think of these as nature's jalapeño poppers, minus the heat.

SERVES 2 TO 4

- **8 small milkweed pods, the size of your thumb or smaller, halved lengthwise**
- **Extra-virgin olive oil**
- **Sea salt and freshly ground black pepper**
- **¼ cup finely chopped lambsquarters or fresh spinach leaves**
- **2 teaspoons nutritional yeast, plus more**
- **1 teaspoon garlic powder**
- **Field Garlic Salt (page 221) (optional)**

Preheat the oven to 375°F. Line a baking sheet with parchment paper and set aside.

Over a medium bowl, gently work open the milkweed pods with your fingers. Pull out the silks and seeds and add them to the bowl. Set aside.

Rub the insides of the milkweed pods with a small amount of oil and sprinkle with a bit of salt. Place the pods on the prepared baking sheet and set aside.

In the bowl with the milkweed silks, add the chopped lambsquarters, nutritional yeast, garlic powder, a sprinkle of Field Garlic Salt (if using), a drizzle of olive oil, and salt and pepper to taste. Mix well.

Using a small spoon, stuff the mixture back into the halved milkweed pods. Set the stuffed pods cut side up on the baking sheet and bake until the filling is fragrant and a bit bubbly, about 15 minutes. Set the oven to broil and broil the pods for another 2 minutes.

Serve immediately.

MULBERR

MULBERRIES ARE ONE of the most recognizable tree fruits in the global North, and yet we don't have an industry for them here in North America! Some old dead people did try, but not for the sake of the fruit—they brought a bunch of Asian mulberry trees to try to start a silk industry here in the States. It . . . did not work. But you know what did work? All those trees easily spreading across the country and rapidly hybridizing with our own native red mulberry to the point that most people have never seen a truly native-to-the-US mulberry tree before! We love dealing with the consequences of our forefathers' choices! JK, they weren't my forefathers. #NotMyForefathers #MineWereBusyFightingRacismNotSilkworms

These days, the fruits of both the non-native and native *Morus* species signal the true beginning of summer, with berries ranging from incredibly tart to nearly sickeningly sweet. A great way to identify a fruiting mulberry is by the slightly boozy-sweet smell of the fallen, fermenting berries beneath the tree. Which means you must also please-pretty-please keep an eye out for tipsy squirrels and birds nearby, and look up before harvesting your mulberries! You think I'm kidding and I am very much not! One year, one of those tipsy birds pooped directly on top of my head. Don't be like me.

Now, it's important to note that while the berries get all the attention, the leaves are also edible—both as greens and as a delicious green tea dupe! The tea is often drunk in Asian countries to help maintain healthy blood sugar levels. There's also a supercool recipe from the Chinese countryside for making tofu with mulberry leaves, wood ash, and water that is not in this book because I've never made it successfully. So, if anyone out there is great at translating different Chinese dialects and wants to be my best friend and help me with that recipe sometime, I would love that!

One of the things I love most about mulberries' particular flavor is how well it accompanies both sweet and savory

IES

dishes. Want to add some interest to a BBQ sauce? Mulberries! Want to add some tasty mystique to a pie? Also mulberries!

The best way, in my opinion, to harvest mulberries is to put a sheet you do not care about under the tree and shake it vigorously. The tree, not yourself (or the sheet, for that matter!). Fewer people will look at you weird if you're shaking the tree as opposed to just shakin' that thang underneath it.

WHAT TO LOOK FOR:

- The **leaves** vary in shape as the tree ages. Younger shrubs have more deeply lobed leaves, often looking like serrated hands with fingers splayed. As they get older, the leaf trends toward a more basic heart shape, but this isn't strictly true across all mulberries.
- The **fruits** are compound berries, like a blackberry, each growing from its own stem. Since the native red mulberry (*Morus rubra*) has become increasingly rare, look out for its non-native-but-hella-widespread cousin, white mulberry (*Morus alba*)! Ripe red mulberries are a deep wine color, while ripe white mulberries can range from golden to a light-pink or light-purple color. During their fruiting stage, red mulberry trees are especially easy to spot if you see a lot of purple stains on the pavement! You can also typically smell the fermenting fallen fruit from a block away, so, uh, beware drunk birds.
- **Lookalike alert!** The only other large tree I could see someone maybe confusing mulberry with is sassafras, but sassafras leaves lack serration on the edges and smell like Froot Loopswhen broken! Mulberry just smells like green!

MULBERRY BBQ SAUCE

Here's where we harness the mulberry's talent for adding a lovely twang, a perkiness, if you will, to savory stuff! You can use any tartish berry for this, like, if you come across some black raspberries on a hike, or if you've got some blackberries or raspberries languishing in the fridge. This recipe is fairly mild in terms of spice, so you're welcome to amp it up with extra cayenne, make it smokier with some chipotles in adobo, or make it sweeter!

MAKES ABOUT 1½ CUPS

2 cups mulberries, rinsed in cold water

1 cup vegetable stock

½ cup ketchup

¼ cup packed brown sugar

2 tablespoons apple cider vinegar

2 teaspoons chili powder

1 teaspoon smoked paprika

¼ teaspoon ground mustard

Pinch of ground cayenne

Sea salt and freshly ground black pepper

In a medium saucepan over medium-high heat, combine the mulberries and vegetable stock. Bring to a simmer and reduce the heat to medium. Cook, stirring occasionally, until the mulberries start to break down, about 7 minutes. Remove the pan from the heat.

Use a wooden spoon to mash the mulberries against the side of the pan to help them release their juices. Using an immersion blender, blend the mixture until smooth. (Alternatively, you can use a regular blender, just give the mixture a little time to cool down and make sure to vent the blender lid to release pressure from the heat!) Transfer the mulberry puree to a medium bowl.

Set a fine-mesh strainer over the same saucepan. Pour the mulberry puree into the strainer, gently pushing down with a wooden spoon or spatula to help the liquid move past the seeds and any other solids, pressing as much liquid as possible back into the pan. Discard the solids.

Return the pan to medium-high heat and add the ketchup, brown sugar, apple cider vinegar, chili powder, smoked paprika, mustard, cayenne, a good pinch of salt, and several cracks of pepper. Whisk thoroughly to combine and bring to a simmer. Reduce the heat to medium-low and cook, stirring frequently, until the sauce has thickened considerably, about 20 minutes. Remove the pan from the heat.

Use immediately or let the sauce cool completely and store in a sealed jar in the fridge for up to 2 weeks.

ROSES

AH, ROSES—THE GLOBAL symbol of love. Except maybe don't use the beautiful manicured rose bouquet from your local florist for these recipes, because you never know what chemicals they're being prepped with! These recipes are better suited for wild roses, be they the roses climbing the side of your home, the multiflora rose slowly eating entire swaths of forest beneath its weedy tendrils, or the beach roses that were brought here to help keep our dunes from falling apart but now are kind of spreading unchecked and they're so pretty that people just pretend they've been here the whole time.

If you needed another reason to please not use store-bought roses, a lot of store-bought roses have also been bred for beauty but not for scent. Wild roses, even the invasive ones, are beautifully scented and make for gorgeous additions to recipes in the form of rosewater, jams, or even just as dried petals applied at the end.

Going out in search of roses is also a wonderful opportunity to get to know the roses that are native to your area! And if you really like what you make with them, you can go back to them during rose hip season and nab some seeds for your own garden. But please, for the love of Pete, do not plant multiflora rose at your home! It will take over your whole garden and I will come and knock on your door and cry at you. That is the Alexis guarantee!

Roses have such a beautiful culinary history around the world, with many of my favorite recipes hailing from northern Africa and the Middle East. They really know their way around rosewater in those parts!

And just to get ahead of this, I do not want to see any letters, emails, or carrier pigeon messages asking why my rosewater method isn't just putting roses in hot water. If you learn nothing else from this book, I do hope you learn from the (many!) tea recipes that dunking plants in hot water makes—you guessed it—tea! Which is quite different from

the aromatic distillation that is rosewater (or orange flower water, another common flower water used in food). If you want to miss out on the most delicate aromatics, that's your business, but I'm out here trying to have your rosewater tasting full-bodied and nuanced like you're walking through a field of flowers. I will answer no further questions on the subject!

WHAT TO LOOK FOR:

You'll likely encounter one of two types of wild roses in your floral frolicking.

- **Beach rose**: If you are near a large body of water, like an ocean or the Great Lakes, congrats, you've got 'em! Grows as a thicket-forming shrub. *VERY POINTY. SO MANY THIN THORNS. RUDE.* Her **flowers** are typically hot pink, but can range from magenta to white. Very fragrant. Large rose **hips** with lil' crowns, available from mid-summer through late winter.
- **Multiflora rose**: Vining and very thorny. Delicate, small, white **flowers** with five petals that emerge in mid-spring. Very small, globe-shaped rose **hips** that are ready for harvest in late summer, but can be collected through winter (as long as the birds don't get them first!).
- **FYI**: I cannot stress enough that these roses are *not* the same type of roses you see in those supermarket long-stem bouquets! Remove that image from your mind before you head out looking for the roses we're talking about.

ROSE PETAL SUGAR

I kind of feel like Rose Petal Sugar should be called fairy dust, but I also feel like fairy dust refers to a drug I've never tried? So, maybe let's just agree to call it Rose Petal Sugar. Excellent for finishing sweet treats as cooking it will decrease its potency, so it's better as a finishing touch. But isn't what we love about floral flavors the fact that they're so fleeting?

MAKES ABOUT 1½ CUPS

1 cup packed fresh rose petals

1 cup granulated sugar

Preheat the oven to its lowest heat, ideally around 150°F and no more than 200°F. (Many ovens have a "keep warm" function that can be adjusted to a lower heat.) Line a baking sheet with parchment paper and set aside.

In a food processor, add the rose petals and sugar and process until the petals are fully broken down.

Spread the mixture out on the prepared baking sheet. Let the sugar dehydrate in the oven until completely dry, about 4 hours, depending on your oven, checking every 30 minutes or so. Let cool completely.

Store the rose petal sugar in an airtight container in a cool, dry place for up to 2 months.

ROSE LEAF SUN TEA

Just the phrase "sun tea" immediately evokes memories of long, sunny summer days. When you ferment leaves from edible plants in the rose family, they do this wild thing where the leaves go from kind of grassy and herbaceous (bleh) to elevated and more like a fruit-infused black tea (oh my!). My friend Michael of @edibleillinois says that his local wild roses yield a tea that tastes like Snapple Peach Iced Tea. I love peach Snapple, so this made the book!

MAKES 1 QUART

1 cup packed rose leaves, rinsed in warm water and dried

Honey or other sweetener (optional)

Lay the rose leaves out on a clean, dry kitchen towel and let them wilt a bit at room temperature for at least 1 hour and up to 3 hours.

Working with one leaf at a time, gently press and roll each leaf between your hands to help break down their cell walls. Add the pressed leaf to a clean quart jar and repeat with the remaining leaves. Seal the jar tightly and set in a cool, dark place (like a kitchen cabinet) to lightly ferment overnight. The leaves should darken a little and will smell a bit fruity.

Preheat the oven to its lowest heat, ideally around 150°F and no more than 200°F. (Many ovens have a "keep warm" function that can be adjusted to a lower heat.) Line a baking sheet with parchment paper.

Spread out the fermented rose leaves on the prepared baking sheet in an even layer. Dehydrate the leaves in the oven until fully dry, about 4 hours, depending on your oven, checking every 30 minutes or so.

At this point, you can store the dried leaves in a clean, sealed jar in a cool, dark place for up to 3 months, or you can go ahead and make your sun tea!

When it's time to make the tea, add ½ cup of the dried leaves (you should end up with about this much after all the fermenting and drying) to a clean quart jar. Fill the jar with water, seal it, and place it in the sun for 1 full day.

Strain out the leaves and sweeten the tea with honey, or as desired. Serve over ice.

ROSEWATER BAKLAVA

Baklava has been one of my favorite desserts since I was five years old, because that's what happens when you go to a Montessori school. Within the baklava recipe is my recipe for homemade rosewater, so you can feel like an alchemist puttering away in your woodland shop. Making the rosewater is easier than you might think, thanks to a setup that includes zero specialized kitchen equipment. The cold lid on top of the pan condenses the aromatic steam that bubbles up from the rose petals, and a bowl sitting on top of the petals in the pan catches the rosewater! Distillation! Just be sure to use the petals from wild roses for maximum aroma and flavor and to avoid the chemicals that are sometimes used on garden roses. If you haven't tracked down wild roses, you can absolutely use store-bought rosewater. Although it's not super-technical, this recipe does take a bit of time, but it's worth it! You can make the rosewater and the syrup up to about a week in advance.

MAKES 1 (13 X 18-INCH) PAN

FOR THE ROSEWATER

1 cup packed rose petals

2 cups ice

FOR THE BAKLAVA

2½ cups granulated sugar

Sea salt

1 tablespoon rosewater (homemade or store-bought)

1 cup extra-virgin olive oil, plus more for the pan

2 cups walnut halves (about ½ pound)

2 cups shelled, unsalted pistachios (about ½ pound)

2 teaspoons ground cinnamon

1 teaspoon ground cardamom

1 (16-ounce) package phyllo dough sheets, thawed

Rose Petal Sugar (page 125), for sprinkling (optional)

MAKE THE ROSEWATER: Add the rose petals to a medium saucepan. Add enough water to cover the rose petals and fill the pan halfway, and set the pan over medium-high heat. Set a small bowl inside the pan, sitting in the water and on top of the petals. The bowl should be wide enough to nearly reach the side of the pan but not touch it, and sit low enough so you can still cover the pan. (If your bowl collection does not include something that fits this admittedly very specific description, choose a bowl that sits low enough as described, then make a little collar for it out of aluminum foil that fits wide enough as described. This way, it will still catch the drops of rosewater that collect under the lid, and they can roll down into the bowl.)

Invert the lid and cover the pan so the handle is facing down into the pan, and the lid forms a little bowl. Place the ice on top of the inverted lid—you may not need all 2 cups to fill it. Bring the water to a light boil, reduce the heat to medium, and simmer undisturbed for 15 minutes. Remove the pan from the heat.

RECIPE CONTINUES

Carefully remove the lid, discarding the melted ice. The liquid inside the bowl is your rosewater! Smell it! Now carefully pour it into a clean jar. You can use this right away (you need only 1 tablespoon for the baklava) or seal the jar and store it in the refrigerator for up to 1 month.

MAKE THE BAKLAVA: In a medium saucepan, combine the sugar and a pinch of salt. Shake the pan a bit so the sugar is in an even layer. Add 1 cup of plain water (not the rosewater yet!), making sure there are no rogue sugar crystals stuck to the side of the pan. Set the pan over medium-high heat and bring to a simmer. Without stirring or shaking the pan, let the syrup simmer until the sugar has completely dissolved, about 2 minutes.

Add the rosewater, cover the pan, and reduce the heat to low. Let cook, undisturbed, for another 2 minutes. Remove the pan from the heat.

Carefully transfer the hot syrup to a jar and let it cool completely at room temperature. Seal the jar and chill if making the syrup ahead of time.

Preheat the oven to 350°F. Grease a baking sheet with a teaspoon of olive oil, using your hand to rub it into an even coating, and set aside.

In a food processor, add the walnuts, pistachios, cinnamon, cardamom, and a pinch of salt. Pulse just until the nuts are finely chopped. Transfer the nut mixture to a medium bowl and set aside.

Divide the phyllo sheets into two stacks. (So, if you've got 16 sheets, make two stacks of 8 sheets each.) Carefully place one stack of phyllo on the prepared baking sheet. Add the nut mixture, spreading it into an even layer. Add the remaining phyllo stack on top, arranging it so the corners mostly line up with the bottom stack.

Using a sharp knife, cut the baklava into small squares or diamond shapes, cutting all the way through both layers of the phyllo. (To make diamonds, cut in straight lines in one direction, then diagonally in the other.) Evenly pour the olive oil all over the top.

Bake until light golden and fragrant, 30 to 35 minutes. Immediately pour the syrup slowly and evenly all over the hot baklava. Some syrup might run out onto the side of the pan—you can spoon it back over on top until all the syrup is absorbed. Let sit for at least 3 hours and up to overnight before serving.

Sprinkle with rose petal sugar (if using), slice through the layers again if needed, and serve.

PURSLAN

PURSLANE, THE TRUE taste of summer! Here in the United States, she's best known for hanging out in the cracks of sidewalks during the warm summer months. But in Central and South America, where she's known as verdolagas, she is a popular green and for very good reason.

Purslane's tender leaves and stems have a tart flavor that increases with exposed sunlight during the day, and make for a perfect summer sandwich topper. She's also the land plant with the highest level of omega-3s! Yet, for some reason, we relegate her here to sidewalks and the edges of parking lots. It's a real princess-disguised-as-a-pauper situation, and I'm here to rectify it!

Purslane is one of the Spanish quelites, which in Mexico and Central America refers to a group of wild greens that frequently make their way onto tables during the summer and fall months. So, if you're looking for additional recipes after trying the ones in this book, I would highly recommend googling "verdolagas" or "quelites," clicking that Translate button if you need to, and seeing where the interwebs take you.

Esteemed New York City forager Marie Viljoen also has a slew of excellent purslane recipes both online and in her book *Forage, Harvest, Feast*. Am I allowed to talk about other books in my book?? I hope so, because I'm going to!

Purslane is truly one of those cosmopolitan girlies, and one of my favorites to teach about because it is found so prolifically all across North America, in cities and rural settings alike!

WHAT TO LOOK FOR:

- You'll find purslane creeping along the ground, with reddish **stems** that grow in little clusters.
- Since it's a succulent, the **leaves** are thick and oval-shaped.
- **Flowers** are very cute, small, and yellow, and show up a little later in the season.
- **Lookalike alert!** Purslane can be differentiated from its lookalike, the creeping spurges (it just sounds bad, doesn't it?), by breaking a stem. Spurges leak a white sap, but purslane has no leaking sap.
- Gather in the morning for milder greens, or in the evening for more tart greens!

EASY PEASY PURSLANE SALAD

The only thing that would make this one-bowl salad easier is just eating the purslane straight out of the ground. Where I live, purslane is a marker of high summer, growing in sidewalk cracks, along curbs, and probably in most people's yards. If you're having trouble finding it in your yard, I guarantee your neighbor is getting rid of it in theirs. Am I suggesting you commit petty theft? No! Please talk to your neighbors! And please eat purslane, you'll thank me later.

SERVES 2

2 tablespoons rice vinegar or the brine from Pickled Magnolia Flowers (page 36)

1 teaspoon Dijon mustard

¼ cup extra-virgin olive oil

Sea salt and freshly ground black pepper

3 cups loosely packed purslane leaves and tender stems, washed and dried

In a medium bowl, whisk together the vinegar and Dijon until combined. Add the olive oil and whisk again, until smooth. Season with salt and pepper to taste. Add the purslane and toss well to coat. Taste and adjust seasoning if needed. Serve immediately!

QUICK-PICKLED PURSLANE

If you're looking for a wild side dish that won't get you uninvited to the next BBQ, this is it! In the immortal words of 50 Cent from his seminal hit "In Da Club," "Switch the style up . . . watch the money pile up." Except, well, you're probably not getting paid to bring something to a cookout, but you will absolutely watch the compliments pile up. There's also a chance that you love this so much, you keep it for your own personal use, adding it to sandwiches or alongside fried things that need something acidic to cut through. Purslane does this cool thing where the acidic compounds inside will actually self-pickle it. (The call is coming from inside the house!) Don't be afraid to get creative with the flavors in your pickle—I sometimes add red pepper flakes or a bit of minced fresh ginger.

MAKES ABOUT 1 CUP

2 cups loosely packed purslane leaves, rinsed and dried

½ teaspoon garlic powder

¼ teaspoon sea salt

In a medium bowl, toss together the purslane, garlic powder, and salt. Let sit for 20 minutes at room temperature. Drain and rinse the purslane. Use immediately or store in an airtight container in the fridge for up to 3 days.

LAMBSQU

WOULD YOU BELIEVE me if I told you a single genus of plants has been feeding people worldwide since the beginning of recorded human history? She's fed ancient Greeks, she was one of the OG North American food crops in the Eastern agricultural complex, she's in the group of wild greens lovingly referred to as quelites in Mexico, and quinoa is a fellow member of her genus! Who is she? She's lambsquarters! And you probably have this gorgeous piece of food history in your very own neighborhood!

It makes sense why chenopodiums (that's botany nerd speak for *lambsquarters*) have been so important culinarily for thousands of years. It boasts several edible parts, and has leafy greens to spare during a time of year when things like spinach, kale, and lettuce are often too tough or too dead to eat!

Its leaves are tender and mild like those of its cousin, spinach, but unlike spinach, due to cute little drops of wax that help it maintain its moisture levels in extreme heat, lambsquarters leaves sparkle! Move over, Edward Cullen.

Also, for future you, and future you's plant-learning endeavors, know that the terms *lambsquarters*, *goosefoot*, and *chenopodium* are often used interchangeably. The name *goosefoot* comes from the leaf's shape, which often does look like a cute lil' goose's foot! I love when names make sense, don't you?

You can use lambsquarters leaves anywhere you'd use spinach, and often to better result. Get outta here, spinach artichoke dip! Her seeds are also tasty in porridge, or cooked like any grain, but unless you're lucky enough to be working with chenopodium quinoa (which is . . . quinoa), the seeds are often a lot of work for not a lot of food. Still fun to winnow, though!

ARTERS

WHAT TO LOOK FOR:

- Lambsquarters is easily recognizable due to the frosted appearance of its top **leaves**. While the sun makes it look covered in glitter, the shine actually comes from tiny bits of wax, which help the lambsquarters retain moisture on hot and dry days. That's one of the reasons the greens stay so tender all summer long! Many plants will also have twinges of purple along their slightly pointed leaves, but not all will.
- The **plants** themselves can get very tall, like five or six feet, if given the opportunity!
- Although we aren't using the **seeds** in this book, they're edible when cooked! Lambsquarters create green and red seed-pod clusters, which have tiny, shiny black seeds inside when you shake them. These seeds can be washed of their saponins and cooked, a reminder that lambsquarters is quinoa's North American counterpart!

LAMBSQUARTERS & ARTICHOKE DIP

Spinach and artichoke who?? We don't know her! One of the many things I love about lambsquarters is that she truly tastes like a green while being neither overly vegetal nor underwhelmingly bland. (Baby spinach, I'm looking at you.) She's approachable! But she brings flavor to the party. And although I get that no one eats spinach and artichoke dip for their health, I do love that lambsquarters also add the kind of nutritional punch to this treat that only a wild green can bring.

SERVES 4 TO 6

Sea salt and freshly ground black pepper

4 cups packed lambsquarters

2 tablespoons unsalted vegan butter

1 (14-ounce) can artichoke hearts, drained, rinsed, and chopped

3 garlic cloves, thinly sliced

2 tablespoons all-purpose flour

¼ cup shredded vegan mozzarella cheese, plus more for topping

¼ cup vegan cream cheese (I like Tofutti, the OG!)

¼ cup unsweetened soy milk

2 tablespoons nutritional yeast, plus more for topping

Crusty bread or crackers, for serving

Fill a large bowl with ice water and set aside.

Bring a large pot of well-salted water to a boil over high heat. Add the lambsquarters and blanch just until vibrant green, 1 or 2 minutes. Remove the pot from the heat.

Using tongs or a slotted spoon, carefully transfer the greens to the ice bath. Let sit for 5 minutes.

In a colander over the sink, drain the blanched greens. Use your hands to squeeze out as much excess water as you can. Set aside.

Preheat the broiler to 500°F.

In a medium oven-safe skillet over medium-high heat, melt the butter. Add the artichokes and cook, stirring, until warmed through, about 3 minutes. Add the garlic and cook until fragrant, another minute. Sprinkle the flour all over the mixture and cook, stirring frequently, until slightly thickened, about 2 minutes. Add the drained greens and stir to warm through, 2 or 3 minutes. Add a big pinch of salt and several grinds of pepper. Remove the pan from the heat.

Add the mozzarella, cream cheese, milk, and nutritional yeast. Stir to combine and let the residual heat in the pan melt the cheeses. Taste and adjust the seasoning if needed. Top with another sprinkle of mozzarella and nutritional yeast.

Broil until bubbly and golden on top, about 3 minutes.

Serve hot with crusty bread or crackers.

LAMB-SPANAKO-PITA

Weirdly enough, spanakopita was one of my favorite dishes to cook as a thirteen-year-old. (What thirteen-year-old makes spanakopita??) There was a project in Latin class where we all had to make Classical-inspired dishes, and I am a showoff and chose spanakopita. I've been eating it ever since! In place of the traditional spinach or chard, this version uses—you guessed it—lambsquarters! People often try to banish lambsquarters from their yards, not knowing it's actually a tasty, healthy powerhouse. But if you only find a handful, you can also just use a combination of greens in this dish. And you'll be making a shockingly easy vegan feta for that salty kick that makes spanakopita so good. If you can find fresh soy milk, it makes for a really great "curd" in this cheese. Check your local Asian or international market!

SERVES 6

FOR THE FETA

2 cups unsweetened soy milk, as fresh and unprocessed as you can find

Juice of ½ lemon

½ teaspoon ground sumac (optional)

Sea salt

FOR THE SPANAKOPITA

2 tablespoons extra-virgin olive oil, plus more for brushing

1 medium onion, chopped

4 garlic cloves, chopped

8 cups packed lambsquarters and/or spinach leaves (about 12 ounces), chopped

Sea salt and freshly ground black pepper

¼ cup chopped fresh dill

¼ cup chopped fresh parsley leaves

1 (16-ounce) package phyllo sheets, thawed

MAKE THE FETA: In a medium saucepan over medium-high heat, add the milk and bring to a rolling boil. Add the lemon juice, reduce the heat to medium-low, and stir to help curds form, cooking for 1 to 2 minutes. Remove the pan from the heat and let cool slightly.

Lay several layers of cheesecloth or a nut milk bag over a colander in the sink. Carefully pour the milk mixture into the cheesecloth and strain the liquid from the curds. Gather up the edges of the cloth or bag, gently twist, and squeeze out the excess liquid. Depending on how dry you'd like your feta, you can use this immediately or let it drain for up to another hour.

Transfer the feta to a large bowl, add the sumac (if using), and season to taste with salt. Set aside.

MAKE THE SPANAKOPITA: Preheat the oven to 350°F. Brush a 9 x 13-inch baking dish with olive oil and set aside.

In a large skillet over medium-high heat, heat the oil. Add the onion and cook, stirring frequently, until translucent and beginning to brown along some edges, about 5 minutes. Add the garlic and cook until fragrant, 1 more minute. Add the lambsquarters, a big pinch of salt, and several grinds of

RECIPE CONTINUES

pepper and cook, stirring occasionally, until vibrant green and just wilted, 5 minutes. Remove the pan from the heat. Stir in the dill and parsley.

Transfer the greens mixture to the bowl with the feta and mix well to combine. Set aside.

Carefully unroll the phyllo sheets into one flat stack. Cut the stack in half so the sheets will roughly fit inside the baking dish.

Lay one phyllo sheet into the prepared baking dish and gently smooth it out. Brush the sheet with olive oil and repeat with another sheet. Repeat with the remainder of one stack (so, half of the total amount of phyllo), brushing a thin coat of oil on each layer. Add the filling, using a spatula to smooth it into an even layer. Repeat with the remaining phyllo sheets, brushing each with oil, including (and especially) the top sheet.

Using a sharp knife, cut the spanakopita into squares—I like to make 12 larger squares, but you can also make 24 cute little ones. Bake until the top is golden, the spanakopita is fragrant, and you can hear the filling sizzling a little, about 1 hour. Let cool slightly and serve.

WILD SUMMER GREENS STOCK

In a shock to probably no one, I am a soup girlie to my core. If you've ever found yourself scouring the house for broth, this magic jar solves that problem. It's also a superb way to use a mishmash of summer greens you've successfully foraged but might not be sure what to do with. You're doing a sort of light salt fermentation on the greens here, "light" in that you're keeping the flavor green and fresh, but not going so far as to make it funky. A spoonful or two in a little pot of water means you get the goodness of summer even after the days start to cool down.

MAKES ABOUT 1 CUP

5 cups packed wild greens, any combination (see Note), about 8 ounces

⅓ cup sea salt, about 4 ounces

Another note about scaling: Your bounty of greens may vary greatly, so don't feel like you need to use these precise amounts. The general ratio of greens to salt is 2:1, meaning whatever your greens weigh (weight, not volume!), use half of that weight of salt. The rest of the process is the same! I give you volume here so it's easier to eyeball things as you're foraging. Just weigh your greens once you're back in the kitchen and go from there.

In a food processor, add the greens and salt and process into a smooth paste.

Transfer the mixture to a clean jar and seal tightly. Let sit at room temperature, sealed and undisturbed, for 3 days.

From here, you can use right away, or store in the refrigerator for up to 3 months.

NOTE: Lambsquarters can factor heavily here, which is why this recipe is in the lambsquarters chapter, but the mix is up to you and what you find. I like to use a combination of three types of wild greens: mild greens like lambsquarters, chickweed, purslane, and young sowthistle (before it gets bitter); tart greens like dock and wood sorrel; and alliums like ramp leaves, field garlic, and nodding onions. Try to use a relatively equal amount of all three groups, but don't sweat it if it's not exact. I'd avoid doing this with majority alliums (like, delicious, but Big Onion Energy, which is not always what you want), and I also leave out any brassicas, such as garlic mustard or burdock.

LINDEN

IF YOU LIVE in the global North, you've got linden! Sometimes they're called *basswood*. Sometimes they're called *lime*, which makes no sense at all, but okay! *Lime* is an alternate form of the Middle English *lind* so I guess it's the same as linden? But we already gave lime to another guy! And the American name *basswood* is not, as I had hoped, inspired by the fish, but instead refers to the fibrous bast of the inner bark.

Lindens are famed worldwide for their flower teas, and this tea has been used to treat anxiety for several hundred years. It also just tastes very good! Also, honey from bees who have pollinated linden has a famously delicate flavor. Also also, lindens are one of the trees with straight-up edible leaves. No prep necessary, just chomp 'em like a lil' giraffe!

They're a member of the mallow family, much like their cousin chocolate, who holds it down in the global South. This makes sense when you consider that linden fruit does have a nutty, chocolatey aroma. In fact, in 1700s France, studies were done to see if linden could be a locally grown, cheaper form of cocoa, but alas, they had trouble getting the flavor to linger. But now we are equipped with technology! Which means you get a beautiful, chocolate-adjacent Chocolate Linden Fruit Ice Cream (page 149) in this chapter too!

So, get out there and get to know your local linden! Cities love them as street trees, so you may have some closer than you think. In this chapter, you'll get to try a recipe for each of the edible parts of the tree: the flowers, the fruit, and the leaves!

WHAT TO LOOK FOR:

- Lindens have heart-shaped **leaves** that start off extraordinarily tender when they first unfurl. Early in the season they'll also sport small, light-green, streamer-like bracts that look like someone glued different leaves to the ends of all the branches.
- Clustered, downward-facing **flowers** that smell *HEAVENLY*. Get ready to negotiate with the bees!
- The flowers become round and downy **fruits** that eventually turn brown and fall from the tree.

LINDEN FLOWER TEA

A tea made with the flowers of linden/bass/lime (why do we keep giving things so many names??) has long been heralded for its antianxiety properties, but I just like it because it tastes good. So, go gather some flowers and pour yourself a cup of chill. When you're foraging for linden blossoms, pull on the stem a few inches back from the blossom so you get a nice cluster with maybe a few lil' leaves on it. You can also make this tea with fresh blossoms; I just like drying them so I have access to their loveliness year-round.

MAKES ABOUT 1 CUP DRIED BLOSSOMS

2 cups fresh linden blossom clusters

Preheat the oven to its lowest heat, ideally around 150°F and no more than 200°F. (Many ovens have a "keep warm" function that can be adjusted to a lower heat.)

Lay the linden clusters on the baking sheet in an even layer. Let them dry in the oven until completely dehydrated, anywhere from 3 to 5 hours, depending on your blossoms and your oven. Let cool completely.

Store the dried blossoms in a sealed jar in a cool, dark place. For 1 cup of tea, use about 1 tablespoon (about 1 blossom) of dried linden flowers or 2 tablespoons of fresh flowers. Steep for 10 to 15 minutes and sweeten if desired.

CHOCOLATE LINDEN FRUIT ICE CREAM

Did you know that linden and cacao are in the same family? They're both in the mallow family, so yes, they're cousins to marshmallows too! The mallow family really does have the dessert scene on lock. Here, we're relying on an oft-forgotten feature of linden fruit, which is that if you roast it, it gets very cacao-y. In the 1700s, the French looked into completely replacing chocolate with linden, but they didn't have refrigeration so it didn't work out. So now, when you make this ice cream, just think: I'm eating the ice cream Marie Antoinette wished she could have had. (I mean, I'm not a historian but that seems right, right?) A few notes: You can dehydrate and grind the linden fruit anytime in advance—just store it in a sealed jar in a cool, dark place until ready to use. Also, nondairy milks can usually be used interchangeably in recipes, but I like soy milk here because the proteins help build a nice ice cream texture.

MAKES ABOUT 1 QUART

4 cups green linden fruit, stems removed

4 cups unsweetened soy milk

2½ cups granulated sugar

1 teaspoon vanilla extract or Nocino (page 199)

Sea salt

5 tablespoons Acorn Starch (page 176) or cornstarch

The night before making the ice cream mixture, place the bowl of an ice cream maker in the freezer.

Preheat the oven to 350°F. Line a baking sheet with parchment paper.

Spread out the linden fruit on the prepared baking sheet. Roast, shaking the pan every 20 minutes or so, until the fruit is browned, fragrant, and feels dry, 45 minutes to 1 hour. Let cool completely.

In a food processor, add the dried linden fruit and grind into a fine powder.

Set a fine-mesh sieve over a medium bowl. Transfer the ground linden fruit to the sieve and, using a spoon or spatula, sift the powder into the bowl, leaving any larger pieces in the sieve. (If you find you have a lot of large pieces left, you can re-grind them and sift again.) You'll need about 1¾ cups of dried linden powder for this recipe, so if you end up with any extra, you can store it in a sealed jar in a cool, dry place for up to 6 months.

RECIPE CONTINUES

In a medium saucepan over medium-high heat, whisk together the milk, ground linden, sugar, vanilla, and a pinch of salt. While the mixture heats, add the Acorn Starch to a glass with ¼ cup of water. Whisk or combine with a fork until smooth and add the starch mixture to the pan. Bring to a simmer and reduce the heat to medium-low. Cook, whisking frequently, until the mixture has thickened and coats the back of a spoon, 7 to 10 minutes. Remove the pan from the heat. Taste and adjust the sweetness or salt. Let the mixture cool to room temperature and chill overnight in a sealed container.

Add the chilled linden mixture to an ice cream maker and churn according to the manufacturer's instructions. Serve the ice cream immediately or store in an airtight container in the freezer for up to 3 months.

LINDEN SHORT-BREAD COOKIES

Historically, during tough times in Western Europe, dried, powdered linden leaves were sometimes used to replace some or all of the wheat flour in a recipe. I wouldn't suggest replacing all of the flour here, unless you like sweet, swampy soup. Get ready for pretty, pale-green cookies and an herbal, matcha-adjacent flavor.

Here, you're cutting the cookies into sort of traditional bars à la Ted Lasso, but you can also roll this dough out and use cutouts if you're feeling frisky! And while I normally don't discriminate among vegan butters, make sure you're using the vegan butter sticks here (as opposed to a "buttery spread") as they behave most like regular butter.

MAKES ABOUT 30 COOKIES

20 to 25 medium linden leaves

1¼ cups (2½ sticks) unsalted vegan butter, softened, plus more for the pan

2¼ cups all-purpose flour

¾ cup powdered sugar

¼ teaspoon sea salt

½ teaspoon vanilla extract or Nocino (page 199)

Preheat the oven to its lowest heat, ideally around 150°F and no more than 200°F. (Many ovens have a "keep warm" function that can be adjusted to a lower heat.) Line a baking sheet with parchment paper.

Lay the linden leaves on the prepared baking sheet and place in the oven until completely dry and crisp, 1 to 2 hours, depending on your oven, checking them every 30 minutes or so. Let cool completely.

In a food processor, add the dried leaves and process into a fine powder. Transfer to a small jar and set aside.

Preheat the oven to 325°F. Grease a 9-inch square baking pan with butter and set aside.

In a large bowl, whisk together the flour, powdered sugar, salt, and ¼ cup of the powdered leaves. Add the butter and vanilla and use a spatula or your hands to form a crumbly dough. (You can also use a hand mixer at medium speed.) The dough may feel a bit dry, but should stick together when pressed.

Transfer the dough to the prepared pan and press it into an even layer. Poke holes all over the dough with a fork. Bake until fragrant and lightly brown along the edges, about 1 hour. While still warm, cut into 18 bars. Let cool completely in the pan.

Serve immediately or store in an airtight container at room temperature for up to 1 week.

GOLDENR

NOTHING HERALDS THE end of summer like the blooming of goldenrod. Its yellow firework flowers punctuate fields, forest edges, and even the sides of highways. Often maligned because people think it's the source of their late summer allergies (Which it's not! That's ragweed! Goldenrod's pollen is too heavy to affect your sweet lil' nose!), goldenrod is not only an excellent food source for lots of native fauna, it's also a great food source for us!

Like many plants in the aster family, goldenrod has a strong, cilantro-adjacent flavor, which you can tone down by blanching. But if you're one of those humans for whom cilantro tastes like soap, I have bad news: Goldenrod greens will taste soapy to you as well. See you in the next chapter, you sweet genetic anomaly, you!

The solidagoes (the genus that all goldenrods fall under) are nearly all native to North and South America and have become a bit invasive in other parts of the world. I recognize this is bad, but the smallest, pettiest part of me smiles knowing that we in the global West haven't only received invasives, we have caused a few too. But also, I am honestly so very sorry to everyone else in the world!!

Goldenrod flowers, while still a little cilantro-adjacent, also bring a lovely floral sweetness to dishes, and ferment into an equally lovely cordial. If you're not into ferments, you can also make a delightful sun tea with goldenrod flowers by simply leaving a bundle of rinsed flowers in a glass jar of water in direct sun for one day! It's refreshing, and contrary to what lots of folks say about allergies and goldenrod, is thought to help with inflammation, not cause it. Woo!

You can also try experimenting with throwing a few goldenrod leaves into dishes in place of cilantro. No more having to run to the grocery when you realize you want tacos, simply go on a neighborhood mosey and gather a few solidago leaves!

WHAT TO LOOK FOR:

- While its golden **flowers** in late summer are known worldwide, goldenrod is much harder to recognize while it's just a shoot.
- **Shoots** will be hairless and green with long, lancelike **leaves** that end in a rounded edge. Some species look decidedly different from others (looking at you, large-leaf goldenrod!), so get to know the species in your area!
- **Lookalike alert!** Goldenrod can be confused with some other asters, but has a very cilantro-adjacent smell when broken.

GOLDENROD SHOOTS

I'm gonna come right out and say that if you're a cilantro person, you will love this recipe. And if you're not, you don't have to make it! This is a simple way to enjoy another wild green, especially one that only gets recognized in the fall for its flowers. To make these, harvest the shoots in mid-spring or early summer when they're knee-height or shorter, and you want them to be tender enough that they snap when you break them. If you see flowers, it's too late! Make the Goldenrod Cordial instead (page 159), or the yummy Goldenrod Leaf Vinaigrette (page 160). If you're worried about a strong flavor, you can blanch the chopped shoots for an extra minute before sautéing.

SERVES 2 TO 4

- **Sea salt and freshly ground black pepper**
- **3 cups chopped goldenrod shoots, cut into 1-inch pieces**
- **1 tablespoon extra-virgin olive oil**
- **1 garlic clove, thinly sliced**
- **Red pepper flakes (optional)**
- **Lemon wedges, for serving**

Fill a medium bowl with a handful of ice and cover with cold water. Set aside.

Fill a medium saucepan with well-salted water and bring to a boil over medium-high heat. Add the chopped goldenrod shoots and cook for 1 minute, just until vibrant green. Remove the pan from the heat.

Using a slotted spoon or skimmer, transfer the shoots to the ice bath. Drain well and set aside.

In a medium skillet over medium heat, heat the oil. Add the garlic and red pepper flakes (if using) and cook until fragrant, 1 or 2 minutes. Add the goldenrod shoots, a pinch of salt, and a few grinds of pepper and cook, stirring frequently, until tender, 3 to 5 minutes. Remove the pan from the heat.

Serve warm with a squeeze of lemon.

GOLDENROD CORDIAL

I love flower ferments. Flowers are so fleeting, so anything that helps them last longer is precious to me. Goldenrod often gets a bad rap because it's confused with ragweed and its pollen, so hopefully this will help get goldenrod back in your good graces. Big ups to my friend Dwight, the flower cordial master of the Midwest, for the tip on how to kick-start the fermentation on this beautiful concoction: His trick is to add a bit of raw apple cider vinegar. So, if it's chilly where you live, or you just love the satisfaction of seeing that Things Are Happening in your ferment, add the apple cider vinegar.

MAKES ABOUT 1 QUART

4 cups distilled or filtered water

Heaping ½ cup goldenrod flowers

Heaping ½ cup granulated sugar

2 tablespoons raw apple cider vinegar (optional)

In a large glass jar or enameled pot, combine the water, goldenrod flowers, sugar, and vinegar (if using). Cover with a thin kitchen towel and secure the towel with a rubber band or string. Stir the mixture twice a day, every day, for 3 to 5 days, depending on the ambient temperature. If it's warm, the mixture will probably only need 3 days. If it's chillier, it'll be closer to 5. Important: You must stir this thing twice a day, every day, or mold will get to it! Ew.

Place a fine-mesh strainer over a medium bowl and strain out the flowers from the cordial. Transfer the cordial to a clean jar, seal it tightly, and store in the refrigerator. The cold environment will slow down the fermentation but not completely stop it, so this keeps well for up to 1 month before it gets too wild in there.

FALL SALAD
with Goldenrod Leaf Vinaigrette

I have been a self-proclaimed salad hater for a lot of my life, but then I figured out I was just eating bad salads with bad salad dressings. Instead, here's a good salad dressing to put on your very good salads! Paired here with kale and roasted squash for full-on fall vibes, it's good for all sorts of salads, especially ones that lend themselves to that cilantro-adjacent flavor.

SERVES 4 TO 6

FOR THE VINAIGRETTE

- 2 cups packed goldenrod leaves
- ½ cup extra-virgin olive oil
- 2 tablespoons rice vinegar
- 2 teaspoons honey
- 1 small shallot
- Sea salt and freshly ground black pepper

FOR THE SALAD

- 1 small delicata squash, cut in half lengthwise, seeded, and cut into ½-inch slices
- 2 tablespoons extra-virgin olive oil
- Sea salt and freshly ground black pepper
- 1 bunch Tuscan or green curly kale, leaves stripped and chopped
- ¼ cup pumpkin seeds

MAKE THE VINAIGRETTE: In a blender, add the goldenrod leaves, olive oil, vinegar, honey, shallot, a big pinch of salt, and several grinds of pepper. Blend until completely smooth. Taste and adjust the seasoning as needed, adding more vinegar, honey, or salt to taste. Transfer the dressing to a clean jar and set aside.

MAKE THE SALAD: Preheat the oven to 400°F. Line a baking sheet with parchment paper and set aside.

In a large bowl, toss together the sliced squash, olive oil, a big pinch of salt, and several grinds of pepper. Lay out the squash on the prepared baking sheet and roast until tender and browned along the edges, 25 to 30 minutes, flipping the slices halfway through. Set aside to cool slightly.

In the same large bowl, add the kale, 1 tablespoon of the vinaigrette, and a pinch of salt. Using your hands or a pair of tongs, massage the dressing into the kale until softened, about 3 minutes. Add the warm squash and a few more tablespoons of the vinaigrette and toss well to coat. Top with the pumpkin seeds and serve.

Store any leftover vinaigrette in a sealed jar in the refrigerator for up to 1 week.

CHICKEN

CHICKEN OF THE WOODS, which I will henceforth refer to as COW, is one of the most recognizable fungi, making it an excellent intro for new foragers. It's neon orange, often tinged with yellow or white, grows in shelves, and has pores (not gills)!

It grows on dead or dying trees, and is actually often what killed the tree in the first place. You can typically find them in the same place year after year, until their mycelium (fungus roots) have used up all the nutrition available to them in the dead wood. So, if you find one that's dried out, mark it in your maps app and visit it again but a smidge earlier in the year next time!

Here in my neck of the woods (Ohio), it is a late-spring through fall mushroom. It boasts a much longer fruiting season than a lot of its edible peers, and I think that contributes to its popularity. When you find it, you want it to be pretty bug-free and very juicy, usually in the first two to four days of fruiting. This will lead to the best-tasting versions of these recipes! You can use slightly older, drier specimens, just know they get tough pretty fast.

Also, a fun general tip for mushrooms but especially handy for COWs because they can sometimes be enormous: If you take home more than you can eat today and want to save some for later, you can either pickle the mushrooms using a basic hot brine, or sauté them in an oil of your choice and freeze them. We'll get into both of those avenues in this chapter! If you freeze them fresh, you will not like the results—their water content makes them gummy and sad when they thaw. They also never seem to bounce back well from being dehydrated whole, but if you do dry them, you can then powder them (as in the Maitake chapter, page 182) and use as an umami seasoning in lots of dishes!

Another important note for COWs in particular is to make sure you cook it all the way through. No quick sears for these guys! No mushroom carpaccio here! There have

OF

E WOODS

been lots of rumors that COWs growing on certain trees will make you sick to your stomach, but what is really happening is the mushrooms weren't thoroughly cooked. In fact, most wild mushrooms, if not cooked thoroughly, will give you tummy troubles—it's too much chitin! For a while it was believed that only certain types of trees caused the reaction, but science reflects that the stomach problems are usually caused by the chitin (pronounced KAI-tin), a type of polysaccharide found in the structures of mushrooms and the exoskeletons of crickets. Cool, and a little gross!

I tend to find COWs on downed hardwoods, usually oaks and maples, but you can find them on lots of other trees too. Every once in a while I get lucky and find them on downed pines!

COW gets its name from the chicken-like consistency of its flesh and how, like chicken, it really takes on the flavor of whatever you cook it with! It's such a fun mushroom, and its bright colors make it a great deal easier to find than some of its counterparts (I'm looking at you, morels). Happy mushroom hunting!

WHAT TO LOOK FOR:

- One of the easiest fungi to identify, chicken (chickens?) of the woods **grows** as neon orange shelves on dead and dying wood.
- Most have a **bright yellow underside**, with the exception of *Laetiporus cincinnatus*, which has a white underside.
- They have **pores, not gills**, and tend to come back in the same spots year after year until they have broken down all the tasty bits from the wood they're growing from. Make sure to leave some behind to spread spores and make new chickens elsewhere! I also love them because in late summer and early fall, before leaves begin changing, you can spot them in the woods from a mile away.
- **Lookalike alert!** I am so for real about the pores-not-gills thing because jack-o'-lantern mushrooms can be mistaken for chicken of the woods. Jack-o'-lanterns have gills (not pores) and they glow in the dark. They will make you poop and throw up. Maybe even at the same time. Don't eat them!

BBQ CHICKEN (of the Woods) SANDWICH

In my neck of the woods, members of the *Laetiporus* genus begin fruiting in early June and typically continue fruiting until fall. So, you don't need to wait until fall to make this recipe, but if you've spent the summer hunting down other things, the COWs are still here for you when the morels and wood ears have long disappeared. And if you happened to make the Mulberry BBQ Sauce (page 121) in summertime, it too could be ready to slather on this sandwich. I love this recipe not only because it's delicious, but also because it's very important to cook chicken of the woods thoroughly, and by shredding it and increasing the exposed surface area, you ensure it gets cooked all the way through when you sauté it. Mushrooms! Sometimes they're like meat!

SERVES 4

2 medium clusters chicken of the woods mushrooms, pulled apart into shreds (about 4 cups shredded)

2 tablespoons extra-virgin olive oil, plus more

Sea salt and freshly ground black pepper

1 cup Mulberry BBQ Sauce (page 121) or store-bought, plus more

4 hamburger buns

On a cutting board, use your hands or two forks to shred the mushrooms into bite-sized pieces. Set aside.

In a large cast-iron or other heavy skillet over medium-high heat, heat the oil. Add the mushrooms and cook, stirring occasionally, until they've released as much liquid as possible and begun to brown along some edges, 10 to 15 minutes. (Yes, it takes this long!) If the mushrooms start to look dry, add a splash more oil. Season with a small pinch of salt and a few grinds of pepper—you can go a little easy here since the mushrooms will also get coated in BBQ sauce, but you do want some seasoning to seep into the mushrooms, not just on the surface. Remove the pan from the heat.

Add a few tablespoons of the BBQ sauce to the mushrooms and toss well to coat. Starting with a little sauce in the hot pan helps it thicken and stick to the mushrooms. Taste and add more sauce to reach your desired consistency.

Divide the mushrooms among the buns, topping with more sauce as desired. Serve hot.

CRAB CAKES OF THE WOODS (Chicken of the Sea Cakes? Wait, no)

There was a restaurant on Cape Cod called Lindsey's and my family was . . . maybe a touch obsessed. At least twice a year we'd find ourselves in the warm dining room or their kitschy fifties' diner and every single visit, from age five to age eighteen, I got the same thing: their phenomenal, family-secret crab cakes. Yes, I broke my vegetarianism twice a summer every summer for a dozen years, knowing full well the guilt I'd feel, because they were that good. With the passing of my eighteenth birthday, I swore them off, crying a bit inside but feeling morally sound. But golly, did I miss them, and I still do! I've dreamed about them as recently as a month ago! Should I, a Filthy Vegan, be ashamed that a closely guarded family recipe for a literal sea bug cake is such a pillar of my identity? Probably!

It was only right to honor those cakes while recently staying at the Massachusetts home that has seen me in so many different iterations, all of whom I love dearly except that one Alexis who spent the entire summer at the beach reading *Twilight*. I opened up some large COWs I'd found and pickled a month before (a neat trick to preserve their texture perfectly that I picked up from @foragerchef's blog), and cooked them with vegan butter, *Codium fragile*, aka dead man's fingers seaweed, sea rocket seedpods, pickleweed, a touch of brown sugar, and a heavy-handed dash of Old Bay. Mixed with bread crumbs, dill, cayenne, and some egg replacer to bind, these guys made me tear up a bit because they took me back to so many summers. So many family dinners. So many Alexises! This was a long intro and if you've made it this far, you're a real one, and I have so much love for you (and crab cakes) in my heart!

RECIPE CONTINUES

SERVES 4 TO 6

- **4 medium clusters chicken of the woods mushrooms**
- **1⅓ cups rice vinegar**
- **2 tablespoons sea salt**
- **1 tablespoon granulated sugar**
- **3 tablespoons unsalted vegan butter, plus more**
- **2 tablespoons fresh seaweed, such as sea lettuce, dead man's fingers, or wakame, finely chopped, or 1 tablespoon crumbled dried nori**
- **2 teaspoons packed brown sugar**
- **1 teaspoon Old Bay seasoning**
- **½ cup panko bread crumbs**
- **1 tablespoon chopped fresh dill**
- **1 tablespoon chopped fresh parsley**
- **1 tablespoon chopped fresh chives or chopped field garlic**
- **Pinch of ground cayenne pepper**
- **¼ cup vegan mayonnaise, plus more for serving**
- **3 tablespoons Just Egg (or other liquid egg replacement)**
- **Arugula, baby lettuces, or lambsquarters, for serving**
- **Lemon wedges, for serving**

Tear the mushrooms into chunks just small enough to fit inside a large jar. Set aside.

In a medium saucepan over medium-high heat, combine the vinegar, salt, sugar, and 1 cup of water and bring to a simmer. Stir to help the sugar and salt dissolve. Remove the pan from the heat.

Carefully pour the hot brine over the mushrooms in the jar. (You may not need all the brine; discard the remainder or save for another use.) Set aside to pickle and let cool completely, about 1 hour. You can use the mushrooms right away after this, or seal and store in the refrigerator for up to 2 months.

Drain the mushrooms and use a clean kitchen towel to gently squeeze out any excess liquid. Tear the mushrooms into small shreds to resemble crabmeat.

In a large skillet over medium-high heat, melt 2 tablespoons of the butter. Add the mushrooms and cook, stirring, until just starting to brown, 5 to 7 minutes. If the pan looks too dry, add a bit more butter. Add the seaweed, brown sugar, Old Bay, and a pinch of salt, and stir to combine. Taste and check for seasoning. Remove the pan from the heat.

In a medium bowl, add the mushroom mixture, bread crumbs, dill, parsley, chives, and cayenne and toss to combine. Add the mayo and egg replacement and mix well. The mixture should hold together when pressed.

Return the pan to medium-high heat and melt the remaining tablespoon of butter. Form a ½-cup portion of the mushroom mixture into a ball and add to the pan, gently pressing down to form a patty. Repeat with another patty or two, depending on your pan. Cook until the underside is golden and the patty holds together when nudged, about 5 minutes. Flip and continue cooking until golden, 3 to 4 minutes. Transfer the cakes to a plate and repeat with any remaining mixture. Remove the pan from the heat.

Arrange the arugula on a serving plate and lay the cakes on top. Serve with lemon wedges and additional mayo, if you like.

ANY DAY GARLICKY MUSHROOM PASTA

One of the hardest things about foraging mushrooms is the rush to cook them before they dry out or go bad. So let me teach you my favorite method for preserving the meaty integrity of chicken of the woods (or any mushrooms): freezing the cooked mushrooms so they're ready to throw into anything, whenever! This is my favorite fast recipe for when I don't have a lot of energy but still want to eat luxuriously. Here, you're using the whole chunk of mushrooms that you froze, but you can also freeze them in smaller portions to use as needed. You can veg this up, if you like, with any leafy cooking greens or even a handful of frozen peas.

SERVES 4 TO 6

FOR THE MUSHROOMS

4 medium clusters chicken of the woods mushrooms (about 1 pound)

3 tablespoons extra-virgin olive oil or unsalted vegan butter

Sea salt

FOR THE PASTA

2 teaspoons extra-virgin olive oil or unsalted vegan butter, plus more

6 garlic cloves, finely chopped

Sea salt and freshly ground black pepper

1 pound pasta, any shape

2 cups packed fresh spinach, lambsquarters, or other cooking greens, or 1 cup frozen peas (optional)

MAKE THE MUSHROOMS: Break the mushroom fronds into bite-sized pieces about 1 x ½-inch in size. Heat a large, dry skillet over medium-high heat. Reduce the heat to medium and add the mushrooms. Stirring frequently, dry-fry the mushrooms until they begin to brown along some edges, 5 to 7 minutes. Add the oil and a pinch of salt and cook, stirring, until the mushrooms have absorbed most of the oil, 3 to 5 minutes. Remove the pan from the heat and let the mushrooms cool completely.

Add the mushrooms to a freezer-safe bag, remove as much air as possible, and freeze for up to 6 months.

MAKE THE PASTA: In a large skillet over medium heat, heat the oil. Add the garlic and cook, stirring, until just fragrant, 1 minute. Add the frozen mushrooms and cook, stirring frequently, to help the mushrooms break up and thaw in the pan. Cook until the mushrooms are soft, warmed through, and fragrant with garlic, about 10 minutes. Taste and season as needed. Remove the pan from the heat.

Bring a large pot of well-salted water to boil over medium-high heat. Cook the pasta just short of the recommended cooking time—it'll finish cooking with the mushrooms. Using a heat-proof measuring cup or mug, reserve 1 cup of the cooking water. Remove the pot from the heat. Drain the pasta and set aside.

Return the pan of mushrooms to medium-high heat. When the mushrooms begin sizzling again, add the pasta and a big splash of pasta water. Shake the pan and toss the pasta to help coat it in the mushrooms and starchy water. Continue cooking and shaking the pan until the pasta is cooked through and has a slightly saucy sheen, 3 to 5 minutes. Add the greens (if using), toss to combine, and cook until the greens are wilted and warmed through, about 3 minutes. Remove the pan from the heat. Taste and adjust seasoning as needed.

Divide the pasta among bowls, and top with a few cracks of pepper and a drizzle of oil.

ACORNS

THE MOST STYLISH beacon of fall, depicted in their little wooden beanies on fall decor worldwide. Acorns have global recognition because, well, oaks do be every-dang-where.

Pliny the Elder called acorns the "first food of mortal man," which, while debatable, speaks to how important the humble oaknut was for humans. So why did they disappear from our collective tables?

Well, the short answer is this: Acorns are a lot of work. Laden with hella bitter tannins to keep the nuts intact until it's time to germinate, acorns need a good long time in water to be rendered palatable by humans. The squirrels win this round. The effort of preparing acorns, paired with their availability since we're competing with squirrels and other critters, made for a situation that spelled P-O-V-E-R-T-Y. And in human history, fewer things fall out of fashion faster than what makes you look poor. We're kinda awful like that!

The longer, and much cooler answer? Acorns never fully left! In South Korea, people can bring acorns to special factories that will leach them and mill them into acorn powder! Acorn powder is used to make the Korean dish dotorimuk, an often savory and always nutty jelly. In North America, several Indigenous groups use these abundant tree nuts for a myriad of dishes, such as soups, breads, flours, stews, and dumplings.

No matter your heritage, somewhere along the line one of your ancestors kept on keeping on because of the abundant energy that oaks so lovingly throw to the ground each year. And for those ancestors, we relearn a few of the old ways.

The nice thing, as far as foraging goes, is we all know what an acorn looks like! A tree nut in a beret! But how to

know where to look for those sweet, hat-wearing nuts? Well, oaks and their leaves come in a variety of shapes and sizes. Some leaf shapes fall into the stereotypical deeply lobed and longer-than-wide category, but you get some screwballs thrown in from the likes of the chestnut and chinkapin oaks. My best recommendation is getting to know the oaks native to and planted in your neck of the woods!

WHAT TO LOOK FOR:

- Before it's even time for the acorns to pop out, get to know your local oaks by identifying the **leaves**! Many, but certainly not all oak leaves, especially in trees in the northern part of North America, share a distinctive shape with two to three lobes and pointed tips. Oak leaves in the southern regions, like the live oak or post oak, tend to be smoother and not as lobe-y.
- Oaks will have heavy **acorn** years, called *mast years*, and light years. This helps control the population of animals who eat acorns. On mast years the over-abundance has a higher likelihood of becoming trees because there aren't enough animals to eat them!
- **FYI**: In my experience, our local bur oaks (*Quercus macrocarpa*) and chinkapin oaks (*Quercus muehlenbergii*) are most worth my harvest time. Bur oaks have huge acorns (with fuzzy hats!), so you get more bang for your buck, and chinkapin oaks have acorns that taste great with barely any leaching, which saves a lot of time!

ABC: ACORN-BASED CARBS (That Aren't Really Carbs?)

Acorn Starch, Acorn Flour, and a Bonus Egg Replacement Formula

Now, there's a trick to eating acorns. All of us who tried as kids to eat one fresh off the ground know this from experience. Acorns contain varying levels of tannins, a set of bitter compounds found in tons of food and drinks that, in high quantities, forms a sort of time-release preservative for acorns. That way they stay so fresh and so clean until it's time for them to germinate.

Acorns from red oaks are often said to contain more tannins as they don't germinate until the spring, whereas acorns from black oaks and white oaks germinate soon after hitting the ground in fall. That being said, tannin levels vary a lot by species and by tree, and the best way to find out how tannic your local nuts are is by seeing how long they take to leach!

Leaching is the process of using cold or hot water to carry the tannins out of your sweet acorns. Cold water leaching takes longer, but, crucially, leaves the starches in the acorn intact. Hot leaching is faster, but leads to a starchless, softer, deep-brown end product.

The hardest part of this recipe is the waiting, but you're rewarded with not one, but two products: acorn flour and starch. (We love a twofer!) Most of the time is inactive, where you're just letting the acorns do their thing, but that being said, the leaching process can take up to 10 days. And from the starch, you've got a bonus product: the ability to make a starch-based egg replacer that works in pretty much any application where you'd otherwise want the binding properties of an egg.

I'm calling for 4 cups of acorns here, but this will truly work for any amount of acorns, so scale up or down depending on your bounty! (I will say, if you've got less than 4 cups, there's a bit of a diminishing return situation once you factor in float-testing and the time and energy to process the nutty buddies. But then again, maybe a small practice batch is good!) The yield will vary depending on your acorns and how much starch they release in the process.

RECIPE CONTINUES

Acorn Starch

4 (or more) cups acorns

TOOLS

1 large bowl or small bucket

1 quart-sized jar

Fill a large bowl with water and add the acorns. Discard any acorns that float.

Fill the quart-sized jar about halfway with cold water. Using your nutcracker of choice, shell the acorns and drop the acorn meat in the water.

In a blender, add the acorn meat and its soaking water. Top off with a bit more water. (You want to help the blender really grind these up.) Blend until the acorn pieces resemble large grits.

Set a fine-mesh strainer over the same quart jar. Carefully pour the blended acorn mixture through the strainer, keeping the grits in the strainer and the cloudy water in the jar. Set the strainer with the grits aside; you'll come back to these soon!

The cloudy water in the jar will settle into a fine starch on the bottom with (probably) brownish water on top. Once the starch has completely settled, pour off the water, keeping the starch in the jar, and replace it with fresh, cold water. Repeat the process—let the starch settle, pour off the old water, add new water—until the water in the jar is clear.

Drain out as much water as you can and spread the starch out on a baking sheet or food dehydrator tray. Set the starch in a sunny spot in your house (do not put this outside, because the squirrels will mock you to your face!) and let it dry, mixing and re-spreading it to dry evenly, for about 2 days. Or dehydrate at 95°F until completely dry, about 24 hours. Let the starch cool completely.

Transfer to a clean jar, seal, and store in a cool, dry place. Use in place of cornstarch and for Starch Egg replacer (recipe follows).

Starch Egg: A Formula

While starches like cornstarch, tapioca starch, potato starch, and arrowroot starch are all excellent as egg replacements in baked goods or other situations where an eggy binding agent is handy, in my kitchen, acorn starch is king! The protein in the starch (because let's remember, the acorn is a nut) does a great job of mimicking the properties of an egg in baked goods.

Anytime I call for a "Starch Egg" in this book, this is what I mean.

3 tablespoons warm water

1 tablespoon Acorn Starch

In a small bowl, add the water and Acorn Starch. Whisk it up and let it sit a minute or two to hydrate, then the starch is ready to use in your recipe.

This formula is equivalent to 1 large egg, so scale up as needed in any situation where you're replacing chicken eggs. (Except for omelets and scrambles. But you knew that.)

Of course, if you don't have Acorn Starch on hand (you mean, you don't have several free days to harvest, process, and babysit a jar of acorn starch and its resident tannins?) this formula, in the same proportions, works for any of the other starches mentioned above!

Acorn Flour

Now we'll do just about the same leaching process with the acorn grits. This process takes a bit longer than the starch since the grits have more surface area for the water to deal with.

Add the reserved grits to a large bowl and cover with cool water. If you live with pets or small children, cover the bowl with a towel and set it somewhere out of their reach. Let the grits sit for at least 8 hours—I like to do this before I head up to bed for the night.

Pour off the soaking water and replace with cool, fresh water. Taste a grit. It will probably taste very gross and bitter the first time! This is the tannic bitterness we're slowly coaxing out of these acorn grits. Repeat this process of soaking, draining, and replacing with fresh water, tasting as you go, until the grits no longer taste gross—in fact, they shouldn't taste like much of anything once they're done leaching out the tannins. This is the part that can take up to 10 days, so try and relax and let the acorns do their thing!

Important note: Your tastebuds are the boss. If your grits taste good but the water still isn't totally clear, it is okay! Remember that a lot of foods and drinks have a certain level of tannins—wine, tea, chocolate, and a host of vegetables and fruits! You're not trying to eliminate them completely, you just want them at a low, tolerable level.

In the same way we dehydrated the starch, we'll dehydrate the grits. Drain out as much water as you can and spread the grits out on a baking sheet or food dehydrator tray. Set the tray in a sunny spot in your house (again, inside! Do not underestimate your neighborhood critters!) and let the grits dry, mixing and re-spreading them to dry evenly, for a few days. Or, if you do have a food dehydrator, dehydrate at 95°F until completely dry, about 24 hours. Let the grits cool completely.

Add the dried acorn grits to a spice grinder (you may need to work in batches) or a food processor and grind into a flour. Transfer the flour to a clean jar, seal, and store in a cool, dry place.

ACORN PANCAKES with Smoked Hickory Syrup

Listen, I love any and all pancakes, but I really love these because they're puffier than your average flapjack and the Acorn Flour brings a nuttiness that's so unique and delicious. It's not often that we think of the role flour plays to impart flavor to a finished dish, so let these babies surprise you in all the best ways. Don't be intimidated by the smoked hickory syrup, either—it's easier than it seems! But if you're short on time (or, well, hickory bark), these pancakes are delicious with any syrup you'd like to use.

SERVES 2 TO 4

FOR THE SYRUP

- 2 to 3 (8-inch) strips hickory or elm bark
- 1 cup lightly packed brown sugar
- Pinch of sea salt

FOR THE PANCAKES

- ¾ cup all-purpose flour
- ¼ cup Acorn Flour (page 179)
- 2 tablespoons granulated sugar
- 1 tablespoon baking powder
- ½ teaspoon sea salt
- 1 cup unsweetened nondairy milk
- 1 tablespoon apple cider vinegar
- 1 teaspoon vanilla extract or Nocino (page 199)
- ½ cup frozen juneberries or other berries (optional)
- Unsalted vegan butter or cooking spray, for the pan

MAKE THE SYRUP: Preheat the broiler to 500°F. Fill a medium saucepan with 4 cups of water and set aside.

Place the hickory bark on a baking sheet and broil until fragrant and just slightly smoky, 5 to 6 minutes. Using tongs, immediately transfer the bark to the pan of water.

Set the pan over medium-high heat and bring to a boil. Reduce the heat to medium-low and simmer until the water is a medium amber color, about 15 minutes. Remove and discard the hickory bark.

Reduce the heat to low and add the brown sugar and salt. Stirring occasionally, gently heat the mixture until the brown sugar has dissolved completely, about 5 minutes. Remove the pan from the heat and set aside. From here you can let the syrup cool completely, and store in an airtight jar in the fridge for up to 2 weeks (unopened). Or you can use it right away!

MAKE THE PANCAKES: In a medium bowl, whisk together the all-purpose flour, Acorn Flour, sugar, baking powder, and salt. Add the milk, vinegar, and vanilla and whisk until smooth. Add the juneberries (if using) and stir to combine. Set aside.

In a large skillet over medium heat, melt a few teaspoons of butter. Working in batches, use about ¼ cup of batter for each pancake and cook, undisturbed, until bubbles form on top, 3 to 4 minutes. Flip and cook until the underside is fully cooked, 1 to 2 more minutes. Transfer to a plate and repeat with the remaining batter. Remove the pan from the heat.

Serve warm with the Smoked Hickory Syrup.

MAITAKE

MAITAKE! ALSO KNOWN as hen of the woods! But I will be using the name *maitake* to avoid confusion with the completely different mushroom that you know and love by now, chicken of the woods! *WHO KEEPS NAMING THESE THINGS!*

Maitake is a late-summer-into-fall mushroom in my area and they grow at the base of oak trees. They also cause something unfortunately (or fortunately, if you like butt jokes) called white butt rot, which is maybe the best phrase I've written in this whole book.

Much more graceful is the Japanese translation of maitake, "dancing mushroom," which is way more rad than calling it a hen, in my humble opinion. Much like COWs, they come back in the same spots year after year! I even find them to be a bit more reliable, as I more often locate them on trees that are still alive. The tree still living means the fungi still has a source of nutrition!

Maitake is a shelf fungus, with its colors ranging from pale khaki to dark-gray, always with white edges and white undersides. They have pores, not gills, and do not bruise like their lookalike cousin, black staining polypore, which is often confused with maitake. (And although the name *black staining polypore* sounds sinister as all get-out, it's not poisonous, so don't worry if you come across any!)

I start looking for maitake when the evenings begin cooling down, usually around mid-September. You will also occasionally find them at the base of trees other than oaks, but they're much rarer. Maitake can get huge and when they do, they get tougher toward the center. But worry not! They'll still be tender around the edges, and some of the tough bits can just be dried and saved for soups.

WHAT TO LOOK FOR:

- It's good that we just covered oaks in the previous section, because maitake love to **grow** at the base of oak trees! In some areas with lots of older oaks, you can find a maitake at the base of nearly every tree come September and October.
- The **fronds** are typically gray with white tips, but can occasionally be brown with white tips.
- They have **pores, not gills**, and grow in shelves, much like the chicken of the woods.
- Maitakes have a **white underside** and they smell quite savory when broken.
- **Lookalike alert!** Maitakes can sometimes be confused with black staining polypore, a sinister-sounding fungus that is actually harmless and also edible (just not terribly visually appealing when cooked). Black staining polypores are much tougher than maitakes, and as the name indicates, will discolor when broken, which maitakes won't do.

MAITAKE BACON

Once, while eating breakfast with my sweet uncle Bill, he looked at me with a mixture of sadness, pity, and I think real genuine apology, and said, "I'm so sorry you can't enjoy God's gift to Earth," and proceeded to put a piece of crispy bacon right into his mouth. If I didn't think it would lead to a lot of bewildered follow-up questions, my response would have been, "Hey, I can also enjoy something that is savory, salty, smoky, and the tiniest bit sweet, just like you're enjoying right now!" Maitake is one of the tastiest and most versatile things you'll find growing at the base of a rotting tree. Here, we're harnessing its savory, meaty qualities to turn it into little hunks of bacon-y goodness.

MAKES ABOUT 1 CUP

2 tablespoons extra-virgin olive oil

1 teaspoon soy sauce

1 teaspoon maple syrup

½ teaspoon nutritional yeast

½ teaspoon garlic powder

¼ teaspoon smoked paprika

Pinch of sea salt

4 cups fresh maitake mushrooms, torn into bite-sized petals (about ½ pound)

Preheat the oven to 350°F. Line a baking sheet with parchment paper and set aside.

In a medium bowl, whisk together the olive oil, soy sauce, maple syrup, nutritional yeast, garlic powder, smoked paprika, and salt. Add the maitake pieces and toss gently with your hands to coat.

Shaking off any excess liquid, transfer the maitake pieces to the prepared baking sheet in an even layer. Bake for about 20 minutes, until the mushrooms have begun to dry out. Flip the mushroom pieces, redistribute into a single layer, and bake for about 10 more minutes, until the edges are just beginning to crisp.

Transfer the maitake bacon to a paper towel to drain and cool slightly. Serve immediately.

CREAM OF MUSHROOM SOUP & SECRET-WEAPON UMAMI POWDER

One of the reasons maitake is such a star is the depth of savory flavor you can get from one mushroom. Most of the time, broths are comprised of lots of different odds and ends to get a full, well-rounded end result. But maitake said, "Nuh-uh," and started singing "I'm Every Mushroom" to the tune of "I'm Every Woman." This recipe is really three (three!) recipes in one: You're making a (1) mushroom stock that's the star of the (2) creamy, comforting soup, and using the not-quite-spent mushrooms from the stock to make my favorite secret kitchen weapon, a single-ingredient (3) umami powder to give depth to your dishes all year long.

MAKES ABOUT 4 CUPS

FOR THE STOCK

3 cups maitake mushrooms, torn into chunks (about 6 ounces)

Sea salt

FOR THE SOUP

Sprig of fresh thyme

Sprig of fresh sage or mugwort

½ cup coconut cream

Sea salt and freshly ground black pepper

¼ cup plain vegan yogurt, for serving (optional)

Chopped cooked maitake or other mushrooms, for serving (optional)

Chopped fresh parsley leaves and or chives, for serving (optional)

MAKE THE STOCK: In a large pot over medium-high heat, add the maitake and 10 cups of water. Bring the mixture to a boil, reduce the heat to barely a simmer, and cook for 2 to 3 hours, until the stock has a rich mushroom flavor. Remove the pot from the heat.

Strain the mushrooms from the stock and reserve them; you'll use them for the umami powder! Salt the stock to taste. Store in an airtight container in the fridge for up to 1 week, or freeze for up to 6 months.

MAKE THE SOUP: In a large saucepan over medium-high heat, add the mushroom stock, thyme, and sage. Bring to a boil and reduce the heat to low. Let the stock simmer with the herbs until very fragrant, about 40 minutes. Remove and discard the herbs and remove the pan from the heat.

Whisk in the coconut cream and season with salt and pepper. You can serve this very simply as-is, or top the soup with dollops of plain yogurt, chopped cooked mushrooms, and fresh herbs for more special occasions.

RECIPE CONTINUES

MAKE THE UMAMI POWDER: Preheat the oven to its lowest heat, ideally around 150°F and no more than 200°F. (Many ovens have a "keep warm" function that can be adjusted to a lower heat.) Set a rack over a baking sheet and line the rack with parchment paper. Set aside.

Very gently squeeze out any excess liquid from the mushrooms. Separate the fronds into relatively uniform pieces so they can dry evenly.

Dehydrate the mushrooms, flipping them every hour or so, until they're completely dry, brittle, and crisp, at least 6 hours and up to 12, depending on your oven. Let cool completely.

In a high-speed blender or spice grinder, add the dried mushrooms and grind to a fine powder. Store in an airtight jar in a cool, dark place for up to 6 months.

MAITAKE STEAK

And just when you thought maitake couldn't take the place of another meaty thing, she says, "Girl, how many times do I have to sing this song??" This is an excellent way to use the hefty bases of any maitakes you've foraged after you've used their fronds in things like bacon, jerkies, cute soup toppings, etc. Important note for any foraged mushrooms: Definitely give them a little brush with a soft pastry brush or something to knock out any dirt before using them. (Store-bought maitakes might not have as big of a chunky base to hold the "steak" together, so depending on what they're looking like, they could be best for another use.)

SERVES 2

Bases from 1 to 2 large maitake mushroom clusters, sliced ½ inch thick

2 tablespoons unsalted vegan butter

1 teaspoon soy sauce, plus more

Using a small paring knife, score both sides of the maitake in a crosshatch pattern, focusing especially on any thicker areas.

Heat a large cast-iron or other heavy skillet over medium-high heat. Add the maitake fronds and dry sear until slightly golden and crisp, about 2 minutes per side. Reduce the heat to medium and add the butter and soy sauce. Flip the mushrooms a few times so they can start soaking up the sauce.

Carefully tilt the pan and use a spoon to baste the mushrooms in the sauce. Continue cooking and occasionally basting the mushrooms until they are a rich, deep golden brown, 5 to 7 minutes, flipping them every few minutes until they've soaked up most of the sauce. Taste a bit of the sauce and add more soy if needed. Remove the pan from the heat.

Serve hot!

BLACK W

UPON LOOKING AT all of the different varieties of black walnut, members of the juglans genus, I realized there are perhaps too many *Juglans* to juggle. So, I will be talking primarily about my hometown juglans: *Juglans nigra*, the eastern black walnut. (How many times can I say juglans in one paragraph??) Rest assured, these recipes work with any black walnut, or honestly, any walnut at all! Just know that if you're not working with one of the black walnuts, it'll lack some of the resinous "weirdness" that black walnuts are best known for. But they'll still be heckin' tasty!

Choosing the season where black walnuts would live in this book was a challenge. At one point there were going to be two different chapters on black walnuts! That's because they can be used at different points in their ripeness cycle: You can pick them right off the tree, green and unripe, to cure in sugar or alcohol to make some very precious potions; you can gather fallen black walnuts, still in their green husk, to ripen and dry out and enjoy in more traditional, nutty applications. The common denominator is, once you factor in all the steeping and curing time, most folks are likely to enjoy black walnuts in fall, heading into winter, depending on where you live.

Growing up in southern Ohio, I first learned about black walnuts not as a tasty treat but as a pesky fall hazard. And I mean that both ways: *fall* as in autumn and *fall* as in booty-on-the-ground. The ends of long summers at camp were signaled with the *THUNK* of their heavy green husks raining down on basketball courts, lawns, cars, sidewalks. Those thick, green, heavily scented husks are probably one of the reasons why black walnuts have fallen by the wayside in terms of casual snacking. They do take a bit of muscle or time to get through!

ALNUTS

Speaking of processing, we do need to acknowledge that getting the nut meat (that's what it's called! I don't make the rules!) out from its fortress of husk and shell takes a bit of doing. But it can be done! And while you will hear some folks suggesting absolutely wild things like putting them in a five-gallon bucket with water and an immersion blender, or elevating your car and spinning its wheels to fire the husked nuts at the back wall of your garage, my favorite method is what I call the Lazy Gal Method:

1. Gather as many fallen, green-husked black walnuts as you can find. Skip any beige ones or any with black spots on the husk.
2. Put your gathered walnuts in a mesh bag and hang up the bag in an area with good air flow where they won't get wet. For me, that's off the corner of one of our bookshelves, directly above an air vent. But it could be a knob on a kitchen cabinet, a corner of a covered porch, o wherever! Get creative!
3. Wait anywhere from 2 weeks to 1 month (sometimes more, depending on how ripe they were when you gathered them). When they're ready, the green husks will have become fully dried, black, and dry or crunchy feeling.
4. Take your bag and thwack it repeatedly on some pavement—this could be a sidewalk, a garage, a patio, wherever! Keep slamming until the husks have broken off the inner shell.
5. At this point your nuts are (1) cured (that's why we waited for them to dry and turn black) and (2) ready to crack! I have a contraption with the absolutely unhinged name of Grandpa's Goody Getter, but you can also use a regular ol' nutcracker or other kitchen weapon of choice.

Now, for those for whom the Lazy Gal Method is still too much, no worries! Some of these recipes call for unripe, green walnuts, husk and all! Wahoo!

One important handling note for the unripe, green walnuts: The green husks will dye black pretty much anything they touch—your fingertips, countertops, clothing, etc.! Very fun if it's on purpose, very unfortunate when it's not. So maybe rock an apron and gloves while processing these guys while unripe.

WHAT TO LOOK FOR:

- Black walnut **trees** tend to be large (even if they're not tall, they just like to take up space!) with dark, deeply furrowed bark.
- Black walnut trees have pinnate, or long and narrow **leaves**.
- Small, tassel-like **flowers** in early spring become **round green husks** with the walnuts developing inside!
- You'll harvest the **nuts** while green and about golf ball–sized for Nocino, or wait for them to turn black and fall to harvest the nuts within. Please don't roll an ankle gathering them. It's a real worry!
- **Lookalike alert!** Because of their similar leaves, black walnut trees sometimes get confused with everyone's least favorite invasive tree: tree of heaven. Thankfully, black walnut doesn't smell like peanut butter that's been left in the rain when you break the foliage, but uh, tree of heaven certainly does.

MAPLE BLACK WALNUT ICE CREAM

A fun bit of food history for my fellow nerds: Before butter pecan, the ice cream flavor was actually black walnut! It was popularized by Black folks in the South soon after Emancipation as we started having the technology to keep things frozen for longer. As it gained in popularity, however, ice cream companies wanted a nut that was cheaper and less difficult to process, thus its cousin pecan stepped up to the plate! Maple black walnut (or just maple walnut) used to be a classic ice cream flavor you could find just about anywhere, so this is my homage to the icon herself.

MAKES ABOUT 1 QUART

- ½ cup black walnut halves, chopped
- 4 cups unsweetened soy milk
- 2 cups granulated sugar
- ½ cup Maple Syrup (page 232) or store-bought
- 1 teaspoon vanilla extract or Nocino (page 199)
- Sea salt
- 5 tablespoons Acorn Starch (page 176) or cornstarch

The night before making the ice cream mixture, place the bowl of an ice cream maker in the freezer.

In a medium skillet over medium heat, toast the walnuts, shaking the pan frequently, until slightly golden and fragrant, 5 to 7 minutes. Remove the pan from the heat.

In a medium saucepan over medium-high heat, whisk together the milk, sugar, Maple Syrup, vanilla, and a pinch of salt. Add the toasted walnuts and let them steep while the mixture warms.

Add the Acorn Starch to a glass with ¼ cup of water. Whisk until smooth and add the starch mixture to the pan. Bring to a simmer and reduce the heat to medium-low. Cook, whisking frequently, until the mixture has thickened and coats the back of a spoon, 7 to 10 minutes. Remove the pan from the heat.

Taste and adjust the sweetness or salt. Let the mixture cool completely and chill overnight in a sealed container.

Strain out and reserve the walnuts.

Add the chilled mixture to an ice cream maker and churn according to the manufacturer's instructions. When the ice cream resembles a soft-serve consistency, add the reserved walnuts and churn until incorporated.

Serve immediately or freeze in an airtight container for up to 3 months.

BLACK WALNUT MUGOLIO/ CHEONG

One of the more expensive things you'll learn to make in this book is black walnut mugolio, something that, in fancy cocktail shops on the internet or IRL, sells for many dollars for a tiny bottle. Black walnut mugolio is a nutty, resinous, sweet, slightly fermented syrup, and making it means you get to witness the magic of sugar pulling all the moisture and flavor out of your unripe walnuts. I'll warn you here and for the Nocino recipe (page 199) to make sure you've gathered those walnuts before late July, or you might break your knife trying to cut through them.

Mugolio is traditionally made using pine cones in Italy, but this is a fun, nutty twist that is equally at home in any cool cook's kitchen. The resulting syrup is delightful added to your morning coffee, drizzled on vanilla ice cream or the Maple Black Walnut Ice Cream (page 196), or even as the sweetener in barbecue sauce! Give yours a taste and see what inspires you.

If you don't recognize the name *mugolio*, you might recognize the name *cheong*, which in Korea is the same syrup and the same process of sugar fermentation. Unripe plum cheong is a popular one if you're looking for a good gateway cheong!

The general formula for mugolio/cheong is equal weights unripe walnuts to sugar, but we know nobody brings a scale with them to the woods! The recipe below calls for a half-dozen green walnuts, because that tends to be about a pound, but you're welcome to weigh yours when you get home, then use that same weight in sugar.

MAKES ABOUT 2 CUPS

About 6 unripe green walnuts, picked off the tree before midsummer, halved

1 pound granulated sugar

In a large, clean jar, layer the walnut halves and the sugar, ending with a layer of sugar. Seal the jar and let sit in a cool, dry place until the sugar has completely dissolved into liquid, 30 to 60 days. For the first week, make sure you "burp" the jar (unseal to release excess carbon dioxide) every day or so. You can also shake the jar while it's sealed to help the fermentation process along. After that, you can burp it every few days, depending on how active your fermentation is.

Strain out and discard the walnut halves and add the syrup to a clean jar. Store in the refrigerator for up to 6 months.

NOCINO

'Tis she, Goth Nut Drink! Nocino is a traditional Italian nut liqueur, fortified with warm spices and high-proof alcohol. While this is one of the lengthier recipes in terms of time before you can try the finished product, I haven't yet met someone who has not deemed it worthwhile at the end. A loving warning: Give this at least six months' steep time before trying it (and honestly, six months is still not enough time for me). Many a forager has, out of natural curiosity, tried a nocino before it's done aging, only to find an acrid liquid reminiscent of a wet burp. Much like all of us, nocino only gets better with time. My best advice, to keep that curiosity at bay, is to let this age somewhere where you'll kind of maybe a little bit forget about it? By the time you unearth it again, it'll be ready.

The fun part is, once you know what a "basic" nocino tastes like, you can play around and make eclectic batches! I've included some flavor suggestions and greatest hits below.

MAKES ABOUT 1 QUART

FOR THE NOCINO

About 20 unripe green walnuts, picked off the tree before midsummer, halved

1 (750mL) bottle of high-quality vodka

Granulated sugar, for simple syrup (you won't need this for at least 6 months)

SPICE COMBINATIONS

Traditional Italian: vanilla beans, citrus peel, whole coffee beans, cinnamon sticks, star anise, fennel seeds

The Forager: dried spicebush berries, smoked hickory bark, roasted dandelion root, prickly ash leaves

The Olivia (named after my friend!): Meyer lemon peel, fresh juniper berries

In a large, clean jar, add the halved walnuts and spices of your choice. Completely cover in the vodka—you may not need all of it, depending on the size of your walnuts and what spices you're using. Seal and store the jar in a cool, dark place for 6 to 8 months.

Set a fine-mesh strainer over a medium bowl. Strain out and discard the solids. Measure the volume of the resulting liquid. You'll want to add that same volume of simple syrup, which we'll make now. Let's say you need about 2 cups of simple syrup.

In a medium saucepan over medium heat, combine 1½ cups of water with 1½ cups of sugar. Bring the mixture to a simmer and stir continuously until the sugar has completely dissolved, 3 to 5 minutes. Remove the pan from the heat and let cool completely.

In a large jar or bottle, combine the nocino with the syrup. Seal and store in a cool, dark place for up to 6 months.

DOCK

DOCK IS ONE of my favorite Old World plant imports. They're tart, they're prolific, and they offer many gifts during different parts of the year. Did you know that the goop from their stems in late spring and early summer can help soothe sunburns and can be used to set jellies? Now you do!

There are a lot of great Central and Eastern European historical recipes using dock leaves. One of my favorites is braiding dock leaves to dry them, which lets them ferment a bit as they dry slowly and makes for an excellent addition to any wintertime soups and stews!

Here in North America, you'll usually find dock in areas that have been disturbed by human movement or construction. A lot of times I'll find them growing alongside newly made paths in our parks. Curly dock will have a signature wavy edge to its leaves, which helps to differentiate them from their less-tasty cousin, bitter dock, who is also edible but usually could use a good blanch before adding to recipes. Curly dock tastes great raw, and I actually prefer it that way!

Dock is a member of the buckwheat family, which I think is most evident in its flower stalk. There, you can find the telltale joints you might see on some of their more distant cousins like lady's thumb or Japanese knotweed.

Dock is also a multi-season wonder, making it extra important in my kitchen. Its dark-green leaves are often some of the first hardy greens I find in the spring, and some of the last I can find in the fall or even early winter after we've already experienced a couple of light frosts.

My favorite way to spot dock in the wild, especially from a distance, is to look for its rusty, brown spikes of dried seeds. These tend to persist well into the growing season after they develop. These seeds are always harvestable, no matter the season you find them, which makes them a very important winter forage for both wildlife and for us!

WHAT TO LOOK FOR:

I'm going to focus on curly dock here, which is tastier than its cousin bitter dock. (I love when names make sense!)

- Long **leaves** have ruffled edges, coming from a central growing point. That center point is where the flower shoot will emerge in spring.
- You probably recognize the **shoots** once they've gone to seed and turned completely brown, as the shoots can persist for multiple seasons, sometimes even into the next spring! The brown shoots really stand out against other green plants in summer, and in late fall and winter when the other plants have dried and died back.
- The **seeds** are small brown ovals with a papery wing, and can easily be stripped from the stalk into a paper bag or jar.

STUFFED DOCK LEAVES

These are reminiscent of a Greek or Middle Eastern dolma, but in place of the often tougher grape leaves, we're using the tart and tender leaves of the dock plant. You don't need to worry about telling the curly and bitter siblings apart for this one, since we're going to blanch the leaves and remove any bitterness. The dried currants here offer a nice sweetness to counter the tartness of the leaves and the nuttiness of the wild rice! A real fall treat.

SERVES 4

Sea salt and freshly ground black pepper

12 curly or bitter dock leaves, rinsed

3 cups cooked wild rice

½ cup dried currants

½ teaspoon Umami Powder (page 190) (optional)

Extra-virgin olive oil, for drizzling

Bring a large pot of well-salted water to a boil over medium-high heat. Add the dock leaves and cook until vibrant green and tender, 2 to 3 minutes. Remove the pot from the heat.

Transfer the leaves to a colander and rinse them under cold water until cool. Dry them well and set aside.

In a medium bowl, combine the wild rice, currants, Umami Powder (if using), and salt and pepper to taste.

Place a dock leaf on a work surface with a short end closest to you. Scoop about ¼ cup of the filling onto the leaf and spread it into a little log across the width of the leaf, leaving about ½ inch on either side. Fold the sides over the filling and carefully roll the leaf into a little packet. Transfer to a plate and repeat with the remaining leaves and filling, nestling the stuffed leaves together on the plate.

Drizzle with olive oil and a bit of salt and pepper. Serve immediately.

DOCK SEED FLOUR

Dock is one of those plants that blends into the scenery until you try looking for it. You spot one, and suddenly you realize it's . . . everywhere! Truly, it is dock's world and we're just living in it. You'll notice their stately brown seed heads pretty much everywhere you look and thankfully, those seed heads provide many a tasty treat. A member of the buckwheat family (along with its much more annoying cousin knotweed), it boasts delicious seeds that give any baked good an earthy, nutty flavor. Gathering the seeds is quite satisfying; you just zip your fingers over those seed heads and poof! You've got a handful of seeds. If you've found a good spot with a lot of dock plants, the foraging process is pretty quick! All they need is a little roast in the oven before blitzing them to smithereens.

Use in any dish where you'd use buckwheat flour, like Dock Crepes (page 206), or swap in a few tablespoons in place of all-purpose flour in dishes like Acorn Pancakes (page 180) or even Garlic Mustard Scissor-Cut Noodles (page 87).

MAKES ABOUT 2 CUPS

4 cups dock seeds, small twigs and leaves sifted out as best you can

Preheat the oven to 350°F.

Spread out the dock seeds on a baking sheet and set aside for about 30 minutes to let any spiders skitter off. (It happens! They'll be tiny if they're in there at all.)

Toast the dock seeds in the oven until dry and aromatic, 10 to 15 minutes. Set aside to let cool completely.

Transfer the toasted seeds to a spice or coffee grinder and grind them—seed, chaff, and all—until evenly ground.

Store in an airtight container in a cool, dark place for up to 6 months.

DOCK CREPES

If you love buckwheat pancakes or crepes, you're going to be utterly obsessed with using dock seed flour for the same dishes. And if you don't love buckwheat in your baked goods (sometimes it can taste sandy, okay?), you will be pleasantly shocked—these are how I want buckwheat crepes to taste.

These crepes work well with both sweet and savory fillings, so they're truly an all-day treat! For sweet, try vegan butter and brown sugar, almond butter and any jam (from this book or your pantry!), or chocolate hazelnut spread. For savory, these are amazing with any cooked mushrooms, like My Plain & Simple Morels (page 62) or a pack of frozen mushrooms like in the Any Day Garlicky Mushroom Pasta (page 170), or any kind of cooked greens like My Best Dandelion Greens (page 49) or Goldenrod Shoots (page 157).

MAKES ABOUT 10 (6-INCH) CREPES

1 cup all-purpose flour

½ cup Dock Seed Flour (page 205)

1½ tablespoons granulated sugar

1 tablespoon Acorn Starch (page 176) or cornstarch

Pinch of sea salt

1 cup unsweetened nondairy milk

1 cup sparkling water

Neutral oil, such as sunflower or grapeseed, for the pan

In a large bowl, whisk together the all-purpose flour, Dock Seed Flour, sugar, Acorn Starch, and salt. Add the milk and sparkling water and whisk again until completely smooth.

In a medium nonstick skillet over medium heat, heat about 1 teaspoon of the oil, swirling it around to evenly coat the pan. Using a small ladle or ¼-cup measure, pour the batter into the hot oil and swirl the pan quickly to spread the batter into a thin layer. Cook the crepe until bubbles have formed and popped, and the surface is nearly matte, 1 to 2 minutes. Flip and cook for another minute and transfer the crepe to a plate. Repeat with the remaining batter. Remove the pan from the heat.

Fill the crepes with the goodies of your choice and serve warm.

QUITE SILLY OF me to combine a native fruit tree and one of the biggest scourges to the American suburban ecosystem in recent memory in the same chapter, but I promise you it makes sense!

To start with: crab apples! They're just tiny apples! I don't care what your mom told you. They're not poison apples, not danger apples, just small apples. As a food history nerd, I find it wild to see how synonymous large apples are with American culture when those are not the apples that were here prior to about four hundred years ago. Many of them, however, are now grafted onto native crab apple species, much like the green apple tree next door to my very own home!

And oh, the Bradford pear, a named cultivar of the Callery pear. Meant to be a clean, low-maintenance street tree, but instead, it's become a mild ecological nightmare. For starters, their branches are very weak and love falling on things you care about. Additionally, although they're not supposed to produce fruit, many of us can somehow take a five-minute jaunt around our city neighborhoods or our hedgerows and see that . . . is not accurate. They do fruit! (It's giving zombie and I do not love it!) Additionally, they crossbreed like a mammajamma and thus frequently escape from their city plantings in the form of their progeny. The least they could do is have the decency to produce fruit we (and the birds) can eat, and thank goodness, they do. In fact, it's pretty tasty when it's ripe.

These two fruits share this chapter because they're diminutive forms of fruits we are very used to eating and

LE &

ORD PEAR

working with, but because of their size we treat them a little differently than we would, say, a full-sized apple or Asian pear. And while they don't taste exactly the same, they can often be used interchangeably, as apples and pears historically have, due to a similar type of sweetness and very similar pectin levels.

And one more time for the folks in the back: Crab apples are not poison, and Bradford pears are edible! I don't know who told several people out on the wild internet otherwise, but please tell them I just want to talk.

WHAT TO LOOK FOR: CRAB APPLES

- Beautiful, sweet-smelling **flowers** that range from white to pink, with five petals each (unless it's an ornamental, which can sometimes be bred for double blooms).
- The flowers become small **fruits**, each on its own stem, with slightly tough skin and multiple seeds inside, like an apple.
- **Lookalike alert!** Sometimes it can be easy to confuse early, tiny crab apples with cherries. But cherry season comes before apple season! And cherries have a singular pit! And delicate skin! Apples and cherries *are* distant cousins (shout-out to the Rosaceae family), but I know you'll be able to tell the difference.

WHAT TO LOOK FOR: BRADFORD PEARS

- The thing you'll probably notice first about Bradford pear trees is their admittedly very pretty white or light-pink **flowers** in early spring. The flowers themselves smell . . . odd and unpleasant. I won't go into it here but if you know you know!
- Bradford pear **fruit** tends to be a sort of light–brownish-green, ranging in size from, say, a small cherry to the size of a golf ball.

BRADFORD PERRY

This delightful play on words—a *perry* is a specific type of hard pear cider—is one of the best ways to use one of the biggest nuisances to plague North America. It also sounds like the name of the token heartthrob in a '90s teen drama!

In my home state of Ohio, it's illegal to sell Bradford pears, but don't worry, they're literally everywhere, and you'll be able to find them just fine in the wild. This is a fun gateway project for foraged fermented beverages, and my favorite reminder that even the most noxious invasives can have redeeming qualities. Nevertheless, I'm still kicking every Bradford pear tree in spring when my whole neighborhood smells like a gym sock.

If your fruit isn't super-ripe, there's an easy alternative for a little assist. Pick up some champagne yeast at a wine or specialty store and follow the instructions on the packet. This way, your foraging timetable is a bit more flexible since it's not dependent on ripe fruit.

You'll notice we're not calling for any sugar here. When you start with super-ripe fruit, you'll end up with a pleasantly sweet bevvie that doesn't need any extra sweetener!

MAKES ABOUT 1 QUART

2½ cups very ripe Bradford pears or crab apples, picked from their stems and rinsed

In a blender, add the Bradford pears and 4 cups of water and blend into a smooth slurry. Transfer the slurry to a large glass jar. Cover the jar with a clean kitchen towel, paper towel, or a few layers of cheesecloth. Secure the towel to the jar with a rubber band. Set aside for 12 hours.

Stir the mixture, replace the towel, and set aside for another 12 hours. Continue this process for 2 to 3 days, stirring the mixture twice a day to keep mold from forming on the top.

Set a fine-mesh strainer over a large bowl and strain out the solids. Discard the solids.

Transfer the liquid to a clean bottle or jar and seal tightly. Let this ferment in a cool, dry place, away from direct sunlight, for 7 to 10 days, burping it (opening the lid to let out excess carbon dioxide) every day or so.

Store the perry in the refrigerator and use within 3 weeks.

CRAB APPLE OR BRADFORD PEAR JELLY

Growing up, my little sister was a proud picky eater, but she has always loved this crab apple jelly. She consistently asks for it, even as her palate has expanded. Also, as someone who will attempt to make a jam from pretty much any edible fruit I find in the wild, I have to express my appreciation for whoever makes Certo. This isn't a paid spot, I just feel comfortable putting my life in their hands! This makes enough for four of those cute little jam jars, so it's great for gifting and sharing your bounty.

As with all of the recipes in this chapter, you can use crab apples or Bradford pears, or a little bit of both! The flavor plays really well with warm spices like cinnamon or nutmeg, so feel free to grate in a little. And in this case it's ideal to use fruit that's still tart and maybe even a bit underripe.

MAKES ABOUT 4 CUPS

- **5 cups firm crab apples or Bradford pears**
- **3¾ cups granulated sugar**
- **1 tablespoon fresh lemon juice**
- **1 (3-ounce) pouch liquid pectin, such as Certo**

In a large saucepan over medium-high heat, combine the crab apples and 3¼ cups of water and bring to a boil. Reduce the heat to medium-low, cover, and simmer until the mixture is slushy and fragrant, about 10 minutes. Mash the fruit using a potato masher or a wooden spoon (you can push the fruit against the sides of the pan). Cover and continue simmering for 5 minutes. Remove the pan from the heat.

Set a fine-mesh strainer over a medium bowl. Pour the fruit mixture through the strainer and reserve in the bowl. Discard the solids.

Return the juice to the pan and stir in the sugar and lemon juice. Return the pan to medium-high heat and bring to a boil. Cook, stirring, until the sugar has completely dissolved, about 1 minute. Remove the pan from the heat and stir in the liquid pectin.

Carefully transfer the jelly mixture into clean jars and seal tightly. From here, you can let the jelly cool completely and store in the fridge for up to 3 months, or preserve using a hot water bath method if that's your jam. (See what I did there?)

CRISPY TOFU IN BRADFORD PEAR SAUCE

I love this sauce because it really shows you the caramelly side of a frost-bletted Bradford pear. Frost breaks down the cell walls and makes the fruit soft, condenses the flavor, and evaporates some of the liquid content. If you've heard of the method of freezing and then thawing an underripe persimmon to coax out its sweetness, it's the same process!

You can absolutely use crab apples here too, but you'll need to plan ahead. Since most of them get eaten by birds before they get super-ripe, you'll use that persimmon method: gather some underripe crab apples, freeze them, then thaw on the counter before using.

SERVES 4

FOR THE SAUCE

3 cups ripe Bradford pears, halved

1 tablespoon sesame oil

2 garlic cloves, minced

1 tablespoon minced fresh ginger

¼ cup soy sauce

2 tablespoons mirin

1 tablespoon sriracha or other chili sauce (optional)

FOR THE TOFU

1 (16-ounce) block extra-firm tofu, cut into 1 x 2-inch blocks

½ cup Acorn Starch (page 176) or cornstarch

1 teaspoon sea salt

½ teaspoon Umami Powder (page 190) (optional)

1 tablespoon neutral oil, such as sunflower or grapeseed, plus more

Hot cooked white or brown rice, for serving

Sliced scallions (green parts), for serving

MAKE THE SAUCE: Working over a medium bowl, push a piece of Bradford pear through a fine-mesh strainer to catch the peel and seeds. (They're so small that peeling them is way too fiddly!) Repeat with the remaining fruit. Discard the peels and seeds and set aside the bowl of pulp.

In a large skillet over medium-high heat, heat the sesame oil. Add the garlic and ginger and cook, stirring, until fragrant, 1 to 2 minutes. Add the reserved fruit pulp, soy sauce, mirin, and sriracha (if using), stir to combine, and bring to a simmer. Reduce the heat to medium-low and cook, stirring, until the sauce is smooth and thickened slightly, 8 to 10 minutes. Taste and adjust the seasoning, adding more soy sauce or mirin as needed. Remove the pan from the heat and set aside.

MAKE THE TOFU: Pat the tofu as dry as you can with a kitchen towel or paper towel. Set aside.

In a medium bowl, add the Acorn Starch, salt, and Umami Powder (if using), and whisk together with a fork. Add the tofu pieces and toss well to coat. Set aside.

In a large nonstick skillet over medium heat, heat the oil. Working in batches, fry the dredged tofu until golden and crispy on all sides, about 2 minutes per side. Transfer the crisped tofu to a rack and continue with the remaining tofu. Remove the pan from the heat.

Toss the tofu briefly with the Bradford pear sauce. Serve immediately with hot rice and top with the sliced scallions.

W
N
I

FIELD GA

THE YEAR WAS 1997. I was five and attempting to help my mom in the garden. Have you met any five-year-olds recently? They're not exactly the most helpful people when it comes to anything requiring fine motor skills. Knowing this from experience, my mom sent me on a little task elsewhere in the yard. What was that task? She pointed me toward a couple tufts of grass that looked mostly like regular ol' grass, but taller and darker green. I'd never noticed it before but when she pointed it out, suddenly I couldn't stop seeing it: all over our lawn, our neighbor's lawn, the little patch of grass across the street, and everywhere!

My mom told me to crush a bit of that grass in my hand, hold it to my nose, and take a whiff. Immediately my nose was filled with the sweetest, sharpest, garlicky smell that, up until that point, I had only associated with boxed pasta dinners or Pizza Hut. I was enchanted. "Who put this here?" I asked my mom. She told me that some weeds, or plants that showed up without anybody inviting them, were sometimes edible and also sometimes delicious and that this grass, onion grass, was one of them. My mom didn't know it yet, but she had sent me on a journey that would go on to define my entire life. And it's all because of onion grass, also known as field garlic!

Field garlic, or *Allium vineale*, isn't native to North America, but you can find it just about everywhere these days. We don't really consider it invasive as it doesn't push out other species with its presence. Instead, it often colonizes spaces where we've tried to install other monocultures, like lawns, but you can also find it among the grasses in slightly wilder green spaces, like forest preserves.

Field garlic is recognizable by its tubular, dark-green fronds, which are often even observable through the snow. They possess a waxy coating that keeps the water in their

RLIC

cells from expanding too much (which would kill the plant) in the winter. This makes them an excellent and delicious winter forage, especially if the ground is soft and wet enough to be able to get out the bulbs along with the fronds.

WHAT TO LOOK FOR:

- Tubular, dark-green **leaves** that look like blades of grass from a distance. A slight waxy coating, which helps them continue to grow and survive even in snowy conditions.
- The leaves get tough by mid-spring, and the plants will produce tiny globe-shaped **bulbs** on top of one of the leaves, a la garlic scapes. Those tiny bulblets are very pungent, and *very* tasty.
- **Lookalike alert!** Grass! LOL but true. If you've ever mowed your lawn and smelled spicy garlic in the air, congrats! Field garlic is hiding in your yard. It grows faster and taller than regular grass, so once you notice it for the first time, you'll start seeing it everywhere.

FIELD GARLIC SALT

Another super easy recipe that just gives any dish a kick in the pants. Wanna make Kraft Mac seem fancy? Field Garlic Salt. Wanna make toast with olive oil just feel decadent? Field Garlic Salt. Odds are, this would make any dish better. And this is a little work you can do to create a staple ingredient that keeps in the pantry!

MAKES ABOUT 1½ CUPS

2 cups chopped field garlic

1 cup sea salt

Preheat the oven to its lowest heat, ideally around 150°F and no more than 200°F. (Many ovens have a "keep warm" function that can be adjusted to a lower heat.) Line a baking sheet with parchment paper.

Spread the chopped field garlic in an even layer on the prepared baking sheet. Dehydrate the field garlic in the oven until fully dried and crisp, 2 to 4 hours, depending on your oven and your greens, checking on them every 30 minutes or so. Shake the pan and flip the greens occasionally to help them dry evenly. Set aside to cool slightly.

In a food processor, add the dried field garlic and salt and pulse until the greens are pulverized into the salt.

Transfer the salt to a clean jar, seal, and store in a cool place for up to 2 months.

FIELD GARLIC PARATHAS

I made these simple parathas and a version of the Violet Leaf Soup (page 57) for my foraging mentor, Sam Thayer, and his daughter a while back. Hoping to impress my guests, I waited anxiously for their reaction. They loved them! Do they require more effort than the Field Garlic Compound Butter (page 225)? Yeah. Is it worth it? Yeah! Will people be so impressed and love them so much? Heck, yes. I'm serving these flaky, beautiful treats to everyone I'm trying to impress from now on, forever.

MAKES 4 PARATHAS

2 cups all-purpose flour, plus more for dusting

1 cup chopped fresh field garlic (or chives, if field garlic isn't in season yet)

⅓ cup vegetable oil, plus more for brushing

1½ teaspoons kosher salt and/or Field Garlic Salt (page 221)

In a large bowl, toss together the flour and field garlic until the field garlic is coated in the flour. Add the oil and salt and use a fork to combine; the mixture will be a sort of bread crumb consistency. Add ½ cup of water and mix again, aiming for a thick dough consistency. If your mixture is still dry, add up to another ¼ cup of water, splashing in a tablespoon or two at a time. Cover the bowl and let the dough rest for at least 20 minutes and up to 1 hour.

Dust a work surface with flour. Cut the dough into 4 equal pieces. Set aside 3 of the pieces under a clean kitchen towel.

Roll out the piece of dough just about as thin as you can, about ⅛ inch. Brush the dough lightly with oil and dust lightly with flour. Fold the sheet of dough like those little accordion fans we made in elementary school, with each fold about ½ inch wide until you have a thick stack of folds. Pinch the ends of the stack, stretch the dough a few inches, and coil it into a little dough spiral.

Roll out the dough spiral to about ¼ inch thick, pinching the ends if they seem to want to run away. Cover the dough with a kitchen towel and repeat with the remaining pieces of dough.

Heat a dry cast-iron or nonstick skillet over medium-high heat. Add a paratha to the hot pan and brush lightly with oil. Cook for 1 minute, flip, and brush the other side with oil. Flip once more to crisp the other side, cooking for just 30 seconds or so. Transfer the cooked paratha to a plate and repeat with the remaining parathas. Remove the pan from the heat.

Serve warm.

FIELD GARLIC COMPOUND BUTTER

Compound butter is my favorite way to seem fancy while using the very lowest-level energy input. Put out a loaf of sourdough bread and compound butter at a dinner? People lose their minds!

MAKES ABOUT 2½ CUPS

2 cups chopped field garlic

2 cups (4 sticks) salted vegan butter, softened

Preheat the oven to its lowest heat, ideally around 150°F and no more than 200°F. (Many ovens have a "keep warm" function that can be adjusted to a lower heat.) Line a baking sheet with parchment paper.

Spread the chopped field garlic in an even layer on the prepared baking sheet. Dehydrate the field garlic in the oven until fully dried and crisp, 2 to 4 hours, depending on your oven and your greens, checking on them every 30 minutes or so. Shake the pan and flip the greens occasionally to help them dry evenly. Set aside to cool completely.

In a large bowl, add the softened butter and the dried field garlic. Use a spatula to fold and mix the field garlic evenly into the butter.

Transfer the butter to a sheet of parchment paper and use the sides of the paper to gently coax it into a log shape. Roll the parchment paper up around the butter, twist the ends of the paper, and continue rolling until the butter forms a smooth log.

Use immediately or store in an airtight container in the refrigerator for up to 2 weeks.

MAPLE

WHAT'S MORE SYNONYMOUS with North American breakfasts than maple syrup? Okay, okay, probably a lot of things, but you know she's up there! One of my favorite things to tell new foragers about is how maple syrup gets made. Much to some folks' surprise, maple sap, when tapped freshly from the tree, is not brown, thick, and sugary sweet, but clear, extremely refreshing, rich in minerals, and almost tastes like water. So how do we get from refreshing tree blood to the syrup we all know and love? Just cook it down! For a long time! That condenses all those minerals and sugars to make the sweet, cozy concoction found in so many pantries across this continent (and around the world!).

If you live in an area that gets a cold winter—and as a Midwesterner raised by a New Englander, when I say "cold winter" I mean y'all are chillin' below 32°F for much of your meteorological winter—then you live in an area where maples can be tapped for sap. Those freezing temperatures are extremely important because in February, March, and April, when the sap starts a-flowing, the alternation of below-freezing temps in the night and above-freezing temps in the day is necessary to trigger the vacuuming of the sap from the roots up to the rest of the tree. Tree physics!

But if you don't live in an area where sugaring—the act of tapping maple trees for sap and cooking them down to syrup or sugar—is possible, worry not! Maples have many wonderful gifts to give beyond their lovely sap.

In Minoh, Japan, there are stands that sell tempura maple leaves after preserving them in salt for a whole year! (We have a riff on the recipe in this very book!)

In the Pacific Northwest, big-leaf maple flowers usher in the spring with their tasty waterfalls of blooms, often used in salads and other dishes!

So, now you know maples are more than just tree blood and pretty fall colors. Hug a maple in your neighborhood today!

Take your drill, set the bit against the trunk, and make a 45-degree angle pointing up. Drill just about ¾ to 1 inch into the trunk. You'll know you've hit it when liquid comes out!

STEP FOUR: Let it flow!

Hammer your spile into the hole and place your bucket where it will catch the falling sap. If you have some of that tubing and a bucket lid, set those up now as well.

Now go do something else for the rest of the day. (You're gonna need a few hours to fill a 5-gallon bucket.)

STEP FIVE: Tree H_2O 4 U

The next morning, go collect your bucket! You can move on to the next section to make maple syrup, but remember you can also enjoy your sap fresh; it's full of minerals and quite refreshing. Some of my friends make their coffee with it this time of year. It's definitely worth a taste, even if you plan to use the bulk of your sap for syrup.

Maple Sugar

You get to go back inside for this one! There aren't a lot of steps here; the main thing to focus on is not burning the Maple Sugar (which is why we're gonna finish ours in the oven). The cooking times here can vary depending on your altitude, so let your senses guide you!

As you'd expect, Maple Sugar brings an earthy, maple flavor that makes anything instantly taste ultra cozy. It doesn't behave exactly the same as granulated sugar, so it's not quite interchangeable, but you'll see I call for it as an option in a couple of recipes in this book.

THINGS YOU NEED:

A pot

A baking sheet

A Silpat (silicone baking sheet liner) or parchment paper

2 cups Maple Syrup (or more; just make sure to use a larger pot if needed)

Preheat the oven to its lowest heat, ideally around 150°F and no more than 200°F. (Many ovens have a "keep warm" function that can be adjusted to a lower heat.) Line a baking sheet with a Silpat and set aside.

Add the Maple Syrup to a medium saucepan and set over medium-low heat. Cook, without stirring, until grainy crystals start to form and the syrup has thickened, about 15 minutes. Remove the pan from the heat.

Carefully spread out the thick syrup on the prepared baking sheet. Bake until the liquid has completely evaporated and you're left with what looks like (and is!) a clump of sugar, about 30 minutes.

From here, you can break the sugar into smallish clumps for things like tea sweetening (this looks a lot cooler and more cottage-y than sugar cubes). Or you can grind the sugar in a food processor or spice grinder. Either way, store the Maple Sugar in an airtight jar in a cool, dark, dry place.

Maple Syrup

Listen. Some folks on the internet will swear you can make maple syrup indoors, but I don't recommend it! The one time I tried it, it took the paint off the walls of our (rental!!) kitchen. So, outside we go!

HERE'S WHAT YOU'LL NEED:

A cord or two of firewood

A fire pit, outdoor fireplace, or makeshift fire pit using stray bricks

A metal grate or anything heatproof on which to securely rest the pot over the fire

A long-handled spoon

5 gallons of maple sap (You can also do this with birch or sycamore sap!)

Patience

STEP ONE: Build the fire and prep the pot

Set up your fire area, including the grate. Get a fire going and let it burn until you've got hot coals, then feed it a bit more wood.

Meanwhile, grab your largest pot and fill it about two-thirds of the way with sap. (The sap can foam and bubble so you want to give it a bit of headroom.) It's likely your sap won't all fit in the pot at once; you'll be adding new sap as liquid cooks off in the pot.

STEP TWO: Game on!

Set your pot of sap on the fire. Check on it periodically, stirring with your long spoon every 30 minutes or so to keep minerally bits on the bottom from burning. Feed the fire as needed to keep a fairly consistent heat.

Now, invite some buddies over because you're gonna be outside for a bit. Sap sessions used to be social gatherings, especially in New England, so why not make a day of it? If there's residual snow on the ground, kids (and grown-up kids) can do fun things like make maple snow cones and maple candy while y'all work!

You'll need at least 3 hours, likely more, depending on the amount of sap you're starting with, the pot size and shape you're working with, and the strength of your fire. (Don't take this as a hint that your fire needs to be hotter! Remember that the maple solids will burn, so let this be a slow, gentle evaporation process.)

Cook the sap down to your desired sweetness. Some folks like a lighter, runnier syrup than what we can get in stores, and some like a syrup that's thicker, almost like molasses. There's no wrong answer!

STEP THREE: Jar it up

Carefully remove the pot from the fire and bring it inside. (I'm assuming that's where you keep your syrup vessels.) Ladle the hot syrup into clean, sanitized jars, leaving as little headroom as possible, and seal tightly. Store in a cool, dark place for up to 1 year. Once opened, keep your syrup in the fridge.

A fun note: I recently took a class from foraged cordials master Dwight Zietlow on making infused maple syrups! You essentially add other flavor or medicinal agents like black cherry bark or dried fruits to the sap before cooking, and then strain them out when you get to a syrup consistency. They're a lot of fun to experiment with and they're very tasty.

MAPLE FLOWER FRITTERS

A quintessential end of winter/beginning of spring recipe. Maples flowering signals to me that my Seasonal Affective Disorder is almost done for the season. And what a delicious celebratory bunch of tree streamers they are! Everyone recognizes maple leaves but not everyone pays close enough attention to the flowers, which come a week or two prior. The flowers are also fun to just straight giraffe off the tree (yes, exactly what it sounds like), but if you're not into that, you can simply make these fritters and impress your friends instead of weirding them out.

If you've missed the window with flowers and your maple tree of choice has started to put out its lil' green helicopters, you can still use the clusters! Just make sure they're really new and young; their color will be a light- or bright-green.

These are great on their own, but they also go great with a tart dipping sauce, like a very mustardy honey mustard!

SERVES 2 TO 4

Heaping ½ cup cornmeal

½ cup all-purpose flour

2½ teaspoons sea salt

¾ cup unsweetened soy milk

1 Starch Egg (page 177)

2 tablespoons Maple Syrup (page 232) or store-bought

Neutral oil, such as sunflower or grapeseed, for frying

20 to 30 maple flower clusters (these can range greatly in size)

In a medium bowl, add the cornmeal, flour, and salt and whisk to combine. Add the milk, Starch Egg, and Maple Syrup and whisk again until smooth. Set aside.

In a high-sided, heavy-bottomed frying pan or skillet, heat about 2 inches of oil to 350°F over medium-high heat. Set a rack over a baking sheet and place near your frying station.

Using the stem ends as little handles, dip the maple flower clusters in the batter, dunking just briefly to coat the flowers. Shake off any excess and add the battered maple cluster to the hot oil. Fry until golden, 1 to 2 minutes per side. Use a slotted spoon or tongs to transfer the fritters to the prepared rack. Repeat with the remaining clusters. Remove the pan from the heat.

Serve warm as is, or with your favorite dipping sauce.

TEMPURA FIRST MAPLE LEAVES

Tempura maple leaves is a seasonal dish that hails from Minoh, Japan. While folks there often choose fallen maple leaves in the autumn (and treat the leaves with a bit of a salt-cure for a few days before frying), you can do the same thing with the first young maple leaves that might start to unfurl around the end of winter. I kind of prefer this method, not because it tastes better—I can confirm from firsthand experience that both versions are delicious—but because when the maple leaves arrive, it's a sign we've been safely handed off from Papa Winter to Mama Spring like the plucky child of divorce the human race is.

You'll be tempted to wait until the leaves are a proper grown-up size, but don't! Make sure you look out for that new leaf shine, like the glossiest little leaf you ever did see. This means they're still tender enough for a quick tempura without the salt-cure.

SERVES 2 TO 4

1 cup all-purpose flour, plus more

1 Starch Egg (page 177)

1 cup ice water

Neutral oil, such as sunflower or grapeseed, for frying

20 young maple leaves

Sift the flour into a small bowl and set aside.

In a medium bowl, whisk together the Starch Egg and ice water. Add the flour and whisk until just combined—a few lumps are okay! Set aside.

In a heavy-bottomed, high-sided skillet, heat about 1 inch of oil to 350°F. Set a rack over a baking sheet and place near your frying station.

Add a bit of flour to a plate. Dredge a maple leaf lightly in the flour, then dip in the tempura batter. Shake off any excess and carefully lay the battered leaf in the hot oil. Fry until just golden, 1 to 2 minutes per side. Use a pair of tongs or a slotted spoon to transfer to the rack. Repeat with the remaining leaves. Remove the pan from the heat.

Serve immediately.

WINTER O

THOUGH THE TITLE of this chapter is indeed winter oysters, *Pleurotus ostreatus* can be found many months in the year, depending on where you live and what your weather's been like. We're nestling them here in the winter chapter because in my neck of the woods, I do tend to find them between December and February more than any other time of year.

Oyster mushrooms are the type of fungi you'll find growing on decaying wood, or even on trees that are still alive (in which case they're actively causing white rot). They're also one of the easiest mushrooms to cultivate, with many people cultivating them in laundry baskets full of straw, or those cute little grow bags you can sometimes get from the farmers' market!

Oyster mushrooms are also secretly metal as hell because they're carnivorous! They paralyze, dissolve, and eat nematodes (tiny wormlike organisms). I don't know what I'm going to do with that information other than savor it! Also, please don't let this factoid scare you, there is no risk of a *Last of Us*–style situation happening. Unless you're a nematode!

Oysters are extremely versatile in the kitchen. Pizza topping, hearty mushroom jerky, sumptuous soup base! If a mushroom can do it, you can probably do it with an oyster. And while *Pleurotus ostreatus* is my hometown oyster, these recipes work with any of the oyster mushrooms growing worldwide, and in any season. So, go find, buy, or grow your own and have fun!

YSTERS

WHAT TO LOOK FOR:

- Winter oysters are really just a hardier oyster mushroom, and **grow** from dead and living wood, just like regular oysters.
- Their **scent** is slightly reminiscent of anise and almonds, they have **curved cap edges** that curl under slightly with age, and their **gills** head all the way down their stipe (which is fungi-person for stem). Oh, and **FYI**: Mind the small black beetles that love hiding from winter weather in the gills.
- **Lookalike alert!** The only lookalikes I see in my area, and I'm using that term very loosely, are mock oysters, or *Phyllotopsis nidulans*, which are smaller in size on average and lack the characteristic oyster mushroom smell.

OYSTER & ACORN SPREAD

I stopped short of calling this a pâté because I am not that girl! Also, anecdotal evidence shows that friends and family members who are suspicious of new foods are more likely to try something if it has a name they understand. Everyone knows what a spread is! This is so savory and versatile and works just as well on a few crackers and eaten over the sink as it does on a snack tray or charcuterie board. There's also something cool about spanning the life cycle of the oak tree in one dish—starch from their acorns and the mushrooms that grow from their decaying remains.

SERVES 2 TO 4

4 cups winter oyster mushrooms (about 8 ounces)

2 tablespoons extra-virgin olive oil

2 garlic cloves, minced

Sea salt or Field Garlic Salt (page 221)

2 tablespoons Acorn Starch (page 176) or cornstarch, plus more

Crackers or crusty bread, for serving

In a food processor, add the mushrooms and pulse just until finely chopped. You can also chop them by hand.

In a large skillet over medium-high heat, heat the oil. Add the mushrooms and cook, stirring frequently, until the mushrooms have given up their liquid and begin to brown on a few edges, about 7 minutes. Reduce the heat to medium-low, add the garlic and salt to taste, and cook until the garlic is fragrant, about 1 minute. Sprinkle the Acorn Starch over the mushroom mixture and toss to coat the mushrooms evenly. You'll notice the starch will start to lightly bind the mushrooms together. If the mixture still looks a bit loose, sprinkle over another tablespoon and toss again. Remove the pan from the heat.

Pack the mushroom mixture into a small bowl or ramekin and serve warm, with crackers or bread.

FRIED OYSTER MUSHROOM PO'BOYS

This sandwich is inspired by my pals at Willowbeez Soulveg in Columbus, Ohio. They're a local, Black-owned vegan restaurant and their oyster mushroom po'boy is so good, I got one any time I was at their stall in the North Market. They transitioned to doing pop-ups, so if I have a craving when they're off, or if I'm traveling and have a decent kitchen where I'm staying, I throw together this very solid approximation to remind me of home!

SERVES 4

FOR THE MUSHROOMS

- 6 tablespoons unsweetened soy milk
- 2 tablespoons Acorn Starch (page 176) or cornstarch
- 1 teaspoon Cajun seasoning
- ½ cup all-purpose flour
- ½ cup cornmeal
- 1 teaspoon sea salt
- ¼ teaspoon freshly ground black pepper
- Neutral oil, such as sunflower or grapeseed, for frying
- 4 cups oyster mushrooms, torn into bite-sized pieces (about 8 ounces)

FOR THE PO'BOYS

- 4 hoagie rolls, split
- Vegan mayonnaise
- Shredded iceberg lettuce
- Sliced pickles (optional)
- Sliced tomatoes (optional)
- Hot sauce (optional)

MAKE THE MUSHROOMS: In a medium bowl, whisk together the milk, Acorn Starch, and Cajun seasoning until smooth. Set aside.

In another medium bowl, whisk together the flour, cornmeal, salt, and pepper. Set aside.

Heat about 2 inches of oil in a high-sided, heavy-bottomed frying pan or skillet over medium-high heat to 350°F. Set a rack over a baking sheet and place near your frying station.

Dip a piece of oyster mushroom in the wet mixture and shake off any excess, followed by the dry mixture, again shaking off any excess. Carefully lay the mushroom in the hot oil and fry until golden and crisp, 1 to 2 minutes per side. Repeat with the remaining mushrooms, working in batches to avoid crowding the pan. Use tongs or a slotted spoon to transfer the fried mushrooms to the prepared rack. Remove the pan from the heat.

ASSEMBLE THE PO'BOYS: Spread the hoagie rolls with mayo and create a nice bed of shredded lettuce on the bottom of the rolls. Nestle several fried mushrooms on the lettuce and top as desired with pickles, tomatoes, and hot sauce. Serve immediately.

OYSTER MUSHROOM CHOWDAH

This is a foraged and veganized homage to the clam and seafood chowders I grew up eating with my mom's family in New England. My nana was raised in Cape Cod and my mom in Boston, so the chowders I grew up with were white and creamy and so good, they haunted my dreams. (Also, I like to live in a world where I pretend Manhattan clam chowder doesn't exist, okay?)

"But, Alexis!" you cry. "New England clam chowder has, well . . . clams in it! And something porky or bacony!" To which I say, "Relax!" In this soup, we're getting our ocean-y flavor from a quick kelp broth, our bacony essence from dulse seaweed, an extra umami boost from miso, and our chewy shellfish texture from giving the mushrooms plenty of cooking time. It's something I can easily make from my pantry in winter, and it definitely scratches that nostalgic itch.

SERVES 4

2 cups fresh oyster mushrooms, cleaned and chopped into bite-sized pieces (about 4 ounces)

½ ounce dried kombu or oarweed kelp

3 tablespoons unsalted vegan butter or extra-virgin olive oil

1 large yellow onion, chopped

¼ ounce dried whole dulse seaweed (the flakes won't work as well here)

Sea salt and freshly ground black pepper

1 pound russet potatoes, unpeeled, cut into ½-inch pieces

1 tablespoon white miso

1 cup unsweetened soy milk

Chopped fresh field garlic or chives, for serving

Oyster crackers or crusty bread, for serving

Add the mushrooms, kombu, and 5 cups of water to a medium saucepan over medium-high heat. Bring to a boil and reduce the heat to medium-low, so the broth simmers gently, and cook for about 30 minutes. It will reduce a bit in that time, which is good—we're concentrating flavors! Remove the pan from the heat. Strain and reserve the broth, and pick out and reserve the mushrooms. Discard the kombu.

Melt the butter in a large, heavy soup pot over medium heat. Add the onion, dulse, and a good pinch of salt and cook, stirring occasionally, until the onions are browned on the edges, 8 to 10 minutes. Add the reserved mushrooms and cook, stirring frequently, until the mushrooms take on a bit of color, 5 to 7 minutes. Add the potatoes, the reserved broth, another pinch of salt, and a few grinds of pepper and stir to combine. Increase the heat to medium-high and bring to a boil. Reduce the heat to medium-low and simmer until the potatoes are tender, about 10 minutes. Remove the pot from the heat, and remove and discard the dulse.

Place the miso in a ladle and dip the ladle in the pot, picking up just enough liquid to cover the miso. Use a spoon to dissolve the miso in the broth, add the mixture back to the pot, and stir to combine. Add the milk and stir again. Taste and check your seasoning—you'll want lots of black pepper here, and probably a bit more salt. Remove the pot from the heat.

Divide the chowder among bowls and top with the field garlic. Stir in oyster crackers if you're old school, or serve the soup with crusty bread.

ROSE HIP

IT'S LIKELY A surprise to no one that I spend a great deal of time in plant identification groups on Facebook. Fewer things bring me a tiny shot of dopamine like helping someone figure out what plant friends have been hiding under their nose! And without fail, every late September or early October, the groups are inundated with people panicking into the ether: "What are these things on my roses now that they're done blooming??" And then I get the joy of letting them know that showy flowers aren't the only gift roses have to give! For after a rose is successfully pollinated, it forms the fruit known as a rose hip.

As a kid running around Cape Cod with my mom, I always knew that school was going to start soon when the beach rose shrubs were more hips than roses. Little did I know at that time, I was being very spoiled by how large the hips of the beach rose, *Rosa rugosa*, are compared to many of its more diminutive cousins. That being said, if you're just wanting your rose hips for something like tea or jelly or cordials like the recipes in this chapter, size does not matter!

Rose hips are extraordinarily high in vitamin C and thus usually find their way into many a cold-weather tea blend. While some rose hips can have tasty flesh, especially after being bletted by a frost or two, I will warn you: The flesh surrounds hairy seeds and if you accidentally eat those seeds, their exit from your body will not be the most pleasant of experiences!

Because of how global and cosmopolitan the *Rosa* genus is, you can find delicious rose hip recipes from across the globe! If you like the ones you find here, I encourage you to look into others from around the world and throughout the centuries. I think the following recipes are an excellent intro to both working with rose hips and really enjoying the tartness that they can bring to a dish or beverage.

S

WHAT TO LOOK FOR:

- If you've read through the Roses section (page 122), you should be good to go.
- Basically, where there were roses, there are now tiny red fruits. And those, friends, are rose hips!

ROSE HIP CORDIAL

If you thought winter killed yeast, you're wrong! The lil' beasties on the outside of those rose hips are still kickin' and are the secret superheroes of this lightly fermented recipe. This cordial tastes really summery for something you gathered in the dead of winter. If you're giving this to wee ones, I'd highly suggest not letting it ferment past the initial two-day stage. And in general, I cap this fermentation at five days, maximum, unless I'm intentionally making a vinegar, which is fine too but it's not this recipe!

MAKES ABOUT 2 CUPS

2 cups warm water

½ cup granulated sugar

½ cup fresh rose hips, unwashed

In a clean quart jar, add the water and sugar and stir until dissolved. Add the rose hips and swirl the jar gently to make sure the rose hips are mostly submerged. Cover the jar with a clean kitchen towel, paper towel, or a few layers of cheesecloth—anything porous. Secure the towel to the jar with a rubber band. Set aside at room temperature for 12 hours.

Using a clean lid, seal the jar tightly and shake the jar to help activate the yeast. You can also use a sanitized stainless steel spoon to stir the mixture if you can't find a suitable lid. (It happens to the best of us!) Replace the towel to cover the jar. Repeat this every 12 hours for 2 days.

After 2 days, the mixture should be active and bubbly. If it still tastes too sweet, let it ferment for another day or so, continuing to shake or stir every 12 hours.

Strain the rose hips from the cordial and discard them. Transfer to a clean jar or bottle and store, sealed, in the refrigerator for up to 3 months.

Drink this cordial over ice, top it with fizzy water, use it as a starter for other fizzy beverages or even wake up a sourdough starter.

ROSE HIP TEA

Sometimes it can feel like winter is a barren wasteland filled with snow and Seasonal Affective Disorder. But look a little closer and you might see some beautiful bursts of red among the dead brambles. Rose hips are actually better if you let them stay on the plant until winter: By then, they will have been bletted by frost (see page 244), which actually makes their flesh sweeter. So, while you might be tempted to start nabbing some in the fall, your patience will be rewarded if you wait. Rose hip tea is the also the perfect concoction to kick up your winter immunity, since it's chock-full of vitamin C!

This recipe serves one with fresh rose hips, but you can also gather a whole bunch, make tea with a few freshies, then dehydrate the rest in a low oven to use for later! If you go that route, use three to four dried rose hips per mug. The sweetener here is totally optional, since winter rose hips are often very sweet on their own because of that frost.

SERVES 1

3 tablespoons fresh dog rose or beach rose hips (3 to 4 large, more if they're small)

Honey or granulated sugar (optional)

In a medium saucepan over medium-high heat, add the rose hips and 1¾ cups of water. Bring to a boil, reduce the heat to medium-low, and simmer gently until the tea reaches a strength you like—I like to go for about 10 minutes. Remove the pan from the heat.

Strain the tea into a mug and sweeten as desired.

ROSE HIP JELLY

This is for the folks who want an easy way to work with rose hips but who might not be dedicated tea drinkers. Or for those who'd rather eat their vitamin C and make it go *wobble*. I like to keep this simple and enjoy it on a piece of sourdough toast, or if I'm feeling decadent, I'll put vegan butter on it first. Basic! But this would be really good in a thumbprint cookie or even a cute lil' hand pie.

If you took it upon yourself to make Linden Flower Tea (page 147) earlier in the year, you can throw a few flower clusters into this mix to add some lovely linden flavor. Just make sure you strain 'em out before adding the pectin!

MAKES ABOUT 4 CUPS

3 cups granulated sugar or Maple Sugar (page 230)

1¾ cups Rose Hip Tea (page 249)

¼ cup fresh lemon juice

1 (3-ounce) pouch liquid pectin, such as Certo

In a medium saucepan over medium-high heat, combine the sugar, tea, and lemon juice. Stirring occasionally to help the sugar dissolve, bring the mixture to a boil. Let it boil hard for 2 minutes and remove the pan from the heat. Stir in the pectin.

Carefully transfer the jelly mixture into clean jars and seal tightly. From here, you can let the jelly cool completely and store in the fridge for up to 3 months, or preserve using a hot water bath method if you'd prefer.

JUNIPER

JUNIPER IS OFTEN thought of in conjunction with Scandinavian or northern European cooking, conjuring images of thinly sliced cured meats, pickled veggies, and tiny, botanical gin-based beverages. So, it's often surprising for people to learn that North America has several native members of the *Juniperus* genus!

Confusingly, American junipers are often referred to as cedars. For example, my particular area's native juniper, *Juniperus virginiana*, is often called the eastern red cedar. Why!? Who did this!? And there are other (actual) cedars that you probably shouldn't eat, which, yes, are in a different genus and not *Juniperus*, but let's face it, the nomenclature does not help! Why did we let old dead dudes name these plants?

While our American juniper berries are not as inherently sweet when fresh off the tree, like many of their northern European counterparts, they are still extraordinarily flavorful and absolutely delicious and are used in many of the same applications!

In addition to being excellent for eating (or I guess I should say "spicing"; you probably don't want to eat too many of these fresh off the tree), simmering juniper branches in water makes your house smell amazing. I often do this when my partner or I are suffering any kind of wintertime sickness.

Junipers are also extraordinarily important to many Indigenous groups across North America because of what they bring to cuisine; because of their extremely rot-resistant wood, which was excellent for building things like chests or fences; and for a myriad of spiritual reasons.

WHAT TO LOOK FOR:

- Unlike some of the conifers, junipers have scaly-looking **leaves, not needles**, which layer on one another.
- Their "cones" are the juniper **berries** you'd recognize from the spice store or maybe a fancy cocktail. You'll want to gather the berries when they've darkened in color, so when you wipe the yeast bloom away with your finger, it's a dusky dark-gray. If they're still green, they're not ready yet!
- **Lookalike alert!** As mentioned before, the cedars in the *Thuja* genus are often conflated with junipers. Thujas are thought to be generally nonedible and you can tell the difference when looking at their fruits. Junipers will look like a standard little round berry, whereas thujas' fruit will look almost like a tiny pine cone. Cute, but not for eating!

JUNIPER CORDIAL

Have you ever thought, "Man, I wish gin tasted like gin but better"? Then the Juniper Cordial might be the fun, fizzy beverage for you! All of the resinous aromatics you love in gin without all of the annoying alcoholic burning taste. (Sorry, gin lovers.) The yeast on juniper berries tends to be quite active—and as we learned in the Rose Hips chapter, just because it's winter doesn't mean wild yeast dies off!—so you can use this to start other fermented beverages like you might use a ginger bug. You can also use this to give sourdough starters a kick in the pants! Much like the Rose Hip Cordial, this shouldn't ferment past about five days. Otherwise, it'll be a vinegar, which you don't want, unless you do!

MAKES ABOUT 2 CUPS

2 cups warm water

½ cup granulated sugar

½ cup fresh juniper berries, unwashed

In a clean quart jar, add the water and sugar and stir until dissolved. Add the juniper berries and swirl the jar gently to make sure they're mostly submerged. Cover the jar with a clean kitchen towel, paper towel, or a few layers of cheesecloth—anything porous. Secure the towel to the jar with a rubber band. Set aside at room temperature for 12 hours.

Using a clean lid, seal the jar tightly and shake the jar to help activate the yeast. You can also use a sanitized stainless steel spoon to stir the mixture. Replace the towel to cover the jar. Repeat this every 12 hours for 2 days.

After 2 days, the mixture should be active and bubbly. If it still tastes too sweet, let it ferment for another day or so, continuing to shake or stir every 12 hours.

Strain the juniper berries from the cordial and discard them. Transfer to a clean jar or bottle and store, sealed, in the refrigerator for up to 3 months.

RYE TARTINES with Juniper Salt

Really hard to go wrong here given my obsession with rye bread, herby salty things, and certain vegan cream cheeses. (Here's lookin' at you, Tofutti!) I'm also a sucker for a dish that looks fancy but is really just things on toast, which is essentially the definition of a tartine. This one is giving quiet Scandi luxury and I love her. If you're making this in late winter or early spring, you can top with some chopped field garlic and it would be *chef's kiss*. You can make these with those crunchy rye crispbreads or with a softer caraway rye thrown in the toaster for a few minutes, and add any toppings you want. You can also consider these an ideal foundation for the Juniper-Cured Carrot Lox (page 257)!

MAKES ABOUT ¼ CUP OF JUNIPER SALT

FOR THE JUNIPER SALT

- 3 tablespoons coarse salt
- 1 tablespoon juniper berries

FOR THE TARTINES

- Rye crispbreads or lightly toasted caraway rye bread
- Vegan cream cheese (I like Tofutti)
- Sliced cucumbers
- Sliced tomatoes
- Sliced red onion
- Chopped fresh dill and/or chives

MAKE THE JUNIPER SALT: In a spice grinder or small food processor, add the salt and juniper berries and blitz until the juniper berries are little specks in the salt. Transfer to a small jar. You can use this right away or seal and store in a cool, dark place for up to 3 months.

MAKE THE TARTINES: Spread the crispbreads with cream cheese. Top simply with a sprinkle of juniper salt, or pile on some toppings first and finish with the juniper salt.

JUNIPER-CURED CARROT LOX

Here it is, folks, in all its chaotic faux-fish glory: how I make carrot smoked salmon. Of all the recipes I've shared on the internet, this one has probably been the most requested. The flavor here squares up to that of a gravlax, which is traditionally cured with salt, sugar, and other spices like dill, plus my sweet baby winter forage: juniper!

I absolutely adored smoked salmon growing up and, much like the gently terrifying spicy chicken sandwiches sold in my junior high cafeteria, it was very hard for me to give up when I went vegetarian as a wee, hungry, angsty teen. In fact, I broke my vegetarianism once as a high school sophomore for a gravlax bagel, eaten while sitting across from the first person I kissed. He remembered I loved it when we "dated" (!) as twelve-year-olds (!), so he bought it for me before I got to the coffee shop, and I was afraid of declining it and being rude to someone who at one point had seen me as cute and desirable. (Who among us hasn't engaged in some people-pleasing behaviors in our youth??)

Eating these carrot lox took me right back to that coffee date, and I laugh in that way we all do looking back on our awkward teen years, like when I think about how often I straightened my hair.

I'm not going to tell you this is a perfect dupe, only because I don't feel like I can confidently say something tastes like something I last ate more than a decade ago (while distracted by a crush, no less). But what I can say is it was close enough to bring me back! If you're looking for a vehicle for these ribbons of smoky, fish-adjacent goodness, layer them on the Rye Tartines with Juniper Salt (page 256). They're also great alongside a tofu scramble or in a breakfast sandwich!

RECIPE CONTINUES

MAKES ABOUT 3 CUPS

3 wide, thick carrots

¼ cup sea salt

2 (3 x 4-inch) nori sheets (the little snack-size guys), crumbled

2 teaspoons fresh or dried juniper berries

1 tablespoon chopped fresh dill

2 tablespoons flaxseed oil

½ teaspoon liquid smoke

½ teaspoon maple syrup

Using a mandoline or vegetable peeler, very thinly slice the carrots into ribbons. Place in a bowl or food storage container and set aside.

In a mortar and pestle, add the salt, nori, juniper berries, and dill and crush into a crumbly, dry paste. Set aside.

In a medium bowl, whisk together the oil, liquid smoke, and maple syrup. Add the salt mixture and whisk again to combine. At this point, you can add more liquid smoke or maple syrup to adjust the flavor.

Pour the mixture over the carrots and toss well to coat. Cover or seal the container and marinate in the fridge for at least 3 hours, ideally overnight.

To serve, shake off any excess curing mixture and add to your dish as desired.

The lox will keep in the fridge for 3 to 5 days. After that, it doesn't go bad so much as it becomes over-cured and too salty!

EVERGREENS

PINES, SPRUCES & FIRS

EVERGREENS: THEY'RE NOT just for Christmas! In fact, they offer a wide breadth of flavors throughout the year, so bookmark this for next spring and summer. Folks often conflate pines, spruces, and firs, but thankfully, while their flavors are each unique in its own way, members of all three groups work well for any of these recipes.

Additionally, like their friends the junipers, pines, spruces, and firs are often covered in a great deal of wild yeast, making them excellent for things like sourdough or fizzy beverage starters. I prefer using the unripe cones, as one would do for Mugolio (page 263), to start fermented beverages as they add a great deal of citrusy goodness, not to mention high levels of vitamin C and lovely resinous notes. Personally, I love spruces the most, as their flavor tends to be a bit lemony!

Of course, evergreens have an association with winter holidays because, well, not many other plants are kickin' this time of year! This also makes them one of the few exciting forages between the months of December and March for those of us who have a snowy, leafless winter. Historically, both in North America and abroad, their needles have often been life-saving sources of vitamin C when other plants were unavailable to us. For many North American Indigenous groups, their inner bark also provided a flour source for breads. So, if anyone ever tries to tell you that there is no foraging in winter, lovingly let them know they're wrong, maybe with a plate of pine shortbread cookies.

Now, just a little reminder: As exciting as it is to learn pine cones are edible, they do have to be completely green all the way through in order to become snacks. I know this is not the state in which we usually find them on the ground, so keep an eye out and search the branches of your friendly neighborhood evergreens to nab the cones before they're too woody.

ENS

WHAT TO LOOK FOR:

- **Pines** have long, softer **needles**. The two most common ones near me, the white and red pine, have bunches of five and two needles, respectively. They make the pine cones (which are where pine nuts come from!) that you'd recognize from the craft store.
- **Spruces** are often recognized for their use in winter holiday decor. My family had spruce Christmas trees when I was growing up. Spruces are quite triangular in shape, have shorter **needles** than pines, and can be a lil' stabby! The needles have white lines set into the underside, differentiating them from yews.
- **Firs** look similar to spruces, but their **needles** are soft. They also give out a pleasant lemony scent when crushed. Spruce might do that too if it didn't stab you first.
- **Lookalike alert!** There is a needle-having evergreen that is not safe for consumption and that is the yew, from the *Taxus* genus. They possess a great deal of glycosides, which, yes, are helpful in the synthesis of life-saving chemotherapy medication, but in their raw state can absolutely be lethal. Luckily, yews usually (not always! but usually) show up as shrubs so you likely won't mistake one for a Christmas tree. And instead of woody cones, yews produce small brown seeds protected by what looks like a cup-shaped red berry.

PINE MUGOLIO

Mugolio, a pine cone syrup with Italian origins, harnesses the power of wild yeast to make a jar of what looks and tastes like liquid gold. A very expensive concoction if you were to buy it in a shop, mugolio is actually the ultimate set-it-and-forget-it recipe. Remember that conifer cones are covered in wild yeast, so when you pack them in with sugar, the yeast eats some of the sugar and liquid is drawn out of the cones. When yeast eats, it burps out CO_2 to produce an acidic (low in pH), very slightly alcoholic, and, best of all, shelf-stable result.

I cannot adequately stress the importance of burping your jar here. That's right, open the lid to release excess carbon dioxide and seal 'er back up. This might not feel like fermentation but it absolutely is! And the last thing you want is a sticky, piney explosion with broken glass and probably tears.

Mugolio's sweet, resinous flavor is incredible drizzled on ice cream, or stirred into tea or sparkling water. If you want to go full cocktail, combine a shot of mugolio with a shot of your favorite herby, botanical gin, and top with ice and sparkling water.

MAKES ABOUT 2 CUPS

2 cups green, unripe pine, spruce, or fir cones

Granulated sugar

Add the cones to a clean 12-ounce jar. Add enough sugar to fill the open space between the cones, plus a thin layer on top. Seal and shake the jar gently to make sure all the cracks are filled, adding more sugar as needed to pack everything in.

Seal the jar tightly and place it in a sunny spot—inside, if it's below freezing outside, or outside if your weather is warmer. Burp the jar each day and let sit for about 2 weeks, until all the sugar has become liquid.

Set a fine-mesh strainer over another clean jar and strain out the pine cones. Discard the pine cones. Seal the jar and store the mugolio in a cool, dark, dry place.

PINE SHORT-BREAD COOKIES

The coziest cookies for a season that, at least for me, requires all the coziness I can get! These are so delightful with tea, tied up in cute bags as gifts, or just as a sweet snack. And this is another instance where the piney-ness isn't overwhelming because you've dried the needles and whizzed them up with sugar! Here, as with the Linden Shortbread Cookies (page 151), make sure you're using the vegan butter sticks (as opposed to a "buttery spread") so your cookies keep their shape.

MAKES ABOUT 30 COOKIES

FOR THE PINE SUGAR

1 cup fresh pine, spruce, or fir needles

1 cup granulated sugar

FOR THE COOKIES

2½ cups all-purpose flour

¼ teaspoon sea salt

1¼ cups (2½ sticks) unsalted vegan butter, softened, plus more for the pan

½ teaspoon vanilla extract or Nocino (page 199)

MAKE THE PINE SUGAR: Preheat the oven to its lowest heat, ideally around 150°F and no more than 200°F. (Many ovens have a "keep warm" function that can be adjusted to a lower heat.) Line a baking sheet with parchment paper.

Lay the pine needles out on the prepared baking sheet and let them dry in the oven until completely dry and crisp, 1 to 2 hours, depending on the needles and your oven. You can start to check them after about 30 minutes to see how things are progressing. Let cool completely.

In a food processor, add the sugar and the dried pine needles and process until the mixture is a fine dust and the pine needles have completely broken down. You're aiming for a consistency between granulated sugar and powdered sugar (similar to superfine sugar).

Transfer the pine sugar to a small jar and set aside. You'll be using ¾ cup of it here, so you can seal and store any extra for another use.

MAKE THE COOKIES: In a large bowl, whisk together the flour, ¾ cup of the pine sugar, and the salt. Add the butter and vanilla and use a spatula to work the ingredients together to form a crumbly dough. (You can also use a hand mixer at medium speed.) The dough may feel a bit dry, but should stick together when pressed.

Turn the dough out onto a large sheet of plastic wrap. Use the sides of the wrap to shape the dough into a log about

RECIPE CONTINUES

12 inches long. Gently pat the ends of the log in to straighten the sides, twist the ends of the wrap to close it, and tuck them under the log. Chill the dough for 1 hour, until firm.

Preheat the oven to 325°F. Line two baking sheets with parchment paper.

Unwrap the dough and slice into thin rounds, no more than ¼ inch thick. Place the rounds 1 inch apart on the prepared baking sheets. Bake for about 15 minutes, until the cookies are just beginning to turn golden on the bottom edges. Let cool for 5 minutes in the pan and transfer to a cooling rack to cool completely. (If you ran out of room on your baking sheets, you can bake off the remaining cookies now.)

Serve immediately or store in an airtight container at room temperature for up to 1 week.

CONIFER NEEDLE CRACKERS

Think of these like a super-seasonal version of a rosemary cracker. There's just a hint of pine here, making these a great way to gently introduce friends or family to the flavor. These absolutely want to be on your wintry snack plates, maybe with Oyster & Acorn Spread (page 239), or alongside some Knotweed Dip (page 73).

You can use conifer needles of any age, just as long as they're fully green (brown spots won't hurt, but they won't bring much flavor). If you see new, bright green tips on a pine, spruce, or fir tree in any season, grab them! They're more tender than mature needles and bring a slight sweetness to anything, including these tasty crackers.

MAKES ABOUT 24 CRACKERS

2 cups all-purpose flour, plus more for dusting

2 tablespoons chopped fresh pine, spruce, or fir needles

1 teaspoon sea salt, plus more for topping

½ teaspoon baking powder

½ cup lukewarm water

¼ cup extra-virgin olive oil, plus more for brushing

Preheat the oven to 400°F. Line two baking sheets with parchment paper and set aside.

In a medium bowl, whisk together the flour, pine needles, salt, and baking powder. Make a well in the middle, add the water and oil, and mix until combined.

Dust a work surface with flour and turn out the dough. Dust the top of the dough with flour as well (it'll be a bit sticky) and knead until smooth, about 3 minutes.

Using your hands, shape the dough into a rough rectangle. Dust once more with flour and use a rolling pin to roll out the dough as thinly as you can, about ⅛ inch or thinner. Try your best to keep the rectangle shape here, but don't stress—homemade crackers look nicer if they're a bit rustic, right?

Using a pizza cutter or knife, cut the dough into roughly 2 x 4-inch rectangles (a superior shape for dipping). Lay the pieces on the prepared baking sheets about ½ inch apart.

Add a few tablespoons of olive oil to a small bowl, and evenly brush the tops of the crackers with oil. Sprinkle lightly with salt and bake until the crackers are slightly puffed up and deep golden along the edges, 12 to 15 minutes. Let cool completely.

Serve immediately or store in an airtight container at room temperature for up to 1 week.

ACKNOWLEDGMENTS

This is such a special list. These people are the reason I am who I am and do what I do, in that weird way I do it. I owe so much to everyone here; more than just their names printed in ink. Please join me in celebrating each one.

To my mother, who's always willing to go for a walk and talk about plants with me. Who gifted me my life and my love of the living world, and the ability to notice unnoticeable things.

To my father, for teaching me the alchemy that is cooking and the stories one can tell with food.

To my sister, who has always been down to gossip over berry picking and is the queen of crabapple jelly.

To my Geoffrey, who learned to love mushrooms for me (truly the grandest gesture of love). Thank you for never letting me give up on myself. I love you to the far edge of the known universe and back.

To Olivia Anand and Rachel Freese, my besties who are willing to try every dish, give all the helpful feedback, dive with me for weird seaweeds, and laugh until our sides stitch. You keep me anchored in the most trying times.

To Jeanelle Hayner Olson, the type A food QUEEN who made working on this book a JOY. It truly couldn't have existed without you and your work ethic, kindness, and endless creativity. Thank you for being my friend and wild plant coconspirator.

To JB Douglas, for somehow managing to be an expert forager, cook, photographer, and basket weaver. Save some talent for the rest of us! Thank you for letting this book borrow your beautiful way of seeing the world.

To my book agent, Nicole Tourtelot, for your endless patience, wisdom, and passionfruit shipments.

To Justin Schwartz and the entire team at Simon Element, thank you for taking a chance on me and creating something beautiful from the bits and bobs in my brain.

To my agency team at UTA. Erik, Sarah, Isaiah, and JB, y'all are truly the dream team.

To Jesse and Laramie Aközbek, for your fantastic photos and friendship. Truly a plant power couple that is a pleasure to know.

To Tyler and John, for the accidentally life-changing anise hyssop. And to Ms. Barbara, for the plant poetry books, the most hilarious story involving beach plums I've ever heard, and your company.

To Elizabeth Andoh, Nicky Buck, Camille George of Wild Glory Bakery, Viviana Matsuda @Mud.Witch, Denise Parker.

To Monica Lyle, for always believing in me (especially when I didn't).

To everyone who ate something you'd never heard of because I said, "No trust me, it's weird and good!"

And to all of my followers across social media. Thank you for the ultimate gift: getting to make a living doing what I truly love. That privilege is never lost on me, and I'm eternally grateful to each of you.

INDEX